Sustainable Bioeconomy Development in the Global South

Matthew Chidozie Ogwu
Sylvester Chibueze Izah
Jorge Gabriel Vazquez-Arenas
Shiferaw T. Feleke • Xun Wei
Editors

Sustainable Bioeconomy Development in the Global South

Volume II Bioeconomy Techniques

Springer

Editors
Matthew Chidozie Ogwu
Department of Sustainable Development
Appalachian State University
Boone, NC, USA

Sylvester Chibueze Izah
Department of Microbiology
Faculty of Science
Bayelsa Medical University
Bayelsa State, Nigeria

Jorge Gabriel Vazquez-Arenas
Instituto Politécnico Nacional
Mexico City, Mexico

Shiferaw T. Feleke
Department of Social Sciences
and Agribusiness
International Institute of Tropical
Agriculture
Dar es Salaam, Tanzania

Xun Wei
Research Institute of Biology
and Agriculture
Shunde Innovation School
University of Science and Technology
Beijing, Beijing, China

ISBN 978-981-96-0304-6 ISBN 978-981-96-0305-3 (eBook)
https://doi.org/10.1007/978-981-96-0305-3

This Springer imprint is published by the registered company Springer Nature Singapore Pte Ltd.
The registered company address is: 152 Beach Road, #21-01/04 Gateway East, Singapore 189721, Singapore

Preface

The global shift toward a sustainable bioeconomy is essential for addressing the environmental, social, and economic challenges of our time. The bioeconomy of the Global South with its abundant biodiversity and untapped natural resources holds significant potential to drive innovation and sustainable growth. However, realizing this potential requires the adoption of tailored techniques that are both regionally relevant and globally impactful.

Sustainable Bioeconomy Development in the Global South – Volume II: Bioeconomy Techniques examines the latest methods and technologies transforming bio-based industries in the Global South. This volume builds on the foundation set by the first volume (*Sustainable Bioeconomy Development in the Global South—Volume I: Status and Perspectives*) with a focus on the potentials and challenges in transitioning toward a sustainable bioeconomy in these regions. The practical applications of bioeconomy techniques such as sustainable raw material extraction, bio-based product production, biorefining, and circular economy approach not only contribute to environmental sustainability but also foster economic inclusivity, offering solutions to pressing issues such as resource depletion, waste management, and renewable energy generation.

Through 17 chapters, this book brings together experts from academia, industry, and policy, highlighting case studies from various regions in the Global South. The topics range from bioenergy generation to advanced wastewater valorization techniques and sustainable food production, showcasing the wide array of bioeconomy applications that are being implemented to address local challenges.

The collaborative nature of this work underscores the importance of cross-border cooperation and knowledge sharing, which is critical for accelerating bioeconomy development globally. The volume aligns with several United Nations Sustainable Development Goals (SDGs), particularly SDG 12 (Responsible Consumption and Production) and SDG 13 (Climate Action). By emphasizing these goals, the book not only highlights technological innovation but also addresses the need for policies that align with long-term sustainability and resilience. These bioeconomy techniques offer promising pathways for mitigating climate change impacts, improving resource efficiency, and fostering sustainable livelihoods in vulnerable regions. This work will inspire the adoption of innovative bioeconomy techniques that support a greener, more equitable future.

Boone, NC, USA — Matthew Chidozie Ogwu
Bayelsa, Nigeria — Sylvester Chibueze Izah
Mexico City, Mexico — Jorge Gabriel Vazquez-Arenas
Dar es Salaam, Tanzania — Shiferaw T. Feleke
Beijing, China — Xun Wei
2024

About the Book

This volume, *Sustainable Bioeconomy Development in the Global South—Volume II: Bioeconomy Techniques*, examines innovative techniques and technologies driving bioeconomy advancements across the Global South. The book focuses on advancing bioeconomy techniques across the Global South for sustainable development, especially those connected to sustainable resource extraction, bio-based processing and production methods, waste-to-value conversion, bioenergy generation, renewable energy solutions, waste management, and the roles of biotechnology.

Through contributions from leading scholars and experts, this book delves into cutting-edge bioeconomy techniques that promote sustainability, resilience, and social innovation while addressing pertinent challenges and opportunities in bioeconomy development through circular economy. The volume aligns with the United Nations Sustainable Development Goals (SDGs), particularly SDG 12 (Responsible Consumption and Production) and SDG 13 (Climate Action).

This book is essential for policymakers, researchers, academics, bioeconomy practitioners, environmental science, sustainable technology, public health, and biodiversity conservation. The book provides a holistic perspective on the role of bioeconomy techniques in supporting the sustainable growth of countries rich in biodiversity but facing environmental and economic challenges.

Contents

Editors and Contributors

About the Editors

Matthew Chidozie Ogwu is Assistant Professor of Integrated Ecology and Sustainable Development at Appalachian State University, USA. He holds a PhD in Biological Sciences from Seoul National University, South Korea. He is an interdisciplinary academic with transdisciplinary skills and diverse convergence research interests (One Health and Eco Health) pertinent to assessing coupled human and natural as well as socio-ecological systems. He has numerous awards, research grants, and scholarships. Dr. Ogwu serves on the board of and as a reviewer for many peer-reviewed journals. He continues to volunteer his time and skills to promote sustainable development.

Sylvester Chibueze Izah is a lecturer at Bayelsa Medical University in Yenagoa, Nigeria, where he also serves as the Assistant Director of Academic Planning, Research, and Innovations. Dr. Izah is a licensed Environmental Health Specialist in Nigeria. He is a multidisciplinary academic with multifaceted abilities relevant to Sustainable Human-Environmental Health Interactions (covering air, soil, and water quality; toxicology; hygiene and sanitation; food science; waste management; biodiversity and their sustainability and risk assessment). He has an impressive track record of over 280 peer-reviewed publications, including journal articles, book chapters, and edited books. Dr. Izah has also collaborated on research projects with colleagues worldwide and actively contributes as an editorial and review board member for several esteemed journals.

Jorge Gabriel Vazquez-Arenas has a PhD in Chemical Engineering from the University of Waterloo (Canada). He is the recipient of numerous awards including the Ontario Graduate Scholarship (OGS) obtained for outstanding research at the PhD level in Ontario, Canada; and the Erasmus Mundus Scholarship 2017 (European Community) in materials for energy conversion and storage, as a visiting professor at "Laboratoire de Réactivité et Chimie des Solides (LRCS)" in Amiens, France. He is an accredited reviewer and monitor of CONAHCYT (RCEA) in different calls and projects in México; and a reviewer in international calls for proposals from Chile, Argentina, Colombia, and Perú.

Shiferaw T. Feleke was born in Ethiopia in 1971. He is an Agricultural Economist at the International Institute of Tropical Agriculture (IITA) based in Dar es Salaam, Tanzania. Before joining IITA in 2014, he worked in the Agricultural Policy Analysis Center (APAC) of the University of Tennessee. He has over 20 years of experience in agricultural research, policy analysis of commodity programs, agricultural productivity analysis, and impact assessments. His research interests include applied circular economy, technology and policy impact evaluation, and international development. He has published several articles in high-impact journals, such as the *American Journal of Agricultural Economics*, *Food Policy*, *Agricultural Economics*, and the *Journal of Agricultural Economics*. Shiferaw holds a PhD in Agricultural Economics from the University of Florida. According to Google Scholar, Shiferaw currently has more than 2500 citations and an h-index of 24.

Xun Wei is a Professor and Vice President at the Research Institute of Biology and Agriculture, University of Science and Technology Beijing. Dr. Wei holds a PhD in Agricultural Economic Management from the Chinese Academy of Social Science and was an experienced project manager in agricultural science and technology programs belonging to the Chinese Ministry of Science and Technology. Dr. Wei's current focus is on biotechnology innovation, sustainable resource allocation, and commercialization policy environment of biotechnology and other cutting-edge agricultural technologies in the context of the bioeconomy landscape. Xun Wei has authored numerous influential publications and books in these areas.

Contributors

B. S. Adeboye Department of Mechanical Engineering, Osun State University, Osogbo, Nigeria

Adeniji Adedeji Adeola Centre for Geography and Environmental Science, University of Exeter, Exeter, UK

Yetunde Eniola Agbeja Department of Aquaculture and Fisheries Management, University of Ibadan, Ibadan, Oyo State, Nigeria

Omotola Fadekemi Ajayi Department of Economics, Chrisland University, Ilugun, Nigeria

Temidayo Oladiran Akinbobola Department of Economics, Obafemi Awolowo University, Osun State, Nigeria

T. O. Akingbade Department of Agricultural Engineering, Adeleke University, Ede, Nigeria

Olamigoke Olawale Akingba Nigerian Centre for Energy and Environment, University of Benin, Edo State, Nigeria

Oluremi J. Akintayo Institute of Agricultural Research and Training (O.A.U), Ibadan, Nigeria

Sulaiman Kajogbola Alimi Department of Environmental Management, Glasgow Caledonian University, Glasgow, UK

Berchie Asiedu Department of Fisheries and Water Resources, University of Energy and Natural Resources, Sunyani, Ghana

R. T. Babalola Department of Food Technology, Osun State University, Osogbo, Nigeria

Feifei Chen Research Institute of Biology and Agriculture, Shunde Innovation School, University of Science and Technology Beijing, Beijing, China

Liang Chen Institute of Animal Sciences, Chinese Academy of Agricultural Sciences, Beijing, China

Ojeleye Abiola Elizabeth Department of Agronomy, Osun State University, Osogbo, Nigeria

Jacob Somorhire Ewutanure Department of Fisheries and Aquaculture, Nigeria Maritime University, Delta State, Nigeria

Waheed Olujide Fawole National Biotechnology Research and Development Agency, Bioresources Development Centre, Ogbomoso, Nigeria

Shiferaw T. Feleke Department of Social Sciences and Agribusiness, International Institute of Tropical Agriculture, Dar es Salaam, Tanzania

S. Montserrat García-Solares Centro Mexicano para la Producción más Limpia, Instituto Politécnico Nacional, Ciudad de México, México

Christy Funmilayo Idowu National Biotechnology Research and Development Agency, Oyo State, Nigeria

O. O. Ilori Department of Mechanical Engineering, Adeleke University, Ede, Nigeria

Adams Ovie Iyiola Department of Fisheries and Aquatic Resources Management, Osun State University, Osogbo, Nigeria

Sylvester Chibueze Izah Department of Microbiology, Bayelsa Medical University, Yenagoa, Nigeria

K. F. Jaiyeoba Department of Agricultural Engineering, Federal University of Technology, Akure, Ondo State, Nigeria

Kokou Kintche Department of Social Sciences and Agribusiness, International Institute of Tropical Agriculture (IITA), Kalambo, DR, Congo

Enoch Akwasi Kosoe Department of Environment and Resource Studies, Simon Diedong Dombo University of Business and Integrated Development Studies, Wa, Ghana

Pius Krütli Department of Environmental Systems Science, ETH, Zurich, Switzerland

W. A. Lamidi Department of Agricultural Engineering, Osun State University, Osogbo, Nigeria

Yanet Lissette Antonio Lopéz Escuela Superior de Ingeniería Química e Industrias Extractivas, Ciudad de México, México

Victor Manyong Department of Social Sciences and Agribusiness, International Institute of Tropical Agriculture, Dar es Salaam, Tanzania

Badmus Mubarak Unidad Profesional Interdisciplinaria en Ingeniería y Tecnologías Avanzadas, Instituto Politécnico Nacional, Ciudad de México, México

Omobolaji O. Obisesan Institute of Agricultural Research and Training (O.A.U), Ibadan, Nigeria

O. E. Ogundahunsi Department of Agricultural Engineering, First Technical University, Ibadan, Nigeria

Olorunfemi O. Ogundele Nigerian Institute of Social and Economic Research, Ibadan, Nigeria

C. A. Ogunlade Department of Agricultural Engineering, Osun State University, Osogbo, Nigeria

Matthew Chidozie Ogwu Goodnight Family Department of Sustainable Development, Appalachian State University, Boone, NC, USA

A. R. Olaniyan Department of Agricultural Engineering, Osun State University, Osogbo, Nigeria

B. O. Oyefeso Department of Agricultural Engineering, University of Ibadan, Ibadan, Nigeria

S. Oyelami Department of Mechanical Engineering, Osun State University, Osogbo, Nigeria

T. F. Oyewusi Department of Agricultural Engineering, Osun State University, Osogbo, Nigeria

Ángel Eduardo Yáñez Río Laboratorio de Electroquímica y Corrosión, Escuela Superior de Ingeniería Química e Industrias Extractivas, Ciudad de México, México

Haruna Sekabira Department of Social Sciences and Agribusiness, International Institute of Tropical Agriculture, Kigali, Rwanda

Guy Simbeko Department of Social Sciences and Agribusiness, Monitoring Evaluation and Department of Natural Resources Management, International Institute of Tropical Agriculture (IITA), Kalambo, DR, Congo

Agricultural Transformation Agenda (ATA-DR Congo), Bukavu, DR, Congo

Johan Six Department of Environmental Systems Science, ETH, Zurich, Switzerland

A. R. Soji-Adekunle Department of Mechanical Engineering, Adeleke University, Ede, Nigeria

Leonhard Späth Department of Environmental Systems Science, Transdisciplinary Lab (TdLab), ETH, Zurich, Switzerland

Samuel Bem Umma Department of Fisheries and Aquaculture, Federal University Wukari, Wukari, Taraba State, Nigeria

Bernard Vanlauwe Department of Natural Resources Management, International Institute of Tropical Agriculture, Nairobi, Kenya

Xun Wei Research Institute of Biology and Agriculture, Shunde Innovation School, University of Science and Technology Beijing, Beijing, China

Benjamin Wilde Department of Environmental Systems Science, ETH, Zurich, Switzerland

Yilin Zhao Zhongzhi International Institute of Agricultural Biosciences, Beijing, China

Sustainable Extraction and Production of Raw Materials for Bio-based Products in the Global South

Samuel Bem Umma, Yetunde Eniola Agbeja, and Adams Ovie Iyiola

Abstract

Nonpathogenic microorganisms are known to be part of the normal gut microflora of animals and specifically play vital roles in fish physiology. They have wide-ranging bioactive molecules with high potential to fish's general health. However, in some instances, it has been reported that irregular gut microbial interactions, pH, and nutrient fluctuations limit the potential activities and benefits of gut microflora. Therefore, as an alternative, the bioactive molecules of this nonpathogenic organism are biotechnologically harnessed and processed for use as additives in fish feeds, by identifying, mass culturing, extracting their extracellular metabolites, and processing them into storable forms for subsequent use. The biotechnology of producing extracellular metabolites from living things, in particular, microorganisms, is now feasible from previously identified and characterized nonpathogenic isolates. Biotechnological applications use biological systems, living organisms, or derivatives thereof, to make or modify products or processes for specific purposes essentially targeted at benefitting humans as end users. Research has shown that optimization of microbial growth and extraction of their metabolites for therapeutic and nutritional purposes are now gaining

S. B. Umma (✉)
Department of Fisheries and Aquaculture, Federal University Wukari, Wukari, Taraba State, Nigeria
e-mail: umma@fuwukari.edu.ng

Y. E. Agbeja
Department of Aquaculture and Fisheries Management, University of Ibadan, Ibadan, Oyo State, Nigeria

A. O. Iyiola
Department of Fisheries and Aquatic Resources Management, Osun State University, Osogbo, Osun State, Nigeria
e-mail: adams.iyiola@uniosun.edu.ng

M. C. Ogwu et al. (eds.), *Sustainable Bioeconomy Development in the Global South*, https://doi.org/10.1007/978-981-96-0305-3_1

more relevance and adding high value to aquaculture production, concerning aquaculture nutritional biotechniques. They can secrete a full complement of secondary metabolic products which are known to be effective therapeutics such as antibiotics, growth promoters such as amino acids, and other biomolecules like enzymes that catalyze metabolism. This chapter therefore focuses on the sources, types, and uses of metabolites. The nutrient requirements of microbial growth, biotechnological, and bioeconomic relevance will also be discussed for sustainable activities in the Global South.

Keywords

Biotechnological applications · Extracellular metabolites · Fish physiology · Nonpathogenic microorganisms

Abbreviations

ATP	Adenosine triphosphate
DNA	Deoxyribonucleic acid
EDTA	Ethylenediaminetetraacetic acid
FAO	Food and Agricultural Organization
PHA	Polyhydroxyalkanoates
PMF	Proton motive force
SSF	Solid-state fermentation

1 Introduction

Microorganisms or microbes are micro creatures that are too small for the naked eyes to see except with the aid of a microscope (Adieze, 2018; Guzik & Mccoy-Simandle, 2017). They are cosmopolitan and seen thriving on both living and non-living things (Shu & Huang, 2021). In their characteristics and widespread nature, they populate any suitable substrate, thereby performing the basic functions of all living things which include but are not limited to movement, respiration, nutrition, irritability, growth, excretion, and reproduction (Sandle, 2016). They are the most populous creatures in terms of biomass and the most prolific organisms in terms of reproduction. They exist in different shapes and forms such as unicellular, multicellular, or clusters (Schopf et al., 2017). It is almost impossible to visualize a single microbial cell with the natural eyes; however, as colonies in plate culture, they can be seen without the aid of a microscope. As single-cell colonies, they are most likely to be differentiated by their shape, size, or color. Microorganisms have a diverse nature, and a unique feature is their very small or tiny size (Adieze, 2018). Microorganisms obtain energy basically from their environment; they have with time succeeded in evolving the various mechanisms necessary for fetching this energy for their use in the areas of digestion and reproduction, etc. The major traits of microorganisms that give them an edge over plants and animals in terms of metabolite production are in their ability to regenerate or grow in a record time of about 15 min for bacteria such as *Escherichia coli*, which require small space for growth, and ease of biological manipulation.

Specific microorganisms have a well-defined relationship that could be symbiotic or parasitic with other microorganisms or higher multicellular organisms (Drew et al., 2021). They maintain a balanced ecosystem through their mutualistic relationships. With this, the microorganisms derive some benefits and at the same time offer some benefits to their host. Bacteria in the gut of fish act on the large molecules of nutrients as substrates by deriving nutrients and shelter, and in turn, secrete extracellular metabolites as by-products of metabolism. These metabolites are essential for the normal physiology and breakdown of macronutrients into smaller bits for easy assimilation into the bloodstream. Most of their secretions are important in maintaining the health and growth of their host (Pinus et al., 2017). An example of such a symbiotic relationship in animals that adds value to their host is one between the fish and the nonpathogenic *Escherichia coli* residing in their gut. In another instance, they could be harmful, that is pathogenic by deriving nutritional benefits and habitat from the host, and at the same time causing harm to them (Bäumler & Sperandio, 2016).

1.1 Classes and Types of Microorganisms

Microorganisms are prokaryotes or eukaryotes. Prokaryotes are a group of organisms whose cells lack a membrane-bound nucleus (karyon). The word prokaryote comes from the Greek word (pro-) meaning before and (karyon) meaning nut or kernel. Prokaryotes do not have a nucleus, mitochondria, or any other membrane-bound organelles. In other words, all their intracellular water-soluble components, that is, protein, deoxyribonucleic acid, and metabolites are located together in the same fluid mix, enclosed by a single membrane known as the cell membranes. Most prokaryotes are unicellular organisms. Prokaryotes are majorly classified into two domains, bacteria, and archaea, while eukaryotes could either be unicellular or multicellular, with a nucleus (Purves et al., 2003; Awuchi et al., 2022).

Microorganisms are divided into seven major types, namely bacteria, archaea, fungi, protozoa, algae, viruses, and helminths (Madigan & Martinko, 2006). Bacteria and archaea are classified under the first and second domains of life respectively. For bacteria, they are usually single-celled or unicellular organisms. They are almost invisible to the natural eyes because of their microscopic nature or characteristics. They are capable of decomposing, fermenting biomass, and fixing nitrogen. In the event of their activities, some are essentially useful to the ecosystem and beneficial to their host, while others could be severely pathogenic and harmful to their host.

1.2 Products of Microorganisms

Microorganisms are economically important to man based on their nature and activity as either nonpathogenic or pathogenic. The derived benefits could be through their mutualistic nature or the secretion of health-important metabolites as by-products of metabolism (Macro, 2021). Sometimes, the nonpathogenic characteristics of microorganisms such as bacteria in the gut of man, animals, and fish specifically are a function of the bioavailability of nutrients for the bacteria to thrive (Toju et al., 2020). Where the needed nutrients are depleted, they can switch their mutualistic roles in other to survive, hence causing harm to their host (Liao et al., 2023). In the agriculture, food, feed, and pharmaceutical industries, the major way of harnessing the products of microorganisms of importance to man such as antibiotics, antibacterial, enzymes, and other bioactive molecules is through culture-based technology of the organisms (Chen et al., 2021). Fungi are cultured using solid-state fermentation, while bacteria are cultured using liquid-state fermentation (Barrios-Gonzalez & Tarrago-Castellanos, 2017). Thus, essential products are derived from the by-products of their metabolic activities. Some examples of microorganisms' products of importance are highlighted in Table 1.

2 Basic Structure, Properties, and Functions of Bacteria and Archaea

2.1 Basic Structure

Bacteria cell structure generally consists of macromolecules such as deoxyribonucleic acid (DNA), proteins, polysaccharides, phospholipids, and the cytoplasm, which is a gel-like substance composed of water, salts, enzymes, nutrients, wastes, and gases. The macromolecules are majorly made up of nucleotides, amino acids, and sugars in various locations within the cells. It is reported that the sequence in which these primary structures are laid within the cells determines their characteristic properties (Ogugbue, 2018; Nelson et al., 2005). Therefore, the genetic code is determined by the specific nucleotides' base sequences in the chromosomal DNA; the amino acids sequence in a particular protein determines the properties and functions of the protein; and the sequence of sugars in the bacterial lipopolysaccharides determines the unique cell wall properties for the bacteria known as peptidoglycan, which differentiates them into Gram-positive or Gram-negative bacteria cells (Britannica, 2020). They could be in different shapes and sizes, from minute spheres, cylinders, and spiral threads, to flagellated rods, and filamentous chains. Under microscope they are sectioned into the following groups: Rod-shape (bacilli) examples are *Escherichia coli*, *Bacillus cereus*; Spherical (coccus) examples are the *Staphylococcus epidermidis*; Curved (Vibroid, Spirochaete) examples are *Vibrio cholera*, *Rhodospirilium rubrum*; Square-shaped (Arcula); Star-shaped (Stella); and Filamentous. Bacterial cells could range from 0.2 μm to 700 μm in diameter. In comparison, bacteria are about 10 times smaller than eukaryotes, with the

Table 1 Some industrially used microorganisms, products, and biological applications

Types of organism source	Types of products	Area of application	References
In agriculture			
Bacillus papillae and *B. thuringiensis*	Toxins	Used in plants to resist insect pest	Khachatourians (2019)
Lueconostoc	Dextrans from sucrose	Used in pickling vegetables and production of other agricultural products	Rosca et al. (2024)
Rhizopus and mucor	Enzymes	Used in enzymes production and in nitrifying the soil	Liu et al. (2024) and Merloti et al. (2019)
In medicine			
Bacillus species	Enzymes	Production of drugs	Danilova and Sharipova (2020)
Lactobacillus sp.	Lactic acid	Production of drugs	Abedi and Hashemi (2020) and Abdul Hakim et al. (2023)
Streptomyces	Secondary metabolites	Production of drugs	Alam et al. (2022) and Zhang et al. (2020)
Aspergillus sp.	Aflotoxins, penicillium	Production of drugs	Shah et al. (2022)
In food and feed industries			
Acetobacter sp. and *Gluconobacter* sp.	Acetic acid	Manufacture of vinegar	Mizzi et al. (2022)
Zymomonas	Alcohol	Production of alcohol	Aminian and Motamedian (2023)
Lactic acid bacteria	Ethanol	Used in brewing industries	Seo et al. (2020) and Abedi et al. (2020)
Lactococcus	Lactic acids	Used as starter in yoghurt manufacture, also as probiotics for gut health	De Chiara et al. (2024) and Maslak et al. (2022)
Ascomycetes (Yeast: *Saccaromyces cerevissiae* var)	Enzymes and alcohol	Used in the production of ethanol and alcoholic beverages	Cason et al. (2020)

advantage of a higher surface-to-volume ratio, higher nutrient uptake rate, rapid excretion of wastes, faster growth rate, and shorter life span, which makes them the most prolific organisms than eukaryotes (Levin & Angert, 2015).

Bacterial cell appendages, envelope, wall, membrane, cytoplasm, chromosome, plasmid, cytoskeleton, gas vacuoles, and ribosome play important functional roles in the general life cycle of the bacterial cell, and in the environment they inhabit. The appendages could either be flagella which facilitate movement or pili and fimbriae which are used for adhesion to surfaces (Haiko & Westerlund-Wikström, 2013). Their movement from one substrate to another and proper adhesion on a particular substrate provide more choice for them on where they can thrive better. The cell envelope functions by housing or enclosing the cytoplasm of the cell. They are layered into the interior cytoplasm membrane, the cell wall, and in some bacterial species the outer capsules (Zerbib, 2017). The cell wall is usually a rigid structure that encloses the bacterial cell thereby giving the bacterial the needed structural integrity, and shape. However the rigidity and strength of the cell wall is characterized by the presence of peptidoglycan, which is located immediately outside of the cytoplasmic membrane. The cell membrane is also known as plasma or cytoplasmic membrane, it is a thin, fragile membrane located just beneath the bacterial cell wall, and it encloses the cytoplasm.

2.2 Basic Functions and Properties

It is reported as the most dynamic structure of the bacterial cell and it functions as a selective permeability barrier that regulates the passage of substances into and out of the cell, while the cytoplasm, also known as the protoplasm, is a gel-like matrix composed of water, salts, enzymes, nutrients, wastes, and gases. It is the major hub of the bacterial cell, where activities such as cell growth, metabolism, and reproduction are carried out. The chromosome is typically a large circular molecule of DNA, localized, but not contained within the membrane, and found floating within a dense region of the cytoplasm called nucleoid region. It consists of genes, which are the basic unit of heredity, needed for growth, survival, and reproduction, while the plasmid are small independent circular rings of DNA in some bacteria that are often encoded for traits that are advantageous but not essential to their bacterial host, and ribosomes are the primary sites of protein synthesis in all living things (Keagle, 2013). Therefore, bacteria being part of living things contain ribosomes, however, they are smaller and are free-standing structures distributed throughout the cytoplasm that are never bound to other organelles, unlike the eukaryotic ribosomes. The gas vacuoles are an important air structure, which is a cylindrical vesicle that allows many bacterial that possess it to adjust their buoyancy, likewise aid vertical movement (Pfeifer, 2012).

They have an independent evolutionary history that makes them stand out from other forms of life hence, constituting a separate domain of life (Woese & Fox, 1977). They are prokaryotes, that is, they do not have a nucleus or any membrane-bound organelles within their cells. They have some similar characteristics with bacteria; however, there is no known species of Archaea that forms spores unlike bacteria and eukaryotes, they usually reproduce asexually by binary fission, fragmentation, or budding (Warren et al., 2019). Archaea could be seen in extreme environments, that is, inhabit environments categorized as being very harsh for survival. It is interesting to note that, they also inhabit the human colon and navel, soil, and aquatic environment, where the play vital roles in the carbon and nitrogen cycle (Warren et al., 2019). Furthermore, there is no known species that is pathogenic, however, they often live as commensals or in mutual relationship with their host (Weber et al., 2016). They have potential usefulness in biotechnology, in terms of their nonpathogenic nature and ability to inhabit extreme environments.

2.3 Economic Importance of Bacteria and Archaea

The economic importance of bacteria to nature and humans cannot be overemphasized. Generally, they play vital roles in the ecosystem, though they may be very small and unnoticeable by the natural eyes, their population tremendously add value and influence the planet's nutrients and compounds ranging from food production and processing to digestion (Jahn et al., 2022). They are essential for the survival of all living things, and the continued sustenance of life, including man, where we derive health benefits from the important microflora that colonize the gut system and other organs. Man has found tremendous help from bacteria, as they are important to man directly or indirectly, being utilized in the agricultural, food, and pharmaceutical industries for the production of raw materials such as fertilizer and pesticides in the agricultural sector, drugs in the pharmaceutical sector, nutritional additives in the food and feed sectors; they are also used for soil and water bioremediation and waste treatment (Akoijam et al., 2022).

Most bacteria secrete important enzymes into their environment which are important in the decomposition of dead organic materials of plant and animal origin, which in turn renew the ecosystem through the release of nutrients needed by other organisms for recycling. Their mode of action is the conversion of macromolecules such as carbohydrates, proteins, fats, nucleic acids, and nitrogenous compounds into smaller molecules in their simplest forms such as CO_2, water, ammonia, hydrogen sulfide, phosphates, nitrates, and simple sugars which then can be used as raw materials by green plants and sugar by animals (Jurtshuk Jr, 1996; Benini, 2020; Egboduku et al., 2024). Saprophytes decompose the organic material and convert it for use directly as nutrients or manual and remove the wasteful components from the earth thereby functioning as the drivers of renewable natural resources. The presence of bacteria is important in virtually every area for the success of human endeavors. In animal husbandry and even in the wild, cattle and other ruminants rely on bacteria to secrete cellulose-digesting enzymes referred to as cellulase for the nutrients contained in the grasses they graze to be broken into absorbable forms in the body. Bacteria are highly involved in the cycling of many important compounds in the ecosystem which is not limited to oxygen, carbon phosphorus, sulfur, and nitrogen. Essentially, their activity closes the gaps between raw material inputs and outputs by bringing back the necessary elements that would sustain the cycle of production.

The nitrogen cycle is a good example of the beneficial roles of bacteria in nutrient cycling; they fixed the inaccessible nitrogen in the air by making it bioavailable to primary producers such as plants (Bernhard, 2010). Nitrogen-fixing bacteria take up the atmospheric nitrogen and convert it into organic nitrogen compounds, with the aid of the nitrogenase enzymes, they combine with hydrogen to form ammonia, which further undergoes conversion by bacteria to an accessible compound. The series of processes is carried out by both free-living and symbiotic bacteria, for example, Azotobacter, Clostridium, Rhizobium, and Bradyrhizobium, respectively. Another important mutual or nonpathogenic interaction of microorganisms with higher multicellular cells is that of bacteria and their leguminous plant host, in which the nitrogen-fixing bacteria, for example, Rhizobium, resides and develops inside the plant tissues, and obtains the energy and carbon needed for their metabolism. In turn, the nitrogen assimilated from the air by bacteria is utilized by the plants as a nutrient (Smercina et al., 2019).

Bacteria are also famously used in the food industries, especially in events where the products obtained from them cannot be chemically produced. For instance, in the dairy industry, lactic acid bacteria are used to convert milk into useful products like cheese, buttermilk, and yoghurt (Coelho et al., 2022). Bacteria are also important in industrial and pharmaceutical production of antibiotics, antibacterial, and antifungal (Katara et al., 2021; Rehman et al., 2020). For instance, Bacillus is used to produce antibiotics such as bacitracin and gramicidin. Bacteria can be used to produce some vitamins such as vitamin B12 from *Bacillus megaterium*, and vitamin K and B-complex from *Escherichia coli* (Kumar et al., 2023). Since bacteria have important and resourceful attributes in the food, agricultural, and pharmaceutical industries, it is also naturally not surprising that they have harmful candidates and effects to humans. Although it is reported that only a small percentage of bacteria cause disease; it is on record that over 90% of human diseases and 10% of plant diseases are caused by the toxins produced by pathogenic bacteria (Kim et al., 2020).

Archaea are known to efficiently recycle the elements surrounding their habitation such as carbon, nitrogen, and sulfur (Pfeifer et al., 2021). They basically can carry out the various steps involved in the nitrogen cycle such as those that remove nitrogen from the ecosystem like nitrate-based respiration and denitrification, and those that introduce nitrogen into the ecosystem like nitrate assimilation and nitrogen fixation (Adam et al., 2018). They can also oxidize Sulphur compounds and release the element to the earth, making it bioavailable for use by other organisms (Chen, 2010). In terms of carbon cycle, they also play important roles in the decomposition of organic matters anaerobically in the ecosystems (David et al., 2020). Because of their ability to inhabit extreme environments they are significantly resourceful industrially in the production of raw materials for agricultural, pharmaceuticals and food processing sectors. For instance, species that are tolerant or can withstand heat, acidity or alkalinity are best candidates to produce enzymes that must be used under such extreme conditions without denaturation (Farkas et al., 2013). They are essential sources of antibiotics, with novel mode of action distinct from bacteria (Chen & Viljoen, 2010; Pfeifer et al., 2021). There are series of potential bioactive products for industrial, pharmaceutical and food processing sectors that can be derived from Archaea. Many more products are derived from other species of Archaea domain or tree of life which have potential in medicine, pharmaceuticals, food, and textile industries that are still not explored. Some bioactive products from Archaea and their functions are:

- Pfu DNA Polymerase compounds which are derived from *Pyrococcus furiosus*, and it is a vital reagent in PCR technique for cloning DNA (Spipida et al., 2018).
- Amylases, Galactosidases, Pullulanases which are derived from *Pyrococcus* species is useful in food processing under high temperatures like in the production of Lactose milk (Spipida et al., 2018).

- Methane which is derived from *Methanosaeta* spp., and *Methanosarcinaceae* spp. is useful in wastewater treatment, and food and dairy production (Vitezova et al., 2020).
- Carotenoids and Canthaxanthin which are derived from *Haloferax alexandrines* are used by feed industries for fish and chicken additives, and as coloring agents (Gaur & Bera, 2023).
- *Polyhydroxyalkanoates* which is derived from *Haloferax mediterranei* is used as a potential alternative to petrochemical-based plastics due to their structural properties and biodegradable nature (Atares et al., 2024).

3 Basic Structure, Properties, and Functions of Eukaryotes (Fungi)

3.1 Basic Structure

The Eukaryotes are microorganisms that make up the third domain classification of life, they are cells that have a nucleus (true nucleus), and other structures called organelles which are enclosed inside the membranes. Hence, its nucleus and other organelles are contained in the cell membrane (Abo Elsoud & El-Kady, 2019). Eukaryotes form the taxon eukaryota, which includes protozoan, algae, fungi, viruses, helminths, multicellular plants, and animals. They are distinct from prokaryotic cells by the presence of a true nucleus, which earns them the name eukaryotes from the Greek word "eu" which means good, and "karyon," which means nut or kernel. The nucleus houses the genetic material such as the DNA, which is responsible for heredity. Most eukaryotic cells also contain membrane-bound organelles such as mitochondria and the Golgi apparatus. The plant and animal components of the eukaryotes contain the chloroplasts, which are responsible for photosynthetic activities (Cavalier-Smith, 1992).

Protozoa are heterotrophic protists; they are single-celled eukaryotes such as amoebas, ciliates, and flagellates, the algae are photosynthetic protists ranging from microscopic unicellular forms to seaweeds, contributing to the oxygen demand of the aquatic ecosystem. The fungi are nonphotosynthetic eukaryotes and play essential roles in decomposition, mycorrhizal associations, and food production. A good example of fungi includes the yeast and mold. Viruses are cellular organisms, meaning they are not truly living organisms but rely on host cells for replication. Helminths are multicellular parasites; they include worms like tapeworms, flukes, and roundworms. Both humans and animals can be their hosts, and they can also exert detrimental effects on their hosts. In general, microorganisms, depending on which one, have a very important role they play in shaping and maintaining the ecosystem (Cavicchioli et al., 2019).

3.2 Functions and Properties

It has been established that fungi are part of the Eukaryotic organism and are essential in the breaking down and recycling of organic matter (Orji et al., 2018). Like other organisms, fungi are involved in the maintenance of the ecosystem through their activities and natural cycles such as carbon, nitrogen, sulfur, and phosphorus cycles (Orji et al., 2018). Descriptively, fungi are heterotrophic and have nuclear material and membrane, mitochondrion, and endoplasmic reticulum; they also pose filamentous structures that are known as hypha. Hypha has chitinous walls, and a network of hypha forms mycelium. Fungi are known to act on varied substrates in nature, with the ability to degrade organic matter into reusable components in nature. They can perform such activities with the aid of some specific enzymes secreted through their metabolic activities on their substrates (Maheshwari, 2016). The ubiquitous nature of fungi in the soil, water, air, food, and some living things gives them the advantage of being a good candidate in biotechnology and agriculture since they are easily manipulated into the production of food products, medicines, additives, antibiotics, and antibacterial substances, and even as entities used in the antagonizing plant pathogens (Reiss & DiSalvo, 2018). However, as important as they are, there are virulent species of these organisms that are dangerous to plants, animals, and humans. Fungi like other living things exhibit the general characteristics of growth and reproduction. They grow through the tips of their hypha and by septation (Maheshwari, 2016). This happens when they have found a suitable substrate to inhabit, which will generally start with colonization, acclimatization, growth, and reproduction. In the event they cannot be acclimatized, the species or colonies are wiped out. The process starts with the lag phase, followed by the log phase or exponential growth window, and then the stationary phase (Vrbal et al., 2019). All these happen as a function of the suitability of the physiochemical components of the environment and nutrient availability (Meyer et al., 2016).

3.3 Reproduction in Fungi

There are two forms of reproduction in fungi, it could be sexual (Telemorph) or asexual (Anamorph). Some fungi species can exhibit both sexual and asexual forms of reproduction; an example of such fungi is the Ascomycetes (McConnaughey, 2014). While sexual reproduction is the fusion of two matured and compatible nuclei, which may not necessarily be gametes, leading to the subsequent meiotic division yielding recombinant progeny, asexual reproduction which is widely known as the common form of reproduction in fungi could be in its simplest form as either budding or fission. Fungi species that undergo meiotic cell division, from fusion of nucleic material and any other form of genetic material exchange can be easily classified (Komluski et al., 2022). There are recognizably four existing classes of fungi based on the type of sexual spores:

- Ascomycetes—Aspergillus and Penicillium are examples.

- Basidiomycetes—Mushroom rust and jelly fungi are examples.
- Deutromycetes—Candida and saccharomyces are examples.
- Phycomycetes—The common black bread mold (Rhizopus) is an example.

3.4 Economic Importance of Fungi

Fungi as food are found to be a good delicacy, for example, the edible mushroom, which is eaten in various regions of the world. They are rich in proteins, and vitamins, and are also important sources of bioactive compounds which constitute a valuable source of supplementary foods (Valverde et al., 2015).

- They are used as medicines are used to produce antibiotics against bacterial infections, example include the novel discovery of Alexander Flemings Penicillium. The antibiotics obtained from fungi are broad spectrum, for example, the β-lactam antibiotics (Ogwu et al., 2022, 2023a, b).
- In agriculture, they are used for various purposes such as nitrifying soil and recycling organic materials in the earth, this assists in balancing the ecosystem through a series of biogeochemical processes, for examples those responsible for these processes include *Penicillium spiculisporus*, *Aspergillus flavus*, and *Penicillium verruculosum* (Akhtar & Amin-ul Mannan, 2020; Ogamba et al., 2023).
- Fungi in food and allied industries are used in the production of food products such as fermented fish, which is a special delicacy in the northeastern part of Nigeria, beverages such as the local beer drink called Buruku, and Kunu which is a sweet and refreshing drink in Nigeria.
- Biotechnologically, several bioactive substances or secretions are derived from the manipulation of some fungi to produce some enzyme products such as xylanase, cellulose, protease, pectinase amylase, and as additives in food and feed for both humans and animals respectively (Komluski et al., 2022).

Microorganisms display an array of relationships during their life cycle in both plants and animals. In plants, a symbiotic relationship is formed and called Mycorrhizae, where they can integrate into the root of the plant. Living cells derive energy from redox reactions through metabolism. The oxidation of carbon sources, e.g., glucose, and electron transfer which involves the processes of biosynthesis and energy metabolism. Only a small part of the electrons is used in reduction reactions to supply new cellular material (i.e., biosynthesis). The fungus derives sugars for their metabolism from the plant roots and in turn, the plant obtains phosphate and other essential minerals to augment the already depleted nutrients from soil through the metabolic activities of the fungus (Moller et al., 2018). The basic beneficial or mutualistic interaction between fungi and plants is by providing nutrients to the plants and then receiving sugar as nutrients from the plants. Essentially, they are the mycorrhizal fungi and are best exemplified by a lichen relationship with a photosynthetic organism, which could be either cyanobacterium or green alga. The fungus grows around the bacterial or algal cells and derives a constant supply of food produced by the photosynthesizing organisms, while the photosynthesizing organisms benefit from the water and nutrients absorbed from the environment by the fungus (McLaughlin Jr & Malik, 2017). In a parasitic relationship, the fungus benefits while the host is harmed. Parasitic fungi live in or on other organisms and get their nutrients using their specialized structures for penetrating a host. They also produce enzymes that break down the host's tissues. However, fungi parasitic activities are not limited to only plants; they are often the major causes of diseases and economic losses in agricultural plants (Iyiola & Valbo-Jørgensen, 2023).

4 Nutrient Requirements of Microorganisms and Growth Media

4.1 Basic Nutrient Requirements

The basic nutrient requirements required for the growth and metabolism of microorganisms are carbon, phosphorus, nitrogen, sulfur, and trace elements which are collectively called macronutrients (Haggag & Abo El Soud, 2012). These minerals can be sourced from bioproducts as illustrated in Table 2.

Microbes need carbon, electrons, and energy to survive. All microorganisms are composed of carbon with macromolecules that contain carbon (lipids, proteins, carbohydrates, nucleic acid). Some organisms may or may not rely in carbon sources, as some can use reduced organic substances as sources of carbon and are called heterotrophs (other feeders) while some can depend on carbon dioxide as a source of carbon and create organic molecules from the inorganic carbon dioxide and are

Table 2 Bioproducts and mineral sources

Nutrients	Functions	Sources	References
Carbon	Building of cellular structures and metabolic processes	Sugars	Wang et al. (2019), Hermsen et al. (2015) and Farrell et al. (2014)
Nitrogen	In metabolism, gene expression, energy production, nucleic acid synthesis, protein synthesis, and metabolism	Amino acids, ammonium salts, or peptides	Berry et al. (2002) and Colla et al. (2015)
Phosphorus	For nucleic acid and synthesis of the membrane	Phosphates	Hallama et al. (2018)
Sulfur	Cofactor and amino acid synthesis	Sulfate salts or sulfur-containing amino acids	Farrell et al. (2014)
Trace elements (magnesium, iron, potassium, and calcium)	Enzymatic functions and reactions	Bioproducts	Schmidt et al. (2019)

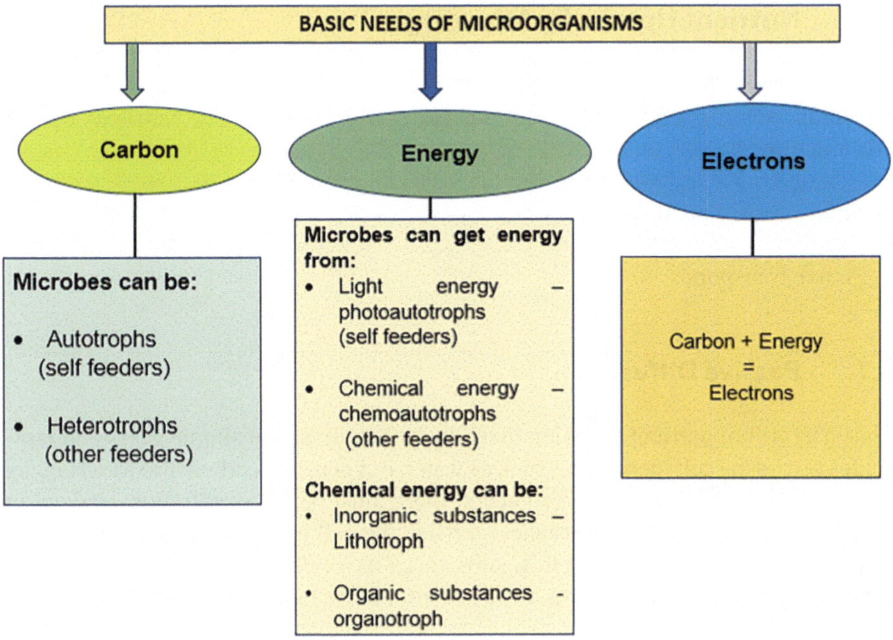

Fig. 1 Basic needs of microorganisms and sources

called autotrophs (self-feeders) (Sharma et al., 2023). In terms of energy for microorganisms, some receive light energy from the sun and are called photoautotrophs (light eaters) while some receive energy from chemicals or compounds that are inorganic and are called chemotrophs (chemical eaters). Chemical energy sources can either be lithotrophs (rock eaters) which derive energy from inorganic bioproducts or organotrophs (organic eaters) which derive energy from organic sources. When all of these are harnessed, electrons are produced for microorganism nutrients (Fig. 1) (Vos et al., 2017).

4.2 Growth Factors

In the presence of carbon sources and inorganic salts, some microbes can build up certain organic molecules from scratch. Some require that the presence of some organic compounds must be present in the environment before they can grow (Umesh & Basheer, 2021). The organic molecules that are essential for the growth of microorganisms are called growth factors (Suman et al., 2015). There are three categories of growth factors:

- Amino acids which are the building blocks of proteins.
- Pyrimidines and purines which are the building blocks of nucleic acid.
- Vitamins (enzyme cofactors).

5 Nutrient Uptake by Microorganisms

Nutrients must be absorbed by the cells from the environment through the cell membrane for support cellular activities (Rouphael et al., 2017). Nutrients can be taken in three ways:

- Passive or simple diffusion.
- Facilitated diffusion.
- Active transport.

5.1 Passive Diffusion

This is a form of nutrient diffusion that allows the passage of simple gases and molecules across the cell membrane, such as water, oxygen, and carbon dioxide (Haggag & Abo El Soud, 2012). For this to occur, there must be a concentration gradient in which the concentration of substances outside the cell must be higher than the concentration inside the cell. With this, substances move easily into the cell due to the concentration gradient decrease, resulting in a slow rate of diffusion of nutrients into the cell.

5.2 Facilitated Diffusion

This process is like passive diffusion as the concentration gradient of substances outside the cell is higher than in the cell, but the difference is the use of permeases which is a carrier protein (Amoo et al., 2021). These proteins are found inside the cell membrane and allow the passage of large molecules due to the channel it creates in the cell. If the concentration gradient reduces, molecules will not move from the environment into the cell. The carrier proteins are action-specific, as they are present in transporting a particular type of molecule or molecules that are closely related to its activities (Enebe & Babalola, 2021; Deng et al., 2019).

5.3 Active Transport

It is the process by which cells transport substances through the carrier proteins in the membrane, utilizing metabolic energy in the cell (Chiaranunt & White, 2023; Pant et al., 2010). This is required as several nutrient uptakes in cells require a means of transportation against a concentration gradient. There are two types of active transport and they both require and utilize carrier proteins:

- Primary active transport.
- Secondary active transport.

Primary Active Transport

It is the process by which chemical energy (Adenosine triphosphate [ATP]) is used to drive the transportation of nutrients around the cell. An example is the ABC system which is a cell transporter that uses ATP-Binding Cassette transporters (Aluko et al., 2023). Each ABC transporter has three different components:

- Membrane spanning proteins that forms the channel that is across the membrane of the cell.
- A region of ATP binding that provides energy by hydrolyzing ATP, thereby providing energy that is needed for the passage of nutrients across the membrane of the cell.
- A protein that is substrate-binding that binds substances to be transporters in the membrane-spanning protein.

In Gram-positive bacteria, the substrate-binding protein is usually attached to the outside region of the cell membrane while in Gram-negative bacteria; it is in the cell periplasm (Compant et al., 2019; Schmidt et al., 2019).

Secondary Active Transport

It is a process by which energy is utilized from a proton motive force (PMF). A PNF is a gradient of ions that occurs when there is transport of electrons within the cell during a process of energy conservation (Sharma et al., 2023). Protons that are positively charged accumulate outside cells that are negatively charged thereby creating a proton gradient between the outside and inside of the cell (Spalvins et al., 2018a, b). For a simple transport system, there are three different types:

- Uniport system.
- Symport system.
- Antiport system.

Each of the mentioned system mechanisms utilizes different protein porter. Unipoters transport a single substance either in or out, symporters transport two substances simultaneously (usually a proton and a molecule), and antiporters transport two substances but in opposite directions (White et al., 2018).

6 Growth Media for Microorganisms from Bioproducts

The culture of microorganisms is a function of the requirement of the organism. This dictates the type of bioproduct to be used as growth media as microbes can be cultured using various types of bioproducts. The commonly used bioproducts for culturing are:

- Residues from agriculture.
- Wastes from food processing.

- By-products from industries.

The bioproducts serve as a substrate that provides nutrients that are essential for the growth of microbes. The nutrients are nitrogen, carbon, phosphorus, and trace elements. Some microbes are action-specific as they require some cofactors to initiate their growth.

6.1 Residues from Agriculture

Residues such as wheat straws, rice husks, corn stover, and bagasse from sugarcane are bioproducts mostly used for the growth of microorganisms (de Bonadio et al., 2018). These residues contain a rich number of lignocellulosic materials, carbohydrates, and some nutrients which makes them good substrates for the growth of microorganisms and the production of metabolites (Da Silva et al., 2019). Bacteria and fungi can grow because of carbon sources found in cellulose, sugars, lignin, and hemicellulose that are embedded in the residues of agriculture (Sadh et al., 2018). They can ferment sugars from the residues to produce bioethanol which is an example of renewable fuel.

- *Enzymes*: The microorganisms can produce enzymes that can be used in industries in the production of textiles, biofuel production, paper, and pulp processing, and in processing food (Hilares et al., 2018). Enzymes, such as ligninases, cellulases, and hemicelluloses, can be produced from microorganisms grown from agricultural residues (Madeira Jr et al., 2017).
- *Organic Acids*: The fermentation of residues can produce organic acids like succinic, citric, and fumaric acid which is used in food production, the pharmaceutical industry, and biodegradable polymers (Venil et al., 2020a, b).
- *Bioplastics*: Through fermentation of sugar from residues, bioplastic polymers such as polyhydroxylalkanoates (PHA) can be produced. PHA are highly biodegradable and are suitable alternatives to conventional plastics (Heredia-Guerrero et al., 2017).
- *Amino Acids*: They can be synthesized from sugars by some microorganisms to produce lysine acid, threonine, and glutamic acid. They are used as food additives, supplements in animal feed, and pharmaceutical applications (Sharma & Ghoshal, 2020).

Agricultural residues are seen as a sustainable way for growth media for microorganisms as it enables the production of enzymes and metabolites which are useful in the processing and application in various industries.

Table 3 Metabolites, functions, and sources from food processing wastes

Metabolites	Sources	Functions	Uses	References
Lactic acid	Lactic acid bacteria—*Lactobacillus* spp.	They ferment sugars in the wastes to produce lactic acid	As a preservative, pH regulator, and enhancer of flavor in the food industry	Sen et al. (2019)
Acetic acid	Acetic acid bacteria—*Acetobacter* spp.	Production of acetic acid through oxidation of ethanol from food processing wastes	Production of vinegar and in industrial processes	Sen et al. (2019) and Sadh et al. (2018)
Bioethanol	Yeast—*Saccharomyces cerevisiae*	Ferment sugars from the waste to produce ethanol	A renewable fuel used as an additive of gasoline or fuel in applications	Panesar et al. (2015)
Enzymes	Fungi and bacteria—Amylases, proteases, lipases	They are cultured on food wastes.	Formulation of detergents, processing of food, and textile industries	Arun et al. (2020) and Madeira Jr et al. (2017)

6.2 Wastes from Food Processing

Wastes such as molasses, whey, and spent grains are commonly used as growth media for microorganisms. They contain organic compounds which serve as nutrients for the growth and metabolite production by microbes. The microorganisms that are cultured on these wastes produce different metabolites which are useful in many industrial applications and have high commercial value (Table 3). Utilization of these wastes for metabolite production is seen as a sustainable way of waste management and economic value.

6.3 By-products from Industries

Microorganisms can be grown from these by-products through fermentation and enzymatic processes in which metabolites can be produced. The highlights in Table 4 present the various products from the industries that can be utilized as a media in microorganisms growth and metabolite production which are useful in several industrial applications.

7 Sources and Uses of Metabolites

7.1 Metabolites

Metabolites are products of microbial metabolism, and it is a large-scale study that involves detecting, identifying, and potentially quantifying small organic molecules

Table 4 Metabolites and their sources from industrial by-products

Metabolites	By-product sources	Constituents	Uses	References
Molasses	Sugar refining processes	Rich in glucose, sucrose, fructose and good for fermentation of microbes	Production of ethanol, lactic acid, citric acid and enzymes	Lopes et al. (2013)
Glycerol	Biodiesel production	Bacteria and yeast can ferment glucose to produce metabolites	Production of metabolites such as citric acid, ethanol and 1,3-propanediol	Embaby et al. (2018)
Whey	Cheese production	Fermentation of lactose in whey to produce lactic acid	Used in industrial and preservation processes	Sash et al. (2018)
Spent grains	Brewing industry	Residual sugars, fibers, and proteins	They produce enzymes, organic acids, and single-cell proteins.	Charalampia et al. (2017)
Citrus peel wastes	Citrus juice production	Sugar, pectin, and essential oils	Production of ethanol, citric acid, limonene, and pectinases which are enzymes	Ahmadi et al. (2015) and Kantifedaki et al. (2018)
Crude glycerol	Biodiesel production	It contains impurities.	Production of butanol, bihydrogen, and biofuels through fermentation by *Clostridium* species	Ahmadi et al. (2015) and Chilakamarry et al. (2021, 2022)

within cells, biofluids, tissues, or organisms (Fiehn, 2002). The goal of metabolomics is to provide a comprehensive print of small metabolic molecules in cells and biofluids without any bias natural to intensive studies of metabolites (Worley & Powers, 2013). Metabolome is the study of the whole set of small metabolites molecules in the body system while metabolites are essentially small molecular weight by-products of metabolic activities of the growing cells or microorganisms (Pinus & Villas-Boas, 2017). Metabolites that are contained in the cell envelope or cell membrane and require the rupture of cells mechanically or chemically by using extracting solvents are referred to as intracellular metabolites (Pinus et al., 2017).

7.2 Sources of Metabolites

Organisms currently used in microbiology and biotechnology for their metabolic products are generally derived from bacteria and eukaryotes. Both fungi and higher multicellular organisms such as plants and animals belong to the eukaryote's domain. Plants and animals, and their cell culture are also used in the biotechnology industry majorly for their products; however, the good criteria that make microorganisms more practical for use are associated with their properties, which enable them to survive and be productive when the environmental conditions is harsh

(Oladipo et al., 2022). They have the following biotechnological advantages over plants and animals for product development:

- *Growth*: They can grow rapidly in a record time of about 15 min under a conducive environment when compared to plants and animals, which could be longer.
- *Space*: It refers to the space required for their growth or culture. Microorganisms require small space and still give higher yields.
- *Environmental conditions*: They can withstand some of the environmental conditions that may not permit other organisms like plants and animals to thrive. They are also not affected by those diseases that affect plants and animals.

Microbial metabolites are either primary or secondary metabolites. However, in some instances, it is quite difficult to differentiate between primary and secondary metabolites since most of their intermediary overlaps (Verpoote et al., 2000). For instance, amino acids can be seen as primary or secondary metabolites (Thirumurugan et al., 2018).

- *Primary Metabolites.*

Primary metabolites are central and present in any autonomously growing organisms or cells, which are directly involved in the basic functions of living organisms such as normal growth, body development, and reproduction. They include lactic acid, amino acids, sugar, etc. which are seen to perform some intrinsic or physiological function in the body of the organism (Mitra et al., 2023). Primary metabolites are products from microbes that are apparent during the exponential growth phase, and whose buildup is a major part in the growth process normally. They are anabolic metabolism intermediates and end-products and are very useful to the cell as building blocks for essential macromolecules (e.g., nucleotides, amino acids), or can be converted to coenzymes (e.g., vitamins). Catabolic metabolism produces some primary metabolites (e.g., citric acid, acetic acid, and ethanol), and they are not useful in synthesizing cellular constituents. Their growth is related to the amount of energy produced and the utilization of substrate. The most important industrial primary metabolites are solvents, amino acids, vitamins, nucleotides, and organic acids (Singh et al., 2017; Awuchi et al., 2022). Primary metabolites are directly involved in normal growth, development, and reproduction. An example of large-scale primary metabolites industrially produced is Ethylene.

- *Secondary Metabolite.*

Secondary metabolites are small metabolic molecules from microbial metabolisms; they have a close connection to primary metabolites through cofactors and precursors (Khetan et al., 1999). Kossel (1891) explained that their presence in the body system is by chance and mostly due to some physiological alterations that are not necessarily important in the day-to-day activity of the body system (Thirumurugan et al., 2018). However, their deficiency over a long time could cause immune impairment and threat to survivability, fecundity, and aesthetics of the organism. The presence and synthesis of the secondary metabolites could be noticed in ecologically deprived organisms even within the same phylogenetic cluster (Tiwari & Rana, 2015). These molecules have atypical makeup, and their constituents are controlled by the organism's nutrition, growth, and enzyme activation. The synthesis of secondary metabolites is generally not common for all bacteria, it is blocked during the growth phase and very low during the inert growth phase. The increasing emergence of microbial secondary metabolism exposes more of the central roles played by secondary metabolites in biological systems (O'Brien et al., 2011), however, there is little knowledge on its metabolism (Damain, 2007). Therefore, there is a need for an increased investigation in improving the synthesis of microbial secondary metabolites (Zothanpuia et al., 2018).

During the growth stages of microorganisms, they produce and secrete diverse primary and secondary metabolites into the surrounding environment. Some specific secondary metabolites, however, require the use of a precursor or cofactor to stimulate either the constructive or destructive microbial metabolism that produces them (Khetan et al., 1999; Thirumurugan et al., 2018). Therefore, in culture media, microbial cells secrete low molecular weight metabolites that is extracellular, and the complete sets of these metabolites are often called exometabolome, which covers both the entire metabolites secreted by the culture cells and the components of the microbial culture media found in the supernatant (Mashego et al., 2007; Pinus & Villas-Boas, 2017). Microbial secondary metabolites are the major bioactive raw materials used in medicine (Blin et al., 2016). Humans consciously or not use secondary metabolites often in medicine, flavorings, pigments, and in recreational drugs (Biologyreference.com 2018). The added value of secondary metabolites is in their specificity to individual species (Picherrky & Gang, 2000). Research has shown that secondary metabolites can affect species in varying ways (Jensen et al., 2014). Consumption of these organic molecules in either way has an impact that is strong on human food, and can be protective against viruses, bacteria, and protozoan parasites (Martin et al., 2018). They are produced by a diverse range of bacteria and fungi and are useful in food, chemical, and nutraceutical industries. Microbial fermentation manufactures most metabolites, rather than synthesizes from chemicals because of the economic competitiveness and the production of isomeric forms of metabolites that is biologically useful (Sharma et al., 2020).

Microbial fermentation produces several industrially important chemicals, for example, glycerol, polyhydroxyl, alcohols, organic acids, and solvents. This is due to their cost-effective production, especially now that the cost implication for manufacturing using petroleum is high (Ruiz et al., 2010). Microbial secondary metabolites secretions are now increasingly being harvested by humans and utilized directly or indirectly to improve their health concerning the potent biological functions of the metabolites as; antiparasitic agents, antitumor agents, antimicrobial against, enzyme inhibitors, immunosuppressive agents, and antibiotics in cases previously treated only with synthetic drugs (Demain & Fang, 2000; Thirumurugan et al., 2018). Several investigations have shown that secondary metabolites from some species of microorganisms are suitably used and have been documented. *Bacillus* sp., Coagulin from *B. coagulans*, and Megacin from *B. megaterium* are some examples (Lisboa et al., 2006).

7.3 Classification of Secondary Metabolites

Based on diversification in structural forms, biosynthesis, and functions, there are about 2,140,000 secondary metabolites presently documented. They are classified into five types, namely (Fig. 2), terpenoids and steroids, fatty acids and their derivatives, and polyketides, nonribosomal polypeptides, alkaloids, and enzyme cofactors (McMurry, 2015).

- *Terpenoids*: They are primarily derived biotechnologically from isopentenyl diphosphates and are presently around 35,000 chemicals identified.
- *Alkaloids*: They are derivatives of amino acids and are around 12,000 chemically known chemicals with an amine structure.
- *Fatty acids and their derivatives, and polyketides*: There are around 10,000 chemically identified and produced biotechnologically from acyl precursors such as propionyl CoA and acetyl CoA.
- *Nonribosomal polypeptides*: They are derivatives of amino acids.
- *Enzyme cofactors*: They are nonprotein and are small molecular-weight enzyme compounds.

Quantitatively, extracts from plants account for about 70% of secondary metabolites, while other sources including bacteria, fungi, and marine organisms follow

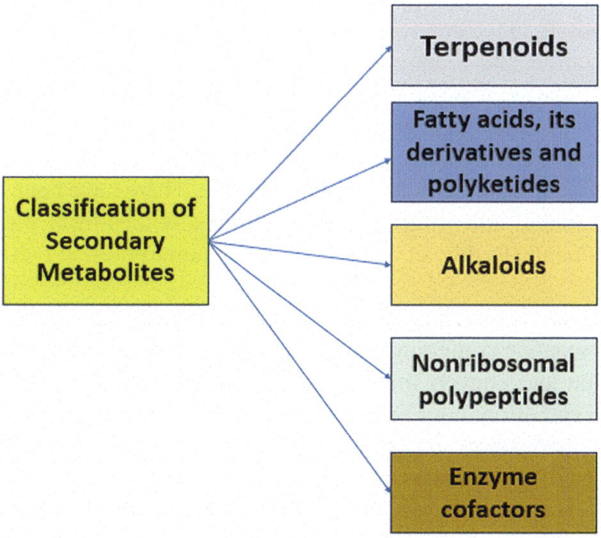

Fig. 2 Classification of secondary metabolites

(Berdy, 2005). Generally, examples of some secondary metabolites are naphthalenes, alkaloids, nucleosides, phenazines, antibiotics, quinolines, peptides, terpenoids, and growth factors. Plants or fungi can be sources of secondary metabolites. Plants are sources of secondary metabolites as they can produce and synthesize diverse groups of organic compounds (Ogwu et al., 2023a, b). Secondary metabolites from plant sources are metabolic intermediates or products that are not essential to the growth and life of the plant producing but are important in plant interaction with their immediate environment and are produced when stress occurs because of bacterial, fungi, and viral infections (Han et al., 2021). Plant secondary metabolites are classified by their chemical structure and can be divided into four major classes namely terpenes, phenylpropanoids (i.e., phenolics), polyketides, and alkaloids. Fungal sources of secondary metabolites are divided into three main classes which are: polyketides, nonribosomal peptides, and terpenes. The most known fungal secondary metabolite is penicillin discovered by Alexander Fleming in 1928.

8 Microorganisms and Sustainable Metabolites Production

There have been several studies on the production of microbial metabolites on an industrial scale. Solid-state fermentation (SSF) is a valorization process for sustainable and profitable product production, but the difficulty faced with this approach is the lack of its basic concepts (Srivastava et al., 2019). The approach is a green approach that converts waste into profitable products. Wastes from agricultural sources are useful substrates for secondary metabolites and enzyme production through SSF because they have a high organic content (Sunish & Thazeem, 2023; Ogwu et al., 2024). The SSF approach is advantageous as it achieves a sustainable breakdown of waste in the environment which has increased due to rapid industrialization, urbanization, and economic progress in the Global South (Guo et al., 2021). As against SSF, forms of waste disposal such as landfills or incineration of wastes contaminate the groundwater by leaching organic pollutants into the groundwater and over 25 times carbon that is destructive into the atmosphere hence resulting in to increase in greenhouse gases and climate change. There have been reports on the valorization of wastes into valued products and metabolites from wastes generated in the environment (Reddy et al., 2021; Chilakamarry et al., 2021; Nizzy & Ogwu, 2024).

SSF is a tool that is useful in transforming agricultural wastes into products with value such as bioactive compounds, biofuels, and metabolites. They use wastes that are low in value and convert them into a resource (Saratale et al., 2020; Ogwu, 2019a, b). SSF can allow microorganisms to grow in environments where water is not free or in limited supply. It can mimic the natural environment of mold and fungi reduce susceptibility to contamination by bacteria and increase efficiency for enzymes. However, the type of substrate dictated the formation of enzymes and metabolites (Chilakamarry et al., 2022). Based on the abundance of resources and biodiversity in the Global South, different microorganisms are used in metabolite production from biobased products. These microbes play a role that is important in the transformation of renewable feedstocks into compounds that are valuable in agricultural, industrial, and healthcare applications. Some microorganisms commonly used for these processes include:

- *Saccharomyces cerevisiae*: It is commonly called yeast and is used in sugar fermentation from biobased feedstocks such as corn, cassava, and sugarcane to produce a biofuel called ethanol (Jouhten et al., 2016; Hilares et al., 2018). It is used in the production of metabolites such as flavor, vitamins, and organic acid compounds from biobased substrates (Wang et al., 2016).
- *Aspergillus* spp.: Numerous species of filamentous Aspergillus which is a fungus from biobased sources in the Global South (Youssef & Singab, 2021). For instance, *A. niger* is used in the fermentation processes of agricultural residues and substances rich in starch to produce citric and gluconic acids which are organic acids and used for different industrial purposes (Vadlapudi et al., 2017).
- *Streptomyces* spp.: They are forms of bacteria that are filamentous, are prolific producers of secondary metabolites, and are principal in different biological activities (Alam et al., 2022). Strains of Streptomyces isolated from Global South soils can be used in the manufacture of antifungals, antibiotics, and bioactive compounds such as tetracycline, erythromycin, and streptomycin (Lacey & Rutledge, 2022).
- *Lactobacillus* spp.: They have used fermentation from biobased products to produce lactic acid. Lactobacillus is a lactic acid bacterium and the acid produced is important in the production of bioplastics, pharmaceuticals, and additives of food (Chamberlain et al., 2022). The strains of Lactobacillus are used in production of probiotics and fermented foods (Lee et al., 2021).
- *Corynebacterium glutamicum*: It is used in industrial production of lysine and glutamic acid which are amino acids from biobased products like cassava, molasses, and corn steep liquor. These acids are useful in food and feed additives, biodegradable polymers, and pharmaceuticals (Arun et al., 2020; Zahoor et al., 2012).
- *Rhizopus* spp.: They are filamentous algae used in fermentation to produce enzymes and metabolites (lipids and organic acids). The strains are used in the production of citric acid which is a useful ingredient in the beverage and food industry, and biodiesel production from feedstocks rich in lipids (Zaveri et al., 2022).
- *Pichia pastoris*: They are yeast species used for recombinant protein production, metabolites, and proteins in the Global South by a process of genetic engineering. *P. pastoris* can secrete high abundance of heterologous proteins and have a robust growth on carbon from biobased sources (Kumokita et al., 2022, 2023).

These microbes are part of the diverse microbes found in the Global South which are harnessed from the production of metabolites in a sustainable manner from biobased feedstocks (Rajeswari et al., 2021). When these are leveraged, biobased industries can be developed in the Global South, thereby contributing to the conservation of resources, economic development, and sustainability of the environment.

8.1 Microorganism Used in Metabolite Production

The production of bacteria from secondary metabolites begins with a stationary phase the lack of nutrients or in response to environmental stress. When bacteria are secreted, they secrete exotoxin, which is a poisonous compound, whereas the prokaryotic cell wall secretes endotoxins. They belong to the following classes: phenazine, polyketides, nonribosomal peptides, ribosomal peptides, glucosides, and alkaloids includes:

- Pyocyanin, from *Pseudomonas aeruginosa.*
- Phenazines from *Pseudomonas* sp. and *Streptomyces* sp.
- Avermectin, from *Streptomyces avermitilis.*
- Epothilones, macrolactones from the soil-welling myxobacterium *Sorangium cellulosum.*
- Erythromycin, *Saccharopolyspora erythraea.*
- Nystatin, from *Streptomyces noursei.*
- Rifamycin, from *Amycolatopsis rifamycinica.*
- Bacitracin, from *Bacillus subtilis* (Tracy strain).
- Gramicidin, from *Brevibacillus brevis.*
- Polymyxin, from *Paenibacillus polymyxa.*
- Ramoplanin, from *Actinoplanes* strain ATCC 33076.
- Teicoplanins, from *Actinoplanes teicomyceticus.*
- Vancomycin, from the soil bacterium *Amycolatopsis orientalis.*
- Microcins, bacteriocins such as microcin V from *Escherichia coli.*
- Thiostrepton, from several strains of streptomycetes, e.g., *Streptomyces azureus.*
- Nojirimycin, an iminosugar from a class of *Streptomyces* species.
- Tetrodotoxin is a neurotoxin produced by *Pseudoalteromonas* and other bacteria living in symbiosis with animals such as, e.g., puffer fish.

8.2 Production Process and Extraction Methods

Certain characteristics are important for microbes as good candidates in industrial microbiology and biotechnology production of raw materials as:

- The organisms must be able to grow in a simple medium and preferably with little or no growth factors aside those in the industrial medium. These certainly will reduce the cost of production (Linder, 2019; Kumar et al., 2022).
- The organisms should be able to grow efficiently at an optimal growth rate in the medium used. This could save production time, input materials, and increase production (Linder, 2019).

- The organisms should also be a good candidate with the desired properties such as being nonpathogenic and producing desired products.
- The products it produces, since it is desired, should be without toxins (Okafor, 2007).
- The microbes should be the one that can be easily manipulated in deriving the right benefits or products.
- The organism should also possess tolerance attributes to extreme environmental conditions.
- The organisms should be the type that can be used to derive a product with ease, using a basic or simple instrument.

With the afore-mentioned the production process of raw materials from microorganisms can be successfully initiated. The production process of metabolites or bioactive substance of industrial relevance from microbes basically starts with the suitable preparation of selected microorganisms, optimizing the growth conditions, such as the selected and suitable media with adequate nutrient compositions, required suitable temperature, and pH for microbial growth, and then inducing the production of desired metabolites or biological molecules (Abo Elsoud & El-Kady, 2019). Selection of suitable isolates or microbes in the industrial production of raw materials is essential to deriving the right products. In specific, only a very small percentage of bacteria is harmful; in other words, they are nonpathogenic, hence harmless to humans. Therefore, it's important to screen and select among other criteria organisms that are nonpathogenic to produce industrial raw materials (Sharma et al., 2020). Besides the nonpathogenic characteristics of microorganisms, other qualities such as tolerance to harsh environments, ease of manipulation and rapid growth are essential for microorganisms in industrial processes (Mazur et al., 2023). Usually, the microbes are first tested biochemically for their pathogenic levels, then their resistance to extreme conditions is tested, and other inert qualities for a specific production process.

The preparation of suitable media is based on the method of optimization. Recently, a robust means such as statistical optimization has been in use which has the capacity to optimize the nutrient requirements, pH, and temperature that is best for the growth and release of extracellular metabolites.

8.3 Substrate or Media for Production of Microbial Products

Different substrates are used for different microbes during industrial production of raw materials or their product. The substrate is adjusted to the required pH and temperature suitable for the microorganisms' growth and metabolites yield. For instance, bacteria are more suitably cultured in a liquid state fermentation, while fungi are mostly cultured in a solid state. Liquid state fermentation process is that is based on the use of a culture media containing water and uses bioreactors which are used intercellularly or extracellularly (Costa et al., 2018). The solid-state fermentation is the process of cultivating microorganisms on materials that are solid, and absence of free liquid (Jorge et al., 2018). Usually, the best range for temperature and pH for most bacteria are 20–45 °C (Lyubertsky et al., 2020) and close to 7.0 (Jin & Kirk, 2018), respectively, and for fungi are 20–30 °C (Dix & Webster, 1995) and 3.0–8.0 (Rousk et al., 2010), respectively. It is important to note the fermenter as well, as on nowadays industrial scale of production; fermenters are made with stainless steel because of its resistance to corrosion. The fermenter is the container containing the medium that microorganisms are grown and as well fermented at regulated conditions (Vassileva et al., 2021).

8.4 Harvesting or Extraction Methods

The products sought for extraction as raw materials for industrial, pharmaceutical or agriculture purposes could be the microbial cells itself like in yeast production, or found lodged in the cells like in streptomycin and enzymes production or free in the medium as penicillin (Gil-Martin et al., 2022). Generally, the procedures employed for extraction depend on the materials being extracted, and examples are solid removal (i.e., insoluble materials) and primary product isolation.

- *Solid removal*: It could be achieved through filtration, whole-broth treatment, coagulation and flocculation, centrifugation, and form fractionation.
- *Primary product isolation*: This is done after the fermented broth are separated into soluble and insoluble fractions, the desired products in the case of bound enzymes or antibiotics may still be bound to the bacterial cells as, therefore, any of the following suitable means such as heat, mechanical and nonmechanic disruption methods can be used to disrupt the cells.

  Some of the means for primary product isolation are:

- *Liquid extraction*: It is also known as solvent extraction or liquid-liquid extraction. It is widely used in the industry, whereby one solute is transferred from one solvent to another where it will be more soluble.
- *Dissociation extraction*: This is where weak bases (fermentation products) are extracted under high pH conditions and weak acids (fermentation products), under conditions of low pH, resulting in an extraction of the solute and materials completely.
- *Ion-exchange adsorption*: This involves the use of chromatography and charcoal adsorption (Millz, 1997; Mielko et al., 2021). The separated products are thereby subjected to a step-by-step purification, usually using chromatography, carbon decolorization, and then crystallization. The final products are then processed and dried for use (Mielko et al., 2021).

The cell disruption method, otherwise known as cell lysis and is a delicate process and it occurs due to the resistance of the cell wall to the high osmotic pressure of the biological molecules inside them (Guaterio et al., 2023). The desired products are carefully extracted in such a way that the intracellular molecules will not be denatured. There are basically two cell disruption methods which are the mechanical cell disruption and nonmechanical cell disruption.

- *Mechanical cell disruption*: The equipment used includes homogenizers which uses the principle of pressure to cause explosion of cells to release the intracellular molecules. The ball mills that cause agitation in suspension with small abrasive particles, breaks the cells because of the shear forces, beads grinding and colliding, ultrasonic disruption is caused through high frequency sound that is produced electronically and transmitted through a metallic tip to an appropriately concentrated cellular suspension (Millis, 1997; Guaterio et al., 2023).
- *Nonmechanical cell disruption*: Cells can be made to disrupt by permeating through various means such as chemical permeabilization, whereby organic solvents are used on the cells by creating canals through the cell membrane for intracellular molecules to leach out; mechanical permeabilization, where cells could be exposed to osmotic shock to damage the cell wall, followed by the release of their intracellular contents like enzymes; enzymatic permeabilization, which is another means where the cells permeate; here, the ethylenediaminetetraacetic acid (ETDA) is used as a destabilizer of the outer membrane of most Gram-negative cells, making the peptidoglycan layer accessible to the enzymes used (Millis, 1997; Mielko et al., 2021).

9 Relevance of Producing Microbial Metabolites for Fish Nutrition

Nowadays, biobased economies such as microbial biotechnologies are highly encouraged in promoting economic growth that is inclusive and sustainable, and employment that is productive (Timmis et al., 2017). As the population of human is increasing, the demand for animal food sources also increases, including farmed livestock and fish, and in turn feed products including fish-derived ingredients (Thiao & Bunting, 2022). A good production of animals is a function of good nutrition, and this is essential to the economic production of a healthy and high-quality product and is also critical since the composition of feed is approximately 50% of the variable cost of production in fish farming (Tadese, 2022). Fish nutrition has tremendously improved in recent years because of the development of commercial diets that are balanced, and produced through biotechnological means that promote the growth of fish optimally and better health of fish (Craig, 2017). The development of new species-specific diet formulation supports the aquaculture industry as it expands to satisfy the increasing demand for affordable, safe, high-quality fish products essential for community livelihoods and food security.

Microbial biotechnology is noted as one of the ancient technological approaches, with an old history of fermentation as a civilization (Linder, 2019). Fish farming involving nutrition has been practiced for millennia as an extensive aquaculture, like the free grazing practice with livestock. This has been able to evolve to semi-intensive and then intensive aquaculture, because of the nutritional needs for fish growth and wellbeing, as well as the economic expectations of fish farmers (Hardy et al., 2022; Umma et al., 2023). Therefore, better nutrition outcomes are obtained when farmed aquatic products are monitored and nutrient contents are improved. This can be achieved through innovative feed additives from a microbial source, which enhances the digestibility and bioavailability of nutrients for fish. Such ingredients are novel and suitable for substituting and/or reducing the dependence on conventional feedstuffs like fishmeal and soya bean meal. Furthermore, probiotics and their products are known for the increase in animal resistance to pathogenic challenges which is an advantage for a successful fish farm business (FAO, 2022).

The initial wide gap between fish and animal nutrition saw fish nutrition lagging, however, this narrative has changed during the past three decades through the rapid research and development in the nutrition of fish which has led to the emergence of suitable semipurified diets for fish growth and prime health (Ronald et al., 2022). This development is obvious in the inland and marine fish production which has increased globally at 76.8% (12.6–54.4 million tons) and 72.2% (9.2–33.1 million tons) live weight equivalent respectively, with a 48.2% (81.6–157.4 million tons) increase in human consumption and 26.9% (5.7–7.8 billion) increase in human population from the 1990s to 2020 (FAO 2022). However, these developments have been overstretched with the huge percentage of resources deployed for nutrition in aquaculture. Therefore, there is a need for changes, in adopting modern diet formulations to maximize production performance, facilitate resilience in immune systems, and achieve an edible product, while also responding to the social licenses and minimizing the footprints in the environment (Rombenso et al., 2022). These new approaches evolving from sound research and development efforts of academic and commercial sectors, and pressures from regulation are on the rise, especially in reassessing the macro and essential nutrients in feed, and importantly, by-products or raw material derived from plants and microbial source for their beneficial purposes in fish nutrition (Rombenso et al., 2022). The application of biotechnology in the production of feed, would either reduce the dependency on the regular ingredients used for fish nutrition or boost their optimal performance in fish.

The use of raw materials or extracts from plants and microbial sources is known for their functionality in areas of biochemistry and physiology as feed additives in fish and other animals. The addition of these beneficial additives in the fish meal daily has the potency of mitigating stress from typical stressors like overcrowding, and handling among others, and serves as precursors of hormones, the substrate of energy, cofactors, and other essential elements (Herrera et al., 2019). These exceptional applications emerging especially out of the advances in microbial biotechnology research are now expanding in diagnostics, pharmaceuticals, agriculture, immune prophylactics, microbial therapies, foods fermented and supplements, bioindustrial products, fish and animal feeds, recycling technologies, and bioremediation. The diverse application of microbial biotechnology has the prospects of profitable business opportunities and is equally on the same level as in the products from chemicals and their chemistry (Timmis et al., 2017).

10 Conclusion

Biobased products which are products from renewable resources have various bioactive and bio properties that are useful for the environment. A major raw material is the metabolites which are produced from microbial metabolism. They are found in the cell membrane and can be extracted by rupturing the cells by mechanical or chemical means. They can be sourced from fungi, bacteria, and multicellular organisms belonging to the eukaryotic group and can appear in primary and secondary forms. The former are associated with the basic functions in living organisms while the latter serves as cofactors or precursors of the former. The classifications of metabolites are based on their structural forms, functions, and biosynthesis. Solid-state fermentation is a sustainable metabolite production approach as it converts the wastes into products that are profitable from renewable resources. This approach is necessary due to the increased urbanization, industrialization, and economic process in Global South countries. This approach saves about 25 times of destructive carbon from sources such as waste disposal in landfills and incineration, which is released into the atmosphere and resulting in climate change. The approach is very useful in transforming wastes from agriculture into valuable bioactive compounds such as biofuels and metabolites. Organisms are usually allowed to grow, and the metabolite produced is a function of the type of microorganism and biobased product. Examples of such are *Saccharomyces cerevisiae, Aspergillus* spp., *Streptomyces* spp., *Lactobacillus* spp., and *Rhizopus* spp., all of which produce different types of raw materials from biobased products. Lastly, the metabolites and other materials are useful in aquaculture industries as the raw materials for feed production. The extracts are known for their biochemical and physiological functionality as feed additives. This inclusion helps to alleviate stress, energy sources, hormone precursors, and cofactors. With these, the prospects of raw materials from biobased products are quite enormous, and therefore sustainable activities should be encouraged in its production and extraction.

References

Abdul Hakim, B. N., Xuan, N. J., & Osian, S. N. H. (2023). A comprehensive review of bioactive compounds from lactic acid bacteria: Potential functions as functional food in diabetes and the food industry. *Food, 12*(15). https://doi.org/10.3390/foods12215850

Abedi, E., & Hashemi, S. M. B. (2020). Lactic acid production-producing microorganisms and substrates sources-state of art. *Heliyon, 6*(10), ee04974.

Abo Elsoud, M. M., & El-Kady, E. M. (2019). Current trends in fungal biosynthesis of chitin and chitosan. *Bulletin of the National Research Centre, 43*, 59.

Adam, P. S., Borrel, G., & Gribaldo, S. (2018). Evolutionary history of carbon monoxide dehydrogenase/acetyl-CoA synthase, one of the oldest enzymatic complexes. *Proceedings of the National Academy of Sciences*, 201716667. https://doi.org/10.1073/pnas.1716667115

Ahmadi, F., Zamiri, M. J., Khorvash, M., Banihashemi, Z., & Bayat, A. R. (2015). Chemical composition and protein enrichment of orange peels and sugar beet pulp after fermentation by two *Trichoderma* species. *Iranian Journal of Veterinary Research, 16*, 25.

Akhtar, N., & Amin-ul Mannan, M. (2020). Mycoremediation: Expunging environmental pollutants. *Biotechnology Reports, 26*, e00452. https://doi.org/10.1016/j.btre.2020.e00452

Akoijam, N., Kalita, D., & Joshi, S. R. (2022). Bacteria and their industrial importance. In P. Verma (Ed.), *Industrial microbiology and biotechnology*. Springer. https://doi.org/10.1007/978-981-16-5214-1_2

Alam, K., Mazumder, A., Sikdar, S., Zhao, Y.-M., Hao, J., Song, C., Wang, Y., Sarkar, R., Islam, S., Zhang, Y., & Li, A. (2022). *Streptomyces*: The biofactory of secondary metabolites. *Frontiers in Microbiology, 13*, 968053. https://doi.org/10.3389/fmicb.2022.968053

Aluko, O. O., Liu, Z., & Sun, X. (2023). The interplay of carbon and nitrogen distribution: Prospects for improved crop yields. *Modern Agriculture, 1*, 57–75.

Aminian, A., & Motamedian, E. (2023). Investigating ethanol production using the *Zymomonasmobilis*crude extract. *Scientific Reports, 13*, 1165.

Amoo, A. E., Enagbonma, B. J., Ayangbenro, A. S., & Babalola, O. O. (2021). Biofertilizer: An eco-friendly approach for sustainable crop production. *Food Security and Safety, 1*, 647–669.

Arun, K. B., Madhavan, A., Sindhu, R., Binod, P., Pandey, A., Reshmy, R., et al. (2020). Re-modelling agro-industrial and food wastes intovalue-added bio actives and biopolymers. *Industrial Crops and Products, 154*, 112621. https://doi.org/10.1016/j.indcrop.2020.112621

Atares, L., Chirait, A., Gonzalez-Martinez, C., & Vargas, M. (2024). Production of polyhydroxyalkanoates for biodegradable food packaging applications using *Haloferax mediterranei* and agrifood wastes. *Food, 13*(6), 950.

Awuchi, C. G., Chukwu, C. N., Iyiola, A. O., Noreen, S., Morya, S., Adeleye, A. O., Twinomuhwezi, H., Leicht, K., Mitaki, N. B., & Okpala, C. O. R. (2022). Bioactive compounds and therapeutics from fish: Revisiting their suitability in functional foods to enhance human wellbeing. *BioMed Research*, 3661866. https://doi.org/10.1155/2022/3661866

Barrios-Gonzalez, J., & Tarrago-Castellanos, M. R. (2017). Solid-state fermentation: Special physiology of fungi. In J. M. Merillon & K. Ramawat (Eds.), *Fungal metabolites. Reference series in phytochemistry*. Springer. https://doi.org/10.1007/978-3-319-25001-4_6

Bäumler, A., & Sperandio, V. (2016). Interactions between the microbiota and pathogenic bacteria in the gut. *Nature, 535*, 85–93. https://doi.org/10.1038/nature1884

Benini, S. (2020). Carbohydrate-active enzymes: Structure, activity, and reaction products. *International Journal of Molecular Sciences, 21*(8), 2727. https://doi.org/10.3390/ijms21082727

Bernhard, A. (2010). The nitrogen cycle: Processes, players, and human impact. *Nature Education Knowledge, 3*(10), 25.

Berry, P. M., Sylvester-Bradley, R., Philipps, L., Hatch, D. J., Cuttle, S. P., Rayns, F. W., & Gosling, P. (2002). Is the productivity of organic farms restricted by the supply of available nitrogen? *Soil Use and Management, 18*, 248–255. https://doi.org/10.1079/SUM2002129

Cason, E. D., Mahlomaholo, B. J., Taole, M. M., Abong, G. O., Vermeulen, J.-G., de Smidt, O., Vermeulen, M., Steyn, L., Valverde, A., & Viljoen, B. (2020). Bacterial and fungal dynamics during the fermentation process of *Sesotho*, a traditional beer of Southern Africa. *Frontiers in Microbiology, 11*, 1451. https://doi.org/10.3389/fmicb.2020.01451

Cavalier-Smith, T. (1992). The origin, losses and gains of chloroplasts. In R. A. Lewin (Ed.), *Origins of plastids*. Springer.

Cavicchioli, R., Ripple, W. J., Timmis, K. N., Azam, F., Bakken, L. R., Baylis, M., Behrenfeld, M. J., Boetius, A., Boyd, P. W., Classen, A. T., Croether, T. W., Danovaro, R., Foreman, C. M., Huisman, J., Hutchins, D. A., Jansson, J. K., Karl, D. M., Koskella, B., Welch, D. B. M., Martiny, J. B. H., Moran, M. A., Orphan, V. J., Reay, D. S., Remals, J. V., Rich, V. I., Singh, B. K., Stein, L. Y., Stewart, F. J., Sullivan, M. B., van Oppen, M. J. H., Weacer, S. C., Webb, E. A., & Webster, N. S. (2019). Scientists warning to humanity: Microorganisms and climate change. *Nature Reviews Microbiology, 17*, 569–586.

Chamberlain, M., O'Flaherty, S., Cobian, N., & Barrangou, R. (2022). Metabolic analysis of *Lactobacillus acidophilus, L. gasseri, L. crispaus* and *Lacticasebacillus rhamnosus* strains in the presence of Pomegranate extract. *Frontiers in Microbiology, 13*, 863228. https://doi.org/10.3389/fmicb.2022.863228

Charalampia, D., Antonios, K. E., Constantina, N., & Haralabos, K. C. (2017). Current trends and emerging technologies in biopigment production processes: Industrial food and health applications. *International Journal of Horticulture and Food Science, 1*, 33–46.

Chen, G.-Q. (2010). Introduction of bacterial plastics PHA, PLA, PBS, PE, PTT, and PPP. In G. G.-Q. Chen (Ed.), *Plastics from bacteria: Natural functions and applications, microbiology monographs* (pp. 1–16). Springer. https://doi.org/10.1007/978-3-642-03287-5_1

Chen, W., & Viljoen, A. M. (2010). Geraniol – A review of a commercially important fragrance material. *South African Journal of Botany, 76*, 643–651. https://doi.org/10.1016/j.sajb.2010.05.008

Chen, Y., Wang, Y., Paez-Espino, D., Poiz, M. F., & Zhang, T. (2021). Prokaryotic viruses impact functional microorganisms in nutrient removal and carbon cycle in wastewater treatment plants. *Nature Communications, 12*, 5398.

Chiaranunt, P., & White, J. F. (2023). Plant beneficial bacteria and their potential applications in vertical farming systems. *Plants, 12*(2), 400.

Chilakamarry, C. R., Sakinah, A. M., Zularisam, A. W., Pandey, A., & Vo, D.-V. N. (2021). Technological perspectives for utilization of wastes glycerol for the production of biofuels: A review. *Environmental Technology and Innovation, 24*, 101902. https://doi.org/10.1016/j.eti.101902

Chilakamarry, C. R., Mimi Sakinah, A. M., Zularisam, A. W., Sirohi, R., Khilji, I. A., Ahmad, N., & Pandey, A. (2022). Advances in solid-state fermentation for bioconversion of agricultural wastes to value-added products: Opportunities and challenges. *Bioresource Engineering, 343*, 126065. https://doi.org/10.1016/j.biortech.2021.126065

Coelho, M. C., Malcata, F. X., & Silva, C. C. G. (2022). Lactic acid bacteria in raw-milk cheeses: From starter cultures to probiotic functions. *Food, 11*(15), 2276. https://doi.org/10.3390/foods11152276

Colla, G., Nardi, S., Cardarelli, M., Ertani, A., Lucini, L., Canaguier, R., & Rouphael, Y. (2015). Protein hydrolysates as biostimulants in horticulture. *Scientia Horticulturae, 196*, 28–38. https://doi.org/10.1016/j.scienta.2015.08.037

Compant, S., Samad, A., Faist, H., & Sessitsch, A. (2019). A review on the plant microbiome: Ecology, functions and emerging trends in microbial application. *Journal of Advanced Research, 19*, 29–37.

Costa, J. A. V., Treichel, H., Kumar, V., & Pandey, A. (2018). Chapter 1 – Advances in solid-state fermentation. In A. Pandey, C. Larroche, & C. R. Soccol (Eds.), *Current developments in biotechnology and bioengineering* (pp. 1–17). Elsevier. https://doi.org/10.1016/B978-0-444-63990-5.00001-3

Craig, S. (2017). *Understanding fish nutrition, feeds, and feeding* (pp. 420–256). Virginia Cooperative Extension. College of Agriculture and Life Sciences, Virginia Tech, Virginia State University Publication.

Da Silva, V. D., Cruz, I. A., Torres, N. H., Figueiredo, R. T., de Melo, L., de Resende, I. T. F., et al. (2019). Agro-industrial wastes: Environmental toxicology, risks, and biological treatment approaches. In R. N. Bharagava (Ed.), *Environmental contaminants: Ecological implications and management* (pp. 1–23). Springer. https://doi.org/10.1007/978-981-13-7904-8_1

Danilova, I., & Sharipova, M. (2020). The practical potential of Bacilli and their enzymes for industrial production. *Frontiers in Microbiology, 11*, 1782. https://doi.org/10.3389/fmicb.2020.01782

David, A., Tripathi, A. K., & Sani, R. K. (2020). Acetate production from cafeteria wastes and corn Stover using a thermophilic anaerobic consortium: A prelude study for the use of acetate for the production of value-added products. *Microorganisms, 8*, 353.

de Bonadio, M. P., de Freitas, L. A., & Mutton, M. J. R. (2018). Carotenoid production in sugarcane juice and synthetic media supplemented with nutrients by *Rhodotorula rubra* l02. *Brazilian Journal of Microbiology, 49*, 872–878. https://doi.org/10.1016/j.bjm.2018.02.010

De Chiara, I., Marasco, R., Gala, M. D., Fusco, A., Donnarumma, G., & Muscariello, L. (2024). Probiotic properties of *Lactococcus lactis* strains isolated from natural whey starter cultures. *Food, 13*(6), 957.

Deng, B., Li, Y., Xu, D., Ye, Q., & Liu, G. (2019). Nitrogen availability alters flavonoid accumulation in Cyclocarya paliurus via the effects on the internal carbon/nitrogen balance. *Scientific Reports, 9*(1), 2370.

Dix, N. J., & Webster, J. (1995). Fungi of extreme environments. In *Fungal ecology*. Springer.

Drew, G. C., Stevens, E. J., & King, K. C. (2021). Microbial evolution and transitions along the parasite–mutualist continuum. *Nature Reviews Microbiology, 19*, 623–638. https://doi.org/10.1038/s41579-021-00550-7

Egboduku, W. O., Egboduku, T., Golohor, O. M., Imarhiagbe, O., & Ogwu, M. C. (2024). Cassava as raw material for sustainable bioeconomy development. In M. C. Ogwu, S. C. Izah, A. A. Cunha Alves, & S. C. Babu (Eds.), *Plant biology, sustainability and climate change, sustainable cassava* (pp. 57–73). Academic. https://doi.org/10.1016/B978-0-443-21747-0.00022-9

Embaby, A. M., Hussein, M. N., & Hussein, A. (2018). Monascus orange and red pigments production by *Monascus purpureus* ATCC16436 through co-solid-state fermentation of corn cob and glycerol: An eco-friendly environmental low cost approach. *PLoS One, 13*, e0207755. https://doi.org/10.1371/journal.pone.0207755

Enebe, M. C., & Babalola, O. O. (2021). Soil fertilization affects the abundance and distribution of carbon and nitrogen cycling genes in the maize rhizosphere. *AMB Express, 11*(1), 24.

FAO. (2022). *The state of world fisheries and aquaculture 2022. Towards blue transformation*. FAO. https://doi.org/10.4060/cc0461en

Farkas, J. A., Picking, J. W., & Santangelo, T. J. (2013). Genetic techniques for the archaea. *Annual Review of Genetics, 47*, 539–561. https://doi.org/10.1146/annurev-genet-111212-133225

Farrell, M., Prendergast-Miller, M., Jones, D. L., Hill, P. W., & Condron, L. M. (2014). Soil microbial organic nitrogen uptake is regulated by carbon availability. *Soil Biology and Biochemistry, 77*, 261–267. https://doi.org/10.1016/j.soilbio.2014.07.003

Gaur, V., & Bera, S. (2023). Microbial canthaxanthin: An orange-red keto carotenoid with potential pharmaceutical applications. *Biotechnologia, 104*(3), 315–328.

Gil-Martin, E., Forbes-Hernandez, T., Romero, A., Cianciosi, D., Giampieri, F., & Battino, M. (2022). Influence of the extraction method on the recovery of bioactive phenolic compounds from food industry by-products. *Food Chemistry, 378*, 131918.

Gopinath, V., & Mampoothiri, K. M. (2014). *Coryneacterium glutamicum. Reference module in food science, encyclopedia of food microbiology* (2nd ed., pp. 504–517). https://doi.org/10.1016/B978-0-12-384730-0.00076-8

Guaterio, G. V., Maciel da Silva, R., Karraz, F. C., Coelho, M. A. Z., Riberio, B. D., & Lemes, A. C. (2023). Cell disruption and permeabilization methods for obtaining yeast products. *Cleaner Chemical Engineering, 6*, 10012.

Guo, H.-N., Wu, S.-B., Tian, Y.-J., Zhang, J., & Liu, H.-T. (2021). Application of machine learning methods for the prediction of organic solid waste treatment and recycling process: A review. *Bioresource Technology, 319*, 124114. https://doi.org/10.1016/j.biortech.2020.124114

Guzik, H., & Mccoy-Simandle, K. (2017). Modeling microorganism transmission with Madagascar hissing cockroaches. *The American Biology Teacher, 79*(5), 365–374.

Haggag, W. M., & Abo El Soud, M. (2012). Production and optimization of *Pseudomonas fluorescens* biomass and metabolites for biocontrol of strawberry grey mould. *American Journal of Plant Sciences, 3*, 836–845. https://doi.org/10.4236/ajps.2012.37101

Haiko, J., & Westerlund-Wikström. (2013). The role of the bacterial flagellum in adhesion and virulence. *Biology (Basel), 2*(4), 1242–1267.

Hallama, M., Pekrun, C., Lambers, H., & Kandeler, E. (2018). Hidden miners – The roles of cover crops and soil microorganisms in phosphorus cycling through agroecosystems. *Plant and Soil, 434*, 7–45. https://doi.org/10.1007/s11104-018-3810-7

Han, M., Zhang, C., Suglo, P., Sun, S., Wang, M., & Su, T. (2021). P-Asartate: An essential metabolite for plant growth and stress acclimation. *Molecules, 26*(7), 1887. https://doi.org/10.3390/molecules26071887

Hardy, R. W., Kaushik, S. J., Mai, K., & Bai, S. C. (2022). Chapter 1 – Fish nutrition – History and perspectives. In R. W. Hardy & S. J. Kaushik (Eds.), *Fish nutrition* (4th ed., pp. 1–16., ISBN:9780128195871). Academic. https://doi.org/10.1016/B978-0-12-819587-1.00006-9

Heredia-Guerrero, J. A., Heredia, A., Dominguez, E., Cingolani, R., Bayer, I. S., Athanassiou, A., et al. (2017). Cutin from agro-waste as a raw material for the production of bioplastics. *Journal of Experimental Botany, 68*, 5401–5410. https://doi.org/10.1093/jxb/erx272

Hermsen, R., Okano, H., You, C., Werner, N., & Hwa, T. (2015). Growth rate competition formula for the growth of *E. coli* on co-utilized carbon substances. *Molecular Systems Biology, 11*, 801.

Herrera, M., Mancera, J. M., & Costas, B. (2019). The use of dietary additives in fish stress mitigation: Comparative endocrine and physiological responses. *Frontiers in Endocrinology (Lausanne), 10*, 447. https://doi.org/10.3389/fendo.2019.00447

Iyiola, A. O., & Valbo-Jørgensen, J. (2023). Intrinsic values of the African environment: A sustainable perspective. In S. C. Izah & M. C. Ogwu (Eds.), *Sustainable utilization and conservation of Africa's biological resources and environment. Sustainable development and biodiversity* (Vol. 888). Springer.

Jin, Q., & Kirk, M. F. (2018). pH as a primary control in environmental microbiology: 1 thermodynamic perspective. *Frontiers in Environmental Science, 6*, 21.

Jouhten, P., Ponomarova, O., Gonzalez, R., & Patil, K. R. (2016). *Saccharomyces cerevisiae* metabolism in ecological context. *FEMS Yeast Research, 16*(7), fow080. https://doi.org/10.1093/femsyr/fow080

Jurtshuk, P., Jr. (1996). Bacterial metabolism. In S. Baron (Ed.), *Medical microbiology* (4th ed.). University of Texas Medical Branch at Galveston.

Kantifedaki, A., Kachrimanidou, V., Mallouchos, A., Papanikolaou, S., & Koutinas, A. A. (2018). Orange processing waste valorisation for the production of bio-based pigments using the fungal strains *Monascus purpureus* and *Penicillium purpurogenum*. *Journal of Cleaner Production, 185*, 882–890. https://doi.org/10.1016/j.jclepro.2018.03.032

Katara, S., Gupta, D. V., Neelam, D., & Rahi, R. K. (2021). Role of bacteria and fungi in antibiotic production. *The Pharma Innovative, 10*(1), 709–714.

Kawal, M., Wattoo, A. G., Khushnood, R. A., Liaqa, A., Igbal, R., & Song, Z. (2023). Chapter 12 – Advancement and challenges in production of biosurfactants. In A. Charles & O. Adetunji (Eds.), *Applications for the next generation biosurfactats in the food sector* (pp. 239–259). Academic. https://doi.org/10.1016/B978-0-12-824283-4.00019-8

Keagle, M. B. (2013). DNA, chromosomes, and cell division. In S. Gersen & M. Keagle (Eds.), *The principles of clinical cytogenetics*. Springer. https://doi.org/10.1007/978-1-4419-1688-4_2

Khachatourians, G. G. (2019). *Insecticidal microbial. Reference module in life sciences*. Elsevier. ISBN:9780128096338. https://doi.org/10.1016/B978-0-12-809633-8.13066-3

Kim, J. S., Yoon, S. J., Park, Y. J., Kim, S. Y., & Ryu, C. M. (2020). Crossing the kingdom border: Human diseases caused by plant pathogens. *Environmental Microbiology, 22*(7), 2485–2495. https://doi.org/10.1111/1462-2920.15028

Komluski, J., Stukenbrock, E. H., & Habig, M. (2022). Non-Mendelian transmission of accessory chromosome in fungi, *30*(2–3), 241–253.

Kumar, M., Poonam, A. S., & Singh, R. P. (2022). Plant growth promoting microbes: Diverse roles for sustainable and ecofriendly agriculture. *Energy Nexus, 7*, 100133.

Kumar, R., Singh, U., Tiwari, A., Tiwari, P., Sahu, J. K., & Sharma, S. (2023). Vitamin B12: Strategies for enhanced production, fortified functional food products and health benefits. *Process Biochemistry, 127*, 44–55. https://doi.org/10.1016/j.procbio.2023.02.002

Kumokita, R., Bamba, T., Inokuma, K., Yoshida, T., Ito, Y., Kondo, A., & Hasunuma, T. (2022). Construction of an l-tyrosine chassis in *Pichia pastoris* enhance aromatic secondary metabolite production from glycerol. *ACS Synthetic Biology, 11*(6), 2098–2107.

Kumokita, R., Yoshida, T., Shirai, T., Kondo, A., & Hasunuma, T. (2023). Aromatic secondary metabolite production from glycerol was enhanced by amino acid addition in *Pichia pastoris*. *Applied Microbiology and Biotechnology, 107*(24), 7391–7401. https://doi.org/10.1007/s00253-12798-5

Lacey, H. J., & Rutledge, P. J. (2022). Recently discovered secondary metabolites from *Streptomyces* species. *Molecules, 27*(3), 887. https://doi.org/10.3390/molecules27030887

Lee, S.-J., Jeon, H.-S., Yoo, J.-Y., & Kim, J.-H. (2021). Some important metabolites produced by lactic acid bacteria originated from Kimchi. *Food, 10*(9), 2148.

Levin, P. A., & Angert, E. R. (2015). Small but mighty: Cell size and bacteria. *Cold Spring Harbor Perspectives in Biology, 7*(7), a019216. https://doi.org/10.1101/cshperspect.a019216

Liao, H., Liu Cai, C., Gao, T., Yang, Q.-E., Yu, Z., Gao, S., Zhou, S., & Friman, V.-P. (2023). Mesophilic and thermophilic viruses are associated with nutrient cycling during hyper thermophilic composting. *The ISMC Journal, 17*, 916–930.

Linder, T. (2019). Making the case for edible microorganisms as an integral part of a more sustainable and resilient food production system. *Food Security, 11*, 265–278.

Liu, L., Cheng, M., Jin, J., & Fu, M. (2024). Effects of different nitrogen applications and straw return depth on straw microbial and carbon and nitrogen cycles in paddy fields in the cool zone. *Scientific Reports, 14*, 6424.

Lopes, F. C., Tichota, D. M., Pereira, J. Q., Segalin, J., De Oliveira, R. A., & Brandelli, A. (2013). Pigment production by filamentous fungi on agroindustrial byproducts: An eco-friendly alternative. *Applied Biochemistry and Biotechnology, 171*, 616–625. https://doi.org/10.1007/s12010-013-0392-y

Lyubertsky, V. A., Zverkov, O. A., Rubanov, L. I., & Seliverstov, A. V. (2020). Optimal growth temperature and intergenic distances in bacteria, archaea, and plastids of Rhodophytic Branch. *BioMed Research International, 2020*, 3465380. https://doi.org/10.1155/2020/3465380

Macro, M. L. (2021). Defining how micro-organisms benefit human health. *Microbial Biotechnology, 14*(1), 35–40. https://doi.org/10.1111/1751-7915.13685

Madeira, J. V., Jr., Contesini, F. J., Calzado, F., Rubio, M. V., Zubieta, M. P., Lopes, D. B., et al. (2017). Agro-industrial residues and microbial enzymes: An overview on the eco-friendly bioconversion into high value-added products. In G. Brahmachari (Ed.), *Biotechnology of microbial enzymes* (pp. 475–511). Elsevier. https://doi.org/10.1016/B978-0-12-803725-6.00018-2

Madigan, M., & Martinko, J. (Eds.). (2006). *Brock biology of microorganisms* (13th ed., p. 1096). Pearson Education.

Maheshwari, R. (2016). *Fungi: Experimental methods in biology. Mycology* (2nd ed., p. 3). CRC Press. ISBN:978-1-4398-3904-1.

Maslak, E., Zloch, M., Aendowski, A., Sugajski, M., Janczura, I., Rudnicka, J., Walczak-Skierska, J., Buszewska Forajta, M., Rafinska, K., Pomastowski, P., Bialczak, D., & Buszewski, B. (2022). Isolation and identification of *Lactococcus lactis* and *Weissella cibaria* strains from fermented beetroot and an investigation of their properties as potential starter cultures and probiotics. *Food, 11*(15), 2257.

Mazur, F., Tjandra, A. D., Zhou, Y., Gao, Y., & Chandrawati, R. (2023). Paper-based sensors for bacteria detection. *Nature Reviews Bioengineering, 1*, 180–192.

McConnaughey, M. (2014). *Physical chemical properties of fungi. Reference module in biomedical sciences*. Elsevier. https://doi.org/10.1016/B978-0-12-801238-3.05231-4

McLaughlin, R. N., Jr., & Malik, H. S. (2017). Genetic conflicts: The usual suspects and beyond. *The Journal of Experimental Biology, 220*(1), 6–17.

Merloti, L. F., et al. (2019). Forest-to-agriculture conversion in Amazon drives soil microbial communities and N-cycle. *Soil Biology and Biochemistry, 137*, 107567.

Meyer, V., Andersen, M. R., Brakhage, A. A., Braus, G. H., Caddick, M. X., Cairns, T. C., et al. (2016). Current challenges of research on filamentous fungi in relation to human welfare and a sustainable bioeconomy: A white paper. *Fungal Biology and Biotechnology, 3*, 6. https://doi.org/10.1186/s40694-016-0024-8

Mielko, K. A., Jablonski, S. J., Lukaszewicz, M., & Miynarz, P. (2021). Comparison of bacteria disintegration methods and their influence on data analysis in metabolomics. *Scientific Reports, 11*, 20859.

Millis, J. R. (1997). Mechanical disruption of cells. In E. Goldberg (Ed.), *Handbook of downstream processing*. Springer. https://doi.org/10.1007/978-94-009-1563-3_1

Mitra, R., Ghosh, S., Mukherjee, G., & Acharya Chiwdhury, A. (2023). Secondary metabolites: Treasure trove for future medicine. In J. M. Merillon & K. G. Ramawat (Eds.), *Plant specialized metabolites. Reference series in phytochemistry*. Springer. https://doi.org/10.1007/988-3-031-30037-0_11-1

Mizzi, J., Gaggia, F., Cionci, N. B., Di Gioia, D., & Attard, E. (2022). Selection of acetic acid bacterial strains and vinegar production from local maltese food sources. *Frontiers in Microbiology, 13*, 897825.

Moller, M., Habig, M., Freitag, M., & Stukenbrock, E. H. (2018). Extraordinary genome instability and widespread chromosome rearrangements during vegetative growth. *Genetics, 210*(2), 517–529.

Narsing Rao, M. P., Xiao, M., & Li, W.-J. (2017). Fungal and bacterial pigments: Secondary metabolites with wide applications. *Frontiers in Microbiology, 8*, 1113. https://doi.org/10.3389/fmicb.2017.01113

Nizzy, A. M., & Ogwu, M. C. (2024). Valorization of cassava processing by-products into biofuel for a sustainable environment. In M. C. Ogwu, S. C. Izah, A. A. Cunha Alves, & S. C. Babu (Eds.), *Plant biology, sustainability and climate change, sustainable cassava* (pp. 291–309). Academic. https://doi.org/10.1016/B978-0-443-21747-0.00010-2

Ogamba, E. N., Iyiola, A. O., Yarkwan, B., & Adetola, B. O. (2023). Potentials, threats, and sustainable conservation strategies of plankton and macrophytes. In S. C. Izah & M. C. Ogwu (Eds.), *Sustainable utilization and conservation of Africa's biological resources and environment. Sustainable development and biodiversity* (Vol. 888). Springer. https://doi.org/10.1007/978-981-19-6974-4_4

Ogwu, M. C. (2019a). Towards sustainable development in Africa: The challenge of urbanization and climate change adaptation. In P. B. Cobbinah & M. Addaney (Eds.), *The geography of climate change adaptation in urban Africa* (pp. 29–55). Springer. https://doi.org/10.1007/978-3-030-04873-0_2

Ogwu, M. C. (2019b). Understanding the composition of food waste: An "-omics" approach to food waste management. In A. P. Gunjal, M. S. Waghmode, N. N. Patil, & P. Bhatt (Eds.), *Global initiatives for waste reduction and cutting food loss* (pp. 212–236). IGI Global. https://doi.org/10.4018/978-1-5225-7706-5.ch011

Ogwu, M. C., Izah, S. C., & Iyiola, A. O. (2022). An overview of the potentials, threats, and conservation of biodiversity in Africa. In S. C. Izah (Ed.), *Biodiversity in Africa: Potentials, threats and conservation, sustainable development and biodiversity* (Vol. 29, pp. 1–20). Springer. https://doi.org/10.1007/978-981-19-3326-4_1

Ogwu, M. C., Iyiola, A. O., & Izah, S. C. (2023a). Overview of African biological resources and environment. In S. C. Izah & M. C. Ogwu (Eds.), *Sustainable utilization and conservation of Africa's biological resources and environment. Sustainable development and biodiversity* (Vol. 888). Springer. https://doi.org/10.1007/978-981-19-6974-4_1

Ogwu, M. C., Osawaru, M. E., Amodu, E., & Osamo, F. (2023b). Comparative morphology, anatomy, and chemotaxonomy of two *Cissus* Linn. species. *Brazilian Journal of Botany*. https://doi.org/10.1007/s40415-023-00881-0

Ogwu, M. C., Odozi, I. P., Ahonsi, O. C., Uleanya, K. O., & Odozi, E. B. (2024). Organic acid production from cassava. In M. C. Ogwu, S. C. Izah, A. A. Cunha Alves, & S. C. Babu (Eds.), *Plant biology, sustainability and climate change, sustainable cassava* (pp. 395–418). Academic. https://doi.org/10.1016/B978-0-443-21747-0.00009-6

Oladipo, A., Enwemiwe, V., Ejeromedoghene, O., Adebayo, A., Ogunyemi, O., & Fu, F. (2022). Production and functionalities of specialized metabolites from different organic sources. *Metabolites, 12*(6), 534.

Panesar, R., Kaur, S., & Panesar, P. S. (2015). Production of microbial pigments utilizing agro-industrial waste: A review. *Current Opinion in Food Science, 1*, 70–76. https://doi.org/10.1016/j.cofs.2014.12.002

Pant, D., Van Bogaert, G., Diels, L., & Vanbroekhoven, K. (2010). A review of the substrates used in microbial fuel cells (MFCs) for sustainable energy production. *Bioresource Technology, 101*, 1533.

Pfeifer, F. (2012). Distribution, formation and regulation of gas vesicles. *Nature Reviews. Microbiology, 10*, 705–715. https://doi.org/10.1038/nrmicro2834

Pfeifer, K., Ergal, I., Koller, M., Basen, M., Schuster, B., & Rittmann, S. K.-M. (2021). Archaea biotechnology. *Biotechnology Advances, 47*, 107668.

Pinus, F. R., Villa-Boas, S. G., & Aggio, R. (2017). A review of analysis of intracellular metabolites from microorganisms: Quenching and extraction protocols. *Metabolites, 7*(4), 53.

Purves, W. K., Sadava, D., Orians, G. H., & Heller, H. C. (2003). Bacteria and archaea: The prokaryotic domains. In *Life: The science of biology* (7th ed., p. 533). Sinauer Associates.

Rajeswari, G., Jacob, S., Chandel, A. K., & Kumar, V. (2021). Unlocking the potential of insect and ruminant host symbionts for recycling of lignocellulosic carbon with biorefinery approach: A review. *Microbial Cell Factories, 20*, 107.

Reddy, C. C., Khilji, A. I., Gupta, A., Bhuyar, P., Mahmood, S., Al-Japairai, K. A. S., & Chua, G. K. (2021). Valorization of keratin waste biomass and its potential applications. *Journal of Water Process Engineering, 40*, 101707. https://doi.org/10.1016/jwpe.2020.101707

Rehman, K., Niaz, S., Tahir, A., Sajid, M., & Akash, H. (2020). Microorganisms and antibiotic production. In M. Z. Hashmi (Ed.), *Advances in environmental pollution research series, antibiotics and antimicrobial resistance genes in the environment* (Vol. 1, pp. 1–6). Elsevier. https://doi.org/10.1016/B978-0-12-818882-8.00001-2

Reiss, E., & DiSalvo, A. (2018). Yeasts. In R. Hunt (Ed.), *Microbiology and immunology online*. University of South Carolina.

Rombenso, A., Araujo, B., & Li, E. (2022). Recent advances in fish nutrition: Insights on the nutritional implications of modern formulations. *Animals (Basel), 12*(13), 1705. https://doi.org/10.3390/ani12131705

Rosca, M. F., Paucean, A., Man, S. M., Chis, M. S., Pop, C. R., Pop, A., & Farcas, A. C. (2024). *Leuconostoccitreum*: A promising sourdough fermenting starter for low-sugar content baked goods. *Food, 13*(1), 96. https://doi.org/10.3390/foods13010096

Rouphael, Y., Cardarelli, M., Bonini, P., & Colla, G. (2017). Synergistic action of a microbial-based bio-stimulant and a plant derived-protein hydrolysate enhances lettuce tolerance to alkalinity and salinity. *Frontiers in Plant Science, 8*, 1–12. https://doi.org/10.3389/fpls.2017.00131

Rousk, J., Baath, E., Brookes, P. C., Lauber, C. L., Lozupone, C., Caporaso, J. G., Knight, R., & Fierer, N. (2010). Soil bacterial communities across a pH gradient in an arable soil. *The ISME Journal, 4*, 1340–1351.

Sandle, T. (2016). *Pharmaceutical microbiology: Essentials for quality assurance and quality control*. Woodhead Publishing. https://doi.org/10.1016/C2014-0-00532-1

Saratale, G. D., Bhosale, R., Shobana, S., Banu, J. R., Pugazhendhi, A., Mahmoud, E., Sirohi, R., Bhatia, S. K., Atabani, A. E., Mulone, V., Uoon, J.-J., Shin, H. S., & Kumar, G. (2020). A review on valorization of spent coffee grounds towards biopolymers and biocatalysts production. *Bioresource Technology, 314*, 123800. https://doi.org/10.1016/j.biortech.2020.123800

Schmidt, J. E., Kent, A. D., Brisson, V. L., & Gaudin, A. C. M. (2019). Agricultural management and plant selection interactively affect rhizosphere microbial community structure and nitrogen cycling. *Microbiome, 7*(1), 146.

Schopf, W., Kitajima, K., Spicuzza, M. J., Kudryavtsev, A. B., & Valley, J. W. (2017). SIMS analyses of the oldest known assemblage of microfossils document their taxon-correlated carbon isotope compositions. *PNAS, 115*(1), 53–58. https://doi.org/10.1073/pnas.1718063115

Sen, T., Barrow, C. J., & Deshmukh, S. K. (2019). Microbial pigments in the food industry – Challenges and the way forward. *Frontiers in Nutrition, 6*, 7. https://doi.org/10.3389/fnut.2019.00007

Seo, S.-O., Park, S.-K., Jung, S.-C., Ryu, C.-M., & Kim, J.-S. (2020). Anti-contamination strategies for yeast fermentations. *Microorganisms, 8*(2), 274.

Shah, Z. A., Khan, K., Rashid, H. U., Shah, T., Jaremko, M., & Igbal, Z. (2022). Insights into metabolic and pharmacological profiling of *Aspergillus ficuum* through bioinformatics and experimental techniques. *BMC Microbiology, 22*, 295. https://doi.org/10.1186/s12866-022-02693-w

Sharma, R., & Ghoshal, G. (2020). Optimization of carotenoids production by *Rhodotorula mucilaginosa* (MTCC-1403) using agro-industrial waste in bioreactor: A statistical approach. *Biotechnology Reports, 25*, e00407. https://doi.org/10.1016/j.btre.2019.e00407

Sharma, R., Garg, P., Kumar, P., Bhatia, S. K., & Kulshrestha, S. (2020). Microbial fermentation and its role in quality improvement of fermented foods. *Fermentation, 6*(4), 106.

Sharma, P., Sangwan, S., Kaur, H., Patra, A., & Anamika, M. S. (2023). Diversity and evolution of nitrogen fixing bacteria. In *Sustainable agriculture reviews 60: Microbial processes in agriculture* (pp. 95–120).

Shu, W.-S., & Huang, L.-N. (2021). Microbial diversity in extreme environments. *Nature Reviews Microbiology, 20*, 219–235.

Singh, R., Kumar, M., Mittal, A., et al. (2017). Microbial metabolites in nutrition, healthcare and agriculture. *3 Biotech, 7*, 15.

Smercina, D. N., Evans, S. E., Friesen, M. L., & Tiemann, L. K. (2019). To fix or not to fix: Controls on free-living nitrogen fixation in the rhizosphere. *Applied and Environmental Microbiology, 85*(6), e02546–e02518. https://doi.org/10.1128/AEM.02546-18

Spalvins, K., Ivanovs, K., & Blumberga, D. (2018a). Single cell protein production from waste biomass: Review of various agriculturalby-products. *Energy Procedia, 147*, 409–418.

Spalvins, K., Zihare, L., & Blumberga, D. (2018b). Single cell protein from waste biomass: Comparison of various industrial by-products. *Energy Procedia, 147*, 409–418.

Spipida, M., Krawczyk, B., Zalewska-Piatek, B., Platek, R., Wysocka, M., & Olszewski, M. (2018). Fusion of DNA-binding domain of *Pyrococcus furiosus* ligase with a TaqStoffel DNA polymerase as a useful tool n PCT with difficult targets. *Applied Microbiology and Biotechnology, 102*(2), 713–721.

Srivastava, N., Srivastava, M., Ramteke, P. W., & Mishra, P. K. (2019). Chapter 23 – Solid-state fermentation strategy for microbial metabolites production: An overview. *New and Future Developments in Microbial Biotechnology and Bioengineering, 2019*, 345–354. https://doi.org/10.1016/B978-0-444-63504-4.00023-2

Suman, G., Nupur, M., Anuradha, S., & Pradeep, D. B. (2015). Single cell protein production: A review. *International Journal of Current Microbiology and Applied Sciences, 4*, 9.

Sunish, K. S., & Thazeem, B. (2023). Microbial biomass. In S. Thomas, M. Hosur, D. Pasquini, & C. Jose Chirayil (Eds.), *Handbook of biomass*. Springer. https://doi.org/10.1007/978-981-19-6772-6_6-1

Tadese, D. A. (2022). Aquatic animal nutrition and feeding for sustainable aquaculture production. *International Journal of Advanced Research in Biological Sciences, 9*(12), 187–194. https://doi.org/10.22192/ijarbs

Thiao, D., & Bunting, S. W. (2022). *Socio-economic and biological impacts of the fish-based feed industry for sub-Saharan Africa* (FAO Fisheries and Aquaculture Circular No. 1236). FAO, Worldfish and University of Greenwich, Natural Resources Institute. https://doi.org/10.4060/cb7990e

Timmis, K., de Lorenzo, V., Verstraete, W., Ramos, J. L., Danchin, A., Brüssow, H., Singh, B. K., & Timmis, J. K. (2017). The contribution of microbial biotechnology to economic growth and employment creation. *Microbial Biotechnology, 10*(5), 1137–1144. https://doi.org/10.1111/1751-7915.12845

Toju, H., Abe, M. S., Ishii, C., Hori, Y., Fujita, H., & Fukada, S. (2020). Scoring species for synthetic community design: Network analyses of functional core microbiomes. *Frontiers in Microbiology, 11*, 1361.

Umesh, M., & Basheer, T. (2021). Microbe mediated bioconversion of fruit waste into value added products: Microbes in fruit waste management. In *Research anthology on food waste reduction and alternative duets for food and nutrition security*. IGI Global.

Umma, S. B., Iyiola, A. O., & Adeshina, I. (2023). Relationship pattern of enteric bacterial load and assessed micronutrients in the gut of *Clarias gariepinus* fish sampled in the Ibadan municipal zone. *The Journal of Basic and Applied Zoology, 84*, 23. https://doi.org/10.1186/s41936-023-00345-7

Vadlapudi, V., Borah, N., Yellusani, K. R., Gade, S., Reddy, P., Rajamanikyam, M., Vempati, L. N. S., Gubbala, S. P., Chopra, P., Upadhyayula, S. M., & Amanchy, R. (2017). *Aspergillus* secondary metabolite database, a resource to understand the secondary metabolome of *Aspergillus* genus. *Scientific Reports, 7*, 7325. https://doi.org/10.1038/s41598-017-07436-w

Valverde, M. E., Hernández-Pérez, T., & Paredes-López, O. (2015). Edible mushrooms: Improving human health and promoting quality life. *International Journal of Microbiology, 2015*, 376387. https://doi.org/10.1155/2015/376387
Vassileva, M., Malusa, E., Sas-Paszt, L., Tracinski, P., Galvez, A., Flor-Peregrin, E., Shilev, S., Canfora, L., Mocali, S., & Vassilev, N. (2021). Fermentation strategies to improve soil bio-inoculant production and quality. *Microorganisms, 9*(6), 1254.
Venil, C. K., Devi, P. R., & Ahmad, W. A. (2020a). Agro-industrial waste as substrates for the production of bacterial pigment. In Z. A. Zakaria, R. Boopathy, & J. R. Dib (Eds.), *Valorisation of agro-industrial residues – Volume I: Biological approaches* (pp. 149–162). Springer. https://doi.org/10.1007/978-3-030-39137-9_7
Venil, C. K., Dufossé, L., & Renuka Devi, P. (2020b). Bacterial pigments: Sustainable compounds with market potential for pharma and food industry. *Frontiers in Sustainable Food Systems, 4*, 100. https://doi.org/10.3389/fsufs.2020.00100
Vitezova, M., Kohoutova, A., Vitez, T., Hanisakova, N., & Kushkevych, I. (2020). Methanogenic microorganisms in industrial wastewater anaerobic treatment. *PRO, 8912*, 1546. https://doi.org/10.3390/pr8121546
Vrbal, P., Schinagl, C. W., Atrmann, D. J., Heiss, B., & Burgstaller, W. (2019). Fungal growth in batch culture- what we could benefit if we start looking closer. *Frontiers in Microbiology, 10*, 2391.
Wang, C. X., Mas, A., & Esteve-Zarzoso, B. (2016). The interaction between *Saccharomyces cerevisiae* and non-saccharomyces yeast during alcoholic fermentation is species and strain specific. *Frontiers in Microbiology, 7*, 502.
Wang, X., Xia, K., Yang, X., & Tang, C. (2019). Growth strategy of microbes from mixed carbon sources. *Nature Communications, 10*, 1279. https://doi.org/10.1038/s41467-019-09261-3
Warren, S. D., St Clair, L. L., & Leavitt, S. D. (2019). Aerobiology and passive restoration of biological soil crusts. *Aerobiologia, 35*, 45–56.
Weber, B., Budel, B., & Belnap, J. (Eds.). (2016). *Biological soil crusts: An organizing principle in drylands* (Ecological Studies 226). Springer.
White, J. F., Kingsley, K. L., Verma, S. K., & Kowalski, K. P. (2018). Rhizophagy cycle: An oxidative process in plants for nutrient extraction from symbiotic microbes. *Microorganisms, 6*(3), 95.
Youssef, F. S., & Singab, A.-N. B. (2021). An updated review on the secondary metabolites and biological activities of *Aspergillus ruber* and *Aspergillus flavus* and exploring the cytotoxic potential of their isolated compounds using virtual screening. *Evidence-based Complementary and Alternative Medicine. Hindawi, 11*, 8860784. https://doi.org/10.1155/2021/8860784
Zahoor, A., Lindner, S. N., & Wendisch, V. F. (2012). Metabolic engineering of *Corynbacterium glutamicum* aimed at alternative carbon sources and new products. *Computational and Structural Biotechnology Journal, 3*, e201210004. https://doi.org/10.5936/csbj.201210004
Zaveri, A., Edwards, J., & Rochfort, S. (2022). Production of primary metabolites by *Rhizopus stolonifera,* causal agent of Almond Hull Rot Disease. *Molecules, 27*(21), 7199. https://doi.org/10.3390/molecules27217199
Zerbib, D. (2017). Bacterial cell envelopes: Composition, architecture, and origin. In D. Miklavčič (Ed.), *Handbook of electroporation*. Springer. https://doi.org/10.1007/978-3-319-32886-7_28
Zhang, L., Song, T., Wu, J., Zhang, S., Yin, C., Huang, F., et al. (2020). Antibacterial and cytotoxic metabolites of termite-associated *Streptomyces* sp. BYF63. *Journal of Antibiotics, 73*, 766–771.

Processing Techniques for Bio-based Products in the Global South

C. A. Ogunlade, A. R. Olaniyan, R. T. Babalola, B. O. Oyefeso, and K. F. Jaiyeoba

Abstract

This chapter delves into the processing techniques for bio-based products in the Global South, emphasizing the intersection of traditional methods and modern advancements. The analysis begins with defining bio-based products and their importance in the context of sustainable development and environmental conservation. It explores the historical background and common traditional methods, particularly focusing on fermentation techniques prevalent in West Africa. The chapter then transitions to modern processing techniques, highlighting their potential to enhance efficiency and product quality while addressing challenges such as economic constraints and technical difficulties. The socioeconomic impacts of these processing activities are examined, noting their influence on small-scale farmers, gender dynamics, and the broader community. The discussion also encompasses the environmental benefits of adopting sustainable practices. The chapter concludes by identifying opportunities for innovation and growth in the bio-based products sector, advocating for policies and investments that support sustainable processing practices. Through this comprehensive analysis, the chapter provides insights into the challenges and opportunities within

C. A. Ogunlade (✉) · A. R. Olaniyan
Department of Agricultural Engineering, Osun State University, Osogbo, Nigeria
e-mail: clement.ogunlade@uniosun.edu.ng

R. T. Babalola
Department of Food Technology, Osun State University, Osogbo, Nigeria

B. O. Oyefeso
Department of Agricultural Engineering, University of Ibadan, Ibadan, Nigeria

K. F. Jaiyeoba
Department of Agricultural Engineering, Federal University of Technology, Akure, Ondo State, Nigeria

M. C. Ogwu et al. (eds.), *Sustainable Bioeconomy Development in the Global South*, https://doi.org/10.1007/978-981-96-0305-3_2

the processing of bio-based products, offering a pathway toward sustainable development in the Global South.

Keywords

Bio-based products · Global South · Traditional processing techniques · Fermentation · Sustainable development

Abbreviations

FAO	Food and Agriculture Organization
GDP	Gross Domestic Product
ICT	Information and Communication Technology
SDGs	Sustainable Development Goals
SMEs	Small and Medium-sized Enterprises
GMO	Genetically Modified Organisms
HPP	High-Pressure Processing
SFE	Supercritical Fluid Extraction
UAE	Ultrasound-Assisted Extraction
R&D	Research and Development

1 Introduction

1.1 Definition and Importance of Bio-based Products

Bio-based products are derived from biological materials and renewable sources such as plants, animals, and microorganisms (Salazar Sandoval et al., 2024). Unlike conventional products that rely on fossil fuels, bio-based products offer a sustainable alternative, reducing dependency on nonrenewable resources and mitigating environmental impacts. These products span a wide range of industries, including agriculture, pharmaceuticals, textiles, energy, and consumer goods (Moreno et al., 2020). Examples include biofuels, biodegradable plastics, natural fibers, and biopharmaceuticals.

Bio-based products play a crucial role in promoting environmental sustainability by significantly reducing carbon footprints and mitigating climate change (Andrew & Dhakal, 2022). Unlike fossil fuel-based products, bio-based products are derived from renewable resources such as plants, animals, and microorganisms. This shift to renewable resources helps conserve biodiversity and maintain natural ecosystems, as it reduces reliance on nonrenewable, polluting resources (Singh, 2024). Furthermore, the use of bio-based products in various industries can lower greenhouse gas emissions, contributing to global efforts to address climate change and fostering a more sustainable future.

The bioeconomy, centered around bio-based products, is a powerful driver of economic development, especially in rural areas of the Global South (Taron & Gebrezgabher, 2024). By creating new markets and job opportunities, it can boost incomes for farmers and entrepreneurs involved in the sector. This economic stimulation can lead to broader rural development, reducing poverty and improving living standards in these communities. The diversification of income sources through the production and sale of bio-based products can also make local economies more resilient to external shocks, such as market fluctuations and environmental changes (Delgado et al., 2024).

In addition to environmental and economic benefits, bio-based products enhance resource efficiency by utilizing agricultural residues, organic waste, and other biomass that would otherwise go to waste (Agoh et al., 2024; Kosoe & Ogwu, 2024; Ogwu et al., 2024). This approach supports a circular economy where materials are reused and recycled, minimizing waste and making the most of available resources. The development and production of bio-based products drive innovation in biotechnology, materials science, and industrial processes, leading to technological advancements with broad applications (Ramchuran et al., 2023). These innovations not only improve the efficiency and sustainability of production methods but also open new possibilities for creating high-value products from renewable resources, further solidifying the importance of bio-based products in addressing critical global challenges.

1.2 Overview of the Global South Context

The term "Global South" refers to regions encompassing Africa, Latin America, Asia, and Oceania, which are generally characterized by lower income levels, less industrialization, and varied socioeconomic development compared to the Global North. The Global South is home to a significant portion of the world's biodiversity and biomass resources, making it a crucial player in the bio-based economy (Rawat et al., 2023).

In the context of the Global South, several key factors shape the landscape for bio-based products (Mandree et al., 2023). Firstly, many countries in this region boast vast agricultural lands and favorable climates for growing a wide range of crops. This agricultural abundance provides a rich supply of raw materials necessary for the development of bio-based products. The diversity of crops grown across these regions supports a variety of bio-based industries, from biofuels to biodegradable materials, offering significant potential for growth and diversification within the local economies (Holland & Shapira, 2024).

Secondly, the Global South is home to a wealth of indigenous and traditional knowledge systems that offer valuable insights into sustainable resource management and processing techniques (Ogwu & Osawaru, 2022). These traditional practices, honed over generations, can enhance the development of bio-based products by providing environment-friendly and culturally appropriate methods. Leveraging this knowledge can lead to innovations that are not only sustainable but also deeply rooted in local contexts, thereby increasing acceptance and effectiveness within communities.

However, despite these advantages, many regions in the Global South face significant economic challenges that hinder the development and commercialization of bio-based products (Mengal et al., 2018). Limited access to capital, inadequate infrastructure, and insufficient market access pose major barriers. Additionally, there is often a technological divide between the Global South and the more industrialized Global North. Bridging this gap requires substantial investments in education, research, and technology transfer. Government policies and regulatory frameworks are also crucial in either supporting or hindering the growth of the bio-based sector (Philippidis et al., 2024). Policies promoting renewable energy, sustainable agriculture, and innovation can drive progress, while unfavorable regulations can stifle development. Social dynamics, including gender roles and community structures, further influence the adoption and impact of bio-based products (Garmendia-Lemus et al., 2024). Empowering marginalized groups, such as women and small-scale farmers, is essential for maximizing the socioeconomic benefits of the bio-based economy.

1.3 Objectives of the Chapter

This chapter aims to provide a comprehensive understanding of processing techniques for bio-based products in the Global South by examining both traditional and modern methods. It highlights the advantages, limitations, and applicability of these techniques in various contexts. Sustainable practices are a key focus, exploring methods that minimize environmental impact, enhance resource efficiency, reduce waste, and conserve energy. By discussing the economic, technical, and social challenges faced in developing bio-based products, the chapter identifies opportunities for innovation, investment, and collaboration. The socioeconomic impacts on local communities, particularly small-scale farmers and gender dynamics, are assessed through illustrative case studies.

Additionally, the chapter reviews the role of government policies, international aid, and partnerships in supporting the bio-based sector, highlighting successful frameworks and initiatives that can be adapted or replicated. It explores emerging technologies and trends in bio-based processing, identifying potential areas for future research and development to drive the growth of the bioeconomy in the Global South. By addressing these objectives, the chapter offers a holistic view of the current state and prospects of bio-based processing techniques, providing valuable insights for researchers, policymakers, and practitioners in the field.

2 Traditional Processing Techniques

2.1 Historical Background

Traditional processing techniques for bio-based products have deep roots in the cultural and historical contexts of the Global South (Cuestas-Caza et al., 2024). These methods have been developed and refined over centuries, often passed down through generations as part of indigenous knowledge systems. Historically, traditional processing techniques were not only a means of producing goods but also a way to preserve food, create materials for everyday use, and maintain cultural practices and rituals (Long, 2024). These methods reflect the ingenuity and adaptability of local communities who, in the absence of modern technology, found effective ways to utilize natural resources for their daily needs.

In many communities, these techniques are intimately linked with local biodiversity and ecological systems, reflecting a deep understanding of and respect for the natural environment (Egboduku et al., 2024; Reyes-García et al., 2024). For instance, in various parts of Africa, Asia, and Latin America, traditional practices are tailored to the specific types of crops and natural resources available in each region. This relationship between traditional processing methods and local biodiversity ensures that these techniques are sustainable and environment-friendly (Onwuka & Adu, 2024). For example, the use of native plant species for food, textiles, and medicinal preparations not only provides for the community's needs but also encourages the conservation of local flora.

The reliance on locally available resources and simple tools makes traditional processing techniques both accessible and sustainable (Kibria et al., 2024). These methods typically require minimal external inputs and are adapted to local conditions, making them resilient in the face of economic and environmental changes. For example, sun drying, a common preservation method, uses the natural heat from the sun to reduce moisture content in fruits, vegetables, and grains. This process requires no electricity or sophisticated equipment, making it an ideal solution for communities with limited access to modern technology.

Moreover, traditional processing techniques often involve communal activities, reinforcing social bonds and cultural heritage (Ardana et al., 2024). The knowledge and skills required for these methods are usually shared among community members, ensuring that everyone, including future generations, can benefit from this collective wisdom. This communal aspect not only helps in the efficient management of resources but also strengthens community ties (Akinola et al., 2020). For example, in many African and Asian communities, the production of fermented foods or the extraction of plant oils is often a collective effort, with different members contributing to various stages of the process. This collaborative approach not only distributes the labor but also fosters a sense of unity and shared purpose.

2.2 Common Traditional Methods

Traditional food processing methods in the Global South encompass a variety of techniques that have been developed over generations to maximize the nutritional value, flavor, and preservation of local foods (Naika et al., 2024). These methods are deeply intertwined with the cultural practices and ecological knowledge of different ethnic groups across the country. They leverage locally available resources and simple, often low-tech tools, making them accessible and sustainable for rural and urban communities alike. The primary goal of these methods is to ensure food safety and longevity, while also enhancing the sensory qualities of food to suit local tastes and dietary preferences. These traditional techniques are not only a testament to the ingenuity of Nigerian culinary practices but also play a crucial role in maintaining food security and cultural heritage (Swinbank & Swinbank, 2021). Some of the common traditional methods are shown in Fig. 1 and explained below.

(i) *Fermentation*: Fermentation is one of the most ancient and widespread traditional processing techniques (Sionek et al., 2023). It involves the biochemical transformation of organic substrates by microorganisms, such as bacteria, yeast, and fungi. In the Global South, fermentation is used to produce a variety of foods and beverages. For instance, in West Africa, traditional fermentation techniques are used to produce products like gari (fermented cassava), *Ogi* (fermented cereal porridge), and palm wine. These methods not only enhance the nutritional value and digestibility of foods but also improve their shelf life.

(ii) *Drying*: Sun drying is another traditional method commonly used to preserve fruits, vegetables, grains, and fish (Arslan & Alibaş, 2024). This technique involves exposing food items to direct sunlight for extended periods, reducing their moisture content and inhibiting the growth of spoilage microorganisms. Sun drying is widely practiced in regions with hot and dry climates, such as parts of Africa and South Asia (Elsamanoudy et al., 2024). It is an energy-efficient method that requires minimal infrastructure and costs.

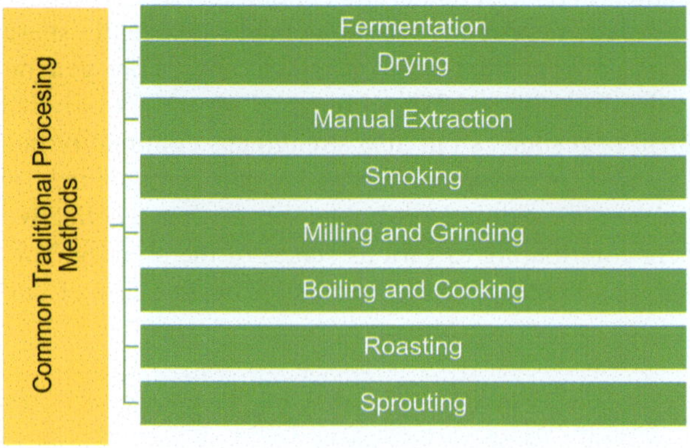

Fig. 1 Common traditional processing methods

(iii) *Manual Extraction*: Manual extraction methods are used to obtain oils, fibers, and other valuable components from plants and animals (Bhatia et al., 2024). For example, traditional methods for extracting coconut oil in Southeast Asia involve grating the coconut, extracting the milk, and then heating it to separate the oil. Similarly, in parts of Africa, shea butter is traditionally extracted by crushing and roasting shea nuts, followed by boiling and skimming off the butter.

(iv) *Smoking*: Smoking is used to preserve meat and fish. This method involves exposing food to smoke from burning wood or other plant materials, which imparts a distinct flavor and helps to prevent microbial growth (Nizio et al., 2023). Traditional smoking techniques vary widely but are commonly used in coastal and riverine communities in the Global South.

(v) *Milling and Grinding*: Manual milling and grinding of grains and seeds have been practiced for millennia. These techniques involve the use of simple tools like stones, mortars, and pestles to process raw materials into flours and pastes. This is a critical step in the preparation of staple foods such as maize flour in Latin America and millet flour in parts of Africa.

(vi) *Boiling and Cooking*: Boiling and cooking are fundamental techniques used to prepare a variety of Nigerian dishes (Omoniyi & Cosmas, 2024). These methods are employed to soften foods, improve their flavor, and make them safe for consumption. For instance, yams, a staple tuber, are often boiled or pounded to make "pounded yam," a popular dish served with various soups. Boiling is also essential in the preparation of "beans," which are cooked until tender and often mashed or stewed (Oyeneye et al., 2023). Cooking techniques in Nigeria often involve the use of traditional earthenware pots and open fires, which impart distinct flavors to the food.

(vii) *Roasting*: Roasting is another traditional method used to enhance the flavor and texture of foods. This technique involves cooking food over an open flame or hot coals (Chobot et al., 2024). In Nigeria, plantains and maize are commonly roasted as street foods. Roasting is also used in the preparation of "groundnuts" (peanuts), which are roasted to create a popular snack and ingredient in various dishes. The roasting process not only improves the taste but also reduces moisture content, thereby extending the shelf life of the food.

(viii) *Sprouting*: Sprouting is a technique used to improve the nutritional value of grains and legumes. By soaking and allowing seeds to germinate, enzymes are activated that increase the availability of vitamins and minerals. In Nigeria, sprouted grains are used to make traditional beverages like "malt drinks," which are enjoyed for their refreshing taste and nutritional benefits. Sprouting also enhances the digestibility of grains, making them more suitable for consumption by children and the elderly (Miyahira et al., 2021).

2.3 Case Study: Fermentation Techniques in West Africa

In West Africa, traditional fermentation techniques are integral to the region's food culture and economy. Fermentation is used to process a variety of staple foods, enhancing their flavor, nutritional value, and shelf life (Murali et al., 2024). One prominent example is the production of *gari*, a fermented and roasted product made from cassava.

***Gari* Production** The process begins with peeling and washing cassava roots, which are then grated into a pulp. The grated cassava is placed in bags and left to ferment for several days. During fermentation, naturally occurring lactic acid bacteria break down the carbohydrates in the cassava, producing lactic acid and other compounds that contribute to the distinctive sour taste of gari. After fermentation, the cassava pulp is pressed to remove excess water, sieved to remove fibrous materials, and then roasted in large pans until it becomes dry and crispy. The final product is a coarse flour that can be stored for long periods and is used to prepare various dishes. The unit operations involved in *gari* production are shown in Fig. 2.

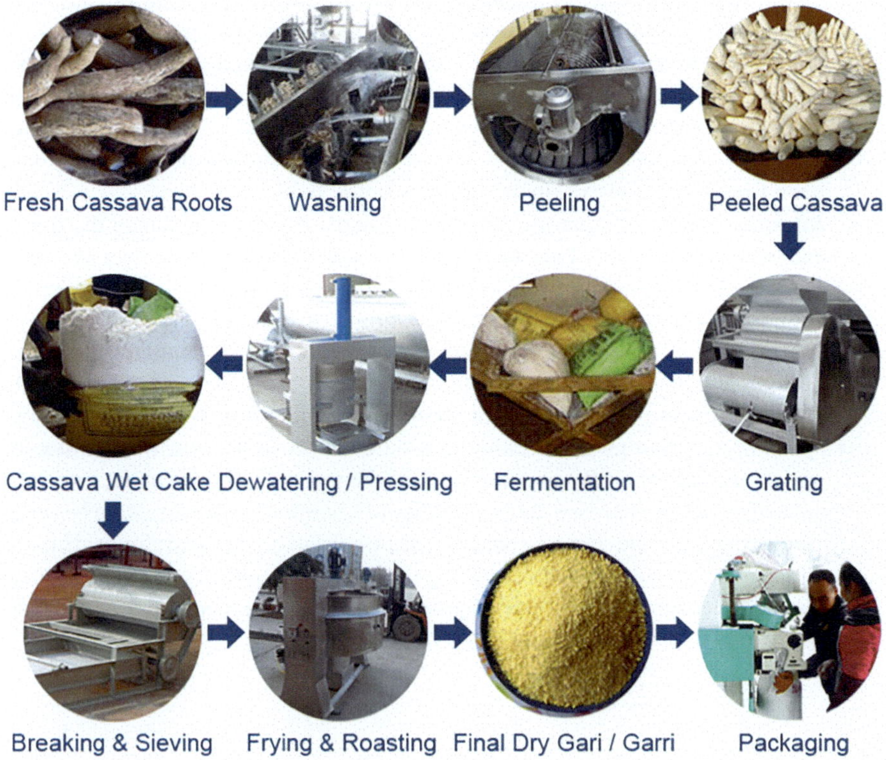

Fig. 2 Gari production processes (Goodwaychina.com)

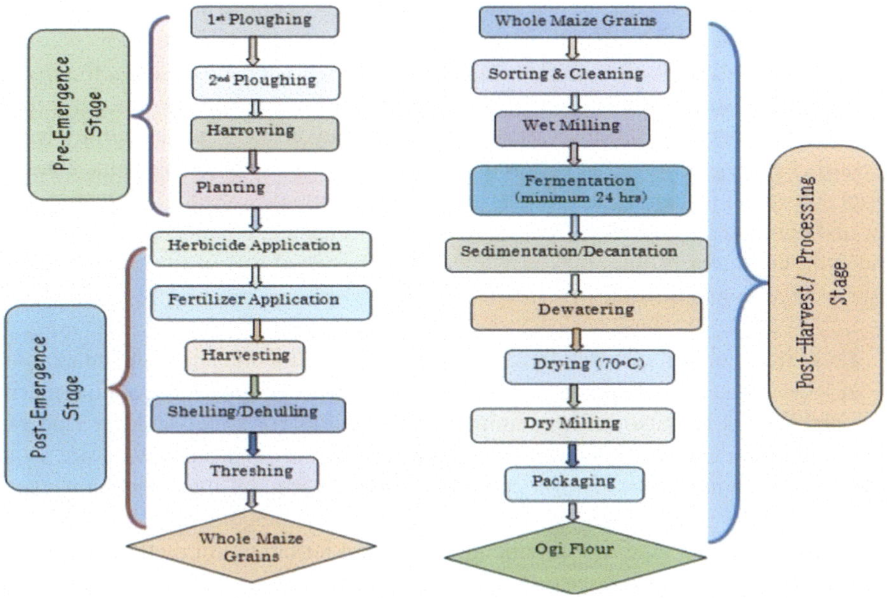

Fig. 3 Production stages of *Ogi* (Tulagha et al., 2021)

***Ogi* Production** Ogi is another traditional fermented product made from grains such as maize, sorghum, or millet. The grains are soaked in water for several days to initiate natural fermentation. After soaking, the grains are wet-milled and the resulting slurry is sieved to remove husks. The liquid obtained is allowed to ferment further until it forms a thick, sour paste. Ogi is typically cooked into a smooth porridge and is a staple food for infants and adults alike. The production line of Ogi is shown in Fig. 3.

Palm Wine Production Palm wine is a traditional alcoholic beverage deeply embedded in the cultural and social fabric of many communities in Nigeria and other parts of West Africa (Adebola, 2024). This unique drink is made from the sap of various species of palm trees, such as the oil palm (*Elaeis guineensis*), date palm (*Phoenix dactylifera*), and raffia palm (*Raphia hookeri*). The process begins with the skillful tapping of the tree, which involves making a careful incision on the trunk or the flower stalk to access the sap. Collectors typically use a sharp, sterilized tool to ensure the sap flows freely into containers, often fashioned from natural materials like calabashes or more modern plastic containers. The sap begins to flow almost immediately, often yielding several liters in a single day. The collected sap undergoes natural fermentation, driven by wild yeasts and bacteria present in the environment and on the tree itself. This spontaneous fermentation starts almost as soon as the sap is collected, causing the sugar-rich liquid to transform into a mildly alcoholic beverage within hours. Palm wine is traditionally consumed fresh due to its rapid fermentation process, which continues to change its taste and alcohol content over time. When freshly tapped, palm wine is sweet, with a relatively low alcohol content and a slightly effervescent quality. As fermentation progresses, the drink becomes more sour and alcoholic, sometimes reaching significant potency if left to ferment for several days. This variability in taste and strength makes palm wine a versatile beverage enjoyed at different stages of fermentation. Fresh palm wine is often consumed in social gatherings, rituals, and ceremonies, symbolizing hospitality, and community. The drink's deep cultural significance, coupled with its unique and evolving flavor profile, makes palm wine a cherished part of the culinary heritage in Nigeria and beyond.

These traditional fermentation techniques illustrate the ingenuity and adaptability of West African food processing methods. They not only provide essential nutrients and flavors but also offer economic opportunities for small-scale producers. The production and sale of fermented products like gari and palm wine are vital sources of income for many rural communities. Furthermore, these techniques are environmentally sustainable, as they rely on locally available materials and natural processes. By preserving and enhancing traditional fermentation methods, there is significant potential to improve food security and economic resilience in West Africa.

3 Modern Processing Techniques

Modern processing techniques represent a departure from traditional methods, incorporating innovative technologies and advanced equipment to improve efficiency, quality, and scalability (Kolasani, 2024). These methods harness the power of science and engineering to address contemporary challenges such as food security, sustainability, and economic development. Unlike traditional techniques, which often rely on manual labor and natural processes, modern methods leverage automation, precision, and data analytics to optimize production processes and ensure consistency in product quality.

3.1 Integration of Technology in Processing

The integration of technology in processing has revolutionized the way bio-based products are manufactured, enabling faster production, better quality control, and reduced environmental impact (Adeleke et al., 2024). Automation and robotics play a significant role in modern processing facilities, streamlining repetitive tasks and minimizing human error. For example, in the production of biofuels, automated systems are used to monitor fermentation processes, control temperature, and pH levels, and harvest microorganisms with precision. Additionally, advances in sensors and monitoring devices allow for real-time tracking of key parameters, ensuring optimal conditions throughout the production process (Arinze et al., 2024).

Furthermore, modern processing techniques often incorporate renewable energy sources and sustainable practices to minimize carbon footprint and resource consumption (Ninduwezuor-Ehiobu et al., 2023). For instance, solar drying technologies are increasingly being adopted in the Global South to preserve agricultural products such as fruits, vegetables, and grains. Solar dryers harness the energy from sunlight to heat and dehydrate food items, reducing moisture content and preventing spoilage. This method not only extends the shelf life of perishable goods but also reduces dependency on fossil fuels and mitigates greenhouse gas emissions. The integration of solar drying technology exemplifies the convergence of modern processing methods with sustainable solutions, highlighting the potential for innovation in addressing pressing global challenges (Ukoba et al., 2024). Some common modern processing techniques and their description are shown in Table 1.

These modern processing techniques represent advancements in science and technology that enable more efficient, sustainable, and innovative approaches to manufacturing and production across different industries.

4 Challenges and Opportunities

4.1 Economic Constraints

Economic constraints present formidable challenges to the advancement of processing techniques for bio-based products in the Global South (Hassan et al., 2019). One of the primary hurdles is the limited access to capital, financing, and investment opportunities, especially for small-scale producers and entrepreneurs. These individuals often lack the collateral or credit history required to secure loans from traditional financial institutions. As a result, they face difficulties in obtaining the necessary funding to establish or expand processing facilities, purchase equipment, or invest in technology upgrades. This financial barrier stifles innovation and growth in the bio-based processing sector, particularly among marginalized communities who are most in need of economic opportunities.

Moreover, the high upfront costs associated with equipment, infrastructure, and technology upgrades exacerbate the financial challenges faced by bio-based product manufacturers in the Global South (Moshood et al., 2021). Investing in modern processing technologies and facilities requires substantial capital investment, which may be prohibitive for many small and medium-sized enterprises (SMEs). Without access to affordable financing options, these businesses struggle to modernize their operations or scale up production to meet growing demand (Gürsu, 2024). Additionally, fluctuating market prices, inadequate market infrastructure, and limited market access further compound the economic constraints faced by bio-based product manufacturers. Uncertainties in market conditions make it challenging for businesses to forecast revenues and plan for long-term sustainability.

Addressing economic constraints in the bio-based processing sector necessitates the implementation of innovative financing mechanisms, public–private partnerships, and supportive policies (Lange et al., 2021). Innovative financing

Table 1 Modern processing techniques

S/N	Modern processing technique	Description
1.	Biotechnology	Biotechnology involves the manipulation of biological systems and organisms to develop products or processes. This can include genetic engineering to create genetically modified organisms (GMOs) for improved crop yields or the production of enzymes for industrial applications
2.	Nanotechnology	Nanotechnology involves the manipulation of materials at the nanoscale (typically 1–100 nanometers) to create new properties or functionalities. In food processing, nanotechnology may be used to enhance packaging materials for better preservation or to develop nanoemulsions for improved delivery of nutrients or flavors
3.	High-pressure processing (HPP)	HPP is a nonthermal processing technique that uses high-pressure water to deactivate microorganisms and enzymes in food products. This method preserves the nutritional quality and sensory attributes of foods while extending shelf life
4.	Extraction technologies	Modern extraction technologies such as supercritical fluid extraction (SFE) and ultrasound-assisted extraction (UAE) use advanced methods to extract bioactive compounds from raw materials. These techniques offer higher extraction efficiencies and reduced processing times compared to traditional solvent extraction methods
5.	Membrane separation	Membrane separation processes, including microfiltration, ultrafiltration, nanofiltration, and reverse osmosis, are used to separate components in liquid or gas streams based on molecular size and properties. These processes are commonly used in the food and beverage industry for concentration, purification, and fractionation of products
6.	Electrospinning	Electrospinning is a technique used to produce nanofibers from polymer solutions or melts by applying an electric field. These nanofibers have a high surface area-to-volume ratio and can be used in various applications, including food packaging, tissue engineering, and drug delivery systems
7.	Microwave and radio frequency heating	Microwave and radio frequency heating technologies use electromagnetic waves to generate heat within food products. These methods offer rapid and uniform heating, making them suitable for thawing, drying, and pasteurization processes in the food industry
8.	Enzyme engineering	Enzyme engineering involves the modification or design of enzymes to improve their catalytic activity, stability, or substrate specificity. Engineered enzymes are used in various industrial processes, including food processing, biofuel production, and pharmaceutical manufacturing, to enhance efficiency and reduce environmental impact

(continued)

Table 1 (continued)

S/N	Modern processing technique	Description
9.	Supercritical fluid technology	Supercritical fluids, such as carbon dioxide or water, are used as solvents in extraction, separation, and purification processes. Supercritical fluid technology offers advantages such as high selectivity, low toxicity, and minimal environmental impact compared to traditional solvent-based methods
10.	Continuous flow processing	Continuous flow processing involves the continuous movement of materials through a series of interconnected processing units, as opposed to batch processing. This approach allows for greater control over reaction conditions, higher throughput, and reduced energy and resource consumption
11.	Green chemistry	Green chemistry, also known as sustainable chemistry, focuses on the design of chemical products and processes that minimize environmental impact and maximize efficiency. Green chemistry principles are applied across various industries to develop safer and more sustainable alternatives to conventional chemical processes
12.	Hydrothermal processing	Hydrothermal processing involves the treatment of materials with hot, pressurized water or steam to facilitate chemical reactions or physical transformations. This technique is used in applications such as biomass conversion, waste treatment, and the production of functional materials
13.	Electron beam processing	Electron beam processing uses high-energy electron beams to irradiate materials for sterilization, cross-linking, or modification. This technique is employed in medical device sterilization, food irradiation, polymer modification, and semiconductor manufacturing
14.	Cold plasma processing	Cold plasma, or nonthermal plasma, is used to treat surfaces or materials for sterilization, decontamination, or surface modification. Cold plasma processing offers advantages such as rapid treatment, low energy consumption, and compatibility with sensitive materials
15.	Cryogenic processing	Cryogenic processing involves the use of very low temperatures, typically below −150 °C (−238 °F), to enhance the properties of materials. Cryogenic processing is used in metallurgy, electronics, and food processing to improve wear resistance, toughness, and shelf life
16.	Hybrid processing technologies	Hybrid processing technologies combine multiple techniques, such as thermal, mechanical, and electromagnetic methods, to achieve synergistic effects and improve processing efficiency. These hybrid approaches are used in materials processing, chemical synthesis, and manufacturing processes to overcome the limitations of individual techniques.
17.	Advanced sensing and control systems	Advanced sensing and control systems integrate sensors, actuators, and automation technologies to monitor and optimize process parameters in real time. These systems enable precise control over manufacturing processes, leading to improved product quality, consistency, and efficiency

mechanisms, such as microfinance initiatives, venture capital funds, and impact investment funds, can provide much-needed capital to SMEs and entrepreneurs in the Global South. Public–private partnerships can facilitate knowledge sharing, technology transfer, and capacity-building initiatives, enabling businesses to access the expertise and resources needed for growth. Supportive policies that incentivize investment in bio-based processing enterprises, particularly those operating in rural and underserved areas, are essential for creating an enabling environment for economic development and sustainable livelihoods (Virgin et al., 2022). By overcoming economic constraints, the bio-based processing sector can unlock its full potential as a driver of inclusive growth, environmental sustainability, and social resilience in the Global South.

4.2 Technical and Infrastructure Challenges

Technical and infrastructure challenges pose significant barriers to the efficient and effective processing of bio-based products in the Global South (Hlangwani et al., 2023). One of the primary challenges is inadequate access to reliable energy, water, and sanitation infrastructure, particularly in remote or off-grid locations where many processing facilities are situated. Without reliable access to these essential resources, processing operations are often hindered, leading to disruptions in production schedules and increased operational costs. Moreover, unreliable energy sources may limit the adoption of energy-intensive processing technologies, further impeding productivity and efficiency (Usman et al., 2024).

Additionally, insufficient technical expertise and skilled labor present significant challenges to the adoption of advanced processing technologies and techniques in the Global South (Miah et al., 2024). Many small-scale producers and entrepreneurs lack the necessary training and expertise to effectively operate and maintain modern processing equipment. This hampers their ability to implement best practices in processing, leading to issues such as inconsistent product quality, low yields, and increased waste. Furthermore, the shortage of skilled labor exacerbates the challenge of adopting innovative processing techniques, hindering the sector's ability to keep pace with technological advancements and market demands (Nizzy & Ogwu, 2024; Sima et al., 2020).

Furthermore, outdated or insufficient processing equipment and facilities constrain the capacity and capability of bio-based product manufacturers in the Global South (Gatica-Neira & Ramos-Maldonado, 2022). Many processing facilities in the region are equipped with outdated or inadequate machinery, which limits their ability to scale up production and meet growing demand. Moreover, the lack of modern processing equipment and facilities hampers the sector's competitiveness in both domestic and international markets, where quality standards and efficiency are paramount.

Addressing technical and infrastructure challenges requires a multifaceted approach that includes targeted investments in capacity building, technology transfer, and infrastructure development (Odeyemi et al., 2024). Governments, international organizations, and development agencies can play a crucial role in providing financial support and technical assistance to upgrade existing processing facilities and build new infrastructure where needed. Moreover, fostering collaboration between industry, academia, and government institutions is essential for promoting knowledge sharing and skills development. By investing in technical capacity and infrastructure development, the Global South can unlock the full potential of its bio-based processing sector, driving economic growth, innovation, and sustainability.

4.3 Potential for Innovation and Growth

Despite the myriad challenges, the bio-based processing sector in the Global South holds significant potential for innovation and growth (Stark et al., 2022). The region's rich biodiversity, abundant natural resources, and diverse agricultural landscapes offer fertile ground for the development of novel bio-based products and processing techniques. With unique raw materials at their disposal, entrepreneurs and researchers have the opportunity to explore innovative solutions that can address local needs while also catering to global markets.

Advancements in biotechnology, nanotechnology, and green chemistry further bolster the potential for innovation in bio-based processing (Wagh et al., 2024). These cutting-edge fields offer new avenues for enhancing the efficiency, sustainability, and value proposition of bio-based products. For instance, biotechnology enables the development of genetically engineered crops with improved traits for processing, while nanotechnology facilitates the creation of nanostructured materials with enhanced properties for various applications.

Moreover, the growing consumer demand for sustainable and eco-friendly products presents favorable market conditions for innovation and growth in the bio-based processing sector (Liu et al., 2024). Consumers are increasingly seeking alternatives to conventional products that are derived from fossil fuels and contribute to environmental degradation. Bio-based materials offer a promising solution, as they are renewable, biodegradable, and often have lower environmental footprints compared to their synthetic counterparts.

Additionally, there is a growing recognition of the environmental and social benefits of bio-based materials, further driving interest and investment in the sector. Governments, businesses, and civil society organizations are increasingly embracing bio-based solutions as part of their sustainability strategies, creating a conducive environment for innovation and growth (Sadhukhan et al., 2023). To fully harness the potential of the bio-based processing sector in the Global South, it is essential to leverage indigenous knowledge, traditional practices, and local resources. Strategic investments in research and development, coupled with supportive policies and partnerships, can catalyze innovation and drive economic development, environmental sustainability, and social inclusion in the region. By embracing innovation and fostering collaboration, the Global South can position itself as a leader in bio-based processing, contributing to a more sustainable and resilient future.

5 Socioeconomic Impacts

5.1 Impact on Small-Scale Farmers

The socioeconomic impacts of bio-based processing in the Global South present a complex interplay of opportunities and challenges, particularly for small-scale farmers (Stark et al., 2022). On the one hand, bio-based processing opens up new avenues for income generation by providing markets for agricultural produce. Through the transformation of raw materials into value-added products like biofuels, bio-based chemicals, or natural fibers, small-scale farmers can effectively capture more value from their crops and augment their incomes. This added value enhances the economic viability of farming activities and contributes to the financial sustainability of rural livelihoods.

Moreover, bio-based processing enables small-scale farmers to diversify their income sources, thereby reducing their dependency on a single crop or commodity (Feleke et al., 2021). By branching out into processing activities, farmers can mitigate the risks associated with fluctuations in market prices and demand for raw agricultural produce. Diversification also serves as a means of hedging against crop failures or adverse weather conditions, enhancing the resilience of small-scale farming enterprises in the face of economic uncertainties.

However, despite the potential benefits, small-scale farmers encounter various challenges in engaging with bio-based processing (Gontard et al., 2018). Access to markets remains a significant hurdle, particularly for those located in remote or underserved areas with limited transportation infrastructure. Many farmers lack the necessary resources and technical knowledge to invest in processing equipment, storage facilities, and transportation networks. Moreover, the transition to bio-based processing often necessitates acquiring new skills and knowledge, which may not be readily available or accessible to small-scale farmers. Additionally, navigating competitive global markets, which prioritize economies of scale and stringent quality standards, poses further barriers for small-scale producers seeking to market their processed products internationally (Mukherjee, 2018).

Addressing these challenges requires a holistic approach that combines targeted investments in infrastructure, capacity building, and market access with supportive policies and interventions. By empowering small-scale farmers with the resources, skills, and opportunities needed to engage in bio-based processing, policymakers, development agencies, and other stakeholders can unlock the full potential of the sector to drive economic development, enhance food security, and promote sustainable livelihoods in the Global South.

5.2 Gender Dynamics in Processing Activities

Gender dynamics play a pivotal role in shaping the landscape of bio-based processing activities in the Global South, influencing the participation, roles, and outcomes of both men and women (Bhujel & Joshi, 2024). While women are often heavily involved in agricultural production and processing activities, their contributions tend to be undervalued and overlooked compared to their male counterparts. Despite being actively engaged in tasks such as harvesting, sorting, and processing crops, women typically receive lower wages and have fewer opportunities for advancement within the agricultural value chain.

Moreover, women in the Global South face additional hurdles in accessing resources critical for engaging in bio-based processing activities (Mandal et al., 2024). Factors such as limited access to land ownership, credit facilities, and training opportunities hinder women's ability to fully participate in and benefit from bio-based processing initiatives. These disparities not only perpetuate economic inequalities but also constrain the overall potential for inclusive growth and sustainable development in rural communities.

Furthermore, entrenched gender norms and stereotypes contribute to the unequal division of labor and decision-making processes within households and communities. Women are often burdened with unpaid care work, including childcare and household chores, which can significantly limit their time and opportunities for engaging in income-generating activities such as bio-based processing (Lightman & Kevins, 2021). Additionally, cultural and social barriers may further restrict women's mobility and participation in public spaces, hindering their access to markets, training, and information relevant to bio-based processing ventures.

Addressing gender disparities and promoting gender equality in bio-based processing necessitates targeted interventions aimed at dismantling structural barriers and empowering women to fully participate in economic activities. This includes initiatives to improve women's access to resources such as land, credit, and training, as well as efforts to challenge traditional gender roles and norms that limit women's economic agency. By ensuring equitable access to resources and opportunities for both men and women, bio-based processing initiatives can contribute to more inclusive and sustainable development outcomes in the Global South.

6 Policy and Support Mechanisms

6.1 Role of Government Policies

Government policies wield significant influence in shaping the landscape of bio-based processing activities across the Global South (Brandão & Santos, 2024). A well-crafted regulatory framework provides the foundation for fostering growth and innovation within the sector while simultaneously addressing broader socioeconomic and environmental objectives. One of the pivotal roles of government policies is the establishment of an enabling regulatory environment conducive to investment and innovation in bio-based processing. This involves the formulation of clear standards, regulations, and certification schemes aimed at ensuring the quality, safety, and environmental sustainability of bio-based products. By delineating guidelines and benchmarks, governments can instill confidence among investors, consumers, and stakeholders in the reliability and integrity of bio-based processing practices.

Furthermore, governments can employ various financial incentives, tax breaks, and subsidies to stimulate private sector investment in bio-based processing infrastructure and technology (Srivastava, 2024). These measures serve to mitigate the upfront costs and risks associated with setting up and operating processing facilities, thereby encouraging businesses to explore and expand their engagement in bio-based processing ventures. Additionally, supportive policies that prioritize sustainable land use, biodiversity conservation, and the protection of natural resources are instrumental in safeguarding the long-term viability of bio-based processing activities (Paleari, 2024). By integrating environmental considerations into policy frameworks, governments can foster a holistic approach to bio-based processing that balances economic growth with ecological sustainability.

Moreover, government policies can play a pivotal role in fostering collaboration and partnership among stakeholders within the bio-based processing ecosystem. By facilitating knowledge exchange, technology transfer, and industry-academia linkages, governments can catalyze innovation and capacity-building efforts within the sector. Furthermore, policies aimed at enhancing market access, promoting research and development, and facilitating international trade can bolster the competitiveness of bio-based processing enterprises on both domestic and global scales. By aligning policies with broader development agendas such as poverty alleviation, food security, and inclusive growth, governments can harness the transformative potential of bio-based processing to drive socioeconomic progress and environmental stewardship in the Global South.

Government policies serve as indispensable tools for steering the trajectory of bio-based processing activities in the Global South. By fostering an enabling regulatory environment, incentivizing investment and innovation, and promoting sustainable practices, policymakers can unlock the full potential of bio-based processing to contribute to the attainment of national development goals and the realization of a more sustainable future.

6.2 International Support and Collaboration

International support and collaboration play indispensable roles in advancing bio-based processing initiatives across the Global South (Tiwari, 2023). Multilateral organizations, development agencies, and international donors serve as crucial partners in providing financial assistance, technical expertise, and capacity-building support to countries within the region. Through various initiatives such as technology transfer programs, knowledge-sharing platforms, and funding schemes, these international partners contribute to bridging the technological and knowledge gaps that often impede the widespread adoption of bio-based processing technologies in the Global South.

One of the primary benefits of international support is its capacity to facilitate technology transfer and knowledge exchange between developed and developing countries. By leveraging the expertise and resources of more advanced economies, countries in the Global South can gain access to cutting-edge technologies, innovative processes, and best practices in bio-based processing. This transfer of knowledge enables local stakeholders to enhance their technical capabilities, improve production efficiency, and adapt sustainable practices tailored to their specific contexts.

Moreover, international collaboration enables countries to leverage each other's strengths and experiences, thereby accelerating the development and deployment of sustainable bio-based processing solutions (Sierra et al., 2021). By sharing lessons learned, exchanging information, and collaborating on joint research and innovation projects, countries can collectively address common challenges and capitalize on emerging opportunities in the bio-based processing sector. This collaborative approach fosters synergies among governments, academia, research institutions, and the private sector, leading to more effective and impactful interventions.

Furthermore, international support plays a vital role in mobilizing financial resources to support bio-based processing initiatives in the Global South. Through funding schemes and investment programs, international donors and development agencies provide critical financial assistance to governments, businesses, and civil society organizations involved in bio-based processing activities. This financial support enables countries to invest in infrastructure development, capacity building, and technology adoption, thereby laying the foundation for sustainable and inclusive growth in the bio-based processing sector.

International support and collaboration are indispensable for advancing bio-based processing initiatives and promoting sustainable development in the Global South (Lange et al., 2021). By leveraging the expertise, resources, and experiences of international partners, countries can accelerate the transition toward a more sustainable and resilient bio-based economy, contributing to poverty reduction, food security, and environmental conservation across the region.

Effective policy and support mechanisms are essential for unlocking the full potential of bio-based processing in the Global South (Ding et al., 2024). By implementing supportive government policies and leveraging international support and collaboration, countries can create an enabling environment for investment, innovation, and sustainable development in the bio-based processing sector. Moreover, by aligning policies with broader development objectives such as poverty reduction, food security, and environmental sustainability, governments and international partners can ensure that bio-based processing initiatives contribute to inclusive and equitable growth in the Global South.

7 Future Directions and Innovations

7.1 Emerging Technologies and Techniques

The future of bio-based processing in the Global South is marked by a promising trajectory of advancements fueled by emerging technologies and innovative techniques (Demirel et al., 2024). The rapid evolution of biotechnology, nanotechnology, and green chemistry is reshaping the landscape of bio-based processing, presenting unprecedented opportunities for enhancing efficiency, sustainability, and product diversification. Biotechnological breakthroughs, such as genetic engineering and metabolic engineering, are empowering scientists to develop novel crops with tailored traits optimized for bio-based processing (Diankristanti et al., 2024). These advancements promise higher yields, improved disease resistance, and optimized biochemical compositions, thus revolutionizing the raw material base for bio-based industries across the Global South.

Furthermore, ongoing progress in fermentation technologies, enzymatic processes, and microbial bioproduction is broadening the horizons of biomass conversion, enabling the transformation of diverse feedstocks into a wide array of valuable products (Lopresto et al., 2018). From biofuels to biopolymers and biochemicals, these cutting-edge techniques offer scalable and sustainable solutions to meet the growing demands for renewable resources and eco-friendly alternatives to conventional products. Moreover, nanotechnology is increasingly being harnessed to enhance the performance and properties of bio-based materials, unlocking new functionalities and applications. Nanostructured materials such as nanocellulose and nanoclays exhibit superior mechanical strength, barrier properties, and biodegradability, rendering them appealing alternatives across various sectors, including packaging, construction, and healthcare.

Additionally, the principles of green chemistry are driving the development of eco-friendly processes and renewable feedstocks for bio-based processing, ushering in an era of sustainable manufacturing practices that minimize waste, energy consumption, and environmental impact (Sharma et al., 2024). By prioritizing the use of renewable resources, reducing hazardous chemicals, and optimizing production processes, green chemistry offers a pathway toward a more sustainable and circular bio-based economy in the Global South. As these emerging technologies and techniques continue to evolve and mature, they hold immense potential to catalyze transformative changes, driving economic growth, environmental stewardship, and social development across the region. Through strategic investments, collaborative research efforts, and supportive policy frameworks, countries in the Global South can harness the full potential of these innovations to chart a sustainable path forward for bio-based processing.

7.2 Prospects for Scaling Up Successful Models

The prospects for scaling up successful models in bio-based processing across the Global South are promising as technologies mature and demonstrate their economic viability (Singh et al., 2022). Leveraging successful case studies and best practices from diverse regions, countries have the opportunity to replicate and adapt proven models to fit their local contexts and development priorities. Community-based biorefineries and cooperative processing units represent exemplary models for decentralized bio-based processing, particularly in rural and underserved areas. These grassroots initiatives empower local communities, create employment opportunities, and add value to agricultural produce, thereby fostering rural development and alleviating poverty.

Moreover, public–private partnerships and cross-sector collaborations have the potential to facilitate the scaling up of bio-based processing initiatives by mobilizing resources, expertise, and networks (Tran et al., 2024). Collaboration among government agencies, research institutions, industry stakeholders, and civil society organizations can foster a conducive environment for innovation and investment, enabling countries to overcome barriers and accelerate the adoption of bio-based processing technologies at scale. By pooling resources and sharing knowledge, stakeholders can collectively address challenges related to technology transfer, infrastructure development, and market access, thus promoting the widespread adoption of sustainable bio-based processing practices.

Furthermore, international cooperation and knowledge-sharing platforms play a pivotal role in disseminating best practices and building technical capacity to support the scaling up of bio-based processing initiatives in the Global South (Sierra et al., 2021). By facilitating exchange programs, collaborative research projects, and capacity-building initiatives, international partners can help countries in the region enhance their capabilities and seize opportunities for sustainable development. By fostering innovation ecosystems and fostering a culture of collaboration and learning, international cooperation contributes to the creation of an enabling environment for scaling up bio-based processing initiatives, thereby unlocking their full potential to drive economic growth, environmental sustainability, and social progress across the Global South.

8 Recommendations for Future Research and Development

Looking ahead, strategic investments in research and development (R&D) are imperative to propel innovation and tackle critical challenges in bio-based processing. Future R&D efforts should prioritize enhancing the efficiency and sustainability of bio-based processing technologies, optimizing feedstock utilization, and valorizing waste streams. By focusing on these key areas, researchers can develop solutions that not only improve the economic viability of bio-based processing but also contribute to environmental conservation and social welfare. Additionally, interdisciplinary research that integrates insights from fields such as agronomy, biotechnology, materials science, and socioeconomic analysis is essential to develop holistic solutions that strike a balance between economic, environmental, and social considerations. By fostering collaboration and knowledge exchange among experts from diverse disciplines, researchers can leverage complementary expertise to address complex challenges and maximize the impact of R&D initiatives.

Furthermore, fostering collaboration between academia, industry, and government agencies is paramount to align R&D priorities with market needs, accelerate technology transfer, and ensure the commercialization of research outcomes. By fostering an ecosystem of collaboration and innovation, stakeholders can collectively identify and address market gaps, regulatory barriers, and technical challenges hindering the adoption of bio-based processing technologies. Moreover, investing in human capital development, education, and skills training is critical to cultivating a skilled workforce equipped with the knowledge and expertise to drive innovation and entrepreneurship in the bio-based processing sector. By investing in education and training programs that nurture talent and foster creativity, countries in the Global South can build a pool of skilled professionals capable of leading the transition toward a more sustainable and resilient bio-based economy.

By prioritizing these recommendations for future research and development, countries in the Global South can position themselves as leaders in the transition toward a more sustainable and resilient bio-based economy. By leveraging R&D investments to drive innovation, foster collaboration, and build human capital, countries can unlock new opportunities for economic growth, environmental stewardship, and social development. Through concerted efforts to advance bio-based processing technologies and practices, countries can harness the transformative potential of bio-based resources to address pressing challenges and create a more prosperous and sustainable future for all.

9 Conclusion

In conclusion, the exploration of processing techniques for bio-based products in the Global South underscores the immense potential of leveraging natural resources, traditional knowledge, and emerging technologies to drive sustainable development. Throughout this discussion, we have highlighted the multifaceted benefits of bio-based processing, ranging from environmental sustainability and economic growth to social empowerment and innovation. By tapping into the rich biodiversity and agricultural abundance of the Global South, countries in the region have the opportunity to unlock new pathways for value addition, job creation, and rural development.

Moreover, our analysis has shed light on the challenges and opportunities inherent in bio-based processing, from economic constraints and technical barriers to gender dynamics and policy imperatives. While these challenges pose formidable obstacles, they also serve as catalysts for innovation, collaboration, and transformative change. By addressing these challenges through strategic investments, supportive policies, and inclusive approaches, countries can harness the full potential of bio-based processing to drive inclusive and sustainable development.

Looking ahead, strategic recommendations for future research and development underscore the importance of continued investment in innovation, capacity building, and collaboration. By prioritizing interdisciplinary research, fostering partnerships between academia, industry, and government agencies, and investing in human capital development, countries in the Global South can position themselves at the forefront of the bio-based revolution. Through concerted efforts to advance bio-based processing technologies and practices, countries can not only enhance their economic competitiveness but also contribute to global efforts toward a more sustainable and resilient future.

In essence, the journey toward a more sustainable and inclusive bio-based economy requires collective action, shared commitment, and visionary leadership. By embracing the principles of sustainability, equity, and innovation, countries in the Global South can chart a course toward a future where bio-based processing catalyzes prosperity, resilience, and well-being for all.

References

Adebola, T. (2024). Branding as a tool to promote geographical indication exports and sustainable development in Africa. In *The Elgar companion to intellectual property and the sustainable development goals* (pp. 499–521). Edward Elgar Publishing.

Adeleke, A. K., Montero, D. J. P., Olu-lawal, K. A., & Olajiga, O. K. (2024). Process development in mechanical engineering: innovations, challenges, and opportunities. *Engineering Science & Technology Journal, 5*(3), 901–912.

Agoh, E. C., Ogwu, M. C., Chukwuemeka, O. S., & Ekeledo, P. I. (2024). Environmental and human health effects of cassava processing and processing waste. In M. C. Ogwu, S. C. Izah, A. A. Cunha Alves, & babu, S.C. (Eds.), *Plant biology, sustainability and climate change, sustainable Cassava* (pp. 203–219). Academic. https://doi.org/10.1016/B978-0-443-21747-0.00001-1

Akinola, R., Pereira, L. M., Mabhaudhi, T., De Bruin, F. M., & Rusch, L. (2020). A review of indigenous food crops in Africa and the implications for more sustainable and healthy food systems. *Sustainability, 12*(8), 3493.

Andrew, J. J., & Dhakal, H. N. (2022). Sustainable bio-based composites for advanced applications: recent trends and future opportunities–A critical review. *Composites Part C: Open Access, 7*, 100220.

Ardana, P. D. H., Suparwata, D. O., Sudrajat, A., Chatun, S., & Harsono, I. (2024). The role of Bali's traditional Subak farming system in the preservation of natural and cultural resources. *West Science Nature and Technology, 2*(01), 31–38.

Arinze, C. A., Ajala, O. A., Okoye, C. C., Ofodile, O. C., & Daraojimba, A. I. (2024). Evaluating the integration of advanced IT solutions for emission reduction in the oil and gas sector. *Engineering Science & Technology Journal, 5*(3), 639–652.

Arslan, A., & Alibaş, İ. (2024). Assessing the effects of different drying methods and minimal processing on the sustainability of the organic food quality. *Innovative Food Science & Emerging Technologies, 94*, 103681.

Bhatia, L., Kaladhar, D. S., Sarkar, T., Jha, H., & Kumar, B. (2024). Food wastes phenolic compounds (PCs): overview of contemporary greener extraction technologies, industrial potential, and its integration into circular bioeconomy. *Energy Ecology and Environment, 9*, 1–31.

Bhujel, R. R., & Joshi, H. G. (2024). Unveiling sustainable agriculture dynamics: a socio-psychological exploration among smallholder farmers in Sikkim, India. *Cogent Social Sciences, 10*(1), 2350118.

Brandão, A. S., & Santos, J. M. (2024). Sustainability from policy to practice: Assessing the impact of European research and innovation frameworks on circular bioeconomy. *Sustainability, 16*(6), 2355.

Chobot, M., Kozłowska, M., Ignaczak, A., & Kowalska, H. (2024). Development of drying and roasting processes for the production of plant-based pro-healthy snacks in the light of nutritional trends and sustainable techniques. *Trends in Food Science & Technology, 149*, 104553.

Cuestas-Caza, J., Toledo, L., & Rodríguez, F. (2024). Transcultural bioeconomy governance in a plurinational state: Sumak Kawsay and bio-based production in two Kichwa territories of Ecuador. *Forest Policy and Economics, 163*, 103227.

Demirel, Ö., Güneş, H., & Can, C. (2024). Sustainable and modern bio-based technologies: new approachs to food safety and security. *Environment, Development and Sustainability*, pp.1–28.

Delgado, A., Rodriguez, D. J., Amadei, C. A., & Makino, M. (2024). Water in circular economy and resilience (WICER) framework. *Utilities Policy, 87*, 101604.

Diankristanti, P. A., Ho, N. H. E., Chen, J. H., Nagarajan, D., Chen, C. Y., Hsieh, Y. M., Ng, I. S., & Chang, J. S. (2024). Unlocking the potential of microalgae as sustainable bioresources from up to downstream processing: A critical review. *Chemical Engineering Journal, 488*, 151124.

Ding, Z., Hamann, K. T., & Grundmann, P. (2024). Enhancing circular bioeconomy in Europe: Sustainable valorization of residual grassland biomass for emerging bio-based value chains. *Sustainable Production and Consumption, 45*, 265–280.

Egboduku, W. O., Egboduku, T., Golohor, O. M., Imarhiagbe, O., & Ogwu, M. C. (2024). Cassava as raw material for sustainable bioeconomy development. In M. C. Ogwu, S. C. Izah, A. A. Cunha Alves, & S. C. Babu (Eds.), *Plant biology, sustainability and climate change, sustainable Cassava* (pp. 57–73). Academic Press. https://doi.org/10.1016/B978-0-443-21747-0.00022-9

Elsamanoudy, G., Abdelaziz Mahmoud, N. S., & Alexiou, P. (2024). Handwoven interior accessories from palm leaves as sustainable elements. *Journal of Cultural Heritage Management and Sustainable Development*. https://doi.org/10.1108/JCHMSD-05-2023-0054

Feleke, S., Cole, S. M., Sekabira, H., Djouaka, R., & Manyong, V. (2021). Circular bioeconomy research for development in sub-Saharan Africa: innovations, gaps, and actions. *Sustainability, 13*(4), 1926.

Garmendia-Lemus, S., Moshkin, E., Hung, Y., Tack, J., & Buysse, J. (2024). European farmers' perceptions and intentions to use bio-based fertilisers: Insights from the theory of planned behaviour and perceived utility. *Journal of Cleaner Production, 434*, 139755.

Gatica-Neira, F., & Ramos-Maldonado, M. (2022). Limits to the productivity in bio-based territorial SMEs. *SAGE Open, 12*(2), 21582440221099294.

Gontard, N., Sonesson, U., Birkved, M., Majone, M., Bolzonella, D., Celli, A., Angellier-Coussy, H., Jang, G. W., Verniquet, A., Broeze, J., & Schaer, B. (2018). A research challenge vision regarding management of agricultural waste in a circular bio-based economy. *Critical Reviews in Environmental Science and Technology, 48*(6), 614–654.

Gürsu, H. (2024). Waste-based vertical planting system proposal to increase productivity in sustainable horticulture; "PETREE". *Sustainability, 16*(8), 3125.

Hassan, S. S., Williams, G. A., & Jaiswal, A. K. (2019). Moving towards the second generation of lignocellulosic biorefineries in the EU: Drivers, challenges, and opportunities. *Renewable and Sustainable Energy Reviews, 101*, 590–599.

Hlangwani, E., Mpye, K. L., Matsuro, L., & Dlamini, B. (2023). The use of technological innovation in bio-based industries to foster growth in the bioeconomy: A South African perspective. *Sustainability: Science, Practice and Policy, 19*(1), 2200300.

Holland, C., & Shapira, P. (2024). Building the bioeconomy: A targeted assessment approach to identifying bio-based technologies, challenges and opportunities. *Engineering Biology, 8*(1), 1–15.

Kibria, M. G., Paul, U. K., Hasan, A., Mohtasim, M. S., Das, B. K., & Mourshed, M. (2024). Current prospects and challenges for biomass energy conversion in Bangladesh: Attaining sustainable development goals. *Biomass and Bioenergy, 183*, 107139.

Kolasani, S. (2024). Revolutionizing manufacturing, making it more efficient, flexible, and intelligent with Industry 4.0 innovations. *International Journal of Sustainable Development Through AI, ML and IoT, 3*(1), 1–17.

Kosoe, E.A., and Ogwu, M.C. (2024). Sustainable cassava processing: processes, techniques, wastes, and waste streams. In: Ogwu, M.C., Izah, S.C., Cunha Alves, A.A. and Babu, S.C. (eds.). Plant biology, sustainability and climate change, sustainable Cassava. Academic. pp. 257–272. https://doi.org/10.1016/B978-0-443-21747-0.00011-4.

Lange, L., Connor, K. O., Arason, S., Bundgård-Jørgensen, U., Canalis, A., Carrez, D., Gallagher, J., Gøtke, N., Huyghe, C., Jarry, B., & Llorente, P. (2021). Developing a sustainable and circular bio-based economy in EU: By partnering across sectors, upscaling and using new knowledge faster, and for the benefit of climate, environment & biodiversity, and people & business. *Frontiers in Bioengineering and Biotechnology, 8*, 619066.

Lightman, N., & Kevins, A. (2021). "Women's work": Welfare state spending and the gendered and classed dimensions of unpaid care. *Gender & Society, 35*(5), 778–805.

Liu, Y., Li, N., Zhang, X., Wei, T., Ma, M., Sun, Q., Li, M., & Xie, F. (2024). Eco-friendly drinking straws: Navigating challenges and innovations. *Trends in Food Science & Technology*, 104511. https://doi.org/10.1016/j.tifs.2024.104511

Long, L. M. (2024). Culinary tourism and contradictions of cultural sustainability: Industrial agriculture food products as tradition in the American Midwest. *Food, Culture & Society, 27*(1), 48–68.

Lopresto, C. G., Verardi, A., Nicoletti, C., Mukherjee, D., Calabro, V., Chakraborty, S., & Curcio, S. (2018). Technological aspects of lignocellulose conversion into biofuels: Key challenges and practical solutions. In *Sustainable biotechnology-enzymatic resources of renewable energy* (pp. 117–154).

Mandal, M., Roy, A., Das, S., Rakwal, R., Agrawal, G. K., Singh, P., Awasthi, A., & Sarkar, A. (2024). Food waste-based bio-fertilizers production by bio-based fermenters and their potential impact on the environment. *Chemosphere, 353*, 141539.

Mandree, P., Thopil, G. A., & Ramchuran, S. (2023). Potential opportunities to convert waste to bio-based chemicals at an industrial scale in South Africa. *Fermentation, 9*(10), 908.

Mengal, P., Wubbolts, M., Zika, E., Ruiz, A., Brigitta, D., Pieniadz, A., & Black, S. (2018). Bio-based industries joint undertaking: The catalyst for sustainable bio-based economic growth in Europe. *New Biotechnology, 40*, 31–39.

Miah, M. T., Erdei-Gally, S., Dancs, A., & Fekete-Farkas, M. (2024). A systematic review of industry 4.0 technology on workforce employability and skills: Driving success factors and challenges in South Asia. *Economies, 12*(2), 35.

Miyahira, R. F., Lopes, J. D. O., & Antunes, A. E. C. (2021). The use of sprouts to improve the nutritional value of food products: A brief review. *Plant Foods for Human Nutrition, 76*(2), 143–152.

Moreno, A. D., Susmozas, A., Oliva, J. M., & Negro, M. J. (2020). Overview of bio-based industries. In *Bio-based products and industries* (pp. 1–40). Elsevier.

Moshood, T. D., Nawanir, G., Mahmud, F., Mohamad, F., Ahmad, M. H., & Ghani, A. A. (2021). Expanding policy for biodegradable plastic products and market dynamics of bio-based plastics: challenges and opportunities. *Sustainability, 2021*, 13.

Mukherjee, S. (2018). Challenges to Indian micro small scale and medium enterprises in the era of globalization. *Journal of Global Entrepreneurship Research, 8*, 1–19.

Murali, R. D., Vishnu Priya, S., Swetha, & Antony, U. (2024). Traditionally fermented foods and beverages for nutritional security and global acceptance. In F. Production (Ed.), *Diversity, and safety under climate change* (pp. 77–87). Springer Nature Switzerland.

Naika, A., Pillay, A., & Paliwal, A. (2024). Indigenous food system for sustainability: South Pacific study. In *Sustainable food systems (Volume I) SFS: Framework, sustainable diets, traditional food culture & food production* (pp. 35–53). Springer Nature Switzerland.

Ninduwezuor-Ehiobu, N., Tula, O. A., Daraojimba, C., Ofonagoro, K. A., Ogunjobi, O. A., Gidiagba, J. O., Egbokhaebho, B. A., & Banso, A. A. (2023). Exploring innovative material integration in modern manufacturing for advancing us competitiveness in sustainable global economy. *Engineering Science & Technology Journal, 4*(3), 140–168.

Nizio, E., Czwartkowski, K., & Niedbała, G. (2023). Impact of smoking technology on the quality of food products: Absorption of Polycyclic Aromatic Hydrocarbons (PAHs) by food products during smoking. *Sustainability, 15*(24), 16890.

Nizzy, A. M., & Ogwu, M. C. (2024). Valorization of cassava processing by-products into biofuel for a sustainable environment. In M. C. Ogwu, S. C. Izah, A. A. Cunha Alves, & S. C. Babu

(Eds.), *Plant biology, sustainability and climate change, sustainable Cassava* (pp. 291–309). Academic. https://doi.org/10.1016/B978-0-443-21747-0.00010-2

Odeyemi, O., Oyewole, A. T., Adeoye, O. B., Ofodile, O. C., Addy, W. A., Okoye, C. C., & Ololade, Y. J. (2024). Entrepreneurship in Africa: A review of growth and challenges. *International Journal of Management & Entrepreneurship Research, 6*(3), 608–622.

Ogwu, M. C., & Osawaru, M. E. (2022). Traditional methods of plant conservation for sustainable utilization and development. In *Biodiversity in Africa: Potentials, threats and conservation* (pp. 451–472). Springer Nature Singapore.

Ogwu, M. C., Odozi, I. P., Ahonsi, O. C., Uleanya, K. O., & Odozi, E. B. (2024). Organic acid production from cassava. In M. C. Ogwu, S. C. Izah, A. A. Cunha Alves, & babu, S.C. (Eds.), *Plant biology, sustainability and climate change, sustainable Cassava* (pp. 395–418). Academic. https://doi.org/10.1016/B978-0-443-21747-0.00009-6

Omoniyi, S. A., & Cosmas, A. (2024). Cooking practice and safety assessment of boiled cassava root sold in streets of Gashua, Yobe state Nigeria. *Food Chemistry Advances, 4*, 100562.

Onwuka, O. U., & Adu, A. (2024). Eco-efficient well planning: Engineering solutions for reduced environmental impact in hydrocarbon extraction. *International Journal of Scholarly Research in Multidisciplinary Studies, 4*(01), 033–043.

Oyeneye, A. I., Emmanuel, O. O., Olusegun, O. O., Morenikeji, O., & Olamide, O. A. (2023). Use of cooking salt in school meal preparation and quality control practices among food vendors engaged in the national home-grown school feeding programme in Southwest Nigeria. *African Journal of Food Science, 17*(1), 1–9.

Paleari, S. (2024). The EU policy on climate change, biodiversity and circular economy: Moving towards a Nexus approach. *Environmental Science & Policy, 151*, 103603.

Philippidis, G., Álvarez, R. X., Di Lucia, L., Hermoso, H. G., Martinez, A. G., M'barek, R., Moiseyev, A., Panoutsou, C., Itoiz, E. S., Sturm, V., & van Leeuwen, M. (2024). The development of bio-based industry in the European Union: A prospective integrated modelling assessment. *Ecological Economics, 219*, 108156.

Ramchuran, S. O., O'Brien, F., Dube, N., & Ramdas, V. (2023). An overview of green processes and technologies, bio-based chemicals and products for industrial applications. *Current Opinion in Green and Sustainable Chemistry, 41*, 100832.

Rawat, Y. S., Singh, G. S., & Tekleyohannes, A. T. (2023). Optimizing the benefits of invasive alien plants biomass in South Africa. *Sustainability, 2024*(16), 1876. https://doi.org/10.3390/su16051876. Academic Editor: Sandra Notaro Received, 15.

Reyes-García, V., García-del-Amo, D., Álvarez-Fernández, S., Benyei, P., Calvet-Mir, L., Junqueira, A. B., Labeyrie, V., Li, X., Miñarro, S., Porcher, V., & Porcuna-Ferrer, A. (2024). Indigenous peoples and local communities report ongoing and widespread climate change impacts on local social-ecological systems. *Communications Earth & Environment, 5*(1), 29.

Sadhukhan, J., Martinez-Hernandez, E., Allieri, M. A. A., Eguía-Lis, J. A. Z., Castillo, A., Dominguillo, D., Torres-Garcia, E., & Aburto, J. (2023). Strategic navigation of world-leading biorefineries and Mexico's policy landscape: A gateway to a sustainable circular bioeconomy. *Journal of Cleaner Production, 434*, 140386.

Salazar Sandoval, S., Amenábar, A., Toledo, I., Silva, N., & Contreras, P. (2024). Advances in the sustainable development of bio-based materials using plant and animal waste as raw materials: A review. *Sustainability, 16*(3), 1073.

Sharma, S., Gallou, F., & Handa, S. (2024). Towards a sustainable tomorrow: Advancing green practices in organic chemistry. *Green Chemistry., 26*, 6289.

Sierra, A. R., Zika, E., Lange, L., de Azúa, P. L. R., Canalis, A., Esteban, P. M., Paiano, P., & Mengal, P. (2021). The bio-based industries joint undertaking: a high impact initiative that is transforming the bio-based industries in Europe. *New Biotechnology, 60*, 105–112.

Sima, V., Gheorghe, I. G., Subić, J., & Nancu, D. (2020). Influences of the industry 4.0 revolution on the human capital development and consumer behavior: A systematic review. *Sustainability, 12*(10), 4035.

Singh, V. (2024). The natural resources: Introduction. In *Textbook of environment and ecology* (pp. 97–105). Springer.

Singh, N., Singhania, R. R., Nigam, P. S., Dong, C. D., Patel, A. K., & Puri, M. (2022). Global status of lignocellulosic biorefinery: Challenges and perspectives. *Bioresource Technology, 344*, 126415.

Sionek, B., Szydłowska, A., Küçükgöz, K., & Kołożyn-Krajewska, D. (2023). Traditional and new microorganisms in lactic acid fermentation of food. *Fermentation, 9*(12), 1019.

Srivastava, P. (2024). Government initiative and policy for agricultural waste utilization as biofuel. In *Emerging trends and techniques in biofuel production from agricultural waste* (pp. 273–304). Springer Nature Singapore.

Stark, S., Biber-Freudenberger, L., Dietz, T., Escobar, N., Förster, J. J., Henderson, J., Laibach, N., & Börner, J. (2022). Sustainability implications of transformation pathways for the bioeconomy. *Sustainable Production and Consumption, 29*, 215–227.

Swinbank, V. A., & Swinbank, V. A. (2021). *Women feed the world: Biodiversity and culinary diversity/food security and food sovereignty* (pp. 187–218). Women's Food Matters.

Taron, A., & Gebrezgabher, S. (2024). Circular bioeconomy: A pathway to sustainable development in an age of global crisis. In *International trade, economic crisis and the sustainable development goals* (pp. 99–117). Emerald Publishing Limited.

Tiwari, A. (2023). Advancement of materials to sustainable & green world. *Advanced Materials Letters, 14*(3), 2303–1724.

Tran, M. H., Paramasivam, P., Le, H. C., & Nguyen, D. T. (2024). Biomass: A versatile resource for biofuel, industrial, and environmental solution. *International Journal on Advanced Science, Engineering & Information Technology, 14*(1), 268.

Tulagha, I., Jaiyeoba, K. F., Ogunlade, C. A., Azeez, N. A., & Adedigba, S. A. (2021). Energy consumption in Processing and Production of Pap (*ogi*) in Nigeria. *Premier Journal of Engineering and Applied Sciences (PJEAS), 2*(2), 24–31. A quarterly publication of the Nigerian Society of Engineers, Ibadan Branch.

Ukoba, K., Yoro, K. O., Eterigho-Ikelegbe, O., Ibegbulam, C., & Jen, T. C. (2024). Adaptation of solar power in the Global south: Prospects, challenges and opportunities. *Heliyon., 10*, e28009.

Usman, F. O., Ani, E. C., Ebirim, W., Montero, D. J. P., Olu-lawal, K. A., & Ninduwezuor-Ehiobu, N. (2024). Integrating renewable energy solutions in the manufacturing industry: challenges and opportunities: A review. *Engineering Science & Technology Journal, 5*(3), 674–703.

Virgin, I., Diaz-Chavez, R., Morris, E. J., Haileselassie, T., Tesfaye, K., De Cliff, S., Njau, K., Munganyinka, E., Muyambi, F., & Otim, M. O. (2022). *State of the bioeconomy in Eastern Africa: 2022*. Stockholm Environment Institute, The East African Science and Technology Commission and BioInnovate Africa.

Wagh, M. S., Sowjanya, S., Nath, P. C., Chakraborty, A., Amrit, R., Mishra, B., Mishra, A. K., & Mohanta, Y. K. (2024). Valorisation of agro-industrial wastes: Circular bioeconomy and biorefinery process–A sustainable symphony. *Process Safety and Environmental Protection, 183*, 708.

Production Techniques for Bio-based Products in the Global South

Toyese Friday Oyewusi, Olayinka Akintola Omotosho,
Busayo Sunday Adeboye, Seun Oyelami,
and Rachael Taiwo Babalola

Abstract

The pursuit to achieve sustainable development has fuelled the desires of all nations to see to the well-being of humans both now and in the future. One of the drivers that can be used to accomplish this sustainable development is a bio-based economy via the use of bio-based products. Bio-based production concept constitutes a new paradigm in a circular bio-based economy for promoting development. The sustainable exploitation of the world's biodiversity from renewable sources to replace fossil raw materials is the main goal of the bioeconomy, and this can be achieved through the production of bio-based products to improve society, the environment, and the economy. The development of a bio-based economy offers a significant opportunity for achieving the Sustainable Development Goals, through the production of sustainable bio-based products. Products made from biological resources show great promise as a substitute for products made from fossil materials and help mitigate negative environmental consequences caused by the latter. Therefore, expanding the market for bio-based products results in the transformation of the industrial sector, thereby fos-

T. F. Oyewusi (✉)
Department of Agricultural Engineering, Osun State University, Osogbo, Nigeria
e-mail: toyese.oyewusi@uniosun.edu.ng

O. A. Omotosho
Land and Water Resources Management Programme, Institute of Agricultural Research & Training, Obafemi Awolowo University, Moor Plantation, Ibadan, Nigeria

B. S. Adeboye · S. Oyelami
Department of Mechanical Engineering, Osun State University, Osogbo, Nigeria

R. T. Babalola
Department of Food Science and Technology, Osun State University, Osogbo, Nigeria

M. C. Ogwu et al. (eds.), *Sustainable Bioeconomy Development in the Global South*, https://doi.org/10.1007/978-981-96-0305-3_3

tering biotechnological innovation and supporting rural development. However, to achieve this, different production techniques are needed to contribute more to increasing the manufacture of many bio-based products from a wide variety of available bioresources. Production techniques are highly desirable and are of critical relevance to the conversion of bio-based resources derived from plants and animals into different categories of bio-products such as biofuels, biochemicals, and biomaterials. Therefore, this chapter provides an overview of bio-based products and the production techniques used in the production of bio-based products in the Global South. The book chapter emphasised bio-based products that fall into the categories of bioenergy, biochemicals, and biomaterials produced from bio-based resources in the Global South using the most common production techniques—biochemical and thermochemical methods. It is imperative that the Global South fully shift to a sustainable bio-based economy considering the expanding population, environmental issues, and the diminishing availability of fossil-based resources. It is anticipated that the data presented in this chapter will offer crucial direction to industrial sector decision-makers or policymakers to accomplish more sustainable social, economic, and environmental goals in the Global South.

Keywords

Bioeconomy · Biodiversity conservation · Bio-products · Bioresources · Global south · Sustainable practices

Abbreviations

AC	Activated Carbon
ANG	Adsorbed Natural Gas
BIOTIC	Biological Traits Information Catalogue
CAGR	Compound Annual Growth Rate
CH_4	Methane
CO	Carbon monoxide
CO_2	Carbon dioxide
EAMs	Electro-active Microorganisms
EU	European Union
EUC	European Union Commission
FAO	Food and Agriculture Organisation
FSRI	Farm Security and Rural Investment Act
IP	Intellectual Property
MFCs	Microbial Fuel Cells
NGOs	Non-Governmental Organisations

NRC	National Research Council
SCP	Single-Cell Protein
SDGs	Sustainable Development Goals
UN	United Nations
UNDP	United Nations Development Programme
US	United States
USDA	United States Department of Agriculture
WCEF	World Circular Economy Forum

1 Introduction

Bio-based products are value-added products made entirely or partially from biological resources. Badgujar and Bhanage (2018), Venkata Mohan et al. (2019), and Ubando et al. (2020) reported that the basic feedstock for producing bio-based products are biological resources, biogenic wastes, and gaseous emissions (CO_2, CO, CH_4, etc.). They offer more features for the product, less resource-intensive production, and effective use of only natural resources. Due to their ability to replace fossil fuels, bio-based products are therefore seen as sustainable and are becoming increasingly significant in the global economy (Oyewusi et al., 2018; Oyewusi et al., 2023a). Based on this, bio-based industries have been able to maximise the use of renewable resources to produce goods like chemicals, materials, and energy. According to Market Analysis Report (2020), the compound annual growth rate (CAGR) for bio-based products is projected to be 12.6% by 2025. At this point, bio-based products are recommended due to their lower emissions and environmental friendliness. Generally, bio-based processing, handling, production, and utilisation have been found to revolve around sustainable development (Muralikrishna & Manickam, 2017). Therefore, the development of a bio-based product offers a significant opportunity for achieving the sustainable development that the worldwide desires.

Sustainable development is expressed as the development that satisfies current needs without jeopardising the capacity of future generations to satisfy their own needs (UN, 2015; Ogwu, 2019a). This makes sustainable development one of the top concerns on the current political and scientific agendas worldwide. Thus, sustainable development addresses major challenges in society, which include resource depletion, climate change, environmental degradation, etc. Three specific pillars are combined to achieve sustainability, and these are presented in Fig. 1. Mensah (2019) reported that these three pillars of sustainability can be connected to the conservation of biodiversity that forms the basis of life from a worldwide standpoint along with a life in dignity as underlying maxims. Hence, sustainability can be defined as the long-term, practically achievable growth of the social, ecological, and economic facets of human existence. This concept relates to fairness in communities, intergenerational relationships, and international relations. The pursuit to achieve sustainable development has fuelled the desires of all nations to see to the well-being

Fig. 1 Framework for assessment of sustainability

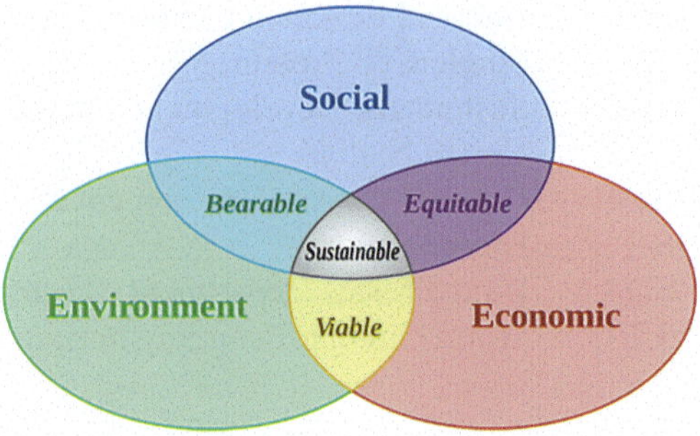

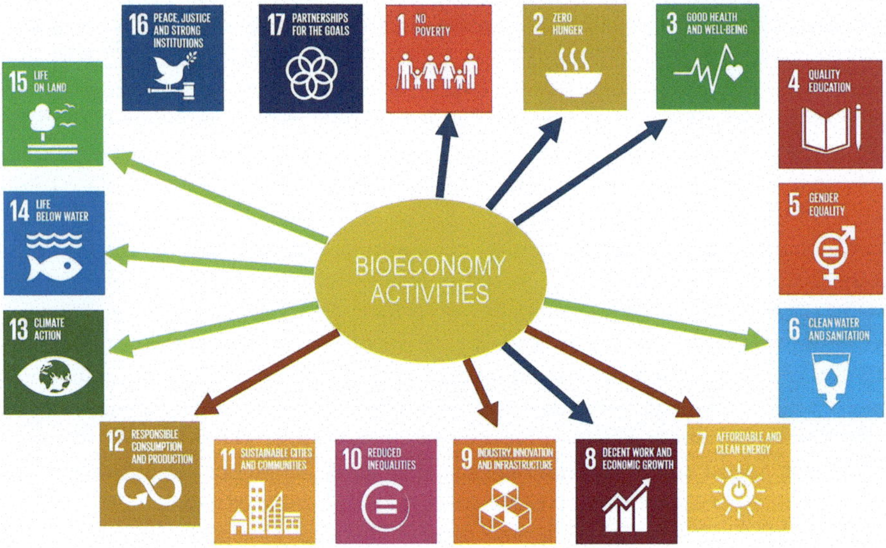

Fig. 2 Contribution of bio-based products to SDGs. Blue arrow: social justice; green arrow: environmental responsibility; red arrow: economic efficiency

of humans both now and in the future. One of the drivers that can be used to accomplish this sustainable development is a bio-based economy via the use of bio-based materials and the production of bio-based products.

The development of a bio-based economy offers an important opportunity to fulfil the intentions of Sustainable Development Goals (SDGs) via the production of sustainable bio-products (UNEP, 2019, 2020). The 2009 concept of planetary boundaries is one of the key guidelines for thinking about sustainability (UN, 2014, 2015). These indicate what constitutes a 'safe margin of manoeuvre' that protects both human livelihoods and the environment (Lamidi & Oyewusi, 2024; Oyewusi & Lamidi, 2024). In this view, bio-based products can be said to directly or indirectly contribute to all the seventeen SDGs in terms of economic efficiency, social justice, and environmental responsibilities as presented in Fig. 2 (UNDP, 2019).

Therefore, to achieve bioeconomy sustainably is the displacement of non-renewable or fossil resources in favour of renewable domestic raw materials and the establishment of robust bio-based products.

The economy of the Global South Countries (GSCs) depends largely on bio-based primary activity sectors like forestry, fisheries, and agriculture (Börner et al., 2017). These sectors frequently offer a great deal of promise for fostering sustainable growth through a transfer of knowledge and technology. Some countries such as China and Brazil use biomass as a modern source of energy and fuel to a greater extent than many industrialised countries (Ahorsu et al., 2018). In fact, the last two decades have been of thorough development for the GSCs with increasing international trade dynamics supplying the European market with products from agriculture (FAO, 2019). This increasing connectedness of bio-based economies comes with opportunities and challenges as a result of rapid innovation in the utilisation of bioresources for bio-based products (Ellabban et al., 2014). Based on this, comprehensive overview of the thrilling highlights and necessary advancements, in addition to a detailed analysis of feedstock, characterisation, bioconversion, and contemporary pre-treatment procedures, appear to be vital. Thus, different production techniques are needed to contribute more to increasing the manufacture of many bio-based products from a wide variety of available bioresources.

Production techniques are highly desirable and are of critical relevance to the conversion of bio-based resources derived from plants and animals into biofuels, biochemicals, and biomaterials (Adams et al., 2018; Nizzy & Ogwu, 2024; Egboduku et al., 2024). Production technique as rightly described is a specific method of doing an activity; in this case, it is a method of converting bio-based resources into the final products. Presently, bio-based production techniques are increasing due to their importance to global economic development. A wide variety of bioresources can now be transformed into value-added bioproducts (Swaminaathan et al., 2024). Although bio-based products have been in existence for the past few years, the production techniques have now received significant improvement (Swaminaathan et al., 2024). To this event, the growing concerns for traditional applications and novel markets for bio-based products have been identified in recent years (Segovia-Hernandez et al., 2021). Shahsavarani et al. (2013) and Leblanc et al. (2022) reported that among the production techniques, turning biomass into bioenergy has proven to be the most ambitious. In addition, Rogers et al. (2018) stated that the pulp and paper industry has discovered that production method can improve process efficiency, yielding savings in money and energy. Therefore, production techniques could help in overcoming barriers encountered in lignin and lignocellulosic biomass for efficient commercial exploitation of the energy.

This chapter, therefore, presents a general idea of bio-based products and the production techniques used in producing these bio-based products in the Global South (GS). To complete this chapter, a comprehensive and methodical literature review of books and online articles was conducted to gather all the reports presented. Based on this, the book chapter emphasised bio-based products that fall into the categories of bioenergy, biochemicals, and biomaterials that were produced from bio-based resources available in the GS regions using the most common production techniques—biochemical and thermochemical methods. It is imperative that the GS fully shift to a sustainable bioeconomy considering the increasing population, environmental challenges, and the diminishing availability of fossil-based resources. It is anticipated that the data presented in this chapter will offer crucial direction to industrial sector decision-makers or policymakers to accomplish more sustainable social, economic, and environmental goals in the GS.

2 Bio-based Products

2.1 An Overview of Bio-based Products

The public occasionally misunderstands the term 'bio-based products' to refer to substances, materials, or objects that are neither harmful to the environment nor have any effect on it. Companies promote their goods by employing the term 'bio-based' to mean 'without affecting the environment'. By general description, the word 'bio-based products' are products produced from renewable bioresources (Pylak et al., 2019). However, based on the definition of the European Union Commission (EUC) (2014), bio-based products are 'value-added products made entirely or partially from biological sources, including animals, plants, microorganisms (fungi, bacteria, and yeast) and enzymes'. They exclude materials that are fossilised or embedded in geological formations (EUC, 2014). In another related development, the United States (US) Secretary of Agriculture in, 2002 Farm Security and Rural Investment Act (FSRI) described bio-based products as 'industrial or commercial value-added products (apart from feed and food) consisting wholly or partially bioresources from agriculture (such as plant, animal, and marine resources), forestry resources or intermediate feedstock' (Busch, 2011). Based on this description, it was discovered that the US Department of Agriculture (USDA) launched the bio-preferred program in 2002 to increase the number of consumers who purchase and utilise bio-based products (Maxwell, 2019). Products that fall under this umbrella include pest control solutions like dethatchers, erosion control, and foliar sprays, as well as agricultural spray adjuvants, compost activators, and energizers (Moreno et al., 2020; Sojka & Saeid, 2022). In view of this, bio-based products are described as 'the products made from plant-related resources and other bioresources in agriculture, forestry, and marine that serve as substitutes for traditional products from fossil-based resources' (USDA, 2020). Therefore, the definition given according to USDA (2020) has been widely accepted and used.

Bio-based products encompass three distinct product categories: bio-based energy (bioenergy), bio-based chemicals (biochemicals), and bio-based materials (biomaterials). They are produced by a biorefinery that integrates the biomass conversion processes. Thus, bio-based products are produced through the integration of biomass conversion processes in a biorefinery; as a result, they are a viable substitute for fossil-based products and help to mitigate the harm that the latter do to the environment (Ruf et al., 2022). In the circular bioeconomy, the ideas and methods of bio-based production offer a new paradigm for fostering growth. The bioeconomy, which is emerging due to new knowledge creation and mentality transformation, is a predictor of the characteristics of the global economy of the future (UN, 2015). The public and private sectors should work together to adopt the new agenda, which will effectively support knowledge across the board and prioritise the future by focusing on a global development strategy that comes from technology and innovation development. Thus, by producing bio-based goods, the bioeconomy aims to replace fossil raw materials consistently with renewable resources, thereby promoting the sustainable use of biodiversity to improve the society, environment, and economy.

2.2 Category of Bio-based Products

Several bio-based products are already produced from biological resources. These bio-based products can be categorised into three primary groups: biochemicals, bioenergy, and biomaterials.

1. *Biochemicals*

Bio-based chemicals (biochemicals) are produced from biomass feedstocks and are utilised as both commercial and industrial bio-based products (Priya et al., 2023). Biochemicals are bio-based chemicals originating from plant, animal, and microbial sources that are substitutes for petroleum-based chemicals, these include lactic and succinic acid and are utilised in various industries like agriculture, food, cosmetics, and pharmaceuticals. Bio-based chemicals are superior to their petroleum-based counterparts in several ways, including cost-effectiveness, environmental impact, and sustainability. Additionally, bio-based chemicals are biodegradable and more eco-friendly. The process of converting bioresources into chemicals with added value through bioconversion has great potential to replace conventional chemicals (Chavan et al., 2022; National Research Council (NRC), 2000). Ogwu et al. (2024) reported that bio-based industrial products would account for significant industrial production, respectively, compared to existing levels for organic compounds, and materials. Products made from biomass and bioenergy will be far more increase in several folds due to its renewable source and biodegradable nature (Mehmood et al., 2021).

2. *Bioenergy*

A sustainable energy source derived from biological resources is known as bioenergy. The fact that it is produced from organic materials, commonly known as biological resources or biomass, qualifies it as a renewable energy source. In this manner, biomass has become a desirable energy source to many nations. Likewise, bioenergy is the most significant renewable energy source in the world, making up more than two-thirds of green energy sources, which account for roughly 6% of the world's energy supply and 55% of all renewable energy (Chong et al., 2021). Also, Dirisu et al. (2024) opined that bioenergy is the most readily and accessible of all forms of renewable energy available globally and accounted for 70% of renewable energy used in 2017. The development of bioenergy has enormous economic, societal, and environmental benefits. Therefore, it has emerged among other sources of energy as a solution to prevailing challenges of greenhouse gas (GHG) emissions when environmental issues like global warming are taken into consideration to advance research into developing alternative energy production methods. Furthermore, compared to conventional energy sources, which have a finite supply, biomass is more widely distributed globally and can be enhanced at all governmental levels, including local, regional, and national; it can also be facilitated with environmentally friendly technology (Priya et al., 2023). Examples include biodiesel, bioethanol, biogas, etc. (Diltz et al., 2011).

3. *Biomaterials*

Bio-based biomaterials (biomaterials) are produced from renewable biological resources like plants, animals, and microorganisms (Haile et al., 2021). They consist of entire plants, plant parts (fruits, seeds, stalks, etc.), some other constituents of plants (proteins, starch, fibre, and lipids), by-products from industrial processing (maize soluble, distiller's grains, etc.), marine resources (seaweed, chitin, etc.), and animal by-products (gelatin, albumen, etc.). It is possible that these resources could be used 'as is' to create new materials but they are always pretreated before used. To extract the required functional elements for biomaterial production, a complete knowledge of the bioresources components—as whole plants, plant parts, or by-products—will be required. This basic knowledge is crucial because it will enable the extraction of additional potentially useful elements from these raw materials (Ali et al., 2020). Biomaterial uses include medical applications, construction materials, etc.; therefore, the modern field of biomaterials combines medicine, biology, physics, chemistry, materials science, etc.

3 Production Techniques for Biomass Conversion into Bio-based Products

Some bio-based industrial products, such as starch, protein, cellulose, lignin, terpenes, oils, etc., are obtained through the physical or chemical processing of biological resources. Others are indirectly produced from carbohydrates by biotechnologies such as microbial and enzymatic processing (Liu et al., 2021). Thermochemical and biochemical conversion are typically the two most widely utilised techniques for converting biomass. When considering thermochemical conversion, high temperatures, and increased pressure are used to break down biomass in

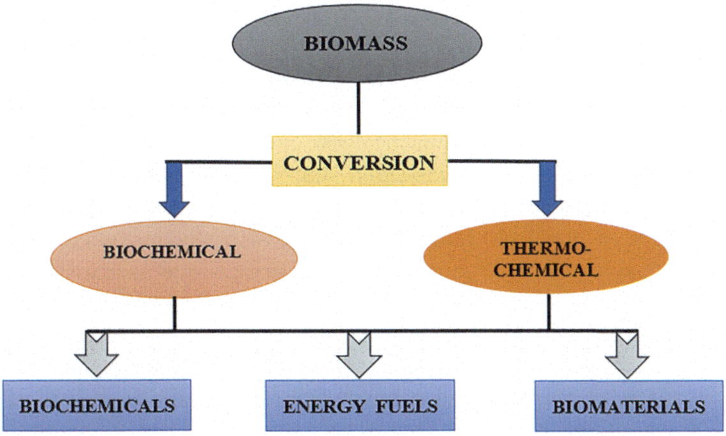

Fig. 3 Production techniques used in the conversion of bioresources

the presence of chemical catalysts to produce fuels, chemicals, and other value-added products (Shen et al., 2015). On the other hand, the biochemical conversion technique uses enzymes to denature and break down biomass materials to produce the desired products (Wainaina et al., 2018). According to Shen et al. (2015), these two conversion techniques can be combined together as a hybrid conversion technique. The hybrid conversion technique utilises thermal disintegration to generate substrates; thereafter, the substrates are then allowed to be consumed by microorganisms to produce chemicals, fuels, and other valuable composites or mixtures.

According to Álvarez-Castillo et al. (2021), both methods of biochemical and thermochemical conversion have been suitably employed for the transformation of biomass into a range of products derived from biomass. Pyrolysis, liquid catalytic cracking, gasification, and combustion are a few techniques under thermochemical production, while examples of biochemical production techniques are anaerobic digestion and esterification. Utilising the thermochemical conversion route as a product technique is a highly sustainable approach because it effectively tackles the fibrous and resistant nature of biomass. On the other hand, Shen et al. (2015) reported that the use of biological production techniques allows for a high degree of product selection with affordable production. Nowadays, great opportunities exist for the conversion of biological resources into value-added products, which is now the focus of most carbon-dependent industries. Although there are several production techniques available, the choice of an appropriate technique is based on the physicochemical properties of the feedstock, stability of the formulation, equipment availability, and the end product. Thus, the two most common production techniques used in the conversion of biomass with their respective categories of bio-based products are illustrated in Fig. 3.

4 Production of Bio-based Products in the Global South

4.1 Bio-based Production in the Global South

Globally, a growing number of industrialised and developing nations are implementing national bioeconomy policies and supporting advancements in bio-based materials, energy, and other value-added products (Stephenson & Damerell, 2022). The development of novel bio-based technologies is reducing fuel emissions and producing renewable chemicals and bio-based materials to replace fossil fuel-based products, which has a major positive impact on climate change and the SDGs (Lange et al., 2021). Beyond agriculture or food production and renewable energy solutions, bio-based production has become major form of bioeconomy that directly or indirectly offers extensive contributions towards low-emission and efficient management of resources in the construction, automotive, and packaging sectors (Stephenson & Damerell, 2022). However, it should be noted that the bioeconomy in developing nations frequently depends largely on bio-based primary sectors like forestry, fisheries, and agriculture (Lange et al., 2021). These sectors frequently have a great deal of potential to advance sustainable growth if knowledge and technology transfer are provided. Biotechnology and digitalisation are two transformative processes that offer the potential to boost the creation of bio-based value in all sectors of the bioeconomy.

Being well-known for its abundance of natural resources—land, water, and biodiversity—the GS is especially well-positioned to both contribute to and profit from the developing bioeconomy. These elements are becoming more and more strategically significant in the global context of a 'bio-based world' (Assan, 2023). The GS has an abundance of biomass and a wide range of methods and technologies for using it, so opportunities for the bioeconomy have a big chance to improve food security, create jobs, and produce low-emission energy solutions (Stephenson & Damerell, 2022). The bioeconomy has the potential to boost economic growth and job creation, encourage regional development, support the use of secure and renewable energy, encourage sustainable industrial development, and enhance environmental quality (Lange et al., 2021). To fully realise this potential, several novel processes that can satisfy performance requirements and promote social, environmental, and economic sustainability must be developed, tested, and implemented. As a result, the GSCs could take the lead in several areas, including the markets for biofuels and the application of new agricultural technologies (Karim et al., 2019).

To transform bio-based industries for sustainable economic growth and development, many nations around the world have made the promotion of a bioeconomy a top priority (Kardung et al., 2021). A fundamental aspect of the bioeconomy is the application of scientific research, knowledge, and innovation to the production of a wide range of agro-industrial and value-added products, in addition to food, feed, fibre, and fuel (Brandão et al., 2021). Creating value around local bioresources by utilising and optimising all parts of primary produce and their products is another essential component of the bioeconomy. A White Paper titled 'Growing the Circular

Bioeconomy, with a Focus on the Global South' was released by the research-based think tank known as Centre for Study of Science, Technology, and Policy (CSTEP) in association with the International Council for Circular Economy (ICCE) (WCEF, 2024). The White Paper, which was unveiled at the 'Unveiling bio-based solutions from the GS' session, advocates for coordinated efforts within the GS to maximise the utilisation of resources derived from forestry, agriculture, and other biological sources to store carbon, improve climate resilience, protect the environment, and boost local economies. It is possible to simultaneously achieve several objectives for creating a circular and regenerative bioeconomy by employing sewage sludge, food wastes, and high-potential products like hemp and innovative plant-based proteins (Ogwu, 2019b; WCEF, 2024).

The UN SDGs are supported by the common global policy agenda, which aims to achieve sustainable bio-based production (UNDP, 2019). This can be primarily increased by making more advancements in the worldwide bio-based production sector that focus on well-being, inclusive transformation, and sustainability (Karim et al., 2019; Akudugu & Ogwu, 2024). The path ahead and the contribution of the bio-based economy can vary greatly depending on the nation and region, given the differences in the circumstances under which national economies function. Africa, Asia, and South America are investing in the creation of new industrial sectors while simultaneously utilising their abundant biological resources (Issa et al., 2019). Because of this, bio-based production is viewed as a means of generating employment and expanding into new markets, which will help to create a low-carbon, resource-efficient economy. It also presents a chance to create novel bio-based products and procedures. Several GSCs have persisted in strengthening and promoting bio-based networks that support sustainable livelihoods and forming alliances for the transfer of technology (El-Chichakli et al., 2016; Börner et al., 2017). The biotechnological development of various processes and bio-based innovations is the primary factor driving the shift towards bio-based production or bioeconomy, irrespective of country and development status. Moreover, it depends on achieving a breakthrough in terms of technical capabilities and economic viability, and in the end, it depends on the supply of sustainable biomass (Scarlat et al., 2015).

Based on this, bio-based production in the GS regions includes products from agriculture, aquaculture, forestry, or agroforestry that are used as food, feed, and energy. Biological processes are added to these, such as those involved in the synthesis of new enzymes and protein products, the recycling of nutrients into fertilisers, and the generation of biofuels. A few instances of creative initiatives are currently in progress in these fields, along with an assessment of their effects (WCEF, 2024). These are some of the areas that were found to have high impact:

1. Place a high value on using agricultural waste and by-products.
2. Utilise all the various products that 'super-plants' like bamboo and hemp can produce.
3. Introduce forest products such as timber, nuts, and fruit to the market.

4. Promote aquaculture and associated goods, such as seagrass.
5. Create biorefineries to generate polymers, chemicals, and enzymes.
6. Grow crops high in protein, such as mycoproteins.
7. Utilise sewage sludge and food waste to create fertilisers and feed.
8. Produce long-term biofuels from leftover materials like food scraps and agricultural waste.

To achieve this change in the GS, several challenges must be overcome, such as the need for technology development, the shortage of skilled labour, the lack of mechanisms to guarantee supply certainty, the high cost of capital and logistics, the lack of consumer awareness and unfavourable perceptions, and the misalignment of regulations.

4.2 Major Bio-based Products in the Global South

Bioresources or biomass can be used to produce a variety of bio-based products. These include:

1. Biofuels

The depletion of fossil fuels, the environmental issues brought on by greenhouse gas emissions, and the rising cost of oil are compelling people to look for alternative sources of energy for heating, cooking, and transportation fuels. In this instance, it may be possible to use organic materials or resources and their inherent chemical compounds that contain valuable resources to produce biofuels. The biomass undergoes various changes during the production before being converted to different forms of bioenergy (Kiran et al., 2014). The following are various forms of biofuel that can be generated from biological resources:

(i) Bioethanol: often referred to as ethyl alcohol. This colourless, less toxic liquid is produced to power automobiles. It is produced via the fermentation process when plants that are sources of sugar (glucose, sucrose, etc.) are fermented by microorganisms (Chin & H'ng, 2013). Bioethanol is used as fuel for power generation in some cosmetics, pharmaceuticals, beverages, and chemical industries. Also, it is used as a cleaning agent, disinfectant, bleaching agent, preservative, raw material for plastics, adhesive in paints and dyes, and solvent for the spirits industry, among other purposes (Chin & H'ng, 2013). Argentina, Brazil, China, India, Mexico, Thailand, and the Gulf States have made significant strides in bioethanol production.

(ii) Biogas: is an inexpensive source of bioenergy, which is created when various microbes break down organic waste anaerobically. During the anaerobic digestion process, hydrogen, carbon dioxide, and acetate are generated, which is attributed to acidogenic bacteria (Oyewusi et al., 2024). As the process continues, methanogens break down the generated hydrogen into water and methane with acetate. Thus, the processes of hydrolysis, methanogenesis, and acidogenesis involved in the production of biogas are finalised by a series of microbial interactions. However, depending on the kind of bacteria involved, the products vary (Prasad et al., 2017). Major countries producing biogas in the GS include China, Thailand, India, Brazil, Kenya, Burkina Faso, Bangladesh, Vietnam, Nigeria, etc.

(iii) Biodiesel: is made from animal fats, used cooking oils, grease or vegetable oils. It is produced through the process of transesterification. However, biodiesel can also be extracted from oilseed using supercritical fluid extraction production techniques (Lee et al., 2014). Thus, biodiesel is a clean-burning, renewable liquid biofuel made up of esters of alkyl groups with high fatty acid content and low aliphatic alcohol content, and due to its zero net carbon dioxide production, biodiesel can be regarded as 'carbon neutral' (Muniraj et al., 2015). Furthermore, because of its high flash point of 100–170 °C and low explosiveness, biodiesel is a safe fuel to store and handle (RRamírez-Arias et al., 2018). In the GS, Brazil, Indonesia, China, Argentina, India, and Thailand have been among the top ten countries producing biodiesel worldwide (Statistical Research Department SRD, 2024), while others include Malaysia, Philippines, Peru, Zambia, Ghana, Liberia, and Nigeria (Adekanbi et al., 2021; Sekoai & Yoro, 2016).

2. Single-Cell Protein

Single-cell protein (SCP) is a type of protein that is produced by bacteria, fungi, algae, and yeast, among other microorganisms (Ritala et al., 2017). Various substrates can be used for the production of SCP through microorganism synthesis (Mondal et al., 2012). These substrates include peels from pineapples, bananas, pomegranates, watermelons, papayas, oranges, beetroot pulp, cucumbers, corncobs, etc. Different microorganisms are used to synthesise SCP from a variety of these substrates that are low-cost or free sources of carbon. For human and animal consumption, SCP is usually made from yeasts and filamentous fungi from specific fungal fermented foods and edible mushrooms (Amara & El-Baky, 2023). Methane is used as a carbon source during the production of SCP, which also lowers greenhouse gas emissions (Ritala et al., 2017). According to Single Cell Protein Products Market (2023), production of SCP can be identified with some regions in the GS such as **Asia** (China, India, Malaysia, etc.), Latin America, the Caribbean (Argentina, Brazil, Columbia, etc.), and Africa (Egypt, Nigeria, South Africa, etc.).

3. Enzymes

Enzymes are proteinaceous biological catalysts that catalyse a variety of metabolic reactions. A vast range of products are produced using enzymes in various industries. For example, the food industry uses pectinases and amylases, and the biofuel industry uses cellulases and tannases to reduce the amount of tannic acid in effluents. According to Ravindran and Jaiswal, the enzymes that are used most frequently are lipases, tannases, hemicelluloses, lignins, pectinases, and amylases. Commercial production of enzymes has been developed with advanced technologies involved in their production. With increases in the number of products in response to expanding markets, the demand for enzymes has increased recently (Bhatia, 2018). The majority of countries in the GS have been producing enzymes in different ways; however, the rate of development observed in the Asia-Pacific area indicates that China, Japan, and India are producing enzymes for industrial use with the potential to become global leaders in this sector soon (Sellami et al., 2022).

4. Activated Carbon

Activated carbon (AC) is a product of destructive distillation techniques that is an amorphous form of elemental carbon from a variety of carbonaceous raw materials, such as lignocellulosic materials, biomass wastes, etc. The thermochemical method is the major production technique used in producing ACs with the support of an experimental rig called an activated carbon reactor (Oyewusi et al., 2019). The AC adsorbent is becoming more and more in demand because it can be used in a variety of industrial fields for purposes like separation and purification (Sajid, 2023). New applications for ACs include batteries and capacitors, as well as storage of renewable and adsorbed natural gas (ANG) as a substitute for light-duty fuel vehicles (S&P Global Commodity Insights, 2023). In the GS, countries such as Indonesia, Malaysia, China, Cameroon, Nigeria, Thailand, and several others have been producing ACs using highly carbon-content raw materials such oil-palm shells, oil-palm fibres, coffee shells, bagasse, coconut shell, corncobs, etc. (Herawan et al., 2013).

5. Antioxidants

Antioxidants are substances that stop molecules from oxidising and becoming free radicals as such, these compounds serve as oxidation inhibitors in fuels and as preservatives in foods and cosmetics. According to Rahaman et al. (2023), certain human diseases are less likely to occur when antioxidants are present in the system. The production of antioxidants from organic matter is due to the possibility of health risks from synthetic antioxidants (Rahaman et al., 2023). To produce bio-based antioxidants, different biological protein sources can be employed such as plant protein (nuts, common beans, cauliflower and their by-products), animal proteins (sheep, goat, milk proteins, egg, fish, and their by-products, skin, etc.) and algae both macroalgae and microalgae (Llauradó Maury et al., 2020; Tadesse & Emire, 2020). Most plants such as fruits, vegetables, and other plant-based, including whole foods cultivated in Asia, Africa, and South America are all sources of vitamins C and E and have been reported to contain effective antioxidants (Gębalski et al., 2023). Cereals such as corn, sorghum, millet, etc., which are essential crops particularly in regions of sub-Saharan Africa, South America, and Asia, have been reported to contain medium to high concentrations of antioxidants (Jideani et al., 2014).

6. Microbial Fuel Cells

Microbial fuel cells (MFCs) are a novel kind of bio-electrochemical reaction that uses the electrons produced by bacterially catalysed biochemical reactions to create electricity (Menandro et al., 2017). Using microorganisms that catalyse reactions, these MFCs can effectively harvest the energy contained in the chemical bonds of organic compounds and convert it directly into electrical energy (Yu et al., 2020). The unique microbial technique used in MFC is an electro-active microorganism (EAM), which is capable of breaking down organic materials using a variety of metabolic pathways to transfer the released electrons to the anodic electrode (Thapa et al., 2022). Thereafter, the electrons produced are moved to the cathode via the external electric circuit to generate electricity (Moradian et al., 2021). This way, the MFCs were used to directly convert the organic materials into electricity. However, MFCs are still in the early phase in many regions of the GS because of the difficulties in streamlining the design and the expense of materials (Mahurede et al., 2023).

7. Dietary Fibre

These days, a healthy lifestyle is more important to people, and they are more interested in eating fruits that are low in calories, sodium, and fats and high in minerals, bioactive compounds, and dietary fibre. The indigestible carbohydrates found in plant cell walls are known as dietary fibres, and they play important roles in both human nutrition and health. Because they are high in fibre, biomass, and agro-industrial by-products are utilised as dietary supplements (Kowalska et al., 2017). Dietary fibres make up nearly half of the calories consumed in some areas, especially in parts of Asia and Africa (FAO, 2022).

8. Pigment.

The production of pigment from microbial sources is becoming more and more interesting due to the growing demand for natural food additives. The natural pigments found in food wastes can be extracted, and they can also be used as a fermentation substrate to help microorganisms produce pigments (Di Salvo et al., 2023). This makes food waste a valuable source of natural pigments. According to Raddatz-Mota et al. (2017), the main producers of pigment in the GS regions include Asia (China, India, Indonesia, Thailand, Philippines, Malaysia, and Vietnam), South America (Brazil, Argentina, Columbia, etc.), and Africa (Egypt, Nigeria, and South Africa).

9. Compost

Compost is described as 'the uniform, stable, odourless, humic-like substance produced from biological process referred to as composting' (Oyewusi et al., 2021). Compost is also referred to as mature organic matter or aerobically stabilised matter. Compost has been applied as land reclamation, soil enhancer, organic fertiliser, highway shouldering, and restorer of wetlands (Oyewusi et al., 2023b). In many nations in the GS, the production of compost has become an effective method of reducing greenhouse gas emissions in the agricultural ecosystem and storing CO_2 as soil carbon (Oyewusi et al., 2021). Also, compost production through biological methods (composting) can aid in soil bioremediation and act as a biological control for any pathogenic plant diseases.

10. Sorbitol.

Sorbitol is a naturally occurring sugar alcohol with huge global commercial interest, with current industrial demand of around 2 million tons per year spanning applications in the food, chemical, textile, cosmetic industries, pharmaceutical, and general health care (Ochoa-Gómez & Roncal, 2017). Considering the fast growth of the food and chemical industries, as well as the wide range of applications for which it is useful, sorbitol production and utilisation present extensive potential. China started manufacturing sorbitol in the late 1950s to make vitamin C and presently, the country has become the largest producer worldwide followed by Indonesia (Phannarangsee et al., 2024). In 2022, a cassava-based sorbitol company was launched in Ado-awaye, Oyo State, Nigeria. This sorbitol factory owned by *Psaltry International* has been confirmed to be the first of its kind in Africa and the second largest in the world with a capacity to produce 25 tons of sorbitol per day, and it has been exporting to other countries in Africa and other regions in GS (Business Day, 2022).

Several products derived from biomass have been introduced and established in recent decades (Ben-Iwo et al., 2016; Wang et al., 2019). Numerous conversion strategies for the manufacturing of various bio-based products are made possible by recent advancements in production processes using biomass as feedstocks. Table 1 shows the general overview of bio-based products that have been produced over time in the GSCs.

5 Benefits, Drivers, and Challenges to the Development of Bio-based Products in the Global South

5.1 Benefits of Bio-based Products in the Global South

Bio-based products have the potential to enhance the national security, environmental quality, and sustainable natural resources while remaining economically competitive thereby contributing to reducing dependency on fossil resources and helping

Table 1 Commercialised bio-based products in the GS

Biochemicals	Biomaterials	Bioenergy
Lactic acid	Propylene	Bioethanol
Formic acid	Polyhydroxy alkanoates (PHA)	Biohydrogen
Ethylene oxide	Acrylic acid derivatives	Biomethane
Acetic acid	Tetrahydrofuran (THF)	Biohythane
Succinic acid	Isoprene	Biogas
Citric acid	Methyl methacrylate	Bio-syngas
Crotonaldehyde	Bio-composites	Propane
Lactic acid	Natural rubber	Biodiesel
Malonic acid	Cellulosic fibres	Bio-oil
Epichlorohydrin	Bio-adhesives	Briquette
Crotonaldehyde	Nanocellulose	Bio-butanol
Xylitol	Compost	Jet fuel
Lysine	Adipic acid (Nylon 6,6)	H-CNG

Table 2 Benefits of bio-based products for the economy, environment, and society

Indicators	Benefits
Economy	Possibility of implementing novel production techniques. New market for innovative bio-based products. Competitiveness and sustainable economic growth. New employment opportunities.
Environment	Reduction of fossil fuel consumption in the economy. Sustainable resource management. Alleviation of ecosystem pressure. Improvement in recycling.
Society	Preserving food security and practicing sustainable resource management. Quality of life and health benefits. Challenges for social innovation.

to accelerate the transition to a green and bioeconomy (Hasenheit et al., 2016). Generally, bio-based products provide various benefits for the nations in the GS, and these benefits have been summarised based on three indicators—economy, environment, and society, and these are presented in Table 2.

The bioeconomy comprises a diverse array of economic activities that are positioned along numerous value chains, each of which includes suppliers, producers, distributors, and buyers (Hasenheit et al., 2016). Moreover, biomass serves as the foundation for all bio-based processes and products, and as such, it serves as the beginning of all value chains related to the bioeconomy. Based on this, it is critical to recognise that the economic, environmental, and social effects of bioeconomy activities are not always confined to the production site of bio-based products (within a regional bioeconomy cluster) but can extend back to the location of biomass production.

5.2 Drivers to Development of Bio-based Products in the Global South

Renewable raw materials have become dominant sources used to produce bio-based products which have continued to experience enormous growth in the last decades. Switching to bio-based resources from conventional resources is affected by several parameters. It differs among countries depending on their conditions or situations, development objectives, and adaption of bioeconomy policies and action plans. Thus, this section identifies some drivers leading to the development of bio-based products, these are somewhat interconnected and complex. Mirzabaev et al. (2014) generated a list of 20 drivers in six categories though, not all of them are included. Discussion on all these drivers is not covered here because it is beyond the scope of this book chapter. Nevertheless, by reviewing the development of bio-based products in several leading countries in the GS, the key factors found as drivers include economy security, economy policies, climate change and friendly environment, financial resources and technology accessibility, the availability of land and yield, and cost of bio-based production and market guarantee (Ferreira et al., 2022). In general, the main factors affecting bio-based production initiatives regarding economic, environmental, and social status in the GS include:

(i) Feedstock accessibility, selection, and efficient application.
(ii) Techniques and tools used in the production of bioenergy and its subsequent application.
(iii) Land tenure system and agricultural practices used to produce feedstock.
(iv) Accessibility and suitability of location for bio-based production; and.
(v) development of a suitable bio-based product life cycle and value chain, as well as adequate production and marketing strategies.

5.3 Challenges to the Development of Bio-based Products

Many obstacles have been identified that impede the growth of bio-based products in the GS, and most of them are related to economic, security, and social issues. Conflicts alongside security, biodiversity loss, food production, commodity prices, and environmental degradation brought on by changes in land usage are a few of these. The GS has a lot of potential for generating bio-based products for both domestic and foreign markets if these problems are resolved and appropriate production technology is put in place (Mohammed et al., 2019). Apart from those abovementioned challenges, the establishment and extension of available bio-based production initiatives in the GS have been impeded by vested interests from a variety of stakeholders, including multi-national agencies, organised private sector, local, state, federal, and international governments, as well as strong but unfavourable government policy imperatives. Another important factor impeding the growth of bio-based products in the GS is the impending socioeconomic and environmental effects, which include deforestation, biodiversity loss, air, soil, and water pollution, GHG emissions, particularly those of methane and carbon monoxide, and soil and water pollution.

Furthermore, the general public is still largely ignorant of the potential advantages of bio-based goods and uses, and as a result, they are not prepared to pay the premium. This implies a roadblock to the adoption and public acceptability of bio-based products, which will undoubtedly impact the growth of the associated sector. According to BIOTIC (2015), other main obstacle to the successful accomplishment of a bio-based scheme is a scarcity of skilled labour coming from the primary production sectors (industries). In addition is the lack of coordination between the numerous parties involved in the value chain at different levels, and inability to link and align the work done by academics or research institutes with the industries. Lastly, lack of standardised laws pertaining to 'intellectual property' (IP) and the expensive cost of patents (BIOTIC, 2015). All these factors have become critical in the GS although, these variables are extremely significant and context-specific, but they differ greatly throughout the GS nations (Mungwe et al., 2016). Therefore, the main barriers to the development of bio-based products in the GS have been identified by researchers and the stakeholders and have been summarised into three important headings as presented in Table 3.

According to Atabani et al. (2012), when considering all of these variables together, it makes sense to draw the conclusion that the acceptance of bio-based products throughout the GS is below expectations. In this regard, modern bioenergy technologies are required to achieve more sustainable energy and economic balance to create more prospects that will be beneficial to the environment and people living in it. When this is achieved, the sustainable development in the GS is certain.

6 Conclusion

Rich bioresources are available in the GSCs for the creation of bio-based products that can replace items derived from fossil fuels. Most of these biomass feedstocks are found in rural areas, since they are produced from agricultural related activities. Production of these value-added products reduces overall production costs and provides substitutes for products derived from fossil resources, hence promoting the growth of a bio-based economy and sustainable development. Consequently, the transition from a fossil-based economy to a circular bioeconomy may be facilitated by the much-needed diversification that bio-based products offer. Based on this potential, bio-based products can be incorporated into several biotechnological processes to produce different kinds of consumer goods. The nature and rate of changes towards sustainability will inevitably involve an increasing reliance on biological processes and bio-based resources. Therefore, the key concerns for policy development and technology innovation revolve around how to positively influence these

Table 3 Barriers to the development of bio-based products in the GS

Barriers	Effects
Feedstock-related barriers	High production cost of raw materials (biomass). Unavailability and inadequate raw materials in needed quantities, qualities, and prices all year round. Seasonal variation in the production of biomass feedstock. Unproductive distribution and transportation schemes of different forms of biomass feedstock. Ineffective bio-waste recovery methods probably utilised as a source of raw materials for bio-based products.
Industry-related barriers	Low level of technological preparedness and low level of commercialisation for numerous bio-based products. Insufficient collaboration among the parties involved in the value chains. Obstacles to the formation of industry-academia collaborations. Insufficient funding to establish new production facilities. Inadequately skilled labour force.
Market-related barriers	The low cost of natural gas and crude oil makes it economically unfeasible to employ biomass feedstock in bio-based production techniques. High price of products made from biomass in comparison to those made from fossil resources. Lack of a specific and comprehensive framework in Global South legislation. Limited functionality of many bio-based products in comparison to their counterparts derived from fossil resources. Tensions between market demands and sustainability objectives. Inconsistent standards and verified labelling for items derived from biotechnology. Deficiencies in the framework of subsidy and policy. Inadequate and unreliable information regarding bio-based products. Poor public sensitisation of the advantages of using items made from bio-based materials. Barriers relating to intellectual property.

changes. In order to ensure that sustainable development goals are accomplished, consideration must be given to economic, environmental, and social factors while evaluating the choice and strategies of a bio-based economy.

Despite the fact that there are many advantages and benefits attached to conversion of bioresources into bio-based products, the transition from the well-established fossil-based economy to the burgeoning bioeconomy is undoubtedly difficult. The lack of large-scale production, technological barriers, decentralisation of the biomass supply chain, and significant initial investment are some of the fundamental difficulties. The primary goal in overcoming these difficulties should be the development of efficient and ideal methods for biomass production, logistics, and monitoring. Furthermore, large-scale operations in biomass utilisation should follow the excellent research efforts in this area that we have already experienced. One approach to implementing research findings for commercialisation is through the establishment of industrial-academic alliances. Thus, focused research, international collaboration across several sectors, safe and responsible biomass consumption, and climate action are necessary for the realisation of the future bio-based economy. In order to contend with conventional, fossil-derived rivals, the industry as a whole should concentrate on the responsible use and effective upgrading of biomass as well as the integration of production techniques for the production of high value-added products. When taken as a whole and in conjunction with the appropriate policies put in place, these strategies have the potential to help shape the future of the bio-based economy and safeguard biodiversity and the environment.

References

Adams, P., Bridgwater, T., Lea-Langton, A., Ross, A., & Watson, I. (2018). Biomass conversion technologies. In *Greenhouse gas balances of bioenergy systems* (pp. 107–139). Academic Press.

Adekanbi, M. L., Ibukun, B. T., & Steve, E. (2021). The potential of biodiesel in mitigating the energy crisis in sub-Saharan Africa energy sectors. *Sustinere: Journal of Environment and Sustainability, 5*(2), 79–89.

Ahorsu, R., Medina, F., & Constantí, M. (2018). Significance and challenges of biomass as a suitable feedstock for bioenergy and biochemical production: A review. *Energies, 11*(12), 3366.

Akudugu, M. A., & Ogwu, M. C. (2024). Sustainable development policies and interventions: A bibliometric analysis of the contributions of the academic community. *Journal of Cleaner Production*, 139919. https://doi.org/10.1016/j.jclepro.2023.139919

Ali, N., Zhang, Q., Liu, Z. Y., Li, F. L., Lu, M., & Fang, X. C. (2020). Emerging technologies for the pretreatment of lignocellulosic materials for bio-based products. *Applied Microbiology and Biotechnology, 104*(2), 455–473.

Amara, A. A., & El-Baky, N. A. (2023). Fungi as a source of edible proteins and animal feed. *Journal of Fungi, 9*(1), 73.

Assan, N. (2023). Socio-cultural, economic, and environmental implications for innovation in sustainable food in Africa. *Frontiers in Sustainable Food Systems, 7*. https://doi.org/10.3389/fsufs.2023.1192422

Atabani, A. E., Silitonga, A. S., Badruddin, I. A., Mahlia, T. M., Masjuki, H., & Mekhilef, S. (2012). A comprehensive review on biodiesel as an alternative energy resource and its characteristics. *Renewable and Sustainable Energy Reviews, 16*(4), 2070–2093.

Álvarez-Castillo, E., Felix, M., Bengoechea, C., & Guerrero, A. (2021). Proteins from agri-food industrial biowastes or co-products and their applications as green materials. *Food, 10*(5), 981.

Badgujar, K. C., & Bhanage, B. M. (2018) Dedicated and waste feedstocks for biorefinery: An approach to develop a sustainable society. In *Waste biorefinery*. Elsevier (pp. 3–38).

Ben-Iwo, J., Manovic, V., & Longhurst, P. (2016). Biomass resources and biofuels potential for the production of transportation fuels in Nigeria. *Renewable and Sustainable Energy Reviews, 63*, 172–192.

Bhatia, S. (2018). Technologies and procedures involved in enzyme production. *Introduction to Pharmaceutical Biotechnology, 2*, 2–42.

BIOTIC. (2015). Biological traits information catalogue. http://www.marlin.ac.uk/biotic/browse.php?sp=6168. Accessed 18 May 2024.

Börner, J., Kuhn, A., & von Braun, J. (2017). *Bio-economy in developing countries*. ZEF Policy Brief No. 25.

Brandão, A. S., Gonçalves, A., & Santos, J. M. (2021). Circular bioeconomy strategies: From scientific research to commercially viable products. *Journal of Cleaner Production, 295*, 126407.

Busch, R. (2011). Bio-based products–market needs and opportunities. In *Renewable raw materials: New feedstocks for the chemical industry* (pp. 169–186).

Business Day. (2022). Nigeria's first cassava-based sorbitol factory for launch. https://business-day.ng/news/article/nigerias-first-cassava-based-sorbitol-factory-for-launch-today/. Accessed 18 May 2024.

Chavan, S., Yadav, B., Atmakuri, A., Tyagi, R. D., Wong, J. W., & Drogui, P. (2022). Bioconversion of organic wastes into value-added products: A review. *Bioresource Technology, 344*, 126398.

Chin, K. L., & H'ng, P. S. (2013). A real story of bioethanol from biomass: Malaysia perspective. In *Biomass now-sustainable growth and use* (pp. 329–346).

Chong, C. T., Loe, T. Y., Wong, K. Y., Ashokkumar, V., Lam, S. S., Chong, W. T., et al. (2021). Biodiesel sustainability: The global impact of potential biodiesel production on the energy–water–food (EWF) nexus. *Environmental Technology & Innovation, 22*, 101408.

Di Salvo, E., Lo Vecchio, G., De Pasquale, R., De Maria, L., Tardugno, R., Vadalà, R., & Cicero, N. (2023). Natural pigments production and their application in food, health and other industries. *Nutrients, 15*(8), 1923.

Diltz, R. A., Luckarift, H. R., & Johnson, G. R. (2011). Sustainable land use for bioenergy in the 21st century. *Industrial Biotechnology, 7*(6), 437–447.

Dirisu, J. O., Salawu, E. Y., Ekpe, I. C., Udoye, N. E., Falodun, O. E., Oyedepo, S. O., Ajayi, O. O., & Kale, S. A. (2024). Promoting the use of bioenergy in developing nations: A CDM route to sustainable development. *Frontiers in Energy Research, 11*, 1184348.

Egboduku, W. O., Egboduku, T., Golohor, O. M., Imarhiagbe, O., & Ogwu, M. C. (2024). Cassava as raw material for sustainable bioeconomy development. In M. C. Ogwu, S. C. Izah, A. A. Cunha Alves, & S. C. Babu (Eds.), *Plant biology, sustainability and climate change, sustainable cassava*. Academic Press. 57–73 pp.. https://doi.org/10.1016/B978-0-443-21747-0.00022-9

El-Chichakli, B., von Braun, J., Lang, C., Barben, D., & Philp, J. (2016). Policy: Five cornerstones of a global bioeconomy. *Nature, 535*(7611), 221–223.

Ellabban, O., Abu-Rub, H., & Blaabjerg, F. (2014). Renewable energy resources: Current status, future prospects and their enabling technology. *Renewable & Sustainable Energy Reviews, 39*, 748–764. https://doi.org/10.1016/j.rser.2014.07.113

European Union Commission EUC. (2014). Bio-based products. The Directorate-General for Internal Market, Industry, Entrepreneurship, and SMEs. https://single-market-economy.ec.europa.eu/sectors/biotechnology/bio-based-products_en#:~:text=Bio%2Dbased%20products%20are%20wholly,geological%20formations%20and%2For%20fossilised. Accessed 18 May 2024.

FAO. (2019). *State of Food Security and Nutrition in the World*. http://www.fao.org/3/ca5162en/ca5162en.pdf. Accessed 20 May 2024.

FAO. (2022). *Agricultural trade in the Global South – An overview of trends in performance, vulnerabilities, and policy frameworks*. https://doi.org/10.4060/cb9120en. Accessed 30 May 2024.

Farm Security and Rural Investment Act. (2002). *Bio-based products; designation guidance for federal procurement*. https://regulations.vlex.com/vid/farm-rural-bio-based-designation-procurement-22668992. Accessed 18 May 2024.

Ferreira, V., Fabregat-Aibar, L., Pie, L., & Terceno, A. (2022). Research trends and hotspots in bioeconomy impact analysis: A study of economic, social and environmental impacts. *Environmental Impact Assessment Review, 96*, 106842.

Gębalski, J., Małkowska, M., Graczyk, F., Słomka, A., Piskorska, E., Gawenda-Kempczyńska, D., et al. (2023). Phenolic compounds and antioxidant and anti-enzymatic activities of selected adaptogenic plants from South America, Asia, and Africa. *Molecules, 28*(16), 6004.

Haile, A., Gelebo, G. G., Tesfaye, T., Mengie, W., Mebrate, M. A., Abuhay, A., & Limeneh, D. Y. (2021). Pulp and paper mill wastes: Utilizations and prospects for high value-added biomaterials. *Bioresources and Bioprocessing, 8*, 1–22.

Hasenheit, M., Gerdes, H., Kiresiewa, Z., & Beekman, V. (2016). *D2. 2: Summary report on the social, economic and environmental impacts of the bioeconomy*. European Union.

Herawan, S. G., Hadi, M. S., Ayob, M. R., & Putra, A. (2013). Characterization of activated carbons from oil-palm shell by CO 2 activation with no holding carbonization temperature. *The Scientific World Journal, 2013*.
Issa, I., Delbrück, S., & Hamm, U. (2019). Bioeconomy from experts' perspectives–results of a global expert survey. *PLoS One, 14*(5), e0215917.
Jideani, A. I., Silungwe, H., Takalani, T., Anyasi, T. A., Udeh, H., & Omolola, A. (2014). Antioxidant-rich natural grain products and human health. In O. Oguntibeju (Ed.), *Antioxidant-antidiabetic agents and human health* (pp. 167–187). InTech Publisher.
Kardung, M., Cingiz, K., Costenoble, O., Delahaye, R., Heijman, W., Lovrić, M., van Leeuwen, M., M'barek, R., van Meijl, H., Piotrowski, S., & Ronzon, T. (2021). Development of the circular bioeconomy: Drivers and indicators. *Sustainability, 13*(1), 413.
Karim, M. E., Karim, R., Islam, M. T., Muhammad-Sukki, F., Bani, N. A., & Muhtazaruddin, M. N. (2019). Renewable energy for sustainable growth and development: An evaluation of law and policy of Bangladesh. *Sustainability, 11*(20), 5774. https://doi.org/10.3390/su11205774
Kiran, E. U., Trzcinski, A. P., Ng, W. J., & Liu, Y. (2014). Bioconversion of food waste to energy: A review. *Fuel, 134*, 389–399.
Kowalska, H., Czajkowska, K., Cichowska, J., & Lenart, A. (2017). What's new in biopotential of fruit and vegetable by-products applied in the food processing industry. *Trends in Food Science & Technology, 67*, 150–159.
Lange, L., Connor, K. O., Arason, S., Bundgård-Jørgensen, U., Canalis, A., Carrez, D., Gallagher, J., Gøtke, N., Huyghe, C., Jarry, B., Llorente, P., Marinova, M., Martins, L. O., Mengal, P., Paiano, P., Panoutsou, C., Rodrigues, L., Stengel, D. B., Van Der Meer, Y., & Vieira, H. (2021). Developing a sustainable and circular bio-based economy in EU: By partnering across sectors, upscaling and using new knowledge faster, and for the benefit of climate, environment & biodiversity, and people & business. *Frontiers in Bioengineering and Biotechnology, 8*. https://doi.org/10.3389/fbioe.2020.619066
Leblanc, F., Bibas, R., Mima, S., Muratori, M., Sakamoto, S., Sano, F., Bauer, N., Daioglou, V., Fujimori, S., Gidden, M. J., & Kato, E. (2022). The contribution of bioenergy to the decarbonization of transport: A multi-model assessment. *Climatic Change, 170*(3), 21.
Lee, K. T., Lim, S., Pang, Y. L., Ong, H. C., & Chong, W. T. (2014). Integration of reactive extraction with supercritical fluids for process intensification of biodiesel production: Prospects and recent advances. *Progress in Energy and Combustion Science, 45*, 54–78.
Liu, J., Wang, S., Peng, Y., Zhu, J., Zhao, W., & Liu, X. (2021). Advances in sustainable thermosetting resins: From renewable feedstock to high performance and recyclability. *Progress in Polymer Science, 113*, 101353.
Llauradó Maury, G., Méndez Rodríguez, D., Hendrix, S., Escalona Arranz, J. C., Fung Boix, Y., Pacheco, A. O., et al. (2020). Antioxidants in plants: A valorization potential emphasizing the need for the conservation of plant biodiversity in Cuba. *Antioxidants, 9*(11), 1048.
Lamidi, W. A., & Oyewusi, T. F. (2024). Soil utilisation patterns in the global south. In: Ogwu, M. C., Izah, S. C., Dessureault-Rompré, J., Gasparatos, D. (eds) *Sustainable soil systems in global south*. Springer, Singapore.
Mahurede, T. P., Chihobo, C. H., Utete, B., & Taru, P. (2023). A review of microbial fuel cell prototypes, their efficacy in wastewater treatment and the contextual situation for Zimbabwe. *Fuel Communications*, 100094.
Market Analysis Report. (2020). https://www.grandviewresearch.com/industry-analysis/bio-based-platform-chemicalsmarket. Accessed 20 May 2024.
Maxwell, J. (2019). Driving the bio-based economy. https://sn.astm.org/features/driving-bio-based-economy-ma19.html
Mehmood, T., Nadeem, F., Qamar, S. A., Bilal, M., & Iqbal, H. M. (2021). Bioconversion of agro-industrial waste into value-added compounds. In *Sustainable bioconversion of waste to value added products* (pp. 349–368).

Menandro, L. M. S., Cantarella, H., Franco, H. C. J., Kölln, O. T., Pimenta, M. T. B., Sanches, G. M., Rabelo, S. C., & Carvalho, J. L. N. (2017). Comprehensive assessment of sugarcane straw: Implications for biomass and bioenergy production. *Biofuels, Bioproducts and Biorefining, 11*(3), 488–504.

Mensah, J. (2019). Sustainable development: Meaning, history, principles, pillars, and implications for human action: Literature review. *Cogent Social Sciences, 5*(1), 1653531.

Mirzabaev, A., Guta, D.D., Goedecke, J., Gaur, V., Börner, J., Virchow, D., Denich, M., & von Braun, J. (2014). Bioenergy, food security and poverty reduction: Mitigating tradeoffs and promoting synergies along the Water-Energy-Food Security Nexus.

Mohammed, I. Y., Abakr, Y. A., & Mokaya, R. (2019). Biofuel and valuable products recovery from Napier grass pre-processing: Process design and economic analysis. *Journal of Environmental Chemical Engineering, 7*(2), 102962.

Mondal, A. K., Sengupta, S., Bhowal, J., & Bhattacharya, D. K. (2012). Utilization of fruit wastes in producing single cell protein. *International Journal of Science, Environment and Technology, 1*(5), 430–438.

Moradian, J. M., Fang, Z., & Yong, Y. C. (2021). Recent advances on biomass-fueled microbial fuel cell. *Bioresources and Bioprocessing, 8*, 1–13.

Moreno, A. D., Susmozas, A., Oliva, J. M., & Negro, M. J. (2020). Overview of bio-based industries. In *Bio-based products and industries* (pp. 1–40). Elsevier.

Mungwe, J. N., Bandiera, L., Accorona, D., & Colombo, E. (2016). Sustainable energization of rural areas of developing countries–A comprehensive planning approach. *Energy Procedia, 93*, 46–52.

Muniraj, I. K., Uthandi, S. K., Hu, Z., Xiao, L., & Zhan, X. (2015). Microbial lipid production from renewable and waste materials for second-generation biodiesel feedstock. *Environmental Technology Reviews, 4*(1), 1–16.

Muralikrishna, I. V., & Manickam, V. (2017). Life cycle assessment. In *Environmental management* (pp. 57–75). Elsevier. https://doi.org/10.1016/b978-0-12-811989-1.00005-1

National Research Council, Commission on Life Sciences, Board on Biology, & Committee on Bio-based Industrial Products. (2000). *Bio-based industrial products: Priorities for research and commercialization.*

Nizzy, A. M., & Ogwu, M. C. (2024). Valorization of cassava processing by-products into biofuel for a sustainable environment. In M. C. Ogwu, S. C. Izah, A. A. Cunha Alves, & babu, S.C. (Eds.), *Plant biology, sustainability and climate change, sustainable cassava*. Academic Press. 291–309 pp. https://doi.org/10.1016/B978-0-443-21747-0.00010-2

Ochoa-Gómez, J. R., & Roncal, T. (2017). Production of sorbitol from biomass. In *Production of platform chemicals from sustainable resources* (pp. 265–309).

Ogwu, M. C. (2019a). Towards sustainable development in Africa: The challenge of urbanization and climate change adaptation. In P. B. Cobbinah & M. Addaney (Eds.), *The geography of climate change adaptation in urban Africa* (pp. 29–55). Springer Nature. https://doi.org/10.1007/978-3-030-04873-0_2

Ogwu, M. C. (2019b). Understanding the composition of food waste: An "-omics" approach to food waste management. In A. P. Gunjal, M. S. Waghmode, N. N. Patil, & P. Bhatt (Eds.), *Global initiatives for waste reduction and cutting food loss* (pp. 212–236). IGI Global. https://doi.org/10.4018/978-1-5225-7706-5.ch011

Ogwu, M. C., Odozi, I. P., Ahonsi, O. C., Uleanya, K. O., & Odozi, E. B. (2024). Organic acid production from cassava. In M. C. Ogwu, S. C. Izah, A. A. Cunha Alves, & babu, S.C. (Eds.), *Plant biology, sustainability and climate change, sustainable cassava*. Academic Press. 395–418 pp. https://doi.org/10.1016/B978-0-443-21747-0.00009-6

Oyewusi, T. F., Aransiola, E. F., Osunbitan, J. A., & Ogunjimi, L. A. O. (2019). A comparative study on the physical properties of briquettes from carbonized and uncarbonized corncobs. *Proceedings of the OAU Faculty of Technology Conference, 7*, 127–135.

Oyewusi, T. F., Osunbitan, J. A., Ogunwande, G. A., & Omotosho, O. A. (2021). Investigation into physico-chemical properties of compost extract as affected by processing parameters. *Journal of Environmental Challenge, 5*, 100370.

Oyewusi, T. F., Osunbitan, J. A., Ogunwande, G. A., Ademiju, T. A., Opadoja, C. O., & Amoo, M. O. (2023b). Effect of compost extract processing parameters on soil physico-chemical properties after harvest. *Suranaree Journal of Science and Technology, 30*(6), 1–9.

Oyewusi, T. F., & Lamidi, W. A. (2024). Alternative land management strategies and their impacts on soil conservation. In: Ogwu, M. C., Izah, S. C., Dessureault-Rompré, J., Gasparatos, D. (eds) *Sustainable soil systems in global south*. Springer, Singapore.

Oyewusi, T. F., Soji-Adekunle, A. R., Oguntunji, M. A., Ayo, V. O., Omotosho, O. A., & Ogundahunsi, O. E. (2024). Biopotential property of maggots as inoculum in anaerobic digestion of substrates for biogas production. *Journal of Engineering in Agriculture and the Environment 10*(2), 43–54.

Oyewusi, T. F., Alebiowu, G., Aransiola. E. F., Soji-Adekunle, A. R. & Adeboye, B. S. (2023a). Effect of Pretreatment on the Physical Properties and Heating Values of Briquettes. *Journal of Renewable Energy and Environment.*

Oyewusi, T. F., Aransiola, E. F., Oyinlola, T. A., & Adesigbin, J. A. (2018). Comparative study of the calorific values and gas emissions of briquettes produced from carbonized and uncarbonized process of corncob. *Journal of the Nigerian Institution of Mechanical Engineers, 8*(2), 20–36.

Phannarangsee, Y., Jiawkhangphlu, B., Thanonkeo, S., Klanrit, P., Yamada, M., & Thanonkeo, P. (2024). Sorbitol production from mixtures of molasses and sugarcane bagasse hydrolysate using the thermally adapted Zymomonas mobilis ZM AD41. *Scientific Reports, 14*(1), 5563.

Prasad, S., Rathore, D., & Singh, A. (2017). Recent advances in biogas production. *Chemical Engineering & Process Technology, 3*(2), 1038.

Priya, A. K., Alagumalai, A., Balaji, D., & Song, H. (2023). The potential of bio-based agro solutions towards sustainable bioeconomy: A review. *RSC Sustainability, 1*, 746–762.

Pylak, M., Oszust, K., & Frąc, M. (2019). Review report on the role of bioproducts, biopreparations, biostimulants and microbial inoculants in organic production of fruit. *Reviews in Environmental Science and Bio/Technology, 18*(3), 597–616.

Raddatz-Mota, D., Pérez-Flores, L. J., Carrari, F., Mendoza-Espinoza, J. A., de León-Sánchez, F. D., Pinzón-López, L. L., Godoy-Hernández, G., & Rivera-Cabrera, F. (2017). Achiote (Bixa orellana L.): A natural source of pigment and vitamin E. *Journal of Food Science and Technology, 54*, 1729–1741.

Rahaman, M. M., Hossain, R., Herrera-Bravo, J., Islam, M. T., Atolani, O., Adeyemi, O. S., Owolodun, O. A., Kambizi, L., Daştan, S. D., Calina, D., & Sharifi-Rad, J. (2023). Natural antioxidants from some fruits, seeds, foods, natural products, and associated health benefits: An update. *Food Science & Nutrition, 11*(4), 1657–1670.

Ramírez-Arias, A. M., Giraldo, L., & Moreno-Piraján, J. C. (2018). Biodiesel synthesis: Use of activated carbon as support of the catalysts. In *Biorefining of biomass to biofuels: Opportunities and perception* (pp. 117–152).

Ritala, A., Häkkinen, S. T., Toivari, M., & Wiebe, M. G. (2017). Single cell protein—State-of-the-art, industrial landscape and patents 2001–2016. *Frontiers in Microbiology, 8*, 300587.

Rogers, J. G., Cooper, S. J., & Norman, J. B. (2018). Uses of industrial energy benchmarking with reference to the pulp and paper industries. *Renewable and Sustainable Energy Reviews, 95*, 23–37.

Ruf, J., Emberger-Klein, A., & Menrad, K. (2022). Consumer response to bio-based products–A systematic review. *Sustainable Production and Consumption, 34*, 353–370.

S&P Global Commodity Insights. (2023). Activated carbon. *Chemical Economics Handbook.* https://www.spglobal.com/commodityinsights/en/ci/products/activated-carbon-chemical-economics-handbook.html

Sajid, M. (2023). Industrial applications of activated carbon. In *Activated carbon: Progress and applications* (p. 24),

Scarlat, N., Dallemand, J. F., Monforti-Ferrario, F., & Nita, V. (2015). The role of biomass and bioenergy in a future bioeconomy: Policies and facts. *Environmental Development, 15*, 3–34.
Segovia-Hernandez, J. G., Sanchez-Ramirez, E., Ramírez-Márquez, C., & Contreras-Zarazúa, G. (2021). *Improvements in bio-based building blocks production through process intensification and sustainability concepts*. Elsevier.
Sekoai, P. T., & Yoro, K. O. (2016). Biofuel development initiatives in sub-Saharan Africa: Opportunities and challenges. *Climate, 4*(2), 33.
Sellami, K., Couvert, A., Nasrallah, N., Maachi, R., Abouseoud, M., & Amrane, A. (2022). Peroxidase enzymes as green catalysts for bioremediation and biotechnological applications: A review. *Science of the Total Environment, 806*, 150500.
Shahsavarani, H., Hasegawa, D., Yokota, D., Sugiyama, M., Kaneko, Y., Boonchird, C., & Harashima, S. (2013). Enhanced bio-ethanol production from cellulosic materials by semi-simultaneous saccharification and fermentation using high temperature resistant Saccharomyces cerevisiae TJ14. *Journal of Bioscience and Bioengineering, 115*(1), 20–23.
Shen, Y., Jarboe, L., Brown, R., & Wen, Z. (2015). A thermochemical–biochemical hybrid processing of lignocellulosic biomass for producing fuels and chemicals. *Biotechnology Advances, 33*(8), 1799–1813.
Single Cell Protein Products Market. (2023). *Single cell protein products market innovations, growth status and technological projections for 2030*. https://www.linkedin.com/pulse/2023-single-cell-protein-products-market-innovations/
Sojka, M., & Saeid, A. (2022). Bio-based products for agriculture. In *Smart agrochemicals for sustainable agriculture* (pp. 279–310). Academic Press.
Statista Research Department. (2024). *Global biodiesel production 2022, by country*. https://www.statista.com/statistics/271472/biodiesel-production-in-selected-countries/#:~:text=Indonesia%20is%20the%20world's%20largest,output%20of%2013.65%20billion%20liters. Accessed 08 April 2024.
Stephenson, P. J., & Damerell, A. (2022). Bioeconomy and circular economy approaches need to enhance the focus on biodiversity to achieve sustainability. *Sustainability, 14*(17), 10643. https://doi.org/10.3390/su141710643
Swaminaathan, P., Saravanan, A., & Thamarai, P. (2024). Utilization of bioresources for high-value bioproducts production: Sustainability and perspectives in circular bioeconomy. *Sustainable Energy Technologies and Assessments, 63*, 103672.
Tadesse, S. A., & Emire, S. A. (2020). Production and processing of antioxidant bioactive peptides: A driving force for the functional food market. *Heliyon, 6*(8).
Thapa, B. S., Kim, T., Pandit, S., Song, Y. E., Afsharian, Y. P., Rahimnejad, M., et al. (2022). Overview of electroactive microorganisms and electron transfer mechanisms in microbial electrochemistry. *Bioresource Technology, 347*, 126579.
Ubando, A. T., Felix, C. B., & Chen, W. H. (2020). Biorefineries in circular bioeconomy: A comprehensive review. *Bioresource Technology, 299*, 122585.
UN. (2014) *The road to dignity by 2030: Ending poverty, transforming all lives and protecting the planet*. https://www.un.org/disabilities/documents/reports/SG_Synthesis_Report_Road_to_Dignity_by_2030.pdf. Accessed 20 May 2024.
UN. (2015). *Transforming our world: The 2030 agenda for sustainable development*. https://sustainabledevelopment.un.org/post2015/transformingourworld. Accessed 20 May 2024.
UNDP. (2019). *The Sustainable Development Goals Report 2019*. https://unstats.un.org/sdgs/report/2019/The-Sustainable-Development-Goals-Report-2019.pdf. Accessed 20 May 2024.
UNEP, United Nations Environment Programme. (2020). *Emissions gap report 2020 executive summary*. United Nations Environment Programme. https://www.unep.org/emissions-gap-report-2020. Accessed 06 April 2024.
USDA. (2020). https://www.biopreferred.gov/BioPreferred/faces/Welcome.xhtml. Accessed 12 May 2024.
Venkata Mohan, S., Dahiya, S., Amulya, K., Katakojwala, R., & Vanitha, T. K. (2019). Can circular bioeconomy be fueled by waste biorefineries—A closer look. *Bioresource Technology Reports, 7*, 100277.

Wainaina, S., Horváth, I. S., & Taherzadeh, M. J. (2018). Biochemicals from food waste and recalcitrant biomass via syngas fermentation: A review. *Bioresource Technology, 248*, 113–121.

Wang, M., Dewil, R., Maniatis, K., Wheeldon, J., Tan, T., Baeyens, J., & Fang, Y. (2019). Biomass-derived aviation fuels: Challenges and perspective. *Progress in Energy and Combustion Science, 74*, 31–49.

World Circular Economy Forum (WCEF). (2024). *Growing the circular bioeconomy, with a focus on the Global South* (p. 56). CSTEP and ICCE.

Yu, Y. Y., Wang, Y. Z., Fang, Z., Shi, Y. T., Cheng, Q. W., Chen, Y. X., Shi, W., & Yong, Y. C. (2020). Single cell electron collectors for highly efficient wiring-up electronic abiotic/biotic interfaces. *Nature Communications, 11*(1), 4087.

Biobased Product Design, Development, and Utilization in the Global South

Haruna Sekabira, Guy Simbeko, Shiferaw T. Feleke, Victor Manyong, Leonhard Späth, Pius Krütli, Bernard Vanlauwe, Kokou Kintche, Benjamin Wilde, and Johan Six

Abstract

The design, development, and use of biobased economy are required to meet the world's increasing demand for food, materials, and energy while maintaining biodiversity and the ecological balance of the planet (Lewandowski, 2018; Sharma et al., 2021). Biobased products are often biodegradable and have a lower environmental footprint. They can help reduce greenhouse gas emissions, decrease dependence on fossil fuels, and minimize pollution (Lewandowski, 2018). The advancement and application of biobased products stimulate innovation and foster economic growth. This opens up new markets and prospects, especially within the agricultural and industrial sectors. For example, biobased plastics, biofuels, and biochemicals are progressively being researched and implemented as sustainable replacements for traditional equivalents. Additionally, biobased products contribute to rural development by providing new income sources for farmers and rural communities (Chandel et al., 2020; Lewandowski, 2018).

H. Sekabira
Department of Social Sciences and Agribusiness, International Institute of Tropical Agriculture, Kigali, Rwanda

G. Simbeko (✉)
Department of Social Sciences and Agribusiness, Monitoring Evaluation and Department of Natural Resources Management, International Institute of Tropical Agriculture (IITA) Kalambo, DR Congo and Agricultural Transformation Agenda (ATA-DR Congo), Bukavu, Democratic Republic of the Congo

Faculty of Economics and Management, Université Catholique de Bukavu (UCB), Bukavu, Democratic Republic of the Congo
e-mail: simbeko.sadi@ucbukavu.ac.cd

M. C. Ogwu et al. (eds.), *Sustainable Bioeconomy Development in the Global South*, https://doi.org/10.1007/978-981-96-0305-3_4

Abbreviations

AFVPE	Anjonus fruit and vegetable processing enterprise
BRICS Countries	Brazil, Russia, India, China, and South Africa
BSF	Black soldier fly
COPED	Company for protection of environment and development
CPCK	Cooperative of planters of coffee in Kabare
DEWATS	Decentralized wastewater treatment system
DIOBASS	*Démarche pour une Intéraction entre les Organisations de Bases et Autres Sources de Savoirs*
ENMCPA	Egnan Naew Mayet Compost Production Association
FESDD	*Femme et Environnement Sain pour le Développement Durable*
GASD	Global action for sustainable development
GHGs	Greenhouse gases
HDDS	Household dietary diversity score
HFIAS	Household food insecurity access scale
LAB	Lactic acid bacteria
NACI	National Advisory Council on Innovation
OECD	Organization for economic co-operation and development
PLA	Polylactic acid
SAIS	Southern Africa innovation support
SANBio	Southern Africa network for biosciences program
SDGs	Development goals
WWTP	Waste water treatment plant

S. T. Feleke · V. Manyong
Department of Social Sciences and Agribusiness, International Institute of Tropical Agriculture (IITA), Dar es Salaam, Tanzania

L. Späth
Department of Environmental Systems Science, Transdisciplinary Lab (TdLab), ETH Zurich, Zurich, Switzerland

Department of Environmental Systems Science, Sustainable Agroecosystems, ETH Zurich, Zurich, Switzerland

P. Krütli
Department of Environmental Systems Science, Transdisciplinary Lab (TdLab), ETH Zurich, Zurich, Switzerland

B. Vanlauwe
Department of Natural Resources Management, International Institute of Tropical Agriculture, Nairobi, Kenya

K. Kintche
Department of Social Sciences and Agribusiness, Monitoring Evaluation and Department of Natural Resources Management, International Institute of Tropical Agriculture (IITA) Kalambo, DR Congo and Agricultural Transformation Agenda (ATA-DR Congo), Bukavu, Democratic Republic of the Congo

B. Wilde · J. Six
Department of Environmental Systems Science, Sustainable Agroecosystems, ETH Zurich, Zurich, Switzerland

1 Introduction

The design, development, and use of biobased economy are required to meet the world's increasing demand for food, materials, and energy while maintaining biodiversity and the ecological balance of the planet (Lewandowski, 2018; Sharma et al., 2021). Biobased products are often biodegradable and have a lower environmental footprint. They can help reduce greenhouse gas emissions, decrease dependence on fossil fuels, and minimize pollution (Lewandowski, 2018). The advancement and application of biobased products stimulate innovation and foster economic growth. This opens up new markets and prospects, especially within the agricultural and industrial sectors. For example, biobased plastics, biofuels, and biochemicals are progressively being researched and implemented as sustainable replacements for traditional equivalents. Additionally, biobased products contribute to rural development by providing new income sources for farmers and rural communities (Chandel et al., 2020; Lewandowski, 2018).

The Global South, comprising developing countries in Africa, Latin America, and Asia, faces unique challenges and opportunities within the framework of the bioeconomy (Taron et al., 2023). The regions described are rich in biodiversity and biological resources, making them well-suited for the advancement of a bioeconomy. However, they face notable challenges such as inadequate infrastructure, limited technical expertise, and financial constraints. The bioeconomy holds great promise for the Global South, as it can stimulate economic expansion, enhance food security, and encourage environmental sustainability. By leveraging their abundant biological resources, these nations can establish industries based on bioproducts, creating jobs, increasing incomes, and supporting sustainable development. Moreover, the bioeconomy can aid the Global South in addressing environmental issues like deforestation, soil degradation, and water scarcity, which are prevalent in many developing countries.

There has been little research on the socio-economic effects of the bio-based, despite the bioeconomy's well-established benefits for the environment. It is also important to point out that even though several processes for biofuel, biogas, biofertilizer, and biostimulant production have already been established in Global South, their development on a larger scale remains limited. It is imperative to comprehend the employment, income distribution, and social justice advantages of the bioeconomy in the Global South to develop inclusive and long-lasting bioeconomic plans. To comprehend customer attitudes, inclinations, and obstacles to embracing biobased products, more study is required. The strategies used to improve customer acceptance and demand can be informed by this data.

The purpose of this chapter is to address the lack of understanding regarding the main challenges and advantages of designing, developing, and using biobased products in the Global South. It aims to provide a thorough analysis of the current state of the bioeconomy in these regions, identify the crucial factors that influence the successful implementation of biobased industries in developing countries, and offer practical solutions and policy recommendations. By doing so, the study aims to contribute to advancing sustainable development and environmental stewardship in the area under study. The research adopts a mixed-methods approach (Feleke et al.,

2021; Sekabira et al., 2022b), combining qualitative and quantitative data collection and analysis. This includes a thorough literature review and case studies in the bioeconomy sector. The study focused on the following objectives:

1. Assess the current state and identify challenges and opportunities of the biobased in the Global South: This involves examining the existing biobased industries, policies, and initiatives in different regions, highlighting the obstacles that developing countries face in advancing their bioeconomy, as well as the potential opportunities that can be leveraged.
2. Examine biobased product design principles and development processes: This involves exploring sustainable material selection, design for end-of-life, and principles of a circular economy critical for biobased product development. It also encompasses the research and development processes, the function of biotechnology, pilot projects, and scaling up production within the Global South.
3. Evaluate the utilization of biobased products and provide policy recommendations: The study investigates market dynamics, consumer acceptance, applications in various sectors, and the environmental and economic benefits of biobased products. Based on the findings, the study will provide recommendations for policymakers, industry stakeholders, and international organizations to bolster the growth of the bioeconomy in developing countries.

Following this introduction, a literature review gives an overview of the historical development of the bioeconomy, key concepts and theories, previous studies, and gaps in current research. The next section focus on the state of the bioeconomy in the Global South, including challenges, opportunities, and case studies from various regions. Subsequent sections explores the design and development of biobased products, examining sustainable material selection, design principles, and technological innovations. In this section, we delves into the exploration of biobased products, focusing on market dynamics, consumer acceptance, and applications across various sectors. The chapter features comprehensive case studies displaying successful biobased product initiatives in the Global South, offering valuable insights and best practices. Additionally, the study outline future prospects and provide recommendations, including a discussion of emerging trends, policy recommendations, and strategies for sustainable growth. Finally, the conclusion succinctly summarize the key findings, as well as the implications for policy and practice, and shed light on future research directions.

2 State of the Bioeconomy in the Global South: Challenges and Opportunities

The Global South's bioeconomy faces several hurdles. These include inadequate infrastructure for waste management, a shortage of expertise in biotechnology and bioengineering, limited access to funding for biobased initiatives, and insufficient policy support (Backhouse et al., 2021; Lewandowski, 2018). There is often a lack of awareness and acceptance of biobased products among consumers and stakeholders. Understanding the challenges and developing advanced skills is crucial for fully realizing the potential of the bioeconomy in developing countries. Most of the current study of bioeconomy has been carried out in developed nations. This underscores the necessity for more studies that are specific to the context and address the unique conditions and challenges of developing countries in the Global South.

Scaling up the circular bioeconomy in developing countries presents significant challenges. Limited institutional capacity and a fragmented private sector hamper progress. Barriers include low-quality technology adoption, as well as lack of access to finance and infrastructure. Additionally, resistance from established industries and a lack of cooperation between sectors complicate efforts (Feleke et al., 2021). It is important to align circular bioeconomy initiatives with existing policies in order Cover a range of political support. Developing countries should tailor their circular bioeconomy policies to fit their own unique contexts, instead of simply adopting models from developed nations. For example, integrating the bioeconomy with agriculture can improve competitiveness in low-income countries. However, there may be span a spectrum of significant informal labor sector in countries like India. Poorly implemented policies can lead to significant investment losses, as exemplified by dysfunctional waste-to-energy plants resulting from inadequate waste segregation systems. Nonetheless, the circular bioeconomy presents opportunities for economic diversification, job creation, and environmental benefits. Developing nations can leverage their young workforce and emerging markets to establish sustainable consumption and production practices (Negi et al., 2021). The transition from a linear economy to a circular model can enhance supply chain resilience, decrease resource dependency, and generate new employment opportunities across different skill levels (Obahiagbon & Ogwu, 2023a, b, 2024a, b). Efficient waste management in the circular bioeconomy can result in the transition from a linear economy to a circular model thereby contributing to improved environmental resilience and poverty reduction. Moreover, it can stimulate advancements in sectors like pharmaceuticals, utilizing local resources for improved health outcomes. Ultimately, the success of circular bioeconomy initiatives depends on a country's resource endowments, development stage, and political environment, as it is illustrated in Fig. 1.

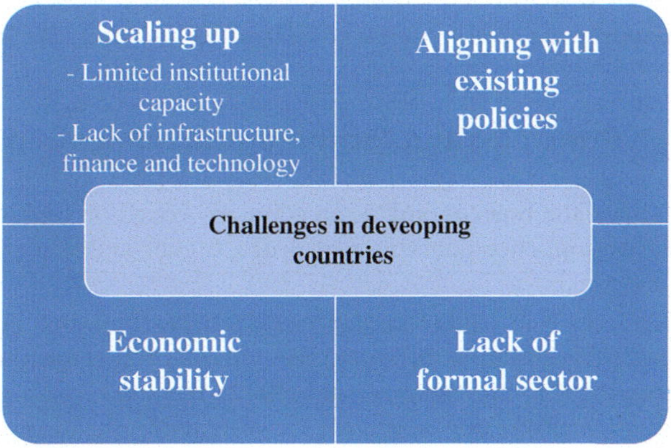

Fig. 1 Challenges in developing nations for moving toward a circular bioeconomy inspired by Negi et al. (2021) study

The shift from a fossil fuel-dependent to a biobased economy is expected to reduce our dependency on fossil fuels and achieve more sustainability, as well as contribute to climate and environmental protection. A study at Universitat Autònoma de Barcelona describes the circular economy as the goal—the desired outcome that aims to separate resource use from natural resource depletion. In contrast, the bioeconomy is seen as the method—the specific biophysical processes that need to be improved to achieve this goal (Chandel et al., 2020; EU, 2020). The study also indicates that the principles of a sustainable bioeconomy and the principles of a circular economy should be in harmony for proper implementation (Lewandowski, 2018).

A circular and sustainable bioeconomy aligns directly with these goals as a revolutionary and restorative system that stimulates industry and the economy while protecting the planet for future generations. It involves replacing plastics and fossil fuels with biological solutions, eliminating toxic chemicals, and reducing waste through innovative materials, products, systems, and business models (Ntwali et al., 2022; Tekman et al., 2020). Advanced technologies for food waste valorization can address resource and environmental challenges, simultaneously addressing capital loss and greenhouse gas accumulation (Ogwu, 2019; Nizzy & Ogwu, 2024). The bioeconomy significantly reduces food waste accumulation, promoting the use of renewable bioresources. It emphasizes the final consumption of bioresources considering agroecosystem limits, resilience, and absorption capacities. A bioeconomy, especially forest-based, supports local communities' socio-economic development and strengthens forest value chains beyond primary forest products (Hetemäki et al., 2023; Rosa & Martius, 2021). Sustainable plantations meet Africa's wood needs, develop related industries, and reuse waste to create goods and services, thus creating jobs. Several studies have briefly explained how circularity is incorporated into the SDGs of the United Nations (Ashukem, 2020). A recent study involves a comprehensive analysis of how these practices support the achievement of SDG goals. The study highlights that circular economy practices have the most substantial impact and create the most synergies with SDG 6 (Clean Water and Sanitation), SDG 7 (Affordable and Clean Energy), SDG 8 (Decent Work and Economic Growth), SDG 12 (Responsible Consumption and Production), and SDG 15 (Life on Land) (Almaqtari et al., 2024; Vargas & Gelcich, 2024) (Fig. 2).

3 Design Principles and Technological Innovation

This strategy defined the bioeconomy as an economy based on biological sciences research and innovation, encompassing new drugs, higher-yielding crops, biofuels, and bio-based chemicals.

Other countries, such as Malaysia and South Africa, have also developed bioeconomy strategies, highlighting the global recognition of the bioeconomy's potential (Bastos Lima, 2021; Bottausci et al., 2022; Hetemäki et al., 2023). The bioeconomy offers a promising approach to sustainable development by using renewable biological resources instead of fossil fuels. This helps address

Fig. 2 Bio-based opportunities and SDGs. (Inspired from Ashukem, 2020; Vargas & Gelcich, 2024 studies)

environmental issues and encourages economic growth. To successfully implement bioeconomy strategies, we need innovative technologies, interdisciplinary research, and supportive policies. These are crucial for overcoming challenges and fully realizing the potential of bio-based industries. By focusing on sustainable practices and integrating environmental considerations, the bioeconomy can significantly contribute to global.

4 Characteristics of Bio-based Value Chains

A central idea of the bioeconomy is cascading use, which involves optimizing the functional and sequential use of biomass to maximize socioeconomic value within the constraints of resource limitations (Casau et al., 2022; Schipfer et al., 2022). Cascading can be understood in different ways. One approach involves efficiently using biomass over time, by reusing and recycling it for various purposes at different stages. For example, paper can be used multiple times and recycled for various applications. Another approach involves prioritizing high-value applications, where plant biomass is first used for the most valuable purposes, such as food production and pharmaceuticals. After these primary uses are met, the remaining biomass can be used for feed, material production, and energy generation from by-products. Biorefineries use biomass as a source for multiple valuable products through various conversion processes, maximizing efficiency (Bastos Lima, 2021). Biorefinery systems differ in their flexibility to process different types of feedstock, the characteristics of their conversion processes, and the range of products they produce (Smit et al., 2022). Grass refineries are designed to use locally sourced meadow grasses and refine them into various products using a cradle-to-cradle approach. This method ensures that the products are fully recyclable without generating waste. The production process maximizes material efficiency by using closed loops and repurposing biogas and heat for heating or drying within the refinery. In bio-based value chains, the focus is on closing material and energy flows, converting linear production processes into circular or closed systems, and thereby minimizing waste. To establish the cascading use of biological resources on a large scale, creating and integrating comprehensive bio-based value chains into broader value networks (Dams et al., 2023; Gatto, 2024). Bio-based value chains exhibit several distinctive features. Firstly, the production of biological resources, especially in forestry and agriculture, is influenced by distinct seasonal patterns.

5 Processing of Bio-based Resources

The core concept of the bioeconomy involves transforming bio-based resources into various products across the food, feed, energy, and material sectors. Due to the unique properties of bio-based resources, suitable conversion methods must be chosen based on the intended application (Gatto, 2024). In developing countries, the economic system heavily depends on finite fossil carbon resources like natural gas, crude oil, and coal. Crude oil is used as the basis for most fuels and is refined into various products such as naphtha, gasoline, diesel, asphalt, heating oil, kerosene, and gas. These products are further processed into intermediate and end products, including plastics, fibers, varnishes, and adhesives. Petroleum (or naphtha) is a liquid raw material composed of hydrocarbons mainly oxidized to produce the desired products (Fattahi, 2017; Nagham & Nuha, 2021). This process uses inorganic, usually metallic, catalysts and requires high temperatures and pressures. It starts with pure and relatively concentrated feedstocks, simplifying product recovery.

Biorefineries investigate various methods for refining renewable resources into fuels, energy, and materials, similar to chemical refining processes (Cuadrado-Osorio et al., 2022). The processes commonly involve using all parts of biomass to create different feed streams that can be turned into basic products. Unlike crude oil, naphtha, and other petrochemical fractions, biomass materials used in biorefineries have lower energy densities, are solid instead of liquid, and are partially oxidized. The most abundant biopolymer and solid raw material, lignocellulose, is made up of three main components: cellulose, hemicellulose (polyose), and lignin. Cellulose and hemicellulose are long chains of sugars, specifically hexoses and pentoses. Lignin is a complex, interconnected polymer made from aromatic alcohol. Fractionation and depolymerization are important steps for further bioconversion. The thermochemical conversion of renewable feedstock, which is less technologically demanding, breaks down biomass into a complex mixture of partially reduced substances. The sugar and lignin fractions are partially oxidized and often need to be reduced to obtain valuable products. To achieve this, CO_2 must be removed from the carbon skeleton, implying that product yields are generally lower than in petrochemical production (Wang et al., 2022). Catalysts, which are highly specific and stop at a particular oxidative step, are used for these reactions. Biocatalysts (whole cells or enzymes) possess these properties but require physiological conditions, performing reactions at moderate temperatures (10–60 °C) under normal pressure.

In forestry, bioenergy from forest biomass is crucial for the energy mix in the Baltic Sea region. The sustainable management and utilization of biomass resources in Sub-Saharan Africa can help stimulate struggling economies. In Kenya, biogas stoves have the smallest environmental footprint compared to other cookstoves. Despite almost entirely green electricity, Costa Rica still relies on fossil fuels for transportation and heating (De Assis Costa et al., 2022). In fisheries and aquaculture, lab-scale experiments in Latvia test the biochemical methane potential of codigesting round goby fish wastes with sewage sludge. Biocarbon/biocoal can substitute nonrenewable fillers in automotive components (Wienhold & Goulao, 2023). Combining the production of biosurfactants, biodiesel, and p-xylene shows how flexible the circular bioeconomy is.

Studies advocate using biocompost over animal feed for rice bran and husk, promoting soil-to-soil recycling. Biochar from agro-wastes serves as an adsorbent in wastewater treatment before returning to fields for soil amendment. Biomethane production from food supply chain biomass feedstock is economically feasible with rationalized natural gas prices and carbon taxes. Biogas and biodiesel production from various waste streams, such as chicken manure and rapeseed straw, demonstrate the potential for renewable energy sources (Denise Assiba Davou et al., 2022). Extracting hydrogen from organic solid wastes and wastewater holds promise for future fuels. Italian studies on dark fermentation and anaerobic digestion of cattle manure and grass silage yield H_2–CH_4 mixtures and bioplastics precursors. Utilizing whey and molasses for biohydrogen and bioplastic production shows the potential of dairy and sugar factory wastes. Combining anaerobically digested sludge with composted sludge as soil amendment reduces health risks and improves nutrient availability (Liew et al., 2022).

6 Case Studies of Successful Bio-based Product Initiatives in the Global South

The study by Feleke et al. (2021) tackles the challenges of natural resource degradation, nutrient-deficient farming systems, hunger, and poverty in Sub-Saharan Africa. It examines how a circular bioeconomy can promote sustainable development by converting agricultural waste into valuable products. The study utilizes a "research for development" (R4D) approach, integrating circular bioeconomy principles into agricultural practices through field experiments, stakeholder interviews, and participatory workshops in order to identify best practices and scalable solutions. The findings indicate that circular bioeconomy initiatives can significantly improve soil health, increase agricultural productivity, and reduce poverty. The research emphasizes the importance of engaging multiple stakeholders and receiving policy support to effectively expand these initiatives.

In Pakistan, Crop et al. (2012) conducted a study focusing on managing agricultural waste, particularly the large volumes of banana waste typically discarded. In this study, the primary objective was to identify sustainable applications for waste materials, with a focus on contributing to the circular bioeconomy. The research utilized a case study methodology to examine the conversion of banana waste into eco-friendly textiles. This involved an in-depth analysis of existing technologies, pilot projects, and an evaluation of the environmental and economic implications of such initiatives. The transformation of banana waste into textiles proved environmentally friendly and economically viable, reducing waste and providing a new source of sustainable textiles that support local economies and minimize environmental impact. IITA (2017) and Sekabira et al. (2022a) explored the need for sustainable food systems in Sub-Saharan Africa, where food security and resource efficiency are critical concerns. The study investigated the potential integration of circular bioeconomy principles into food systems to improve sustainability. It assessed existing practices and discovered opportunities to include circular bioeconomy principles. The research revealed that these integrations can greatly enhance sustainability by minimizing waste, improving resource efficiency, and bolstering food security.

Various authors (Hetemäki et al., 2023; Jindo et al., 2020; Kigo, 2021; Marenya & Barrett, 2007) investigated the environmental and waste management challenges associated with fishery waste in Kenya. The study aimed to examine the possibilities of converting fishery waste into biogas as a sustainable energy source through field experiments, laboratory analysis, and economic feasibility studies. Fishery waste was collected and processed in anaerobic digesters to produce biogas, with findings indicating that this method is effective for waste management and renewable energy production (Akshaya & Jacob, 2020; Al-Wahaibi et al., 2020; Sarker, 2020).

In Kenya, Hetemäki et al. (2023) and Jindo et al. (2020) addressed the need for sustainable and renewable energy sources, focusing on using algal biomass for biofuel production. The research involved cultivating algae, extracting biofuels, and evaluating the efficiency and sustainability of the process through laboratory experiments and pilot projects. The findings suggest that algal biomass is a promising source of biofuel with significant potential to reduce reliance on fossil fuels (Ahmad & Ashraf, 2023). In 2022, Namibia endorsed its National Bioeconomy Strategy for 2023–2028 to promote innovation, conservation, and sustainable use of its biological resources. This plan addresses the country's challenges while preserving its natural environment (Abegunde et al., 2019; Bottausci et al., 2022). The role of policy in transitioning to a sustainable bioeconomy is critical. The global bioeconomy is attracting attention from public officials, corporate leaders, social and biophysical scientists, and the public. Governments adopting bioeconomy concepts in their policies aim to resolve environmental degradation and unsustainable resource use while promoting employment, rural value creation, economic growth, proper use of bioresources, renewable energy security, climate change mitigation, circularity, and environmental protection (Bojago & Abrham, 2023).

South Africa's bioeconomy-strategy policy involves an interagency effort, including departments of science and technology; commerce, industry, and competition; agriculture, land management, and rural advancement; environment, forestry, and fisheries; and health (Lange et al., 2022; Runres-South- Africa, 2023; Simbeko et al., 2023). However, the policy framework has been described as moderately incoherent due to incomplete strategies despite having policy elements on the bioeconomy. Efforts are ongoing to advance policy capacity and adapt the regulatory system to anticipated changes, with the National Biodiversity Economy Strategy (NBES) aiming to grow while ensuring sustainability and conservation of indigenous biological resources.

The South African government promotes bioentrepreneurship within bioeconomic sectors, including the bioprospecting field (Aymard et al., 2021; Hlangwani et al., 2023; Runres-South- Africa, 2023). The institute of sciences of South Africa and the South African Medical Research Council conducted a review of bioethics regulation and assessed the policy impact of genome editing. The National Advisory Council on Innovation (NACI) suggested assessing the effectiveness of the bioeconomy in alignment with the bioeconomy strategy. South Africa has established foreign-supported regional initiatives like the Southern Africa Network for Biosciences (SANBio) and the Southern Africa Innovation Support Program (SAIS). Although there is no global policy for the adoption and implementation of bioeconomy strategies, numerous countries are pursuing measures and policies to advance their bioeconomies, with macro-regional policy approaches, multilateral policy processes, intergovernmental dialogue, and stakeholder-driven initiatives emerging as trends (Hlangwani et al., 2023). The bioeconomy strategy for Eastern Africa has spurred national efforts to develop specific bioeconomy policy initiatives, emphasizing the need for a comprehensive and well-defined development strategy to enhance the effectiveness and impact of national action plans.

Nevertheless, Malaysia and South Africa have extensive bioeconomy strategies aligned with frameworks of developed nations. The transition to a sustainable bioeconomy can take various forms, leading to different societal types. Hence, the shift toward small-scale bioeconomic activities is frequently sidelined in political discussions, and the move toward reduced consumption faces opposition due to conflicting interests and goals. Failures in implementing bioeconomic activities may have adverse effects on natural resources and the environment, such as unsustainable biomass usage, competition for land, intensified agriculture, threats from invasive species, and negative ecological impacts. The ECONOBIO project in Cameroon aims to safeguard biodiversity while fostering economic prospects for local communities. It is an innovative model based on developing sustainable economic sectors such as agroforestry, ecotourism, and the valorization of nontimber forest products (Ashukem, 2020; Assan, 2023). The goal is to reduce local populations' dependence on destructive logging and preserve fragile ecosystems (Acobta et al., 2023).

In Nigeria, where 60 million hectares of primary tropical forests were lost during the twentieth century, the organization Reforest'Action partnered with the local company Green Sahara Farms (ESG, 2024) to counter deforestation. They reintroduce trees into farmers' daily lives by developing agroforestry systems that incorporate trees alongside and within fields. These trees play a vital role in providing ecosystem services such as diversifying agricultural production through fruit trees, enriching the soil, and protecting crops. The presence of a diverse range of species is essential in these agroforestry systems. For instance, in a cashew field, various tree species like papaya or pomegranate can be cultivated to yield a secondary fruit harvest. Additionally, leguminous species like acacia or cassia not only enrich the soil with nitrogen but also provide feed for livestock, thereby contributing to farmers' food security and income stability.

The bioeconomy in Rwanda is making significant progress due to a strong partnership involving multiple stakeholders. Recently, at the Bio Africa Convention, UVU Bio revealed an impactful project in collaboration with the University of Rwanda, the Ministry of Education of Rwanda, and the Deutsche Gesellschaft für Internationale Zusammenarbeit (GIZ) GmbH on behalf of the German Federal Ministry for Economic Collaboration and Development. This project aims to reshape the bioeconomy landscape in Rwanda, paving the way for sustainable growth, job creation, and innovation (Johnson et al., 2022; Sekabira et al., 2022a). In the Democratic Republic of the Congo (DRC), Ethiopia, Rwanda, and South Africa, the results showed that sorting organic from inorganic waste and using organic waste as compost had a 31% likelihood of contributing to successful engagement in CBE practices while sorting waste and using organic waste as livestock feed contributed to such success by 17%. Using organic waste as compost and livestock, feed had the lowest success rate of 11%. Therefore, community-based innovations that encourage the integration of circular bioeconomy practices among smallholders—especially those involving waste sorting and the use of organic waste as compost—are more likely to succeed in establishing circular food systems.

In DR Congo, the *Femme et Environnement Sain pour le Développement Durable* (FESDD) organization has convinced 400 households to sort their household waste in two bags (inorganic and organic). Organic waste is sorted and transported to rural areas for composting. Over 430 coffee farmers have used 749 tons of compost in the past 2.5 years. The effectiveness of this innovation stems from its ability to not only enhance coffee yields and soil fertility in Kabare but also reduce urban unsanitary conditions in the city of Bukavu (RUNRES-DRC, 2023). Furthermore, the *Démarche pour une Intéraction entre les Organisations de Bases et Autres Sources de Savoirs* (DIOBASS) organization is experimenting circular with economy approaches by trying to restore soil nutrients by creating compost with organic household waste. The organic waste is collected from 300 households in Bukavu City and from the Nyawera market, where a lot of waste accumulates. Forty-eight tons of compost are expected to be produced every 6 months. Most of the farmers hope to increase their production from 0.6 tons to 1 ton of coffee per hectare. GASD (Global Action for Sustainable Development) is another organization producing co-compost by recovering organic waste from different sources. (1) household waste; 200 households from Bukavu city, (2) waste from Katana and Kabamba markets, (3) coffee pulps from Cooperative of Planters of Coffee in Kabare (CPCK); a coffee farmer cooperative washing stations, as well as (4) human waste from ecological toilets built on the two markets: Katana and Kabamba (RUNRES, 2023; RUNRES-DRC, 2023). The co-composting innovation depicted in Fig. 3 offers a significant opportunity to enhance agricultural production and offer sustainable solutions to the health and sanitation challenges stemming from the accumulation of urban waste.

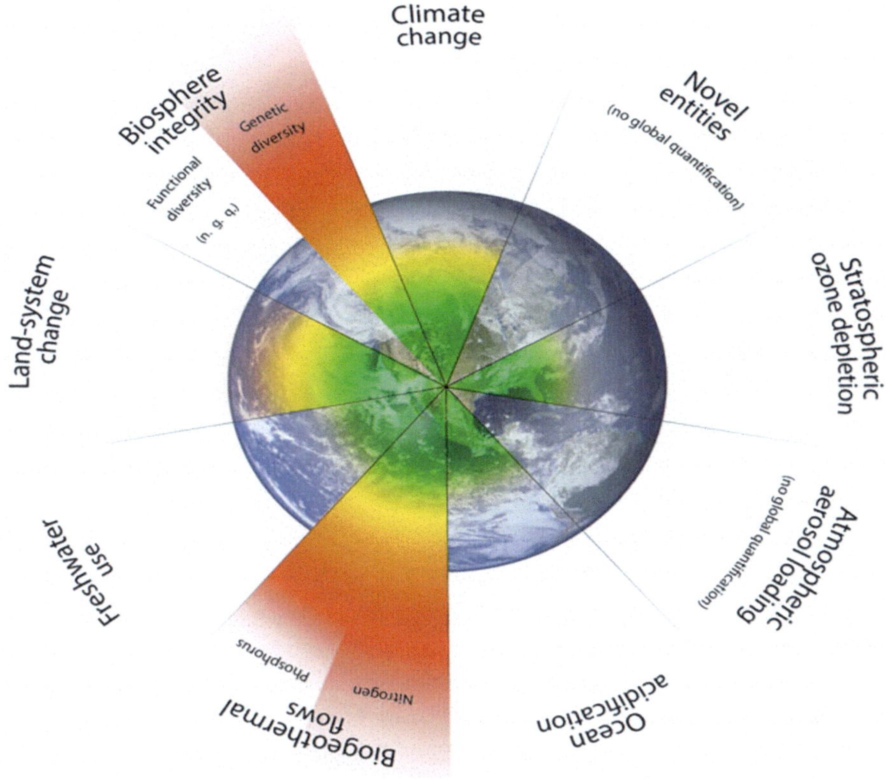

Fig. 3 The green-shaded area represents the safe operating space. (Adapted from Lewandowski, 2018 study)

Several innovations were introduced (Fig. 4) in Ethiopia. ENMCPA (Egnan Naew Mayet Compost Production Association) association serves as a bridge between waste collectors and compost users, the local banana farmers. Once the composting facility is operational, the enterprise in the city region of Arba Minch plans to process and recycle 1100 tons of compost per year. Currently, ENMCPA receives municipal biowaste from four women's associations. This innovation leads to the recycling of municipal organic solid waste and creates job opportunities for women. The Anjonus Fruit and Vegetable Processing Enterprise (AFVPE) produces banana flour from physiologically matured unripen bananas (RUNRES-Ethiopia, 2023). Runres has introduced a recycling system for urine, where a Urine-diverting A portable toilet has been set up in high-traffic areas, such as bus terminals, public parks, and markets, to gather source-separated human urine. The collected urine is then transformed into struvite fertilizer, which is seen as a sustainable alternative to urea fertilizer. Additionally, it is trying to modify the existing public toilets so that they can separate at the source human urine from feces. Lastly, the collected urine

Fig. 4 Bioeconomy policies and strategies established. (Inspired by Bastos Lima, 2021)

is transported daily to the treatment facility, where it is converted into struvite (RUNRES, 2023; RUNRES-Ethiopia, 2023).

Figure 4 illustrates some key bioeconomic practices introduced in Rwanda. In Rwanda, Akanoze is a small-scale cassava-processing unit in Kamonyi that recycles the nonutilized cassava peels, which account for approximately 30% of the weight of the tubers. Cassava peels were traditionally discarded through rotting or burning because of their high cyanide content. However, a new method has been introduced to make use of these peels. They are now sliced, pressed to remove moisture, sun-dried, and then ground to produce flour. This process makes the peels safe for animal consumption. This flour is sold to a gross retailer as a basis for the manufacturing of animal feed (RUNRES, 2023; RUNRES-Rwanda, 2023). Due to the high cost of imported animal feed in Rwanda, which is often unaffordable for subsistence farmers, this innovation provides a cost-effective local alternative that is rich in fiber for feeding animals. Maggot Company operates a Black Soldier Fly (BSF) facility that repurposes green waste, brewery waste, and household waste from the local community. These organic wastes are chopped finely so that they can be fed to BSF larvae (Fig. 5). For approximately 13–18 days, the larvae consume waste before being sold to fish and livestock farmers as a source of high-protein feed for animals, mainly poultry and pigs. Since animal feed is frequently imported into Rwanda and sold at an exorbitant cost for subsistence farmers, this innovation presents an affordable, local option that also creates economic opportunities within the community. COPED (Company for Protection of Environment and Development) is a solid waste management company that operates in the larger Kigali region, is piloting a program with the support of RUNRES to sort waste at the household level in the

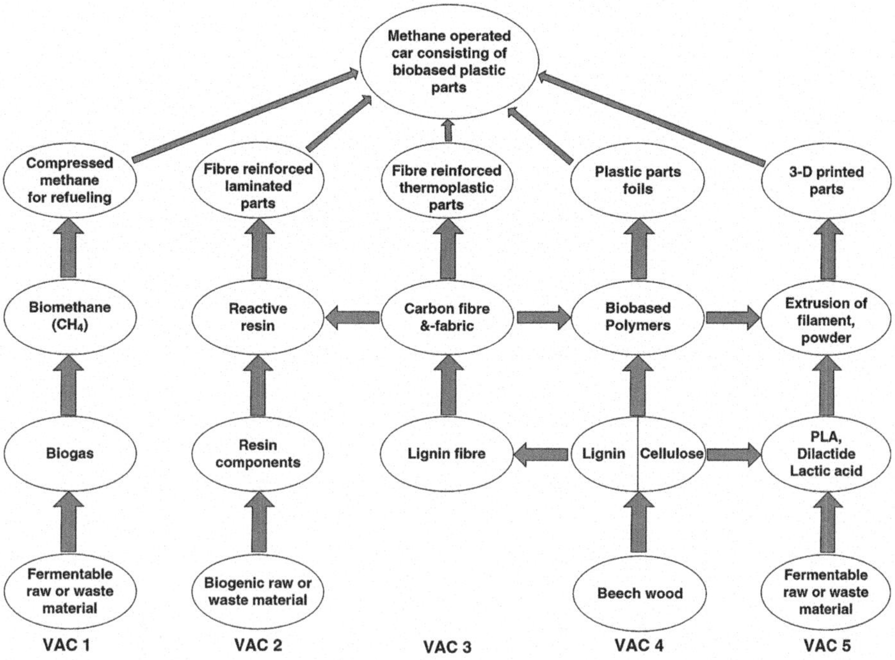

Fig. 5 Bio-based value chain. (Adapted from Dahmen et al., 2019; Lewandowski, 2018 works)

local community. Once sorted, the organic waste is transported to the COPED composting facility (RUNRES-Rwanda, 2023). Three distinct grades of compost are offered to meet the needs of various clients: grade 1 for gardeners, grade 2 for annual crop farmers, and grade 3 for perennial crop farmers. Hence, in this local area, sorting waste at the household level is a new and needed practice to produce high-quality compost.

In South Africa, a public-private A collaboration was formed with a private entity (Duzi Turf), a public utility (Umgeni Water), and a local government (Msunduzi Municipality) to investigate the potential for co-compost production on a municipal scale. The private company co-composts garden and green waste and dewatered sewage sludge provided by the municipality and public utility, respectively. The private company is conveniently located next to the landfill site and the public utility to facilitate the collection of the materials (RUNRES South- Africa, 2023). The garden waste is then (1) chipped to the required structural consistency, (2) combined with sludge from a Waste Water Treatment Plant (WWTP), and (3) composted in windrows. In addition to supporting the production of turf grass, this co-compost will benefit the farmer-cooperatives from the nearby Sobantu community by improving soil fertility management and yields. Umgeni Water, a state-owned entity, envisions installing an integrated Decentralized Wastewater Treatment System (DEWATS) and pour-flush urine diversion system as an eco-friendly onsite sanitation and resource recovery solution in a rural school. The innovation seeks to: (1) tackle sanitation management issues at a school accommodating around 500 students (Sikhululiwe School) and (2) contribute to improved livelihoods of people

Fig. 6 (**a**) Enclosed compost heaps at the Mudaka waste management facility, (**b**) screening and packaging of the compost produced at Mudaka, (**c**) waste unloading and categorization at the Mulungu waste management facility. (The figure was adapted from Sekabira et al., 2023 work)

in the Vulindlela community. At present, many technological solutions in the ecological sanitation sector encounter bureaucratic obstacles related to the reuse of human waste (RUNRES South-Africa, 2023). Therefore, this pilot-scale innovation will be useful for building data-based evidence to help influence the regulations/policies that stymie the reuse of human waste and water/wastewater treatment sludge at a large scale (RUNRES, 2023). This innovation (Fig. 6) has the potential to alleviate the sanitation challenge presented by rapidly filling pit latrines and create several employment opportunities while at the same time improving the productivity of local farming operations. The enhanced biochar produced would also be sold to the local community and in the nearest urban centers.

The study provided intracity, intercity, city region, and overall sample descriptions. However, due to space constraints, we focused on detailing the results of the overall sample and key variables across city regions. Figs. 7, 8, 9, and 10 illustrates that the use of CB practices is more likely to improve food security for the HFIAP and HFIAS 2 score and HDDS dietary diversity) among households engaged in CB practices (Figs. 3 and 4) Either 1283 for HFIAS score and 0,277 for HDDS, with all other things being equal.

In Fig. 4, The results indicate that approximately 37% of the sampled households were severely food insecure, 30% were moderately food insecure, 10% were mildly food insecure, and only 23% were food secure. Kamonyi had the highest percentage of food-secure households (49%), followed by Arba, Minch (32%), then Msunduzi (19%), and least was Bukavu (3%). However, severe food insecurity was highest in Arba Minch (51%), followed by Bukavu (39%), then Msunduzi (36%), and at least Kamonyi (25%) (Fig. 11).

7 The Current State of Design and Use of Biological Products

Designing a biological product requires the use of biobased raw materials and adherence to strict specifications on the manufacturing process and additives. This includes limited use of additives and manufacturing aids, avoidance of ionizing treatments, and environmentally friendly, nonpolluting processes. As a bioproduction pioneer, Généthon established its pharmaceutical-grade bioproduction

Fig. 7 (**a**) Filtered and packaged organic compost prepared for sale to local farmers, (**b**) cleaning and separating of matured, unripe banana fingers, (**c**) the end product available on the market: packaged banana flour, (**d**) Struvite reactor for the portable toilet used for source. (The figure was adapted from Sekabira et al., 2023 study)

Fig. 8 (**a**) A cooperative of women is employed daily to peel cassava, (**b**) employees bring the organic waste collected from the market to a vegetable cutter machine, (**c**) the Black Soldier Fly (BSF) larvae are fed on the chopped organic waste (and brewing waste sourced from a local brewery), (**d**) waste is sorted at the household level between organic (green bag) and inorganic (blue bag), (**e**) at the compost production site, men are employed to remove remaining inorganic components and to prepare and maintain the compost heaps. (The figure was adapted from Sekabira et al., 2023 study)

Fig. 9 (**a**) The windrows are composed of a mixture of organic waste and sewage sludge. For this reason, the process is called co-composting and not only composting, (**b**) filled up toilet in Vulindlela community, (**c**) Garden waste is chipped before mixing with the sewage sludge. (The figure was adapted from Sekabira et al., 2023 study)

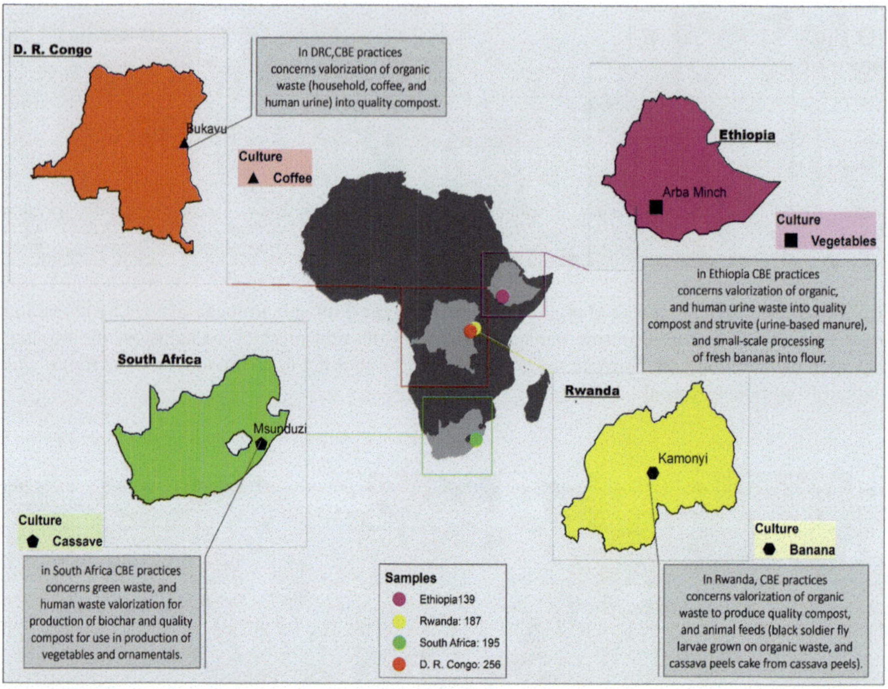

Fig. 10 Mean values of the Household Dietary Diversity Score (HDDS) and the Household Food Insecurity Access Scale (HFIAS) score across different city regions. (The figure was adapted from Sekabira et al., 2023 study)

capabilities in 2008. Nowadays, the application of biotechnology and bioengineering, including recombinant DNA technologies, spans a broad spectrum of applications such as enhancing human and animal food sectors, biofuels, materials, chemicals, and pharmaceuticals gain wide domains in Global South. Genetic engineering is expected to be instrumental in the advancement of nonfood applications, but its use in modern biotechnology is controversial. Policies related to modern biotechnology applications may have diverse implications that need to be considered when assessing their impacts. The debate on modern biotechnology use was reignited by new plant breeding technologies (NPBT) for genetic editing and associated regulatory issues (Feleke et al., 2021). Two popular current biological products among amateur and expert gardeners, according to Premier Tech Home and Garden, include compost, recognized as an effective way to transform some waste into usable material for gardening and food production. Various compost types include composted manure and products based on mushrooms or shrimp. All ingredients nourish the soil naturally and promote vigorous growth of stronger roots and larger flowers. Even if plants are not grown for human or animal consumption, it's important to fertilize them with fertilizers derived from natural sources rather than laboratory-made products (Peng et al., 2023; Seleiman et al., 2020). In Table 1, we

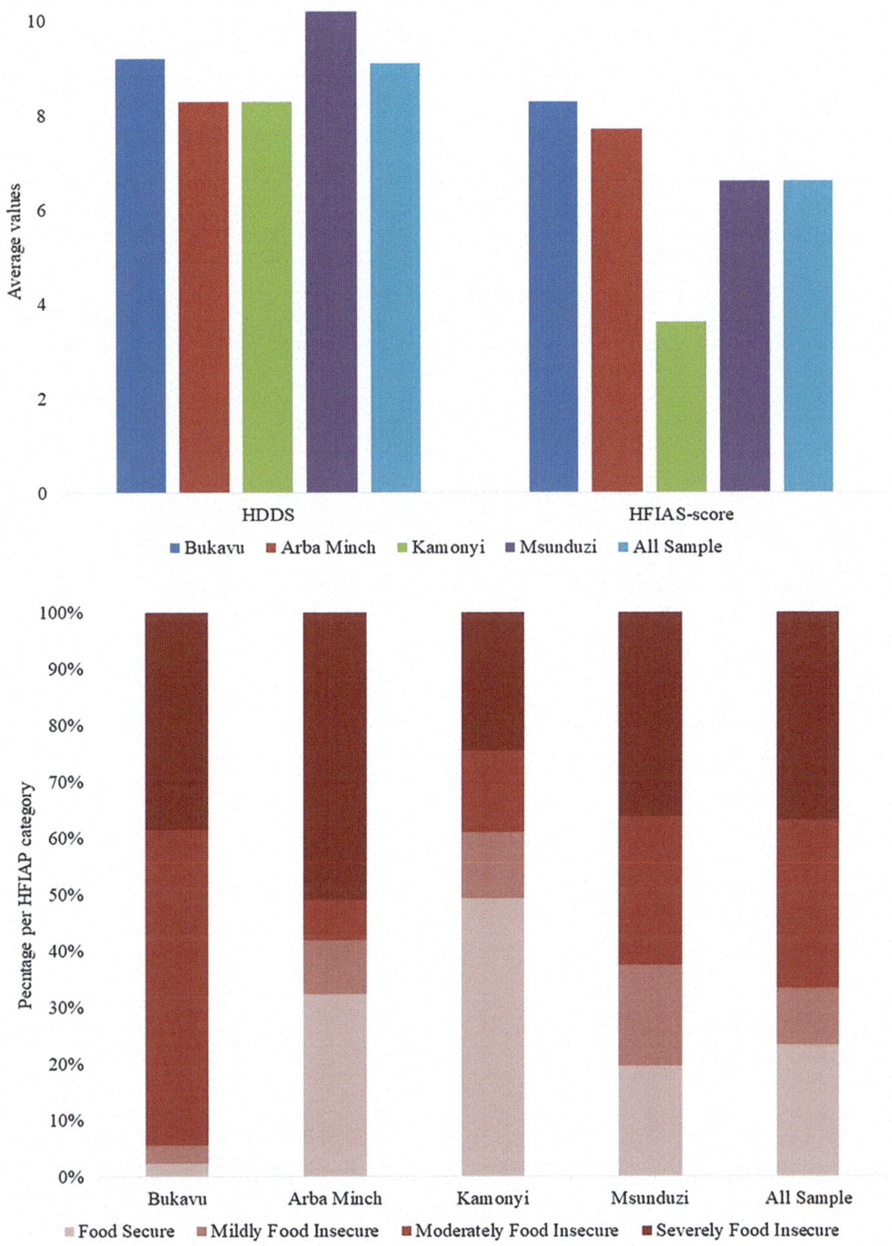

Fig. 11 Proportions of households per Household Food Insecurity Access Prevalence (HFIAP) category per city-region. (Adapted from Sekabira et al., 2024)

Table 1 The current state of design and use of biological products after 2020 (authors compilation)

Countries	Title of the document (s)	Biobased product design	Biobased product development, and utilization	Focus domains	Problem solved	References
South Africa and Development countries	Review of current and future biobased stabilization products enzymatic and polymer-based road construction materials	StabilizedAdditives to enhance weak subgrade soils, such as dispersive, erodible, and collapsible types and soft or expansive clays)	Road base, sub-base, and subgrade materials Utilization of bio-derived enzymes and microbes polymeric additives to avert road failure caused by water penetration and/or erosion	Implementation of biobased materials for infrastructure development Soil stabilization techniques, Application of biobased stabilization products in the road construction sector and using modernized techniques in the road construction industry	Improve the primary engineering properties of the material to meet a road construction standard	Ramdas et al. (2021)
Africa	A critical review of the current status and challenges facing biomass biorefineries in Africa and future perspectives for bioeconomy development	Biomass residues and marine biomass, Biofuels and biogas production and syngas	Digestate and compost, biofertilizers and biostimulants	The production of bioenergy; and the production of agricultural biologicals (biofertilizers and biostimulants) for agronomic purposes	The adaptation of existing valorization of sustainable environment pathways to the African context.	Fertahi et al. (2023)
Iran	Biobased Business Models: specific and general learnings from recent good practice cases in different business sectors.	Bioconversion of residual organic streams using Black Soldier Fly (BSF) Larvae	Production of feed quality protein meal and oil with high nutritional value	Biobased good practices on the market	The competitiveness and sustainability potential of Biobased Business Models for biobased solutions.	Hatvani et al. (2022)

Developing countries	Qualitative investigation of sewage sludge composting: effect of aerobic/anaerobic pretreatments	Aerobic and anaerobic digestions and physicochemical properties	Wastewater treatment plant	Waste sewage sludge	Management and recycling of organic waste within an integrated waste management system	Naserian et al. (2021)
B.R.I.C.S countries	Green finance, renewable energy investment, and Environmental protection: empirical evidence from B.R.I.C.S. countries	Reducing carbon emissions for sustainable environment	Renewable energy (R.E.U)	The role of R.E.U. and green finance in environmental quality	Cost of environmental degradation	Hailiang et al. (2023)
Gabon, Kenya, Nigeria, South Africa, and Tanzania	Outlook for Sustainable forest Bioeconomy in Gabon, Kenya, Nigeria, South Africa, and Tanzania	Wood products and fuel wood development	Forestry, sawnwood, plywood, paper and wood products, biotextiles, biochemicals, biopharma, and services related to those products	Forest Bioeconomy	Large forest resources and stable institutions and population growth	Hetemäki et al. (2023)
Developing countries	Production of gasolines and monocyclic aromatic Hydrocarbons: From fossil raw materials to green processes	Applications of monocyclic aromatic hydrocarbons Manufacture of monocyclic, aromatic, hydrocarbons, and gasolines from fossil raw materials	Pharmaceuticals, detergents, cosmetics, plastics, rubbers, etc.	Renewable raw materials for the production of fuels and chemicals	Limited availability of fossil raw materials in the next 50 years	Busca (2021)

(continued)

Table 1 (continued)

Countries	Title of the document (s)	Biobased product design	Biobased product development, and utilization	Focus domains	Problem solved	References
East Africa region	Challenges, and opportunities of transnational innovation systems using cassava as the case study crop in the East Africa region	Bio-based cassava products	Cassava-based biorefineries, novel cassava food and feed products as well as starch-based bio packaging products. Bio-based products	Biomass production, bioprocessing, and value addition to the rich bioresources	Production, processing, and market challenges for cassava East Africa	Lutta et al. (2024)
Ethiopia, Rwanda, DR. Congo, South Africa	Determinants and success of engagement in circular bioeconomy practices in African food systems	Circular bioeconomy practices	Sorting waste and using organic waste as compost and as livestock feed	Circular bioeconomy	Circular, flexible, and sustainable food systems	Sekabira et al. (2023)
Ghana, Ethiopia, Nigeria, Tanzania, South Africa	Circular Bioeconomy Research for Development in Sub-Saharan Africa: Innovations, Gaps, and Actions	Biotechnology	Biochar production, BiomassWeb, integral use of oil palm, Seaweed value addition, from farmer to Pharma, Insect-based feed (black soldier fly larvae), Soybean inoculant (Nodumax), Biodiesel, High-quality cassava peels, Organic fertilizer and Edible insects (e.g., termites); edible mushroom	Circular bioeconomy	Depletion of natural resources, nutrient-deficient farming systems, hunger, and poverty in sub-Saharan Africa (SSA)	Feleke et al. (2021)

South Africa	Cost–Benefit Analysis of Municipal Sludge as a Low-Grade Nutrient Source: A Case Study from South Africa	Using sludge as a source of nutrients in agriculture	The application of sludge on cropland.	Crop growth, groundwater, soil, and food value chain	Risks posed by inorganic and organic contaminants to plant growth, soil, groundwater, and consequently the food chain and environment related to the utilization of digested sludge as a fertilizer on cropland	Seleiman et al. (2020) and Tesfamariam et al. (2020)
Asia and Africa	Evidence of uptake of different pollutants in plants harvested from soil treated and fertilized with organic materials as source of soil nutrients from developing countries	Pollutants from treated waste material	Absorption of pollutants from treated waste material treated	Applications of organic materials in agriculture and use of polluted organic materials in agriculture	Misconception about the use of organic materials (sludge, urine, human waste and urban waste) in agriculture and organic farming	Olowoyo and Mugivhisa (2019)
India	Unlocking the nutrient value of sewage sludge	Sewage sludge	Pretreated Sludge	Used nutrient value of sewage sludge	Associated challenges in using the sludge	Singh et al. (2022)
Irak	Review on Chemical Separation of Crude Oil and Analysis of its Components.	Crude oil	Hydrocarbons	Hydrocarbure	Conditions and the phase diagram of the subsurface oil mixture	Nagham and Nuha (2021)

(continued)

Table 1 (continued)

Countries	Title of the document (s)	Biobased product design	Biobased product development, and utilization	Focus domains	Problem solved	References
Developing countries	Waste biorefinery toward a sustainable circular bioeconomy: a solution to global issues	waste biorefinery as bioenergy and biomaterials	biopolymers and biolipids	Circular bioeconomy	environmental problems and food security.	Leong et al. (2021)
Africa	Global fossil fuel reduction pathways under different climate mitigation strategies and ambitions	Global coal, oil, and natural gas supply (intended for all uses)	Adopt unambiguous near- and long-term reduction benchmarks within coal, oil, and gas production	Energy transitions	Viability of various pathways for reducing fossil fuel use	Achakulwisut et al. (2023)
Oman	Techno-economic evaluation of biogas production from food waste via anaerobic digestion	Biogas production from food waste via anaerobic digestion	Biogas production	Food waste	Environmental Problems causing by umping sites	Al-Wahaibi et al. (2020)

Arabian Gulf Region (Bahrain, Kuwait, Oman, Qatar, Saudi Arabia, and the United Arab Emirates)	Harnessing the energy transition from total Dependence on fossil to renewable energy in the Arabian Gulf region, considering population, climate change impacts, ecological and carbon footprints, and United Nations' Sustainable Development Goals	Fossil energy sources (oil and natural gas) domestically, industrially, commercially, economically, and transportation-wise	Photovoltaic (PV) solar energy	Renewable energy sources	Per capita ecological and carbon footprints, including oil consumption and its environmental impacts with SDGs, and ecological and carbon footprints	Salem et al. (2023)
China	Striving toward environmental sustainability: how natural resources, human capital, financial development, and economic growth interact with ecological footprint in China	Natural resources and financial development	Ecological footprint	Sustainable development.	The connection between economic growth, natural resources, human capital, and financial development, environment, and ecological footprint	Zia et al. (2021)
India	Unification of Waste Management from Fish and Vegetable Markets Through Anaerobic Co-digestion	Anaerobic digestion of fish waste (FW) as sole substrate	Codigestion with vegetable wastes (VW): three trials viz., T1 (mono-digestion of VW), T2 (mono-digestion of FW) and T3 (codigestion of VW and FW)	Waste management	Management of fish waste quantity originating from fish markets and harbors in coastal regions	Akshaya and Jacob (2020)

(continued)

Table 1 (continued)

Countries	Title of the document (s)	Biobased product design	Biobased product development, and utilization	Focus domains	Problem solved	References
Costa Rica	The Future of Fossil Fuels: Challenges and Opportunities in a Low-Carbon World	Natural gas as a bridge fuel, and the use of fossil fuels in nonenergy applications	Use of geothermal energy in oil production	Fossil fuels in a world with low carbon emissions	Reduce greenhouse gas emissions, declining costs of Renewable energy, and political and social pressure to reduce fossil fuel use in low carbon world	Alagoz and Alghawi (2023)
The Middle East and North Africa (MENA) countries	Natural resources abundance, economic globalization, and carbon emissions: Advancing sustainable development agenda	Natural resource	Economic globalization	Natural resource and economic globalization	Impact of natural resources and economic globalization on the environmental degradation	Xiaoman et al. (2021)
China India Japan Nigeria Ethiopia Egypt Brazil Mexico Colombia	A Global Assessment: Can Renewable Energy Replace Fossil Fuels by 2050?	renewable energy production	Hydroelectric, wind, solar power, biopower and biothermal, biofuels, geothermal, and tidal	Renewable energy	Effectiveness of using renewable energy pathways for a complete to shift from fossil fuels to renewable energy by 2050	Holechek et al. (2022)

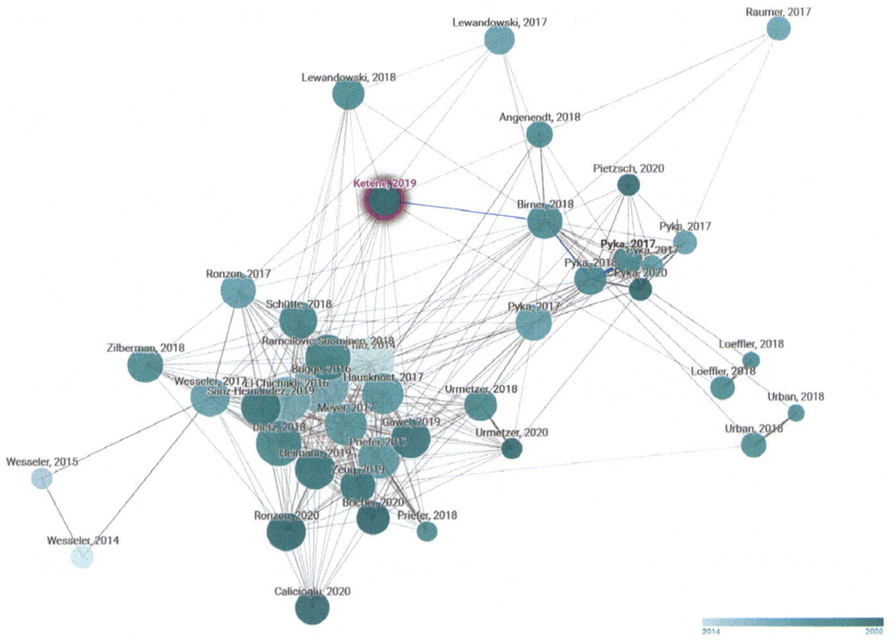

Fig. 12 Connected papers graph before 2020 (author conception)

present the current state of design and bio-based products used in Global South in the world after 2020, and in Fig. 12, we present the connected papers graph on bio-based related to the Ketene (2019) and Lewandowski (2018) works.

8 Applications and Challenges Associated with Biobased Products

Biobased materials have great potential for packaging fast foods such as salads, as well as fresh or minimally processed fruits and vegetables. The high oxygen (O_2) and carbon dioxide (CO_2) permeability of many biobased materials makes them suitable for packaging respiring foods, such as fruits and vegetables. Joint Research Centre (2016) examined the use of biobased polymers as direct edible coatings for some food products. These biobased polymers, including certain polysaccharides, can be applied by immersing, spraying, or brushing the solution onto the food. The edible coating is a barrier to mass transfer without contaminating the food (Lewandowski, 2018). Lewandowski (2018) highlighted that GMO-free starch-based plastic is utilized for manufacturing food service items such as plates and cutlery. Cellophane, a natural biopolymer, is also employed as a packaging film for wrapping bakery and confectionery products. It's essential to consider the durability of bio-based primary packaging and its impact on product shelf life in the food packaging industry. The limited applications of bio-based materials in food packaging are often attributed to their poor properties. De Assis Costa et al. (2022) and Sharma et al. (2021) noted that factors like performance, processing, and cost limit the application of most biobased materials in food packaging. Analyses of the physical and mechanical properties of some biobased materials, such as PLA, starch, and PHB, indicated that water vapor permeability (WVP), a crucial parameter for starch, needs improvement before it can be used for food packaging (Versino et al., 2023). Blending biobased materials with other materials can create composites that significantly enhance overall performance compared to nonblended materials. For example, nanoparticles fused into biopolymers yield composites that are heat and cold-stable, impermeable to gases, and possess strong, firm, and antimicrobial properties (Sherwood et al., 2016). A study by Tekman et al. (2020) and Tumwesigye et al. (2019) demonstrated higher water vapor permeability in biobased composite materials compared to the unmodified materials tested.

Furthermore, Dahmen et al. (2019) emphasized the enhanced mechanical and physical properties of composite materials obtained from PLA and cellulose. Most biobased polymers, including starch, cellulose, and proteins, are renewable and abundant agriculture-derived materials (Chandel et al., 2020). Due to their biodegradability, biobased materials are suitable for producing compost, which carries nutrients and organic materials to the soil. They are also adaptable to many plastic processing methods, requiring little or no adjustments.

Biobased materials are anticipated to continue being important sources for food packaging, with ongoing research aimed at enhancing their functionality and future applications (Hasan et al., 2018). Utilizing plant fibers as reinforcement has the potential to facilitate the development of high-performance, renewable, and sustainable composite materials. These composites are envisioned to bridge the property–performance gap between renewable and petroleum-based polymers (Lewandowski, 2018; Norli et al., 2017). Increased research on blends from polymers and nanoclays (nanocomposites) is expected to result in better barrier and mechanical properties, potentially leading to the commercialization of these composite materials as substitutes for conventional packaging materials (Versino et al., 2023).

9 Conclusion

Policies have a significant impact on the shift to a sustainable bioeconomy. The public, government officials, business executives, social and biophysical scientists, and others are becoming more and more interested in the global bioeconomy. Consequently, the OECD took the lead in developing an international framework to handle the goals and problems associated with the bioeconomy. The incorporation of bioeconomy principles into policy frameworks seeks to address issues such as degrading environmental conditions, decreasing the use of unsustainable resources, creating jobs, promoting rural development, stimulating economic expansion, guaranteeing the sustainable utilization of bioresources, securing renewable energy, mitigating climate change, improving circularity, and safeguarding the environment. There isn't currently a global policy in place to embrace and put bioeconomy concepts into practice. However, many nations are implementing policies to advance their bioeconomies, characterized by macro-regional policy approaches, multilateral processes, intergovernmental dialogues, and stakeholder-driven initiatives. The Regional Bioeconomy Strategy for Eastern Africa has spurred national efforts to develop specific bioeconomy-policy initiatives. The transition to a sustainable bioeconomy is becoming a global trend. Developing countries are employing different methods to encourage growth in their bioeconomies. However, unlike developed countries, most developing nations lack a comprehensive perspective in their bioeconomy strategies. This is evident in the challenges they face when implementing clear regional policies and innovation strategies.

Without a comprehensive and clearly defined bioeconomy development strategy, the effectiveness and impact of national action plans may be limited. Despite this, Malaysia and South Africa have developed holistic bioeconomy strategies that are aligned with those of developed countries. There are numerous ways to transition to a sustainable bioeconomy, leading to various socioeconomic models. However, political marginalization of small-scale bioeconomic enterprises is common, and competing interests and goals hamper the shift toward lower consumption levels. An insufficient application of bioeconomic activities can have detrimental repercussions on the environment and natural resources, such as unsustainable biomass consumption, competition for land, intensification of agriculture, hazards from invading species, and negative ecological effects. Future policy goals must incorporate ecological and social considerations in line with the Sustainable Development Goals (SDGs) to avoid these negative impacts. Numerous reports criticize the lack of measurement and sustainability in most global bioeconomy frameworks. To achieve systemic change, new methods must be developed to identify, measure, and communicate the bioeconomy's contributions to environmental sustainability.

By applying an outcome-tracking approach to data from secondary sources, we demonstrated that circular bioeconomy initiatives led to economic, social, and environmental benefits. These benefits include higher crop yields, improved food security, increased resource efficiency, job creation, and reduced greenhouse gas emissions. Our study's contributions are limited to examining in large manner the small-scale circular bioeconomy interventions aimed at producing food, feed, fuel, and fertilizers. Further research is needed to assess their feasibility on a commercial scale. Furthermore, there is a need for research to explore the technical feasibility of producing high-value products, such as biobased chemicals and polymers. These products can be used as building blocks for biobased products in the manufacturing industry. The technical and commercial feasibilities of high-value products, including herbicides, biopesticides, food additives, detergents, and cosmetics, have been demonstrated in various African countries. A study identified numerous value pathways for secondary organic resources beyond energy and fertilizers on technologies, applications, and products for biobased wastes.

Additionally, our study focused on independent value chains rather than value webs. There are no links within or between the considered value chains based on cascading use and joint biomass use principles. Future research should explore how a biomass value web can be created for system optimization following the BiomassWeb project implemented in Ethiopia, Ghana, and Nigeria. The study employed a qualitative analytical framework to assess the economic, environmental, and social impacts of circular bioeconomy interventions. Relying solely on qualitative analysis makes it challenging to advocate for the expansion of promising circular bioeconomy innovations. Future research efforts should generate quantitative evidence to persuade policymakers to endorse the advancement of sustainable bioeconomies. Currently, global shifts toward sustainable development encompass the transition from a linear to a circular economy and the energy transformation advocating for decarbonization, digitalization, distributed energy generation, and energy efficiency.

References

Abegunde, V. O., Sibanda, M., & Obi, A. (2019). The dynamics of climate change adaptation in Sub-Saharan Africa: A review of climate-smart agriculture among small-scale farmers. *Climate, 7*(11), 132. https://doi.org/10.3390/cli7110132

Achakulwisut, P., Erickson, P., Guivarch, C., Schaeffer, R., Brutschin, E., & Pye, S. (2023). Global fossil fuel reduction pathways under different climate mitigation strategies and ambitions. *Nature Communications, 14*(1), 5425. https://doi.org/10.1038/s41467-023-41105-z

Acobta, A. N., Ayompe, L. M., Wandum, L. M., Tambasi, E. E., Muyuka, D. S., Egoh, B. N. (2023). Greenhouse gas emissions along the value chain in palm oil producing systems: A case study of Cameroon. *Cleaner and Circular Bioeconomy, 6*, 100057. ISSN 2772-8013. https://doi.org/10.1016/j.clcb.2023.10005

Ahmad, A., & Ashraf, S. S. (2023). Sustainable food and feed sources from microalgae: Food security and the circular bioeconomy. *Algal Research, 74*, 103185. https://doi.org/10.1016/j.algal.2023.103185

Akshaya, N. B., & Jacob, S. (2020). Unification of waste management from fish and vegetable markets through anaerobic co-digestion. *Waste and Biomass Valorization, 11*(5), 1941–1951. https://doi.org/10.1007/s12649-018-0438-z

Alagoz, E., & Alghawi, Y. (2023). The future of fossil fuels: Challenges and opportunities in a low- carbon world. *Journal of Earth Sciences Knowledge and Applications, 5*(3), 381–388.

Aljamali, N. M., & Salih, N. S. (2021). Review on chemical separation of crude oil and analysis of its components. *Journal of Petroleum Engineering & Technology, 11*(2), 35–49.

Almaqtari, F. A., Elsheikh, T., Hussainey, K., & Al-Bukhrani, M. A. (2024). Country-level governance and sustainable development goals: Implications for firms' sustainability performance. *Studies in Economics and Finance, 41*(3), 684–723. https://doi.org/10.1108/SEF-05-2023-0272

Al-Wahaibi, A., Osman, A. I., Al-Muhtaseb, A. H., Alqaisi, O., Baawain, M., Fawzy, S., & Rooney, D. W. (2020). Techno-economic evaluation of biogas production from food waste via anaerobic digestion. *Scientific Reports, 10*(1), 15719. https://doi.org/10.1038/s41598-020-72897-5

Ashukem, J.-C. N. (2020). The SDGs and the bio-economy: Fostering land-grabbing in Africa. *Review of African Political Economy, 47*(164). https://doi.org/10.1080/03056244.2019.1687086

Assan, N. (2023). Socio-cultural, economic, and environmental implications for innovation in sustainable food in Africa. *Frontiers in Sustainable Food Systems, 7*, 1192422. https://doi.org/10.3389/fsufs.2023.1192422

Aymard, C., Brulhart, F., & Vieu, M. (2021). *Management international International Management Gestiòn Internacional.*

Backhouse, M., Lehmann, R., Lorenzen, K., Lühmann, M., Puder, J., Rodríguez, F., & Tittor, A. (2021). *Bioeconomy and global inequalities: Socio-ecological perspectives on biomass sourcing and production*. Springer International Publishing. https://doi.org/10.1007/978-3-030-68944-5

Bastos Lima, M. G. (2021). Corporate power in the bioeconomy transition: The policies and politics of conservative ecological modernization in Brazil. *Sustainability, 13*(12), 6952. https://doi.org/10.3390/su13126952

Bojago, E., & Abrham, Y. (2023). Small-scale irrigation (SSI) farming as a climate-smart agriculture (CSA) practice and its influence on livelihood improvement in Offa District, Southern Ethiopia. *Journal of Agriculture and Food Research, 12*, 100534. https://doi.org/10.1016/j.jafr.2023.100534

Bottausci, S., Midence, R., Serrano-Bernardo, F., & Bonoli, A. (2022). Organic waste management and circular bioeconomy: A literature review comparison between Latin America and the European Union. *Sustainability, 14*(3), 1661. https://doi.org/10.3390/su14031661

Busca, G. (2021). Production of gasolines and monocyclic aromatic hydrocarbons: From fossil raw materials to green processes. *Energies, 14*(13), 4061. https://doi.org/10.3390/en14134061

Casau, M., Dias, M. F., Matias, J. C. O., & Nunes, L. J. R. (2022). Residual biomass: A comprehensive review on the importance, uses and potential in a circular bioeconomy approach. *Resources, 11*(4), 35. https://doi.org/10.3390/resources11040035

Chandel, A. K., Garlapati, V. K., Jeevan Kumar, S. P., Hans, M., Singh, A. K., & Kumar, S. (2020). The role of renewable chemicals and biofuels in building a bioeconomy. *Biofuels, Bioproducts and Biorefining, 14*(4), 830–844. https://doi.org/10.1002/bbb.2104

Crop, A., Society, S., Mtambanengwe, F., Mapfumo, P., Chikowo, R., Chamboko, T., & Pleasant, M. (2012). Climate change and variability: Smallholder farming communities in zimbabwe portray a varied understanding over 75% of rural populations within sub-Saharan Africa rely on rainfed smallholder agriculture for subsistence and livelihoods (FAO, 2006). *African Crop Science Journal, 20*, 227–241.

Cuadrado-Osorio, P. D., Ramírez-Mejía, J. M., Mejía-Avellaneda, L. F., Mesa, L., & Bautista, E. J. (2022). Agro-industrial residues for microbial bioproducts: A key booster for bioeconomy. *Bioresource Technology Reports, 20*, 101232. https://doi.org/10.1016/j.biteb.2022.101232

Dahmen, N., Lewandowski, I., Zibek, S., & Weidtmann, A. (2019). Integrated lignocellulosic value chains in a growing bioeconomy: Status quo and perspectives. *GCB Bioenergy, 11*(1), 107–117. https://doi.org/10.1111/gcbb.12586

Dams, B., Maskell, D., Shea, A., Allen, S., Cascione, V., & Walker, P. (2023). Upscaling bio-based construction: Challenges and opportunities. *Building Research & Information, 51*(7), 764–782. https://doi.org/10.1080/09613218.2023.2204414

Davou, D. A., Gbegnide, H. A., Aïna, M. P., & Ouendo, E.-M. (2022). Evaluation of solid biomedical waste management practices in six health facilities in southern Benin. *GSC Advanced Research and Reviews, 12*(2), 144–163. https://doi.org/10.30574/gscarr.2022.12.2.0223

De Assis Costa, F., Nobre, C., Genin, C., Medeiros Rocha Frasson, C., Araújo Fernandes, D., Silva, H., Vicente, I., Takaes Santos, I., Feltran-Barbieri, R., Ventura Neto, R., & Folhes, R. (2022). *Bioeconomy for the Amazon: Concepts, limits, and trends for a proper definition of the tropical forest biome*. World Resources Institute. https://doi.org/10.46830/wriwp.21.00168

ESG. (2024). *Green Sahara Farms Unveils ESG Report, Targets Net- Zero Emission | FINANCIAL TRUST.* https://financialtrust.com.ng/green-sahara-farms-unveils-esg-report-targets-net-zero-emission/

EU. (2020). *Agroecological practices supporting food production and reducing food insecurity in developing countries: A study on scientific literature in 17 countries*. Publications Office. https://doi.org/10.2760/82475

European Commission. Joint Research Centre. (2016). *A global view of bio-based industries: Benchmarking and monitoring their economic importance and future developments: EU Brazil Sector Dialogues Workshop*, 18–19 February 2016, JRC Seville. Publications Office. https://doi.org/10.2788/153649

Fattahi, B. (2017). Sustainability: A business imperative, not a moral sacrifice. *Journal of Petroleum Technology, 69*(05), 46–48. https://doi.org/10.2118/0517-0046-jpt

Feleke, S., Cole, S. M., Sekabira, H., Djouaka, R., & Manyong, V. (2021). Circular bioeconomy research for development in Sub-Saharan Africa: Innovations, gaps, and actions. *Sustainability, 13*(4), 1926. https://doi.org/10.3390/su13041926

Fertahi, S., Elalami, D., Tayibi, S., Taarji, N., Lyamlouli, K., Bargaz, A., Oukarroum, A., Zeroual, Y., El Bouhssini, M., & Barakat, A. (2023). The current status and challenges of biomass biorefineries in Africa: A critical review and future perspectives for bioeconomy development. *Science of the Total Environment, 870*, 162001. https://doi.org/10.1016/j.scitotenv.2023.162001

Gatto, A. (2024). *Towards sustainable global food and biomass systems: Interactions between food loss and waste reductions, dietary shifts, and transitioning to a circular bio-based economy*. Wageningen University. https://doi.org/10.18174/648608

Hailiang, Z., Iqbal, W., Chau, K. Y., Raza Shah, S. A., Ahmad, W., & Hua, H. (2023). Green finance, renewable energy investment, and environmental protection: Empirical evidence from B.R.I.C.S. countries. *Economic Research-Ekonomska Istraživanja, 36*(2), 2125032. https://doi.org/10.1080/1331677X.2022.2125032

Hasan, M. K., Desiere, S., D'Haese, M., & Kumar, L. (2018). Impact of climate-smart agriculture adoption on the food security of coastal farmers in Bangladesh. *Food Security, 10*(4), 1073–1088. https://doi.org/10.1007/s12571-018-0824-1

Hatvani, N., Van Den Oever, M. J. A., Mateffy, K., & Koos, A. (2022). Bio-based business models: Specific and general learnings from recent good practice cases in different business sectors. *Bio-Based and Applied Economics, 11*(3), 185–205. https://doi.org/10.36253/bae-10820

Hetemäki, L., Tegegne, Y. T., & Ochieng, R. M. (2023). *Outlook for sustainable forest bioeconomy in Gabon, Kenya, Nigeria, South Africa and Tanzania*. Circular Bioeconomy Alliance. https://doi.org/10.62164/20241

Hlangwani, E., Mpye, K. L., Matsuro, L., & Dlamini, B. (2023). The use of technological innovation in bio-based industries to foster growth in the bioeconomy: A South African perspective. *Sustainability: Science, Practice and Policy, 19*(1), 2200300. https://doi.org/10.1080/15487733.2023.2200300

Holechek, J. L., Geli, H. M. E., Sawalhah, M. N., & Valdez, R. (2022). A global assessment: Can renewable energy replace fossil fuels by 2050? *Sustainability, 14*(8), 4792. https://doi.org/10.3390/su14084792

IITA. (2017). *2017 Annual report: Serving the African farmers and communities*, 124.

Jindo, K., Schut, A. G. T., & Langeveld, J. W. A. (2020). Sustainable intensification in Western Kenya: Who will benefit? *Agricultural Systems, 182*(September 2019), 102831. https://doi.org/10.1016/j.agsy.2020.102831

Johnson, F. X., Canales, N., Fielding, M., Gladkykh, G., Aung, M. T., Bailis, R., Ogeya, M., & Olsson, O. (2022). A comparative analysis of bioeconomy visions and pathways based on stakeholder dialogues in Colombia, Rwanda, Sweden, and Thailand. *Journal of Environmental Policy & Planning, 24*(6), 680–700. https://doi.org/10.1080/1523908X.2022.2037412

Ketene, M. B. (2019). Bioeconomy: Shaping the transition to a sustainable, biobased economy., Iris Lewandowski (Springer, 2018). *Corvinus Journal of Sociology and Social Policy, 10*, 165. https://doi.org/10.14267/CJSSP.2019.1.9

Kigo, F. (2021). Research in business & social science the moderating effect of gender on the contribution of social enterprises to social transformation in Kiambu County, Kenya. *International Journal of Research in Business and Social Science, 10*(8), 257–265.

Lange, S. L., Senekane, M. F., & Naicker, N. (2022). Understanding the socio-demographic profile of waste re-users in a Suburban setting in South Africa. *Resources, 11*(5), 45. https://doi.org/10.3390/resources11050045

Leong, H. Y., Chang, C.-K., Khoo, K. S., Chew, K. W., Chia, S. R., Lim, J. W., Chang, J.-S., & Show, P. L. (2021). Waste biorefinery towards a sustainable circular bioeconomy: A solution to global issues. *Biotechnology for Biofuels, 14*(1), 87. https://doi.org/10.1186/s13068-021-01939-5

Lewandowski, I. (Ed.). (2018). *Bioeconomy: Shaping the transition to a sustainable, biobased economy*. Springer International Publishing. https://doi.org/10.1007/978-3-319-68152-8

Liew, C. S., Yunus, N. M., Chidi, B. S., Lam, M. K., Goh, P. S., Mohamad, M., Sin, J. C., Lam, S. M., Lim, J. W., & Lam, S. S. (2022). A review on recent disposal of hazardous sewage sludge via anaerobic digestion and novel composting. *Journal of Hazardous Materials, 423*, 126995. https://doi.org/10.1016/j.jhazmat.2021.126995

Lutta, A. I., Bößner, S., Johnson, F. X., Virgin, I., Trujillo, M., & Osano, P. (2024). Transnational innovation systems for bioeconomy: Insights from cassava value chains in East Africa. *Frontiers in Sustainable Food Systems, 8*, 1205795. https://doi.org/10.3389/fsufs.2024.1205795

Marenya, P. P., & Barrett, C. B. (2007). Household-level determinants of adoption of improved natural resources management practices among smallholder farmers in western Kenya. *Food Policy, 32*(4), 515–536. https://doi.org/10.1016/j.foodpol.2006.10.002

Nagham, M., & Nuha, S. S. (2021). Review on chemical separation of crude oil and analysis of its components. *Journal of Petroleum Engineering & Technology*. https://www.researchgate.net/profile/Nagham-Aljamali/publication/355595793_Review_on_Chemical_Separation_of_Crude_Oil_and_Analysis_of_Its_Components/links/617800bea767a03c14b60c9e/Review-on-Chemical-Separation-of-Crude-Oil-and-Analysis-of-Its-Components.pdf

Naserian, E. S., Cheraghi, M., Lorestani, B., Sobhanardakani, S., & Sadr, M. K. (2021). Qualitative investigation of sewage sludge composting: Effect of aerobic/anaerobic pretreatments. *Arabian Journal of Geosciences, 14*(10), 836. https://doi.org/10.1007/s12517-021-07232-x

Negi, S., Hu, A., & Kumar, S. (2021). Circular bioeconomy: Countries' case studies. In *Biomass, biofuels, biochemicals* (pp. 721–748). Elsevier. https://doi.org/10.1016/B978-0-12-821878-5.00008-8

Nizzy, A. M., & Ogwu, M. C. (2024). Valorization of cassava processing by-products into biofuel for a sustainable environment. In M. C. Ogwu, S. C. Izah, A. A. Cunha Alves, & S. C. Babu (Eds.), *Plant biology, sustainability and climate change, sustainable cassava* (pp. 291–309). Academic Press. https://doi.org/10.1016/B978-0-443-21747-0.00010-2

Norli, I., Fazilah, A., & Mohamad Pazli, I. (2017). Agricultural biomass utilisation as a key driver for Malaysian bioeconomy. In I. S. Dabbert, I. Lewandowski, J. Weiss, & A. Pyka (Eds.), *Knowledge-driven developments in the bioeconomy* (pp. 141–159). Springer International Publishing. https://doi.org/10.1007/978-3-319-58374-7_8

Ntwali, V. M., Cishesa, T. H., Mirindi, M. N., Bafurume, D. B., Balungwe, E. B., Musale, D. K., & Bahananga, J. B. M. (2022). Evaluation of plastic waste in pasture areas and frequency of foreign bodies in the rumen after slaughter in urban areas: Case of Bukavu and its hinterland. *Journal of Entomology and Zoology Studies, 10*(5), 50–61. https://doi.org/10.22271/j.ento.2022.v10.i5a.9061

Obahiagbon, E. G., & Ogwu, M. C. (2023a). Sustainable supply chain management in the herbal medicine industry. In S. C. Izah, M. C. Ogwu, & M. Akram (Eds.), *Herbal medicine phytochemistry* (Reference series in phytochemistry). Springer. https://doi.org/10.1007/978-3-031-21973-3_64-1

Obahiagbon, E. G., & Ogwu, M. C. (2023b). Consumer perception and demand for sustainable herbal medicine products and market. In S. C. Izah, M. C. Ogwu, & M. Akram (Eds.), *Herbal medicine phytochemistry* (Reference series in phytochemistry). Springer. https://doi.org/10.1007/978-3-031-21973-3_65-1

Obahiagbon, E. G., & Ogwu, M. C. (2024a). Organic food preservatives: The shift towards natural alternatives and sustainability in the Global South's markets. In M. C. Ogwu, S. C. Izah, & N. R. Ntuli (Eds.), *Food safety and quality in the Global South* (pp. 299–329). Springer. https://doi.org/10.1007/978-981-97-2428-4_10

Obahiagbon, E. G., & Ogwu, M. C. (2024b). The Nexus of business, sustainability, and herbal medicine. In S. C. Izah, M. C. Ogwu, & M. Akram (Eds.), *Herbal medicine phytochemistry* (Reference series in phytochemistry). Springer. https://doi.org/10.1007/978-3-031-21973-3_67-1

Ogwu, M. C. (2019). Understanding the composition of food waste: An "-Omics" approach to food waste management. In A. P. Gunjal, M. S. Waghmode, N. N. Patil, & P. Bhatt (Eds.), *Global initiatives for waste reduction and cutting food loss* (pp. 212–236). IGI Global. https://doi.org/10.4018/978-1-5225-7706-5.ch011

Olowoyo, J. O., & Mugivhisa, L. L. (2019). Evidence of uptake of different pollutants in plants harvested from soil treated and fertilized with organic materials as source of soil nutrients from developing countries. *Chemical and Biological Technologies in Agriculture, 6*(1), 28. https://doi.org/10.1186/s40538-019-0165-0

Peng, X., Jiang, Y., Chen, Z., Osman, A. I., Farghali, M., Rooney, D. W., & Yap, P.-S. (2023). Recycling municipal, agricultural and industrial waste into energy, fertilizers, food and construction materials, and economic feasibility: A review. *Environmental Chemistry Letters, 21*(2), 765–801. https://doi.org/10.1007/s10311-022-01551-5

Ramdas, V. M., Mandree, P., Mgangira, M., Mukaratirwa, S., Lalloo, R., & Ramchuran, S. (2021). Review of current and future bio-based stabilisation products (enzymatic and polymeric) for road construction materials. *Transportation Geotechnics, 27*, 100458. https://doi.org/10.1016/j.trgeo.2020.100458

Rosa, S. F., & Martius, C. (2021). *Forest-based bioeconomy in sub-Saharan Africa: Looking at benefits and burdens from a social sustainability standpoint*. Center for International Forestry Research (CIFOR). https://doi.org/10.17528/cifor/007951

RUNRES. (2023). Runres approach. *RUNRES*. https://runres.ethz.ch/approach/

Runres-drc. (2023). Runres innovations introduced in DR Congo. *RUNRES*. https://runres.ethz.ch/innovations/drc/

Runres-Ethiopia. (2023). Inovations introduced by runres in Ethiopia. *RUNRES*. https://runres.ethz.ch/innovations/ethiopia/

Runres-Rwanda. (2023). Innovation introduced by runres in Rwanda. *RUNRES*. https://runres.ethz.ch/innovations/rwanda/

Runres-South- Africa. (2023). Innovation introduced by Runres in South Africa. *RUNRES*. https://runres.ethz.ch/innovations/south-africa/.

Salem, H. S., Pudza, M. Y., & Yihdego, Y. (2023). Harnessing the energy transition from total dependence on fossil to renewable energy in the Arabian Gulf region, considering population, climate change impacts, ecological and carbon footprints, and United Nations' Sustainable Development Goals. *Sustainable Earth Reviews, 6*(1), 10. https://doi.org/10.1186/s42055-023-00057-4

Sarker, S. (2020). By-products of fish-oil refinery as potential substrates for biogas production in Norway: A preliminary study. *Results in Engineering, 6*, 100137. https://doi.org/10.1016/j.rineng.2020.100137

Schipfer, F., Pfeiffer, A., & Hoefnagels, R. (2022). Strategies for the mobilization and deployment of local low-value, heterogeneous biomass resources for a circular bioeconomy. *Energies, 15*(2), 433. https://doi.org/10.3390/en15020433

Sekabira, H., Nijman, E., Späth, L., Krütli, P., Schut, M., Vanlauwe, B., Wilde, B., Kintche, K., Kantengwa, S., Feyso, A., Kigangu, B., & Six, J. (2022a). Circular bioeconomy in African food systems: What is the status quo? Insights from Rwanda, DRC, and Ethiopia. *PLoS One, 17*(10), e0276319. https://doi.org/10.1371/journal.pone.0276319

Sekabira, H., Tepa-Yotto, G. T., Djouaka, R., Clottey, V., Gaitu, C., Tamò, M., Kaweesa, Y., & Ddungu, S. P. (2022b). Determinants for deployment of climate-smart integrated pest management practices: A meta-analysis approach. *Agriculture, 12*(7), 1052. https://doi.org/10.3390/agriculture12071052

Sekabira, H., Simbeko, G., Feleke, S., Manyong, V., Späth, L., Krütli, P., Vanlauwe, B., Kintche, K., Wilde, B., & Six, J. (2023). Determinants and success of engagement in circular bioeconomy practices in African food systems. *Cleaner and Circular Bioeconomy, 6*, 100065. https://doi.org/10.1016/j.clcb.2023.100065

Sekabira, H., Feleke, S., Manyong, V., Späth, L., Krütli, P., Simbeko, G., Vanlauwe, B., & Six, J. (2024). Circular bioeconomy practices and their associations with household food security in four RUNRES African city regions. *PLOS Sustainability and Transformation, 3*(4), e0000108. https://doi.org/10.1371/journal.pstr.0000108

Seleiman, M. F., Santanen, A., & Mäkelä, P. S. A. (2020). Recycling sludge on cropland as fertilizer – Advantages and risks. *Resources, Conservation and Recycling, 155*, 104647. https://doi.org/10.1016/j.resconrec.2019.104647

Sharma, P., Gaur, V. K., Sirohi, R., Varjani, S., Hyoun Kim, S., & Wong, J. W. C. (2021). Sustainable processing of food waste for production of bio-based products for circular bioeconomy. *Bioresource Technology, 325*, 124684. https://doi.org/10.1016/j.biortech.2021.124684

Sherwood, J., Clark, J., Farmer, T., Herrero-Davila, L., & Moity, L. (2016). Recirculation: A new concept to drive innovation in sustainable product design for bio-based products. *Molecules, 22*(1), 48. https://doi.org/10.3390/molecules22010048

Simbeko, G., Nguezet, P.-M. D., Sekabira, H., Yami, M., Masirika, S. A., Bheenick, K., Bugandwa, D., Nyamuhirwa, D.-M. A., Mignouna, J., Bamba, Z., & Manyong, V. (2023). Entrepreneurial potential and agribusiness desirability among Youths in South Kivu, Democratic Republic of the Congo. *Sustainability, 15*(1), 873. https://doi.org/10.3390/su15010873

Singh, V., Phuleria, H. C., & Chandel, M. K. (2022). Unlocking the nutrient value of sewage sludge. *Water and Environment Journal, 36*(2), 321–331. https://doi.org/10.1111/wej.12739

Smit, A. T., Van Zomeren, A., Dussan, K., Riddell, L. A., Huijgen, W. J. J., Dijkstra, J. W., & Bruijnincx, P. C. A. (2022). Biomass pre-extraction as a versatile strategy to improve biorefinery feedstock flexibility, sugar yields, and lignin purity. *ACS Sustainable Chemistry & Engineering, 10*(18), 6012–6022. https://doi.org/10.1021/acssuschemeng.2c00838

Taron, A., Majumder, A., Bodach, S., & Agbefu, D. (2023). *Public-private partnerships for the circular bioeconomy in the Global South: Lessons learned.* International Water Management Institute (IWMI). https://doi.org/10.5337/2023.205

Tekman, M. B., Wekerle, C., Lorenz, C., Primpke, S., Hasemann, C., Gerdts, G., & Bergmann, M. (2020). Tying up loose ends of microplastic pollution in the Arctic: Distribution from the sea surface through the water column to deep-sea sediments at the HAUSGARTEN observatory. *Environmental Science and Technology, 54*(7), 4079–4090. https://doi.org/10.1021/acs.est.9b06981

Tesfamariam, E. H., Ogbazghi, Z. M., Annandale, J. G., & Gebrehiwot, Y. (2020). Cost–benefit analysis of municipal sludge as a low-grade nutrient source: A case study from South Africa. *Sustainability, 12*(23), 9950. https://doi.org/10.3390/su12239950

Tumwesigye, W., Aschalew, A., Wilber, W., Atwongyire, D., Mary Nagawa, G., & Ndizihiwe, D. (2019). Climate smart agriculture for improving crop production and biodiversity conservation: Opportunities and challenges in the 21st century – A narrative review. *Journal of Water Resources and Ocean Science, 8*(5), 56. https://doi.org/10.11648/j.wros.20190805.11

Vargas, C. A., & Gelcich, S. (2024). Integrated actions across multiple sustainable development goals (SDGs) can help address coastal ocean acidification. *Communications Earth & Environment, 5*(1), 319. https://doi.org/10.1038/s43247-024-01485-6

Versino, F., Ortega, F., Monroy, Y., Rivero, S., López, O. V., & García, M. A. (2023). Sustainable and bio-based food packaging: A review on past and current design innovations. *Food, 12*(5), 1057. https://doi.org/10.3390/foods12051057

Wang, M., Liu, K., Yu, J., Zhang, C.-C., Zhang, Z., & Tan, Q. (2022). Recycling spent lithium-ion batteries using a mechanochemical approach. *Circular Economy, 1*(2), 100012. https://doi.org/10.1016/j.cec.2022.100012

Wienhold, K., & Goulao, L. F. (2023). The embedded agroecology of coffee agroforestry: A contextualized review of smallholder farmers' adoption and resistance. *Sustainability, 15*(8), 6827. https://doi.org/10.3390/su15086827

Xiaoman, W., Majeed, A., Vasbieva, D. G., Yameogo, C. E. W., & Hussain, N. (2021). Natural resources abundance, economic globalization, and carbon emissions: Advancing sustainable development agenda. *Sustainable Development, 29*(5), 1037–1048. https://doi.org/10.1002/sd.2192

Zia, S., Rahman, M. U., Noor, M. H., Khan, M. K., Bibi, M., Godil, D. I., Quddoos, M. U., & Anser, M. K. (2021). Striving towards environmental sustainability: How natural resources, human capital, financial development, and economic growth interact with ecological footprint in China. *Environmental Science and Pollution Research, 28*(37), 52499–52513. https://doi.org/10.1007/s11356-021-14342-2

Circular Economy Approaches in Bioeconomy in the Global South

Matthew Chidozie Ogwu, Enoch Akwasi Kosoe, Adams Ovie Iyiola, and Waheed Olujide Fawole

Abstract

The intersection of the circular economy (CE) and bioeconomy represents a transformative approach to addressing sustainable development challenges, particularly in the Global South. This chapter highlights the synergies between CE and bioeconomy, highlighting their combined potential to foster resource efficiency, waste reduction, and sustainable growth. The bioeconomy is centered on the sustainable utilization of biological resources, while CE focuses on optimizing resource use through recycling, regeneration, and the minimization of waste. Together, these frameworks offer significant opportunities for addressing pressing environmental and socioeconomic issues, especially in regions such as Africa, Southeast Asia, and Latin America, where resource scarcity, waste mismanagement, and socioeconomic disparities persist. Despite the potential bene-

M. C. Ogwu (✉)
Goodnight Family Department of Sustainable Development, Appalachian State University, Boone, NC, USA
e-mail: ogwumc@appstate.edu

E. A. Kosoe
Department of Environment and Resource Studies, Faculty of Integrated Development Studies, Simon Diedong Dombo University of Business and Integrated Development Studies, Wa, Ghana
e-mail: ekosoe@ubids.edu.gh

A. O. Iyiola
Department of Fisheries and Aquatic Resources Management, Faculty of Renewable Natural Resources Management, College of Agriculture and Renewable Natural Resources, Ejigbo Campus, Osun State University, Ejigbo, Osun State, Nigeria
e-mail: adams.iyiola@uniosun.edu.ng

W. O. Fawole
National Biotechnology Research and Development Agency, Bioresources Development Centre, Ogbomoso, Oyo State, Nigeria

M. C. Ogwu et al. (eds.), *Sustainable Bioeconomy Development in the Global South*, https://doi.org/10.1007/978-981-96-0305-3_5

fits, the implementation of circular bioeconomy initiatives in the Global South faces numerous challenges, including economic constraints, limited technological infrastructure, and inadequate waste management systems. Additionally, social barriers and the lack of robust policy frameworks hinder the widespread adoption of CE principles. This chapter explores these barriers while presenting case studies of innovation hubs and startups leading the way in developing sustainable solutions. Technological advancements in biotechnology, waste-to-energy systems, and bioproduct innovations are also examined, demonstrating the role of innovation in driving the circular bioeconomy forward. The chapter further outlines the importance of policy and governance frameworks in promoting CE and bioeconomy integration. It emphasizes the role of governments, private sectors, and international organizations in facilitating the transition to sustainable systems through collaborative efforts, policy coherence, and investment in research and development. By integrating CE into bioeconomic strategies, the Global South can achieve environmental benefits, such as reduced emissions and enhanced biodiversity, and economic gains, including job creation and rural development. In conclusion, the chapter underscores the critical need for collaborative, multistakeholder efforts to scale up circular bioeconomy solutions in the Global South. Future research should focus on addressing existing barriers, fostering innovation, and developing comprehensive policy frameworks that promote sustainability and inclusivity.

Keywords

Circular economy · Bioeconomy · Environmental sustainability · Global South · Waste management · Technological innovations

Abbreviations

CE	Circular Economy
CBE	Circular Bioeconomy
SDGs	Sustainable Development Goals
BECCS	Bioenergy with Carbon Capture and Storage
PPP	Public–Private Partnership
IRENA	International Renewable Energy Agency

1 Introduction

In recent decades, two interrelated concepts—Circular Economy (CE) and Bioeconomy—have gained prominence as essential frameworks for sustainable development. The circular economy (CE) is a systems-based approach that focuses on minimizing waste and maximizing resource efficiency through the continuous use of materials. This contrasts with the traditional linear economic model of "take, make, dispose," which typically results in high levels of resource extraction, energy use, and waste generation (Kirchherr et al., 2017). CE emphasizes the need to design out waste, keep products and materials in use, and regenerate natural systems through strategies like reuse, remanufacturing, and recycling (Ellen MacArthur Foundation, 2013). On the other hand, bioeconomy refers to the production, utilization, and conservation of biological resources, including biomass and bio-based processes, to provide sustainable solutions in sectors such as agriculture, forestry, energy, and industry (Bugge et al., 2016). Bioeconomy prioritizes the sustainable management of natural resources, integrating biotechnology to develop bio-based products, bioenergy, and materials. The concept of bioeconomy is seen as an essential pathway for reducing the dependence on fossil fuels, decreasing greenhouse gas emissions, and mitigating climate change impacts (D'Amato et al., 2017). While these two paradigms are distinct, their integration—the circular bioeconomy—represents a powerful tool for achieving environmental, economic, and social sustainability (Ghisellini et al., 2016). The circular bioeconomy applies circular economy principles to the bioeconomy, emphasizing the efficient use of biological resources and reducing waste throughout the value chain, from resource extraction to production and end-of-life management (Carus & Dammer, 2018).

The Global South, encompassing regions such as Africa, Latin America, and parts of Asia, faces unique socioeconomic and environmental challenges, including rapid population growth, urbanization, resource scarcity, and environmental degradation (Patterson et al., 2021). Many countries in these regions rely heavily on natural resources for economic development, which has resulted in unsustainable practices such as deforestation, overfishing, and the exploitation of mineral resources. These practices have often contributed to biodiversity loss, ecosystem degradation, and climate vulnerability. The adoption of circular economy principles in the Global South is particularly relevant because it provides an opportunity to transform these challenges into sustainable growth models. By reducing waste, promoting resource efficiency, and fostering innovation in renewable energy and materials, CE can create more resilient economies that are less reliant on finite resources (Korhonen et al., 2018). The circular economy also promotes social inclusivity by creating new job opportunities in the recycling, waste management, and remanufacturing sectors, which are critical for addressing poverty and unemployment (Schroeder et al., 2019). Several countries in the Global South are beginning to recognize the importance of CE approaches. For example, South Africa has made strides in waste management policies, promoting recycling and the creation of secondary markets for materials (Meyer, 2020). Similarly, India has adopted strategies for resource recovery from waste, particularly in the energy and water sectors, highlighting the potential of CE in transforming urban and rural economies (Rawat et al., 2019).

The integration of CE principles into the bioeconomy represents a transformative shift toward sustainable development, especially in the Global South. This integration, often referred to as the circular bioeconomy, seeks to optimize the use of biological resources by closing resource loops, reducing waste, and regenerating natural ecosystems. In a circular bioeconomy, biological materials are utilized in a manner that maximizes their value over multiple cycles, from food production to bio-based materials and bioenergy, before they are returned to the environment in a regenerative way. The circular bioeconomy provides a critical framework for addressing some of the most pressing global challenges, such as food security, climate change, and resource scarcity. For example, agricultural residues that are traditionally discarded as waste can be converted into bioenergy or compost, reducing both the environmental footprint and the economic costs associated with waste management (Stegmann et al., 2020). By applying CE principles to bioeconomy sectors, such as agriculture, forestry, and fisheries, the Global South can develop more sustainable production systems that enhance resource efficiency and promote environmental conservation. Moreover, the circular bioeconomy aligns with the United Nations' Sustainable Development Goals (SDGs), particularly those related to responsible consumption and production (SDG 12), climate action (SDG 13), and life on land (SDG 15). The adoption of circular bioeconomy strategies can thus play a pivotal role in achieving sustainable development targets in the Global South, ensuring that economic growth is decoupled from environmental degradation (Basso et al., 2021).

This chapter aims to highlight circular economy approaches in the bioeconomy of the Global South, discussing opportunities and challenges associated with their implementation. The chapter will provide an overview of the circular economy and bioeconomy concepts and their interlinkages and analyze the relevance of circular economy approaches in the Global South and how they contribute to sustainable development. Challenges and barriers to adopting circular bioeconomy practices, including economic, technological, and governance-related constraints will be presented as well as policy recommendations and strategies to scale up circular bioeconomy practices.

2 The Circular Economy: Concepts and Frameworks

The circular economy (CE) is a sustainable economic model that aims to decouple economic growth from resource consumption and environmental degradation. Unlike the traditional linear economy, which follows a "take, make, dispose" approach, CE emphasizes the continuous use of resources by creating closed-loop systems that minimize waste, pollution, and resource extraction (Geissdoerfer et al., 2017). At its core, CE seeks to extend the lifecycle of materials by promoting the reuse, refurbishment, remanufacturing, and recycling of products and resources (Kirchherr et al., 2017).

CE is built upon the understanding that natural resources are finite and that current patterns of consumption and production are unsustainable. By keeping products and materials in circulation for as long as possible, the circular economy reduces the demand for raw materials and the energy needed to process them, while simultaneously addressing environmental issues such as waste and pollution (Ellen MacArthur Foundation, 2013). In this model, waste is seen as a valuable resource, and the entire value chain—from production to disposal—can be optimized to reduce environmental impacts.

2.1 Principles of Circular Economy: Reduce, Reuse, Recycle

The principles of CE are often summarized by the "3Rs" framework: Reduce, Reuse, and Recycle (Fig. 1). These principles guide the development and implementation of CE practices in various sectors, from manufacturing to agriculture and waste management (Ghisellini et al., 2016).

1. *Reduce*: The first principle aims to minimize resource consumption and waste generation by designing products that require fewer materials, energy, and natural resources. This includes reducing the demand for virgin resources and increasing the efficiency of production processes (Kirchherr et al., 2017). For example, in agriculture, reducing the use of synthetic fertilizers and pesticides can lead to more sustainable farming practices.
2. *Reuse*: The second principle involves reusing products and materials in their current form, either for their original purpose or in a new application. This extends the lifecycle of goods and reduces the demand for new products. In the case of packaging, reusable systems (e.g., glass bottles that can be returned and refilled) offer a tangible example of reuse in action (Stahel, 2016).
3. *Recycle*: The third principle focuses on recovering and recycling materials that would otherwise be discarded as waste. Recycling involves processing materials to create new products, reducing the need for virgin raw materials. Commonly

Fig. 1 Circular economy principles. (Source: WalletMax, n.d.)

recycled materials include plastics, metals, paper, and organic waste, which can be transformed into compost or bioenergy.

According to Geissdoerfer et al. (2017), these principles are interconnected and form the foundation of a circular system where resource input is minimized, and waste output is reduced, ensuring that natural ecosystems are preserved and restored.

2.2 Historical Context of Circular Economy

The concept of the circular economy is not entirely new but has evolved over decades, drawing inspiration from multiple disciplines and frameworks. Its historical roots can be traced back to the industrial ecology movement, which viewed industrial systems as part of broader ecological systems (Frosch & Gallopoulos, 1989). In the 1960s and 1970s, early environmental thinkers like Kenneth Boulding proposed the idea of a "closed-loop economy" in which waste would be minimized and resources would be reused (Boulding, 1966). Boulding famously described Earth as a "spaceship" with finite resources, emphasizing the need for a sustainable economic model that respects ecological boundaries.

During the 1990s, the concept gained momentum as environmental concerns over waste, pollution, and resource depletion became more pronounced. Concepts such as "cradle-to-cradle" design, popularized by McDonough and Braungart, advocated for products to be designed with the entire lifecycle in mind so that materials could be repurposed at the end of their use without generating waste. This thinking aligned closely with the principles of the circular economy, and their work has been influential in CE's development.

By the twenty-first century, the circular economy gained institutional support, particularly in Europe, where governments and organizations like the Ellen MacArthur Foundation began promoting it as a solution to environmental challenges (Ellen MacArthur Foundation, 2013). Today, CE has become a global movement, with countries and regions across the world adopting policies and practices that prioritize circularity in resource management.

2.3 Global and Regional Efforts to Promote Circular Economy

Countries around the world are rapidly adopting the circular economy (CE) concept as they aim to achieve long-term economic growth, resource efficiency, and environmental sustainability (Table 1).

Efforts to promote the circular economy have gained traction globally, with regional initiatives being spearheaded in Europe, Asia, and the Global South. These initiatives are often driven by both environmental concerns and the economic benefits that CE offers. Europe has been at the forefront of circular economy policy

Table 1 Success stories of some countries in relation to circular economy

Countries	Successes
China	China launched a CE strategy address limited resources and excessive energy consumption in 2015. The policy focuses on "reduce, reuse, and recycle"to achieve high resource efficiency productivity by 15% over a five-year period
Denmark	Companies design their products to have a longer lifespan and allow for reuse and recycling of material with high economic value
Carlsberg group	A company based in Denmark that developed the world's first biodegradable beer bottle
United States of America	In 2016, a US chapter of the Ellen MacArthur Foundation launched a Circular Economy 100 (CE100) program, to accelerate the realization of their goal and create new circular economy opportunities in the United States
Netherlands	The manufacturing, food and biomass industries, consumer products, plastics, and construction are the economic sectors that have transited to circular economy
Germany	The rationale is that Germany uses incineration to produce massive amounts of electricity, but in the process, it burns material that could be recovered and utilized again

development. The European Commission adopted the Circular Economy Action Plan in 2015, to transition the European Union (EU) toward a more sustainable economic model (European Commission, 2015). This plan focuses on key sectors such as plastics, food waste, critical raw materials, and construction. The European Green Deal, launched in 2019, further strengthens the EU's commitment to circular economy principles, aiming to make Europe the first climate-neutral continent by 2050 (European Commission, 2019). China has also made significant progress in promoting the circular economy. Since the early 2000s, China has adopted policies aimed at improving resource efficiency, waste management, and environmental sustainability. The Circular Economy Promotion Law, enacted in 2009, encourages the development of eco-industrial parks and the integration of recycling into the production process (Yuan et al., 2006). China's focus on CE is closely tied to its broader environmental goals, including reducing pollution and improving energy efficiency.

In the Global South, circular economy efforts are emerging, though they face unique challenges due to limited infrastructure, technological constraints, and socioeconomic factors. Countries such as South Africa and Kenya have implemented circular economy principles in waste management and agriculture (Velis et al., 2017). For example, South Africa's National Waste Management Strategy promotes recycling and the creation of green jobs through the development of recycling industries (Meyer, 2020). In India, circular economy efforts focus on reducing plastic waste and promoting resource recovery in urban areas (Rawat et al., 2019). In Latin America, countries like Brazil have adopted circular economy strategies in the agricultural sector, focusing on bioenergy production and waste valorization

(Stegmann et al., 2020). The circular economy is relevant globally, but its adoption and impact vary across regions. Developed economies, particularly in Europe, have advanced policies and infrastructure to support CE, with a focus on decoupling economic growth from resource consumption. In contrast, the Global South faces more complex challenges, including economic instability, limited access to technology, and weaker regulatory frameworks. However, CE holds immense potential in these regions, particularly for addressing issues such as poverty, unemployment, and environmental degradation (Schroeder et al., 2019). In the Global South, waste management presents a critical area where circular economy principles can be applied. Many countries in Africa, Asia, and Latin America are grappling with rapid urbanization and the consequent generation of large amounts of waste. The application of CE principles, such as recycling, resource recovery, and waste-to-energy technologies, offers opportunities for economic development while reducing the environmental impacts of waste (Velis et al., 2017). Additionally, sectors such as agriculture can benefit from circular practices by promoting sustainable land use, reducing chemical inputs, and valorizing agricultural waste (Basso et al., 2021). The circular economy also provides opportunities for innovation in the Global South. For example, in Kenya, companies are developing innovative recycling and upcycling solutions for plastic waste. In India, circular economy initiatives focus on reducing resource intensity in industries such as construction, textiles, and electronics (Rawat et al., 2019). Despite these efforts, challenges remain, including the need for better infrastructure, financing, and regulatory support.

3 Bioeconomy: Concepts and Relevance to the Global South

The bioeconomy refers to the production, utilization, and conservation of biological resources—such as plants, animals, and microorganisms—to generate food, bio-based products, and energy through sustainable processes (Bugge et al., 2016). The bioeconomy focuses on replacing fossil-based resources with biological ones, aiming to create a more sustainable and circular use of resources in various industries. As global populations increase and climate change pressures mount, the bioeconomy has emerged as a key pillar of sustainable development, offering pathways to reduce environmental degradation while promoting economic growth (Langeveld et al., 2010). At the core of the bioeconomy concept is the notion of utilizing biological resources in a manner that respects ecological limits and replenishes natural systems. This approach includes promoting resource-efficient production processes, using renewable biomass, and minimizing waste by creating circular material flows (McCormick & Kautto, 2013). A sustainable bioeconomy thus integrates economic development with the regeneration of natural ecosystems, which is particularly relevant in regions such as the Global South, where agriculture, forestry, and fisheries play significant roles in national economies.

The bioeconomy is multidimensional and spans several key sectors that are critical for both environmental sustainability and economic growth. These sectors include agriculture, forestry, fisheries, and bioenergy, all of which rely on biological resources for production and can benefit from circular bioeconomy principles.

Agriculture Agriculture forms the backbone of the bioeconomy in many countries, particularly in the Global South. It involves the cultivation of plants and animals for food, fiber, and other products. Sustainable agricultural practices, such as organic farming, agroecology, and precision farming, aim to optimize resource use while minimizing environmental impacts like deforestation, soil degradation, and water scarcity (Schader et al., 2015). The agricultural bioeconomy also promotes the valorization of agricultural waste and byproducts, turning what would typically be discarded into bio-based products or energy (FAO, 2018).

Forestry Forestry is another crucial component of the bioeconomy, especially in regions with significant forest cover, such as parts of Africa, Southeast Asia, and Latin America. Sustainable forest management aims to harvest timber and nontimber forest products without depleting biodiversity or forest ecosystems (Pülzl et al., 2014). Additionally, forestry can contribute to carbon sequestration and bioenergy production, making it a vital sector for climate change mitigation.

Fisheries and Aquaculture Fisheries and aquaculture are central to the bioeconomy in coastal regions and countries with rich marine resources. Sustainable fisheries focus on harvesting aquatic resources without overexploitation, while aquaculture provides a controlled environment for the cultivation of fish, mollusks, and algae (Pauly & Zeller, 2016). The bioeconomy encourages the use of fish waste and byproducts in the production of bio-based products, such as fertilizers and fishmeal, contributing to a circular economy in the sector.

Bioenergy The bioenergy sector utilizes biomass to produce energy in the form of biofuels, biogas, and bioelectricity. Bioenergy is essential for reducing dependence on fossil fuels and mitigating greenhouse gas emissions (Faaij, 2006). In the Global South, where access to clean energy is often limited, bioenergy offers a promising solution for rural electrification and energy security. For instance, biogas systems, which convert organic waste into methane, are becoming increasingly popular in rural areas for cooking and heating.

While the bioeconomy holds great promise for sustainable development in the Global South, it also faces significant challenges. These challenges stem from a combination of economic, environmental, and governance-related factors that limit the capacity of countries to fully harness the potential of biological resources.

Economic Development Many countries in the Global South are highly dependent on the extraction and export of raw materials, such as agricultural commodities, timber, and fish, which makes them vulnerable to global market fluctuations (Pang et al., 2020). Furthermore, the transition to a bio-based economy requires significant investments in research, infrastructure, and technology, which may be difficult to achieve in regions where financial resources are scarce. The high upfront costs associated with bio-based industries, such as bioenergy production or biotechnological innovations, can also act as barriers to entry for smallholders and rural communities (Bracco et al., 2018).

Environmental Sustainability The Global South faces critical environmental challenges, including deforestation, biodiversity loss, land degradation, and water scarcity. Unsustainable agricultural practices, such as slash-and-burn farming and overfishing, have led to the degradation of ecosystems that are vital for sustaining biological resources. Without proper management, the expansion of bioeconomy sectors could exacerbate these environmental pressures (Haberl et al., 2019). For instance, large-scale bioenergy production requires significant land and water resources, potentially displacing food crops and leading to land-use conflicts.

Technological and Institutional Gaps The transition to a sustainable bioeconomy requires advanced technologies and strong institutional frameworks, which are often lacking in the Global South (Pülzl et al., 2014). Developing bio-based industries, such as bioplastics or biofuels, rely on cutting-edge biotechnology and innovation, yet many countries lack the necessary research infrastructure and skilled workforce to support such advancements. Moreover, weak governance structures and inconsistent regulatory frameworks can hinder the development of policies that promote sustainability in the bioeconomy.

Social Equity and Inclusion Ensuring that the benefits of the bioeconomy are equitably distributed is a major challenge in the Global South. Marginalized communities, such as smallholder farmers, indigenous peoples, and women, often have limited access to resources, technology, and markets (Altieri, 2009). Without targeted policies that promote social inclusion and empower vulnerable groups, the bioeconomy could widen existing inequalities and fail to deliver inclusive growth.

Several studies have explored the opportunities and challenges of developing bioeconomy sectors in the Global South. A review of the literature reveals that the bioeconomy is seen as a key driver of sustainable development, particularly in regions with abundant biological resources but limited access to fossil fuels and industrial infrastructure (D'Amato et al., 2017). However, the literature also highlights the unique constraints faced by developing countries, such as inadequate funding, limited technical expertise, and the need for policy coherence. In Africa, the bioeconomy has been promoted as a solution to food insecurity, rural poverty, and environmental degradation. For example, the African Union's Agenda 2063 outlines the importance of sustainable agriculture and bio-based industries for achieving inclusive growth on the continent (African Union, 2015). Similarly, countries like Kenya have adopted bioeconomy strategies to improve the management of natural resources and create green jobs (Janssen & Rutz, 2011). However, challenges such as land tenure insecurity, lack of access to credit, and weak governance have limited the effectiveness of these initiatives. In Latin America, the bioeconomy has gained traction in the bioenergy sector, particularly in Brazil, where sugarcane-based ethanol production has become a key component of the country's renewable energy strategy (Goldemberg et al., 2014). Brazil's experience demonstrates the potential of the bioeconomy to drive economic growth while reducing greenhouse gas emissions. However, concerns remain about the social and environmental impacts of large-scale biofuel production, particularly in terms of land use, water consumption, and food security (Pang et al., 2020). In Southeast Asia, the bioeconomy has been integrated into national development plans to promote sustainable forestry, fisheries, and agriculture. In countries like Malaysia and Thailand, bio-based industries such as palm oil production and aquaculture have become critical components of national economies (Abolhosseini & Heshmati, 2014). However, these industries face significant challenges, including deforestation, loss of biodiversity, and conflicts over land use.

4 Interlinking the Circular Economy and Bioeconomy

The circular economy (CE) and bioeconomy represent two complementary paradigms aimed at achieving sustainability, resource efficiency, and environmental stewardship. The bioeconomy emphasizes the use of biological resources to produce food, energy, and materials, whereas the circular economy seeks to eliminate waste and promote the reuse, recycling, and regeneration of resources (Geissdoerfer et al., 2017). When integrated, these concepts form a circular bioeconomy, wherein the production and consumption of bio-based resources follow circular principles, enhancing both environmental and economic sustainability.

The alignment between the circular economy and bioeconomy lies in their shared goal of fostering sustainable resource management. The bioeconomy focuses on utilizing renewable biological resources, while the circular economy promotes closing resource loops to reduce waste and extend the life cycle of materials. Integrating CE principles into the bioeconomy enhances resource efficiency, ensuring that biological resources are used in a regenerative, rather than linear, fashion. Research highlights that by integrating CE practices—such as reducing material waste, promoting resource recovery, and reusing by-products—industries can minimize environmental impact and improve resource efficiency (D'Amato et al., 2017). For example, in agriculture, crop residues can be repurposed for bioenergy production or organic fertilizers, contributing to both circularity and sustainability (Ogwu et al., 2023a, b, 2024; Egboduku et al., 2024).

The circular bioeconomy encompasses several key concepts that drive its implementation across different sectors:

Biomass Valorization Circular bioeconomy frameworks emphasize the valorization of agricultural and industrial biomass, converting it into high-value products such as biofuels, bioplastics, and fertilizers. The valorization of biomass by-products, such as crop residues and organic waste, is critical for achieving a zero-waste system (Bugge et al., 2016; Nizzy & Ogwu, 2024).

Bioenergy and Carbon Capture Circular bioeconomy models support technologies such as bioenergy with carbon capture and storage (BECCS), which help mitigate greenhouse gas emissions. BECCS processes not only generate renewable energy but also capture carbon emissions, contributing to negative emissions and climate change mitigation.

Sustainable Food Systems By integrating circular bioeconomy practices, food systems can reduce waste, regenerate soils, and promote nutrient cycling. For instance, composting food waste and crop residues helps enhance soil fertility while reducing the need for synthetic fertilizers (Ghisellini et al., 2016).

Biorefineries Circular bioeconomy models advocate for biorefineries that convert biomass into a wide range of products, following a zero-waste approach. These facilities process every part of the biomass, creating biofuels, chemicals, and bio-based materials (Cherubini, 2010).

Circular economy principles offer solutions to several challenges inherent in the bioeconomy, such as resource inefficiency, waste generation, and environmental degradation. Incorporating CE practices into bioeconomic systems can enhance sustainability by promoting resource optimization, reducing waste, and regenerating biological inputs.

Resource Efficiency The circular economy enhances resource efficiency by optimizing the use of raw materials through recycling, repurposing, and reusing. In the context of bioeconomy, CE ensures that biological resources are utilized to their full potential, minimizing waste and maximizing productivity (Stahel, 2016).

Waste Reduction and Valorization One of the key challenges in the bioeconomy is the management of waste streams. CE frameworks address this by promoting waste valorization—converting waste into valuable products—thereby reducing the environmental impact of waste disposal (D'Amato et al., 2017).

Sustainable Production and Consumption Circular bioeconomy models support sustainable production and consumption patterns by minimizing reliance on nonrenewable inputs and encouraging the use of renewable, bio-based materials. This contributes to the long-term sustainability of bio-based industries (Lieder & Rashid, 2016).

Innovation in Bio-based Materials CE drives innovation in the development of bio-based materials, such as biodegradable plastics, which offer sustainable alternatives to conventional fossil-based products. Circular approaches ensure that these materials are reused or recycled, reducing their environmental footprint (European Commission, 2018).

In the Global South, the traditional bioeconomy often faces significant challenges related to resource depletion, environmental degradation, and socioeconomic inequalities. The traditional bioeconomy relies heavily on the extraction and use of natural resources without sufficient emphasis on resource regeneration or waste management. This can lead to deforestation, loss of biodiversity, and soil degradation, as seen in many developing regions (Borras et al., 2018; Ogwu, 2019a, b). In contrast, the circular bioeconomy offers a more sustainable approach by integrating CE principles to create closed-loop systems that regenerate resources and minimize waste. In the Global South, this approach could be particularly beneficial in addressing issues of food security, resource scarcity, and environmental resilience. For example, circular bioeconomy models can promote the use of agricultural waste for bioenergy, reducing dependence on nonrenewable energy sources while creating local employment opportunities (Sillanpää & Ncibi, 2019). Furthermore, the circular bioeconomy fosters resilience by reducing reliance on external inputs, such as chemical fertilizers and fossil fuels, and promoting the use of locally available bio-based resources. This can enhance the self-sufficiency of communities in the Global South, while also contributing to climate change adaptation and mitigation efforts (Murray et al., 2017).

5 Challenges in Implementing Circular Economy in Bioeconomy in the Global South

The implementation of a circular economy (CE) within the bioeconomy offers substantial benefits, especially in the Global South. However, various socioeconomic, technological, and policy-related challenges hinder the full adoption of CE principles in these regions. These barriers include economic and policy constraints, limited technological advancements, insufficient infrastructure for recycling and waste management, social barriers, and challenges in stakeholder engagement.

Economic and Policy Constraints In the Global South, economic limitations pose one of the most significant barriers to implementing circular economy practices. Many developing nations prioritize immediate economic growth over long-term sustainability, leading to insufficient investments in the circular bioeconomy. The cost of transitioning from a linear to a circular model can be prohibitive, particularly for small and medium enterprises (SMEs) that dominate the agricultural and industrial sectors in these regions. Moreover, the absence of strong policy frameworks to support circular bioeconomic activities further exacerbates these challenges. A lack of clear regulatory guidelines, incentives, and subsidies to promote CE practices makes it difficult for businesses and industries to adopt circular strategies (Alaerts et al., 2019). In addition, the bioeconomy in these regions often relies heavily on the extraction of natural resources, without sufficient attention to sustainability. Weak enforcement of environmental regulations and policies leads to the continued exploitation of these resources, hindering efforts to shift toward a regenerative, circular model (Levänen et al., 2018).

Limited Technological Advancements Technological limitations are another critical challenge in implementing CE within the bioeconomy in the Global South. The lack of access to advanced technologies for resource recovery, waste recycling, and sustainable production processes hampers the transition to a circular bioeconomy. This technological gap is particularly pronounced in the agricultural and industrial sectors, where outdated practices continue to dominate (Kumar & Singh, 2020). For instance, advanced biorefineries and waste-to-energy technologies, which are crucial for converting biomass and organic waste into value-added products, remain underdeveloped in many Global South countries. Additionally, technological advancements in areas such as precision agriculture and bio-based material production have been slow to reach these regions due to financial constraints, limited research capacities, and inadequate knowledge transfer from developed nations (Stahel, 2016).

Lack of Infrastructure for Recycling and Waste Management Infrastructure deficits represent a major challenge in transitioning to a circular bioeconomy in the Global South. Effective waste management and recycling systems are essential for closing resource loops and minimizing waste. However, many developing countries struggle with inadequate infrastructure for waste collection, sorting, and recycling (Velenturf et al., 2019). In several African and Southeast Asian countries, for example, only a fraction of waste is properly managed, with the majority ending up in landfills or informal dumping sites. This leads to environmental degradation and hinders efforts to recover valuable materials from waste streams. Furthermore, the lack of formal recycling industries results in lost opportunities for economic value generation from bio-based waste, such as agricultural residues and organic waste (da Silva Veras et al., 2020).

Social Barriers and Stakeholder Engagement The successful implementation of a circular bioeconomy requires active engagement from a wide range of stakeholders, including governments, businesses, local communities, and civil society. However, social barriers, such as lack of awareness, resistance to change, and limited public participation, often impede progress. In many Global South countries, there is a limited understanding of the benefits of circular bioeconomy practices, both at the individual and institutional levels (Korhonen et al., 2018). Furthermore, cultural attitudes toward waste management and recycling vary across regions, with some communities showing reluctance to adopt circular practices due to traditional practices and social norms. Without the active participation of stakeholders, it is challenging to implement circular economy principles effectively. Governments and institutions need to foster a culture of sustainability by promoting education and awareness campaigns that highlight the environmental, economic, and social benefits of circular bioeconomic activities (Yuan et al., 2006).

In Africa, countries such as Kenya and South Africa have made strides in integrating circular economy principles into their bioeconomies, particularly in waste management and agriculture. Kenya's ban on plastic bags and the promotion of sustainable agricultural practices offer a promising example of how policy can drive circular economy initiatives. However, challenges such as limited infrastructure, inadequate funding, and reliance on imported technology continue to hinder large-scale implementation (Ndiritu & Nyambane, 2022). Southeast Asia, home to major bio-based industries such as palm oil production, faces significant challenges in adopting circular bioeconomic practices. The region struggles with issues such as deforestation, unsustainable land use, and pollution from agricultural waste. However, countries like Indonesia and Malaysia have begun exploring circular bioeconomy models, focusing on converting agricultural residues into biofuels and bioplastics (Yong et al., 2016). In Latin America, Brazil and Argentina have made efforts to promote circular bioeconomy practices, particularly in agriculture and forestry. Brazil's biofuel industry, powered by sugarcane-based ethanol production, exemplifies a successful circular bioeconomic initiative. However, the region faces challenges related to deforestation, land degradation, and unequal access to technology and infrastructure.

6 Technological and Innovation Pathways

Technological advancements and innovations are key drivers in transitioning toward a circular bioeconomy. These innovations not only enhance the efficiency of resource use but also contribute to the development of sustainable products, processes, and services that address the challenges associated with waste, pollution, and resource depletion. Technology plays a pivotal role in facilitating the shift from a linear to a circular bioeconomy by enabling the development of efficient processes that minimize resource input, reduce waste, and promote recycling. Advanced technologies in materials science, digitalization, and automation are critical in optimizing resource flows across the bioeconomy sectors, including agriculture, forestry, and manufacturing. Technologies such as precision agriculture and smart farming systems, which leverage sensors and data analytics, have revolutionized agricultural practices by reducing inputs and waste, thus enhancing productivity and sustainability (De Clercq et al., 2016). Moreover, digital platforms and blockchain technology provide transparency and traceability in supply chains, ensuring sustainable sourcing of bio-based products and fostering consumer confidence in circular bioeconomic practices (Fraga-Lamas & Fernández-Caramés, 2019; Obahiagbon & Ogwu, 2023a, b, 2024a, b).

Innovations in recycling technologies are central to the circular bioeconomy, particularly in converting waste into valuable products. Advances in recycling methods, such as chemical recycling and pyrolysis, have made it possible to recover materials from complex waste streams, including plastics and agricultural residues. These methods not only reduce the amount of waste sent to landfills but also help in producing bio-based chemicals, fuels, and bioplastics (Agoh et al., 2024; Kosoe & Ogwu, 2024; Meys et al., 2020). Waste-to-energy technologies, such as anaerobic digestion and gasification, provide additional pathways for converting organic waste into bioenergy. These solutions are especially relevant in the Global South, where agricultural and municipal waste is abundant, and energy access remains a critical issue. For example, anaerobic digestion plants in India and biogas production from organic waste in Kenya have demonstrated the potential of waste-to-energy systems to address both waste management and energy needs (Kampman et al., 2021).

Biotechnology and bioengineering are at the forefront of innovation in the circular bioeconomy. The use of genetically engineered microorganisms and enzymes in industrial biotechnology has revolutionized the production of bio-based products, including biofuels, bioplastics, and biochemicals (Hatti-Kaul et al., 2020). Advances in synthetic biology have also enabled the development of engineered microbes capable of converting waste biomass into high-value products such as bio-based chemicals, pharmaceuticals, and nutraceuticals (Sillanpää & Ncibi, 2019). In agriculture, biotechnological innovations, such as the development of biofertilizers and biopesticides, are reducing the reliance on synthetic inputs and promoting sustainable farming practices. These innovations are crucial in addressing the environmental impacts of traditional agriculture, such as soil degradation and chemical pollution, while enhancing crop yields and food security (Ray et al., 2022).

The Global South has seen a surge in innovation hubs and startups that are advancing the circular bioeconomy through technological solutions. In Africa, for instance, startups like Recycle Points in Nigeria and Sanergy in Kenya have developed innovative recycling and waste management solutions that address urban waste challenges while creating jobs and supporting local economies (Asante, 2020). Recycle Points incentivizes households to sort and recycle waste by offering points that can be redeemed for household goods, while Sanergy converts organic waste from informal settlements into fertilizers and renewable energy. In Southeast Asia, innovation hubs like the Thailand Institute of Scientific and Technological Research (TISTR) have been instrumental in promoting circular bioeconomic initiatives. TISTR has facilitated the development of bio-based products and waste management solutions in sectors such as agriculture, food, and energy (Singh et al., 2020). Similarly, in Latin America, startups like Biome Makers in Argentina are using biotechnology to analyze soil microbiomes and optimize agricultural productivity, demonstrating how innovation can contribute to both environmental sustainability and economic development.

The advancement of the circular bioeconomy requires strong partnerships between governments, industry, academia, and civil society. Collaborative efforts foster innovation by enabling the sharing of knowledge, resources, and technologies across sectors and regions. Public–private partnerships (PPPs) have played a crucial role in driving innovation in the circular bioeconomy, particularly in areas such as waste management, bioenergy production, and sustainable agriculture (Velenturf & Purnell, 2017). International collaborations, such as the European Union's Horizon 2020 program, have also provided funding and support for circular bioeconomic initiatives in the Global South. Additionally, platforms such as the Circular Economy Lab in Africa and the Latin American Circular Economy Coalition facilitate knowledge exchange and best practice sharing among countries working toward circular bioeconomic goals (Geissdoerfer et al., 2017). Technology and innovation are essential to realizing the potential of a circular bioeconomy. From advances in recycling and waste-to-energy technologies to breakthroughs in biotechnology and bioengineering, these innovations offer sustainable solutions to the challenges of resource efficiency and waste management. The Global South has emerged as a hub for circular bioeconomic innovation, with startups and innovation hubs demonstrating how local solutions can contribute to global sustainability efforts. To fully realize the benefits of the circular bioeconomy, continued investment in technological development and strong partnerships will be essential.

7 Policy and Governance Frameworks

The successful implementation of a circular bioeconomy (CE) in the Global South depends on robust policy and governance frameworks. These frameworks should promote sustainable practices, address challenges specific to the region, and foster international cooperation. In the Global South, several countries have begun developing policies to promote CE and bioeconomy practices. Countries such as South Africa, Kenya, and Colombia have implemented national strategies that focus on enhancing resource efficiency, reducing waste, and transitioning to bio-based economies (De Jesus & Mendonça, 2018). For instance, South Africa's National Waste Management Strategy integrates circular economy principles to promote recycling, waste reduction, and the sustainable management of natural resources. Similarly, Kenya's Green Economy Strategy and Implementation Plan emphasizes the role of sustainable agriculture, renewable energy, and waste management in achieving long-term economic growth (Kosoe et al., 2023).

These policies focus on enhancing sustainable resource management, fostering innovation, and encouraging the development of circular value chains. However, while progress is being made, significant challenges remain, particularly in terms of policy coordination, enforcement, and scaling up these initiatives across various sectors (Akudugu & Ogwu, 2024; IRENA, 2020). International frameworks and regional cooperation play a critical role in advancing the circular bioeconomy. The United Nations' Sustainable Development Goals (SDGs), particularly SDG 12 on responsible consumption and production, provide a global framework for promoting sustainable resource use and reducing waste. Additionally, frameworks such as the African Union's Agenda 2063 and the European Union's Circular Economy Action Plan provide guidelines and action plans to support regional collaboration on circular bioeconomy initiatives (African Union, 2015). Regional cooperation initiatives, such as the African Circular Economy Alliance (ACEA) and the Latin American and Caribbean Circular Economy Coalition, aim to facilitate knowledge sharing, policy alignment, and capacity building among member states. These platforms enable countries in the Global South to share best practices, collaborate on regional projects, and attract international funding for sustainable development initiatives (Geng et al., 2019).

Despite the growing interest in CE and bioeconomy policies, there are significant barriers to their effective implementation in the Global South. One of the key challenges is the lack of financial resources and investment to support large-scale projects. Many countries in the region face economic constraints that limit their ability to allocate funds for infrastructure development, research, and innovation in the CE and bioeconomy sectors (Haas et al., 2020). Additionally, limited institutional capacity and governance issues, such as weak regulatory enforcement and corruption, hinder the effective implementation of policies. In some cases, conflicting priorities between economic development and environmental sustainability further delay the adoption of CE practices (Schroeder et al., 2019). Furthermore, the informal sector, which plays a significant role in waste management and recycling in many countries in the Global South, is often excluded from formal policy discussions and support systems, leading to fragmented implementation efforts.

Governments, the private sector, and international organizations each have distinct but complementary roles in promoting the circular bioeconomy. Governments are responsible for creating enabling policies, providing incentives, and ensuring that regulatory frameworks are in place to support CE initiatives. For example, governments can introduce tax breaks, subsidies, and grants to encourage businesses to adopt circular practices, invest in green technologies, and engage in sustainable resource management (Borrello et al., 2020). The private sector is critical in driving innovation and scaling up circular bioeconomy solutions. Companies can play a leading role in developing sustainable products, improving waste management processes, and integrating CE principles into their operations. Public–private partnerships (PPPs) offer opportunities to leverage private sector expertise and resources for the benefit of circular bioeconomy projects (UNIDO, 2018). International organizations, such as the United Nations, the World Bank, and the International Renewable Energy Agency (IRENA), provide technical assistance, funding, and capacity-building support to help countries in the Global South develop and implement CE policies. These organizations also facilitate knowledge exchange and collaboration across borders, ensuring that countries can learn from global best practices and tailor solutions to their unique contexts (IRENA, 2020).

To promote the successful implementation of circular bioeconomy practices in the Global South, the following recommendations for policymakers are proposed:

Strengthen Policy Coordination and Enforcement Governments should prioritize the development of integrated policies that align CE and bioeconomy goals with national development plans. Additionally, strengthening regulatory frameworks and improving enforcement mechanisms is crucial for ensuring compliance.

Enhance Financial Support and Investment Governments and international organizations should create innovative financing mechanisms, such as green bonds, to support CE projects. Public–private partnerships can also be leveraged to attract investment and scale up sustainable initiatives.

Build Institutional Capacity Capacity-building programs should focus on improving the technical and managerial skills of government officials, policymakers, and industry stakeholders to ensure effective policy implementation.

Engage the Informal Sector The informal sector plays a significant role in waste management and recycling in many countries in the Global South. Policymakers should incorporate informal workers into formal policy frameworks, providing them with access to resources, training, and protection under labor laws.

Foster International Collaboration Governments should actively participate in regional and international platforms, such as the African Circular Economy Alliance, to share knowledge, collaborate on projects, and access global funding opportunities.

Promote Stakeholder Engagement Engaging local communities, businesses, and civil society in the development and implementation of CE policies is essential for ensuring broad-based support and promoting behavior change toward sustainability.

8 Sustainability and Circular Bioeconomy

The circular bioeconomy (CBE) integrates principles of sustainability by promoting the regeneration of natural systems, minimizing waste, and reducing reliance on nonrenewable resources.

8.1 Environmental Benefits of Circular Bioeconomy

One of the primary environmental advantages of the circular bioeconomy is the significant reduction in greenhouse gas (GHG) emissions. By shifting from a linear "take-make-dispose" model to a circular system, CBE reduces waste generation and promotes recycling and resource efficiency. This mitigates the environmental impacts associated with waste disposal, such as methane emissions from landfills and air pollution from incineration (Geissdoerfer et al., 2017). Additionally, by replacing fossil-based materials with bio-based alternatives, the CBE fosters the use of renewable resources, contributing to climate change mitigation. Enhanced biodiversity is another critical environmental benefit. Circular bioeconomy practices promote sustainable land use and forestry management, which can help preserve habitats and restore ecosystems. Sustainable agriculture practices, such as agroforestry and organic farming, also contribute to biodiversity conservation (Morone & Imbert, 2020). Moreover, CBE initiatives play a vital role in water conservation. Many bioeconomy sectors, including agriculture and forestry, are water intensive. Circular systems that reuse water, prevent contamination, and enhance soil moisture retention through organic farming can significantly reduce water consumption and improve overall water resource management (Ronzon & M'barek, 2018).

8.2 Economic Benefits of Circular Bioeconomy

The circular bioeconomy has the potential to generate significant economic benefits. By promoting resource efficiency, CBE reduces the cost of raw materials, minimizes production waste, and encourages innovation in product design and material recovery, leading to more competitive and resilient business models (McCormick & Kautto, 2013). Job creation is another major economic benefit of the circular bioeconomy. As CBE involves diverse sectors such as agriculture, forestry, fisheries, and waste management, it has the potential to create new jobs, especially in rural areas. The focus on local resource cycles, recycling industries, and bio-based products fosters rural development by creating employment opportunities and supporting local economies (Wijkman & Skånberg, 2015). Resource efficiency in CBE also supports long-term economic sustainability. By optimizing the use of renewable resources and reducing dependence on imported fossil-based materials, countries can decrease their vulnerability to global market fluctuations and supply chain disruptions. This resilience is particularly crucial for countries in the Global South, where economic stability is often intertwined with resource security (Carus & Dammer, 2018).

8.3 Social Benefits of Circular Bioeconomy

The circular bioeconomy also has the potential to contribute to poverty reduction by creating jobs in sectors such as agriculture, forestry, and bio-based industries. This can provide livelihoods for vulnerable populations, particularly in rural communities where economic opportunities are often limited (OECD, 2018). By improving income security, CBE can play a significant role in reducing poverty levels and promoting social equity. Improved health outcomes are another key social benefit. Sustainable agricultural practices reduce the use of harmful chemicals, leading to safer food production and cleaner air and water. Furthermore, by promoting access to bio-based products, such as pharmaceuticals and healthcare supplies derived from renewable resources, CBE contributes to better public health (Poirier et al., 2017).

While the benefits of CBE are clear, there are challenges to ensuring sustainability in its initiatives. One significant challenge is the uneven distribution of resources and infrastructure needed to implement circular bioeconomy systems, particularly in the Global South. Many countries face limitations in technological advancements, skilled labor, and capital investment, which can hinder the development of bio-based industries (Ladu & Blind, 2017). Another challenge lies in balancing economic development with environmental conservation. In some cases, the expansion of bioeconomy sectors such as bioenergy or agriculture may lead to land-use changes that negatively impact ecosystems and biodiversity. This trade-off highlights the need for careful planning and regulation to ensure that CBE initiatives do not exacerbate environmental degradation (El-Chichakli et al., 2016). Stakeholder engagement and public acceptance of bio-based products also present challenges. Promoting awareness and understanding of the benefits of CBE, while addressing concerns related to product safety, pricing, and availability, is crucial for gaining widespread support (Hobbs & Goddard, 2015).

To ensure the long-term sustainability of CBE initiatives, it is essential to monitor and evaluate key sustainability indicators. These indicators may include metrics related to resource efficiency, such as material recycling rates, energy consumption, and water usage (Bocken et al., 2016). Environmental indicators, such as GHG emissions reductions and biodiversity impacts, must also be regularly assessed to determine the success of CBE strategies in mitigating environmental harm. Economic indicators, including job creation rates, income distribution, and rural development, help gauge the socioeconomic benefits of CBE initiatives. Monitoring these indicators enables policymakers and stakeholders to identify areas for improvement and ensure that CBE contributes to broader societal goals. Social sustainability indicators, such as poverty reduction, public health improvements, and stakeholder engagement, should also be measured to assess the inclusivity and equity of bioeconomy practices. Regular evaluations of these indicators help to ensure that CBE initiatives benefit all sectors of society and align with sustainable development objectives (OECD, 2018).

9 Conclusion

A circular economy offers a promising framework for sustainable development by minimizing resource extraction, reducing waste, and promoting the efficient use of materials. Its relevance extends across global regions, including the Global South, where circular practices can address socioeconomic and environmental challenges. As more countries adopt CE principles, it is crucial to tailor these approaches to the specific needs and constraints of different regions, ensuring that circular economy initiatives contribute to inclusive and sustainable development worldwide. The integration of CE principles into bioeconomy strategies presents a promising pathway for sustainable development, particularly in the Global South. This conclusion highlights key trends, future research directions, and collaborative opportunities that can drive innovation and scalability in the circular bioeconomy. As the global focus on sustainability intensifies, several trends are emerging within the circular bioeconomy. These include the increasing role of bio-based products in reducing dependence on fossil fuels, advancements in waste-to-energy technologies, and a growing emphasis on resource efficiency. Future research should explore the scalability of these innovations and their applicability to different regions, particularly in the context of developing countries. Additionally, there is a need to investigate the socioeconomic impacts of circular bioeconomy initiatives, such as job creation, rural development, and community resilience, ensuring that solutions are inclusive and equitable.

Technological advancements continue to shape the future of the circular bioeconomy. Innovations in biotechnology, bioengineering, and artificial intelligence are poised to enhance waste-to-value processes, improve resource recovery, and optimize supply chains. Technologies like precision agriculture, advanced recycling, and bio-based energy solutions hold immense potential for addressing resource scarcity and environmental degradation. Future research should focus on the development and deployment of these technologies in low-resource settings to ensure their accessibility and affordability. International collaboration will be crucial for scaling up circular bioeconomy solutions. North-South cooperation can facilitate knowledge transfer, capacity building, and financial support for implementing circular bioeconomy practices in the Global South. Likewise, South–South collaborations provide opportunities for sharing context-specific solutions, particularly in regions with similar socioeconomic and environmental challenges. Such collaborations can accelerate the adoption of circular bioeconomy models and promote sustainable development on a global scale. To scale up circular bioeconomy initiatives, it is essential to develop a clear roadmap that integrates innovation, policy frameworks, and multistakeholder partnerships. Governments must create enabling environments through supportive policies and incentives for private sector investment in bio-based industries. The establishment of innovation hubs, demonstration projects, and community-led initiatives can provide proof-of-concept examples, helping to scale circular bioeconomy solutions across regions. Educational programs and skill-building initiatives will also play a critical role in equipping future generations with the expertise needed to drive the circular bioeconomy forward.

The Global South is uniquely positioned to benefit from circular bioeconomy strategies, given its rich biodiversity, agricultural base, and growing demand for sustainable development. Integrating CE into bioeconomy strategies can help these regions transition to more sustainable production systems, reduce resource dependency, and address pressing environmental issues such as deforestation, soil degradation, and water scarcity. Governments, businesses, and civil society organizations need to collaborate on creating a circular bioeconomy that prioritizes both economic development and environmental sustainability. To realize the full potential of the circular bioeconomy, a collaborative approach involving governments, private sector actors, academia, and nongovernmental organizations is needed. Multistakeholder platforms should be established to facilitate dialogue, share best practices, and foster innovation. International organizations and financial institutions must also play a pivotal role in providing technical assistance and funding for circular bioeconomy projects in the Global South. Public–private partnerships will be crucial in driving investment and innovation.

Future research should focus on identifying region-specific challenges and solutions for implementing circular bioeconomy models in the Global South. This includes exploring the potential of local bio-based resources, developing low-cost technologies, and understanding the social and economic impacts of circular bioeconomy initiatives. Additionally, research should investigate the role of policy frameworks in incentivizing circular practices and overcoming barriers to implementation. Continued collaboration between academia, industry, and governments is essential for advancing circular bioeconomy innovations and achieving sustainable development goals.

References

Abolhosseini, S., & Heshmati, A. (2014). The development of biofuels in developing countries: Current status and policies. *Renewable and Sustainable Energy Reviews, 38*, 677–690.

African Union. (2015). *Agenda 2063: The Africa we want*. African Union.

Agoh, E. C., Ogwu, M. C., Chukwuemeka, O. S., & Ekeledo, P. I. (2024). Environmental and human health effects of cassava processing and processing waste. In M. C. Ogwu, S. C. Izah, A. A. Cunha Alves, & S. C. Babu (Eds.), *Plant biology, sustainability and climate change, sustainable cassava* (pp. 203–219). Academic. https://doi.org/10.1016/B978-0-443-21747-0.00001-1

Akudugu, M. A., & Ogwu, M. C. (2024). Sustainable development policies and interventions: A bibliometric analysis of the contributions of the academic community. *Journal of Cleaner Production*, Article # 139919. https://doi.org/10.1016/j.jclepro.2023.139919

Alaerts, L., Van Acker, K., & Rousseau, S. (2019). Impacts of policies and societal transitions on bio-based and circular economies. *Sustainability, 11*(1), 106.

Altieri, M. A. (2009). Agroecology, small farms, and food sovereignty. *Monthly Review, 61*(3), 102–113.

Asante, L. A. (2020). Urban waste management and job creation in the circular economy: The case of recycle points in Nigeria. *Waste Management, 102*, 379–388.

Basso, L., Carus, M., & Dammer, L. (2021). Circular bioeconomy: A critical analysis of its concepts and evolution. *Journal of Cleaner Production, 298*, 126810.

Bocken, N., de Pauw, I., Bakker, C., & van der Grinten, B. (2016). Product design and business model strategies for a circular economy. *Journal of Industrial and Production Engineering, 33*(5), 308–320.

Borras, S. M., et al. (2018). The bioeconomy and global land use. *Journal of Peasant Studies, 45*(5–6), 789–804.

Borrello, M., Lombardi, A., Pascucci, S., & Cembalo, L. (2020). *The seven challenges for transitioning into a bio-based circular economy in the Agri-food sector*. Recent Developments in Bioeconomy.

Boulding, K. (1966). The economics of the coming spaceship earth. In H. Jarrett (Ed.), *Environmental quality in a growing economy* (pp. 3–14). Johns Hopkins University Press.

Bracco, S., Calicioglu, O., Gomez San Juan, M., & Flammini, A. (2018). Assessing the contribution of bioeconomy to the total economy: A review of national frameworks. *Sustainability, 10*(6), 1698.

Bugge, M. M., Hansen, T., & Klitkou, A. (2016). What is the bioeconomy? A review of the literature. *Sustainability, 8*(7), 691.

Carus, M., & Dammer, L. (2018). The circular bioeconomy—Concepts, opportunities, and limitations. *Industrial Biotechnology, 14*(2), 83–91.

Cherubini, F. (2010). The biorefinery concept: Using biomass instead of oil for producing energy and chemicals. *Energy Conversion and Management, 51*(7), 1412–1421. https://doi.org/10.1016/j.enconman.2010.01.015

D'Amato, D., Droste, N., Allen, B., Kettunen, M., Lähtinen, K., Korhonen, J., et al. (2017). Green, circular, bio-economy: A comparative analysis of sustainability avenues. *Journal of Cleaner Production, 168*, 716–734.

da Silva Veras, G., Sousa, A. D., & Nascimento, A. M. (2020). Circular bioeconomy and waste management: An integrated approach. *Waste Management & Research, 38*(7), 657–668.

De Clercq, M., Vats, A., & Biel, A. (2016). Agriculture 4.0: The future of farming technology. *World Government Summit Report, 23*, 42–53.

De Jesus, A., & Mendonça, S. (2018). Lost in transition? Drivers and barriers in the eco-innovation road to the circular economy. *Ecological Economics, 145*, 75–89.

Egboduku, W. O., Egboduku, T., Golohor, O. M., Imarhiagbe, O., & Ogwu, M. C. (2024). Cassava as raw material for sustainable bioeconomy development. In M. C. Ogwu, S. C. Izah, A. A. Cunha

Alves, & S. C. Babu (Eds.), *Plant biology, sustainability and climate change, sustainable cassava* (pp. 57–73). Academic. https://doi.org/10.1016/B978-0-443-21747-0.00022-9

El-Chichakli, B., von Braun, J., Lang, C., Barben, D., & Philp, J. (2016). Five cornerstones of a global bioeconomy. *Nature, 535*(7611), 221–223.

Ellen MacArthur Foundation. (2013). *Towards the circular economy: Economic and business rationale for an accelerated transition*. Ellen MacArthur Foundation.

European Commission. (2015). *Closing the loop – An EU action plan for the circular economy*. https://eur-lex.europa.eu/legal-content/EN/TXT/?uri=CELEX%3A52015DC0614

European Commission. (2018). *A sustainable bioeconomy for Europe: Strengthening the connection between economy, society and the environment*. European Commission.

European Commission. (2019). *The European green deal*. https://eur-lex.europa.eu/legal-content/EN/TXT/?uri=COM%3A2019

Faaij, A. (2006). Modern biomass conversion technologies. *Mitigation and Adaptation Strategies for Global Change, 11*(2), 343–375.

FAO. (2018). The future of food and agriculture—Alternative pathways to 2050.. Food and Agriculture Organization of the United Nations.

Fraga-Lamas, P., & Fernández-Caramés, T. M. (2019). A review on blockchain technologies for an advanced and cyber-resilient automotive industry. *IEEE Access, 7*, 1750–1778.

Frosch, R. A., & Gallopoulos, N. E. (1989). Strategies for manufacturing. *Scientific American, 261*(3), 144–152. https://doi.org/10.1038/scientificamerican0989-144

Geissdoerfer, M., Savaget, P., Bocken, N. M. P., & Hultink, E. J. (2017). The circular economy–A new sustainability paradigm? *Journal of Cleaner Production, 143*, 757–768.

Geng, Y., Sarkis, J., & Ulgiati, S. (2019). Sustainability, circular economy, and industrial ecology: A strategic roadmap for economic transformation. *Resources, Conservation & Recycling, 146*, 1–4.

Ghisellini, P., Cialani, C., & Ulgiati, S. (2016). A review on circular economy: The expected transition to a balanced interplay of environmental and economic systems. *Journal of Cleaner Production, 114*, 11–32.

Goldemberg, J., Coelho, S. T., & Guardabassi, P. (2014). The sustainability of ethanol production from sugarcane. *Energy Policy, 39*(11), 7399–7407.

Haas, W., Krausmann, F., Wiedenhofer, D., & Heinz, M. (2020). How circular is the global economy? An assessment of material flows, waste production, and recycling in the global economy. *Journal of Industrial Ecology, 24*(5), 1091–1101.

Haberl, H., Erb, K. H., Krausmann, F., Gaube, V., Bondeau, A., Plutzar, C., & Gingrich, S. (2019). Quantifying and mapping the human appropriation of net primary production in earth's terrestrial ecosystems. *Proceedings of the National Academy of Sciences, 104*(31), 12942–12947.

Hatti-Kaul, R., Nilsson, L. J., Zhang, B., Rehnberg, N., & Lundmark, S. (2020). Designing biobased recyclable polymers for plastics. *Trends in Biotechnology, 38*(1), 50–67.

Hobbs, J. E., & Goddard, E. (2015). Consumer willingness to pay for locally produced food: A case of cattle production and product attributes. *Canadian Journal of Agricultural Economics, 63*(3), 377–402.

IRENA. (2020). *Power to the people: Innovation and technological change in the global south*. International Renewable Energy Agency.

Janssen, R., & Rutz, D. (2011). Sustainability of biofuels in developing countries: New approaches for trade and certification. *Sustainability, 3*(5), 644–663.

Kampman, B., van Grinsven, A., Dinkel, F., & Blom, M. (2021). Waste-to-energy options in Kenya: An assessment of waste technologies and their impacts. *Frontiers in Environmental Science, 9*, 770685.

Kirchherr, J., Reike, D., & Hekkert, M. (2017). Conceptualizing the circular economy: An analysis of 114 definitions. *Resources, Conservation and Recycling, 127*, 221–232.

Korhonen, J., Honkasalo, A., & Seppälä, J. (2018). Circular economy: The concept and its limitations. *Ecological Economics, 143*, 37–46.

Kosoe, E. A., & Ogwu, M. C. (2024). Sustainable cassava processing: Processes, techniques, wastes, and waste streams. In M. C. Ogwu, S. C. Izah, A. A. Cunha Alves, & S. C. Babu

(Eds.), *Plant biology, sustainability and climate change, sustainable cassava* (pp. 257–272). Academic. https://doi.org/10.1016/B978-0-443-21747-0.00011-4

Kosoe, E. A., Achana, G. T. W., & Ogwu, M. C. (2023). Regulations and policies for herbal medicine and practitioners. In S. C. Izah, M. C. Ogwu, & M. Akram (Eds.), *Herbal medicine Phytochemistry* (Reference series in phytochemistry). Springer. https://doi.org/10.1007/978-3-031-21973-3_33-1

Kumar, A., & Singh, E. (2020). Circular economy and bioeconomy for sustainable development. *Journal of Environmental Management, 267*, 110646.

Ladu, L., & Blind, K. (2017). Developing an innovation indicator for a circular bioeconomy: Mapping patents in green chemistry. *Journal of Cleaner Production, 163*, 52–62.

Langeveld, H., Sanders, J., & Meeusen, M. (Eds.). (2010). *The biobased economy: Biofuels, materials and chemicals in the post-oil era*. Earthscan.

Levänen, J., Lyytimäki, J., Lahcen, B., & Anbarasan, S. (2018). Challenges in implementing circular economy in the global south. *Journal of Cleaner Production, 198*, 1565–1577.

Lieder, M., & Rashid, A. (2016). Towards circular economy implementation: A comprehensive review in the context of the manufacturing industry. *Journal of Cleaner Production, 115*, 36–51.

McCormick, K., & Kautto, N. (2013). The bioeconomy in Europe: An overview. *Sustainability, 5*(6), 2589–2608.

Meyer, J. (2020). Waste management in South Africa: The role of the private sector and civil society. *Waste Management & Research, 38*(1), 46–56.

Meys, R., Kätelhön, A., Bachmann, M., Winter, B., Zibunas, C., Suh, S., & Bardow, A. (2020). Achieving net-zero greenhouse gas emission plastics by a circular economy. *Science, 370*(6515), 620–626.

Morone, P., & Imbert, E. (2020). Circular bioeconomy: what's in a name? *Sustainability, 12*(5), 1709.

Murray, A., Skene, K., & Haynes, K. (2017). The circular economy: An interdisciplinary exploration of the concept and application in a global context. *Journal of Business Ethics, 140*(3), 369–380.

Ndiritu, S. W., & Nyambane, G. (2022). Transitioning to a circular economy in Africa: Challenges and opportunities. *Journal of Environmental Economics and Policy, 11*(1), 54–71.

Nizzy, A. M., & Ogwu, M. C. (2024). Valorization of cassava processing by-products into biofuel for a sustainable environment. In M. C. Ogwu, S. C. Izah, A. A. Cunha Alves, & S. C. Babu (Eds.), *Plant biology, sustainability and climate change, sustainable Cassava* (pp. 291–309). Academic. https://doi.org/10.1016/B978-0-443-21747-0.00010-2

Obahiagbon, E. G., & Ogwu, M. C. (2023a). Sustainable supply chain Management in the Herbal Medicine Industry. In S. C. Izah, M. C. Ogwu, & M. Akram (Eds.), *Herbal medicine phytochemistry* (Reference series in phytochemistry). Springer. https://doi.org/10.1007/978-3-031-21973-3_64-1

Obahiagbon, E. G., & Ogwu, M. C. (2023b). Consumer perception and demand for sustainable herbal medicine products and market. In S. C. Izah, M. C. Ogwu, & M. Akram (Eds.), *Herbal medicine phytochemistry* (Reference series in phytochemistry). Springer. https://doi.org/10.1007/978-3-031-21973-3_65-1

Obahiagbon, E. G., & Ogwu, M. C. (2024a). Organic food preservatives: The shift towards natural alternatives and sustainability in the global South's markets. In M. C. Ogwu, S. C. Izah, & N. R. Ntuli (Eds.), *Food safety and quality in the global south* (pp. 299–329). Springer. https://doi.org/10.1007/978-981-97-2428-4_10

Obahiagbon, E. G., & Ogwu, M. C. (2024b). The nexus of business, sustainability, and herbal medicine. In S. C. Izah, M. C. Ogwu, & M. Akram (Eds.), *Herbal medicine phytochemistry* (Reference series in phytochemistry). Springer. https://doi.org/10.1007/978-3-031-21973-3_67-1

OECD. (2018). *Biotechnology for sustainable growth and development: A synthesis report*. Organisation for Economic Co-operation and Development.

Ogwu, M. C. (2019a). Towards sustainable development in Africa: The challenge of urbanization and climate change adaptation. In P. B. Cobbinah & M. Addaney (Eds.), *The geogra-*

phy of climate change adaptation in urban Africa (pp. 29–55). Springer Nature. https://doi.org/10.1007/978-3-030-04873-0_2

Ogwu, M. C. (2019b). Understanding the composition of food waste: An "-Omics" approach to food waste management. In A. P. Gunjal, M. S. Waghmode, N. N. Patil, & P. Bhatt (Eds.), *Global initiatives for waste reduction and cutting food loss* (pp. 212–236). IGI Global. https://doi.org/10.4018/978-1-5225-7706-5.ch011

Ogwu, M. C., Osawaru, M. E., Amodu, E., & Osamo, F. (2023a). Comparative morphology, anatomy, and chemotaxonomy of two *Cissus* Linn. species. *Brazilian Journal of Botany*. https://doi.org/10.1007/s40415-023-00881-0

Ogwu, M. C., Iyiola, A. O., & Izah, S. C. (2023b). Overview of African biological resources and environment. In S. C. Izah & M. C. Ogwu (Eds.), *Sustainable utilization and conservation of Africa's biological resources and environment* (Sustainable development and biodiversity) (Vol. 888, pp. 1–34). Springer. https://doi.org/10.1007/978-981-19-6974-4_1

Ogwu, M. C., Odozi, I. P., Ahonsi, O. C., Uleanya, K. O., & Odozi, E. B. (2024). Organic acid production from cassava. In M. C. Ogwu, S. C. Izah, A. A. Cunha Alves, & S. C. Babu (Eds.), *Plant biology, sustainability and climate change, sustainable cassava* (pp. 395–418). Academic. https://doi.org/10.1016/B978-0-443-21747-0.00009-6

Pang, Y., Mortimer, S., & Lynch, J. (2020). Biofuel and bioenergy production: Environmental impacts and implications for land use and food supply. *Environmental Research Letters, 15*(8), 084021.

Patterson, J., Schulz, K., Vervoort, J., Van Der Hel, S., Widerberg, O., Adler, C., et al. (2021). Transforming urban systems: Towards sustainable development in cities in the global south. *Sustainability, 13*(10), 5532.

Pauly, D., & Zeller, D. (2016). Catch reconstructions reveal that global marine fisheries catches are higher than reported and declining. *Nature Communications, 7*(1), 1–9.

Poirier, B., Lapointe, G., Gagné, S., & Tanguay, P. (2017). Bioeconomy of bioproducts: A sectoral and territorial approach. *Journal of Cleaner Production, 143*, 84–95.

Pülzl, H., Kleinschmit, D., & Arts, B. (2014). Bioeconomy–an emerging meta-discourse affecting forest discourses? *Scandinavian Journal of Forest Research, 29*(4), 386–393.

Rawat, M., Ramanathan, A. L., & Subramanian, V. (2019). Circular economy strategies in waste management in India. *Environmental Science & Policy, 101*, 205–214.

Ray, D. K., Ramankutty, N., Mueller, N. D., West, P. C., & Foley, J. A. (2022). Yield trends are insufficient to double global crop production by 2050. *PLoS One, 7*(6), e66428.

Ronzon, T., & M'Barek, R. (2018). Socioeconomic indicators to monitor the EU's bioeconomy in transition. *Sustainability, 10*(6), 1745.

Schader, C., Muller, A., El-Hage Scialabba, N., Hecht, J., Isensee, A., Erb, K. H., & Niggli, U. (2015). Impacts of feeding less food-competing feedstuffs to livestock on global food system sustainability. *Journal of the Royal Society Interface, 12*(113), 20150891.

Schroeder, P., Anggraeni, K., & Weber, U. (2019). The relevance of circular economy practices to the sustainable development goals. *Journal of Industrial Ecology, 23*(1), 77–95.

Sillanpää, M., & Ncibi, M. C. (2019). *The circular economy in global perspective* (pp. 1–382). Academic.

Singh, R., Bajar, S., & Wani, S. P. (2020). Advances in bioeconomy and circular economy. *Global Circular Bioeconomy, 2*, 151–160.

Stahel, W. R. (2016). The circular economy. *Nature, 531*(7595), 435–438.

Stegmann, P., Londo, M., & Junginger, M. (2020). The circular bioeconomy: Its elements and role in European bioeconomy clusters. *Resources, Conservation and Recycling, 155*, 104720. https://doi.org/10.1016/j.resconrec.2019.104720

UNIDO. (2018). *Circular economy: A new paradigm for industrial development?* United Nations Industrial Development Organization.

Velenturf, A. P., & Purnell, P. (2017). Resource recovery in the circular economy: Identifying opportunities and barriers for resource recovery from waste. *Resources, Conservation & Recycling, 117*, 20–31.

Velenturf, A. P., Archer, S. A., Gomes, H. I., Christgen, B., Lag-Brotons, A. J., & Purnell, P. (2019). Circular economy and the waste hierarchy. *Journal of Industrial Ecology, 23*(1), 96–106.

Velis, C. A., Cook, E., & Fischer, K. (2017). Circular economy and global secondary material supply chains. *Waste Management & Research, 35*(4), 427–429.

WalletMax. (n.d.). *Circular economy: Repair, reuse, recycle*. Retrieved August 15, 2024, from https://www.getwalletmax.com/post/circular-economy-repair-reuse-recycle

Wijkman, A., & Skånberg, K. (2015). The circular economy and benefits for society. *Club of Rome Report, 5*(1), 23–30.

Yong, J. Y., Klemeš, J. J., & Varbanov, P. S. (2016). Advances in sustainable development and the circular economy. *Renewable and Sustainable Energy Reviews, 61*, 1171–1186.

Yuan, Z., Bi, J., & Moriguichi, Y. (2006). The circular economy: A new development strategy in China. *Journal of Industrial Ecology, 10*(1–2), 4–8.

Bio-Based Industries and Their Products in the Global South

A. R. Olaniyan, C. A. Ogunlade, R. T. Babalola, and T. O. Akingbade

Abstract

In this chapter, we discuss the essence of bio-based industries in driving sustainable economic growth in the Global South. Resources of the traditional fossil fuel become increasingly depleted and environmental concerns rise; the promising alternative is revealed when there is a shift toward bio-based industries. These industries leverage renewable biological resources to produce a variety of goods, including biodegradable polymers, biofuels, and bio-based chemicals, which serve as ecologically friendly alternates for petrochemical products. The chapter also discusses the types of feedstocks utilized by bio-based industries, such as plant, animal, and microbial sources, and emphasizes the importance of establishing sustainable and efficient supply chains. The technological advancements in converting biomass into valuable products and energy forms, such as biodiesel and ethanol, are explored, stressing the potential economic and biological benefits. The Global South faces substantial challenges in adopting bio-based industries, even though it has a promising outlook. These challenges include technological barriers, resource management issues, and the impact of global trade dynamics. The perceptions' importance of market acceptance for bio-based products is also addressed in this chapter. This is crucial for driving demand and fostering industry development. Global South economic strength can improve because of adoption of bio-based solutions, create jobs opportunities, and reduce

A. R. Olaniyan (✉) · C. A. Ogunlade
Department of Agricultural Engineering, Osun State University, Osogbo, Nigeria
e-mail: adetutu.olaniyan@uniosun.edu.ng

R. T. Babalola
Department of Food Science and Technology, Osun State University, Osogbo, Nigeria

T. O. Akingbade
Department of Agricultural Engineering, Adeleke University, Ede, Nigeria

M. C. Ogwu et al. (eds.), *Sustainable Bioeconomy Development in the Global South*, https://doi.org/10.1007/978-981-96-0305-3_6

carbon footprint. In this chapter, there is a comprehensive analysis of the current state and future projections of bio-based industries in the Global South. It examines the visions into how these industries can enhance sustainability and a prosperous future.

Keywords

Feedstock · Bioeconomy · Bio-product · Bio-industries · Biopolymer · Biomass · Bio-products

Abbreviations

GS	Global South
LC	Life Cycle
LCA	Life Cycle Assessment
AI	Artificial Intelligence
TPS	Thermoplastic Starch
3D	Three Dimension

1 Introduction

Economic development and sustainability of the Global South can be vastly influenced by bio-based industrialization. The quest for clean energy sources, renewable resources, favorable climatic conditions, and acceptable bio-material products needed to foster a bio-circular economy necessitated the increased insurgence of bio-based industries globally. Production of goods and energy from biological resources has added significant economic benefits to the Global South countries (GBA, 2024). Over-reliance on fossil-based materials contributed to the depletion of natural resources, global warming, and pollution (Moreno et al., 2020). Globally, bio-based industries are cautiously replacing fossil-based industries to attain a sustainable economy by harnessing biological resources for the manufacturing of invaluable products. The emergence of bio-based industries can present fresh opportunities for job creation and economic growth in the Global South. Countries can address societal challenges and promote sustainable practices by implementing innovative bio-based approaches and optimizing resource efficiency (Neykov et al., 2021). In areas with a sizable rural population, the bioeconomy can also be used to advance social inclusion and rural development (Altoukhov et al., 2021). A commitment to bioeconomic development can result in the development of new value chains and the production of bio-based products that are sold in domestic and international markets. Bio-based product utilization cuts across various sectors, aiding in expanding the workforce, economic development, and enhancing the concept of circular economy (Moreno et al., 2020). This is evident in biomedical application of bio-based materials for healing wounds and tissue engineering (Seddiqi et al., 2021), biofuel for automobiles, bio-based chemicals for production of lubricants, cosmetics, construction materials such as bio-fiber, bio-based polymers for textile industries, inks from plants, enzyme propelled detergent, bio composite from agricultural waste for constructions (Agoh et al., 2024; Kosoe & Ogwu, 2024; Vinod et al., 2020) which are all embraced by the current global market. Countries in the GS are already keying into the concept of bio-based products. For instance, Bangladesh now produces 73% of the total energy used in the country from biomass (Huda et al., 2014).

However, the GS faces many difficulties in its transition to a bio-based economy. Factors militating against bio-based economies in developing and under-developed countries are highlighted in Fig. 1. These factors highlight the difficulties and complexities involved in making the shift to a bio-based economy, emphasizing the necessity of resolving concerns, encouraging customer acceptance, managing resources sustainably, and gaining political support to achieve the potential advantages of a bio-based economic model. Having addressed these, the GS can benefit greatly from bio-based industries in terms of job creation, sustainability, and economic growth. Countries in the GS can establish themselves as major players in the

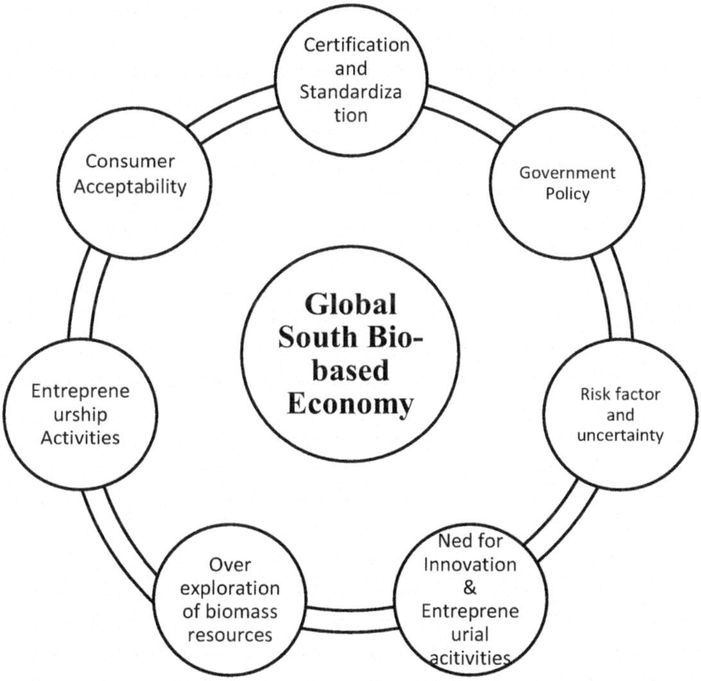

Fig. 1 Factors militating against bio-based economy in GS

global bioeconomy scene by utilizing their natural resources and adopting bioeconomic principles.

2 Feedstock for Bio-Industries

Bio-based feedstock is the bedrock for global sustainability goals by offering renewable substitutes for conventional fossil fuel-based resources. It is imperative to conduct a comprehensive life cycle assessment of biochemical production systems to guarantee that the utilization of bio-based feedstock is consistent with sustainable objectives (Ögmundarson et al., 2019). Establishing dependable and cost-effective supply chains for premium biomass feedstock is essential in meeting the challenging goals outlined for bio-based industries development. According to Moreno et al. (2020), bio-based industries utilize feedstock from natural sources such as sugar, rapeseed, linseed, animal fats, cellulosic material, plant residue, municipal waste, processing waste, for example, sludge palm oil, and wood shavings, to produce an array of bio-based products such as biodiesel, biopolymer, bio-fiber, bio-based cosmetics, and detergent. Comparing the life cycles of bio-products derived from fossil and bio-based feedstock can shed light on the environmental effects of various production routes and help guide decision-makers toward more environment-friendly choices (Dunn et al., 2015). Considering that some developing nations already contribute substantial amounts of biofuel, biomaterials, and biochemical to international markets, investing in bio-industry offers chances for ethical banking and long-term sustainable rural development (Gelder et al., 2012). The financial sustainability of bio-based product processes depends critically on the choice of appropriate biomass feedstock and the optimization of production scales (Nguyễn et al., 2015). By lowering dependency on finite fossil fuels, minimizing environmental effects, and promoting economic development—particularly in areas like the GS—leveraging bio-based feedstock provides a route toward a more sustainable future. Bio-based feedstock can significantly contribute to the advancement of the global sustainability agenda by tackling issues with feedstock availability, production efficiency, and environmental sustainability. The premise of harnessing agricultural resources to produce a bio-based product that is eco-friendly renders potential benefits in environmental pollution and bioeconomy. A wide variety of resources are necessary for the bio-based industry to function; Table 1 shows categories of bio-based feedstock used in bio-based industries.

2.1 Techno-Economic Assessment of Bio-Based Feedstock for Bio-Industries

Bio-based feedstock utilization mitigates against carbon emission and reduces energy consumption while serving its optimal purpose (Yadav & Agarwal, 2021). Bio-based feedstocks are more advantageous in terms of reduction in dependence on nonrenewable resources, biodegradability, and carbon-dioxide footprint

Table 1 Categories of bio-based feedstock for bio-based industries

Bio-based feedstock	Plant-based	Animal-based	Microbial-based
	Carbohydrates: glucose, cellulose, cellulosic fiber, alginates	Carbohydrate: chitin, hyaluronic	Carbohydrates: xanthan gum, curdlan, hyaluronic
	Lipids: corn, soy, castor, rapeseed, linseed, cotton seed, coconut, almond, algae	Lipids: animal fat	Microbial polyesters: polyhydroxyalkanoates, poly(lactic acid)
	Proteins: soy protein, wheat gluten	Proteins: collagen, elastin, keratin, silk	
	Lignin: dry land plants		

reduction (Lovett et al., 2022). Sustainability benefits over their fossil-based counterparts in thc commercialization of bio-based feedstock in a variety of industries are highly promising due to their abundance. Techno-economic assessment of bio-based feedstock gives insight into the feasibility and viability of bio-based feedstock for commercial purposes. The bio-based feedstock is assessed from three focal points.

2.1.1 Availability of Feedstock

Lignocellulosic biomass such as forest and agricultural residues is abundant and primarily made up of cellulose, hemicelluloses, and lignin (Szwarc & Głowacka, 2021). Santhanam et al. (2011) state that since lignocellulosic material makes up most plant cells, the percentage of plants with lignocellulosic content is significant and as such promises to be a reliable feedstock for commercialization in biogas plants, bioethanol production, and other bio-based industries. Lignocellulosic is thought to be the largest biomass resource accessible worldwide (Kamusoko et al., 2022; Zhao et al., 2022). Additionally, the diversity of plant sources with lignocellulosic content is further highlighted by the fact that lignocellulosic biomass from weedy plants represents an alternate feedstock for profitable bioethanol production (Djali et al., 2021). Due to its low ecological and agronomic requirements, lignocellulosic biomass has been recognized as a leading sustainable nonfood crop for lignocellulosic biofuels. As a result, there are numerous commercialization opportunities in bio-energy, biogas, bioethanol, bio-based polyols, biochemical, and other bio-based industries. These opportunities will support environmentally responsible and sustainable practices.

Cellulose a major component of lignocellulose has been identified as the most prevalent polysaccharide on Earth (Seddiqi et al., 2021). It can be sourced from biomass such as cell walls of wood and plants, some species of bacteria, and algae, as well as tunicates. The inherent abundance and diversification of sources of cellulose have created the opportunity to find new uses for this adaptable substance. Hence, the sustainability of lignocellulosic materials is undebatable (Seddiqi et al., 2021).

On the other hand, fiber, sugar, wood, and fat are natural feedstock domesticated in different regions across the global south; this is based on agricultural factors which include climate and soil conditions. There are unique agricultural raw materials in each region; Bangladesh is known for its forest resources, while countries like Thailand and Brazil are known for sugarcane production. To produce bioproducts and address issues related to competitive land use, bio-based industries utilize agricultural resources, primarily plant residues and waste that are available and accessible in most regions (Ogwu et al., 2023, 2024).

The availability of lipids as a bio-feedstock for fossil-based products is evident in various investigations into the use of different sources of lipids to produce biofuel, bio-product, and sustainable materials. Examples of sources of lipid are lipid production from lignocellulosic hydrolysate, lipid fermentation using a mix-medium of corncob acid hydrolysate and soybean oil by *Trichosporon dermatis* (Fei et al., 2016; Huang et al., 2017). Lipids can also be sourced from animal fat which is quite undesirable for consumption but can be harnessed for the production of bio-based products as well as from waste cooking oil.

Also, huge bio-feedstock can be supplied from the forest, aside the conventional forest products, chemicals can be derived from nonwood resources in the forest. Resins, Nano-crystalline cellulose, Betulin, and tannins are chemicals derived from materials such as beeswax, nuts, mushrooms, honey, and aromatic plants. Cellulose from wood is now to produce ethanol, and naphthalene as there is a slash in paper production as printing of the magazines, newspapers, and memos has dropped drastically as a result of digitization (Verkerk et al., 2022).

2.1.2 Mode Collection

Bio-based feedstock is characterized by its dispersed nature around the world. For commercial purposes, feedstock must be gathered from its various sources and regions for successful materialization in the bio-industrial world. The financial impact of the methods used to get the feedstock to the processing stage (sorting, packing, transportation, and selection) referred to as feedstock costs must be assessed. The economic effectiveness of the collection mode is highly dependent on the location of the bio-industry. Also, the feedstock preparation or pretreatment needs to be considered. In Brazil, the cost of feedstock accounts for up to 60–70% of bioethanol production from sugarcane (Macrelli et al., 2012). There are significant regional variations in feedstock varieties, costs, and availability as well as in capital and operating expenses, tax rates, soil characteristics, and composition.

2.1.3 Transformation Technology

Efficiency, technicality, and cost of producing biofuel, biochemical, biopolymers, and other bio-based materials are essential factors in assessing the viability of bio-based industries The conversion process of bio-based feedstock to eco-friendly products has paved the way for modern biomass conversion technology with low-emission rate, expansion of the scope of the agricultural sector, rehabilitation of regarded areas, watershed protection and habitats through dedicated cropping system (Feehan & Petersen, 2003). Popa (2018) identifies the use of thermochemical, biochemical, and chemical ways of processing biomass to produce chemicals and energy depending on available feedstock and the efficiency of the process. Processes such as transesterification, hydrolysis, fermentation, and hydrotreating of waste materials also result in the production of biofuel.

3 Bio-Based Product

Bio-based products are transforming industries and opening the door to a more sustainable economy. These products range from everyday necessities to cutting-edge inventions. Bio-based products have the potential to improve the sustainability of natural resources, environmental quality, and national security while competing economically. Agricultural and forest crops may serve as an alternative feedstock to fossil fuels to moderate price and supply disruptions in international petroleum markets and help diversify feedstock sources that support the nation's industrial base (Egboduku et al., 2024). Bio-based products tend to be environment-friendly because they are produced by less polluting analogous processes than in the petrochemical industry (NRCC, 2000). Growing consumer demand for eco-friendly products and rising oil prices are driving the shift toward bio-based chemicals and polymers, opening new avenues for sustainable development (Jong et al., 2012). The use of bio-based products has the potential to lessen the effects of climate change, encourage the circular economy, and cut greenhouse gas emissions. As an example, biodegradable plastics and bio-based composites offer sustainable substitutes for traditional materials, resulting in decreased environmental effects and carbon footprints (Ögmundarson et al., 2019; Islam et al., 2023). Bio-based materials offer biodegradability, lower carbon footprints, and reduced toxicity when compared to synthetic counterparts; this is demonstrated by the development of bio-based adhesives, bio-composites, and bio-based polymers (Arias et al., 2021; Caferra et al., 2022; Patel et al., 2018). The potential for bio-based products to be more environmentally sustainable than fossil fuel-based alternatives is highlighted by LCA and environmental evaluations; these findings underscore the significance of integrating economic and environmental performance in the development of renewable products (Majer et al., 2018; Meramo et al., 2022; Merighi et al., 2020). To comprehend the LC impacts of bio-based products and to promote sustainable production practices, environmental assessments of bio-based products, such as bio-based plastics and biofuels, are crucial (Broeren et al., 2017; Ghasemi et al., 2020; Isola et al., 2017).

Bio-based materials represent a significant market with a wide range of products. These products and processes aid in creating employment opportunities and new markets. A wide range of products can be produced from different biological resources.

3.1 Categories of Bio-Based Products

(a) Energy: Ethanol, biodiesel, biogas, naphthalene
(b) Chemicals: Cleaning chemicals, surfactants, soaps, detergents, agricultural chemicals, fuel additives, and feed additives. Other forms of chemical are known as specialty chemicals, such as fatty acids and lipids.
(c) Materials: These can be:
  (i) Agricultural materials: Fertilizers, herbicides, pesticides.
  (ii) Construction materials: Composite materials, niopolymers.
  (iii) Pharmaceuticals products: Biopolymers, textiles, fibers, bio-plastics.
  (iv) Other materials: Adhesives, absorbents, paints, ink, dyes, packaging materials, lubricants, paper and paper products, alternative fibers/bonded fabrics/textiles, oils, waxes, binders.

3.2 Assessment of Environmental Impact of Bio-Based Product

Growing evidence points to the bio-based economy as a means of addressing ecological and social issues, including resource depletion, environmental degradation, and climate change through sustainable practices (Bennich & Belyazid, 2017; Nakhate & Meer, 2021). Sustainability principles need to be respected and put into practice if the bio-based economy is to have the anticipated effects on the environment and society. To prove that a product satisfies required sustainability standards, along with associated certification programs, ecolabels are useful in lowering doubts about its characteristics and advantages.

Evaluations of bio-based products' life cycles and environmental impacts point to their potential for greater environmental sustainability than fossil fuel-based alternatives, highlighting the significance of combining environmental and economic performance in the development of renewable products (Majer et al., 2018; Meramo et al., 2022; Merighi et al., 2020). Environmental sustainability of bio-based products measured in terms of how it affects natural resources, ecosystems, climate change, and human health can be assessed based on the LCA.

LCA of bio-based products assesses the cumulative potential environmental impact of these products. With the ever-growing exploration of bio-feedstock for producing bio-products, sources of the feedstock mostly agricultural waste and forest resources need to be replenished. Given their renewable nature, bio-based material resources outperform mineral- and fossil-based resources in many important ways.

4 Bio-Based Industries in the Global South

In the GS, bio-based industries have the potential to have a big impact on sustainability and economic growth. The bioeconomy, with the quest to utilize biological resources to produce goods and energy, offers the GS opportunities to make the most of their natural resources and advance sustainable practices (GBA, 2024). Sustainable biomass production and conversion into a range of bio-based products, chemicals, and energy sources are essential steps in the shift to a bio-based economy (Besi & McCormick, 2015). The emergence of bio-based industries can present fresh opportunities for job creation and economic growth in the GS. Through the implementation of creative bio-based strategies and resource efficiency optimization, nations can tackle societal issues while advancing sustainable practices (Neykov et al., 2021). By embracing bioeconomic development, new value chains and bio-based products that serve both domestic and international markets can be established.

Countries with Agricultural Systems that are highly productive will benefit from the bioeconomy, as agriculture has the potential to be a significant source of energy, chemicals, and materials in the twenty-first century (Singh et al., 2003). Due to the level of development, the GS is characterized by an agrarian economy that favors biomass generation. The GS is endowed with a wide variety of renewable biological materials that can be used for a range of applications, promoting resource efficiency and sustainable development. Bio-based Industries in the GS are based on the renewable biological materials found in the region and the technological amenities available.

4.1 Factors Influencing Bio-Industries Development in the Global South

Some factors influence the fostering and sustainability of bio-based industries in the GS. These factors include:

1. Class and availability of feedstock: The feedstock's price and availability have a significant impact on what can be produced because they serve as the raw material for the creation of chemicals, fuels, and bio-based materials.
2. Market demand: although a wide range of products can be made from biomass, the market for biobased products is diverse, ranging from specialty chemicals that are needed in small quantities to bulk energy and chemicals. Identifying products with high demand is crucial in deciding the products to be produced. However, the demand for certain products—like biofuels—depends heavily on policy decisions; for other products, it depends on consumer demand for environment-friendly products; and for still other products, it depends on the need for specialized molecules from the chemical industry from a business standpoint.
3. The investment and production decisions taken on the ground. There is a wide gap between the products that can, in theory, be produced from biomass, and

what is or will be produced in reality. This, in turn, depends upon two factors: the maturity of the technology and its economic viability.

4. Technological advancement: enhanced technology has a great impact on the development of bio-based industries in the GS. Innovation, sustainability, and the growth of bio-based industries are all greatly aided by technological developments, which make it possible to create new materials, methods, and products that support an increasingly environment-friendly and sustainable industrial landscape. The production of second-generation biofuels, various bio-based chemicals, and renewable jet fuels has been made possible by technological advancements, demonstrating the influence of improved technology development on bio-based industries (Tsiropoulos et al., 2018).
5. Government policy: The regulatory framework, financial incentives, and strategic direction provided by government policies play a pivotal role in molding the growth of bio-based industries. Governments can drive economic growth and industrial development by implementing industrial policies, which in turn encourage innovation and sectoral transformation in bio-based industries (Chen & Xie, 2019; Akudugu & Ogwu, 2024). Governments can create a conducive environment for industrial growth by developing and enacting industrial policies that support the development, research, and commercialization of bio-based products.

4.2 Categories of Bio-Based Industries in the Global South

Industries in the Global South can be broadly categorized in conformity with alternative products to petroleum, petrochemical products, and new products for biomedical and other materials.

1. *Bio-Energy Industries*

   Energy is the major driver of most nation economies. Bio-energy in the form of electricity, gas, and biofuel (liquid or solid) is essential in addressing challenges relating to energy insecurity, global climatic change, and a declining economy. GS is harnessing the potential inherent in the available biomass for the production of energy. Most economies of developing countries are agrarian in nature which supports bioeconomy. Agricultural produce and forest resources, industrial and municipal waste are reliable feedstock for energy generation. GS countries (Brazil, Thailand, Bangladesh, Zimbabwe, Malawi, and Kenya) are producing bio-energy in the form of bioethanol mostly from sugarcane, biodiesel from oil-bearing seeds, palm oil sludge, waste vegetable cooking oil, production of biogas from crop residues, municipal waste, and animal dungs. Bio-energy industries can be:

   (i) Liquid fuel, that is, bioethanol and biodiesel.
   (ii) Biogas industries: Cooking gas
   (iii) Bioelectricity generating industries

Maximizing the potential of bio-energy to address energy challenges and promote global sustainable development requires cooperation, innovation, and supportive policies.

2. *Biomaterials and Bio-Chemicals*
   (a) Biomedical industry: In biomedical engineering applications, where wood-based nanotechnologies have been developed to support sustainability, wood is seen as a potential replacement for petroleum-based materials (Jiang et al., 2017). Wood's historical significance as a primary material for a variety of applications highlights its potential to support sustainable growth in the biomedical engineering industry. The use of wood-derived biopolymers in 3D printing processes for biomedical applications has gained traction due to their biodegradability, biocompatibility, and noncytotoxicity (Xu et al., 2018). The utilization of wood-based biopolymers in tissue engineering and customized drug delivery systems demonstrates the industry's exploration of novel biomaterials for a variety of biomedical applications. Further evidence of the potential of wood-based structures containing responsive polymers for biomedical engineering applications comes from the integration of wood-based materials into functional materials for biomedicine and water management (Berglund & Burgert, 2018). Some subsets of the biomedical industry are:
      (i) Biomedical engineering: This is a fast-moving field that uses a variety of materials and technologies to solve different medical problems. With their use in tissue engineering, therapy, diagnosis, and other biomedical applications, magnetic nanoparticles have drawn a lot of attention and shown promise for transforming healthcare procedures (Reddy et al., 2012).

      The versatility of hydrogels, particularly self-healing hydrogels, has been demonstrated by their extensive applications in biomedicine, soft electronics, sensors, and actuators (Zhang & Khademhosseini, 2017). The potential of photo-cross linkable hydrogels to revolutionize medical treatments is evident in the extensive research being conducted on their applications in tissue engineering, regenerative medicine, drug delivery, cancer therapies, and biosensing (Choi et al., 2019). The growing field of biomedicine, sensing, advanced energy technology, and environmental applications is highlighting the wide range of applications for nanoparticle hydrogel composites, which include carbon dot-based hydrogels (Wang et al., 2023).

      According to Wang (2024), there is a growing interest in investigating new materials for medical applications. One such material is polyoxometalate-based metal-organic frameworks, which are emerging as novel materials with significant potential in biomedical research. Innovative developments in robotics, biomedicine, and industry are being made possible by polyurethane-based multifunctional materials, underscoring the revolutionary potential of material science across a range of industries (Sikdar et al., 2022). Additionally, catalase's industrial uses are progressing in the fields of biosensor development, bioremediation, food processing, and biomedicine, demonstrating the wide range of uses for enzymatic materials (Zhang et al., 2019).

(ii) Tissue engineering: The goal of developing polymeric systems for tissue engineering and biomedical applications is to produce materials that are adaptable to different medical requirements. These materials may include smart polymeric materials for sophisticated biomedical applications (Ravichandran et al., 2012). The design and production of biological tissue and organ replacements are being revolutionized by computer-aided tissue engineering, opening up new avenues for tissue transplantation and regenerative medicine (Sun et al., 2004). Reproductive medicine is being developed at the forefront by biomedical engineering, which integrates biosensors, biomaterials, and AI to address biological and medical challenges (Zhu et al., 2022).

(b) The pharmaceutical industry has grown to be one of the most significant and lucrative industries in the world thanks to its reliance on advanced technology, highlighting the crucial role that technology plays in advancing biomedical research (Sulik, 2011). Pharmaceutical sciences heavily rely on lipid nanocapsules with surface-active characteristics, which emphasize the importance of interfacial phenomena in different formulations (Mouzouvi et al., 2017). According to Di et al. (2013), carbon nanotubes, a commonly used nanomaterial, exhibit dual proinflammatory and antifibrotic effects that highlight their potential in biomedical applications. Further, the development of crystalline micro- and nanomaterials is offering important new perspectives for the pharmaceutical industry's production of commercially valuable nano-cocrystals (Tan et al., 2021). This industry makes use of biopolymers to create biodegradable materials for use in medicine. Pectin, chitosan, starch, and cellulose are examples of naturally occurring biopolymers.

Cellulose: A case of availability leads to invention. It has proven its versatility as a feedstock for useful biomedical materials (Sediqq et al., 2020). Its capacity for chemical modification has proven advantageous in terms of application for a variety of products. Because of their low cytotoxicity, biocompatibility, and biodegradability, cellulosic materials in particular hold out a lot of promise as reasonably priced materials for biomedical applications in the future. Biomedical industries use cellulose and its derivatives in wound dressing, drug delivery, and cell immobilization.

Starch: Due to their renewable source and suitability for use in composites, starch-based biopolymers, such as TPS, have shown promise in packaging applications (Temesgen et al., 2021). These biopolymers provide a sustainable way to lessen reliance on traditional plastics because they can be tailored to match specific needs for packaging materials.

Pectin: Natural plant cell wall polysaccharide pectin has drawn interest as a flexible biopolymer with a range of uses in the food packaging, pharmaceutical, and biomedical sectors. The pectin-based biopolymers in food packaging applications are due to their unique properties, such as high porosity, specific surface area, and thermal conductivity (Mellinas et al., 2020). Pectin has also been investigated for several uses outside of food packaging, such as in the pharmaceutical and biomedical sectors.

Chitin: This is a naturally occurring biopolymer present in the exoskeletons of fungi, insects, and crustaceans, and has drawn a lot of interest due to its numerous and varied uses. After processing, chitin can be converted into chitosan, a derivative with special qualities that can be used in a variety of ways in agriculture, medicine, and the food business (Kumirska et al., 2010). Chitin can be processed to create chitosan, a derivative with unique properties that have numerous applications in the food industry, medical field, and agriculture (Kumirska et al., 2010).

(c) Bioagriculture: It is a field that combines agriculture and biology and is experiencing notable progress driven by innovative research and technologies. Using cutting-edge cell culture techniques, livestock intestinal organoids have been successfully cultured in a variety of farm animals, including pigs, cattle, rabbits, horses, sheep, and chickens (Beaumont et al., 2021). This shows the potential to improve animal health and productivity. Bioagriculture has been identified by China's bioeconomy development plan as a critical area for growth, emphasizing the role that sustainable agricultural methods play in guaranteeing food security (Chen, 2024).

Sustainable farming practices are looking into encapsulated organic-mineral fertilizers with nanoporous structures, a sign of the trend toward ecologically friendly agricultural innovations (Yanovska et al., 2021). Resources available for this industry are:

- Forest resources: Wood-based resources have grown to be significant assets in the bioagricultural industry due to sustainable solutions proffer for a range of applications. For instance, wood-derived nanocellulose has shown promise in boosting plant growth, soil quality, and water retention in agricultural practices. Using bioagricultural techniques and technologies has helped developing nations create sustainable food production systems and increase agricultural productivity. To improve nutritional value and promote health benefits through novel food products, the bioagricultural sector has investigated biotechnological applications in functional foods (Reynolds & Martirosyan, 2016). Useful resources for bioagricultural industries are.

(d) Bio-manufacturing: Bioenvironmental protection is an essential component of sustainable developments, and biomanufacturing is a key player in the advancement of environment-friendly practices. To produce necessary chemical materials with less of an impact on the environment, biomanufacturing provides a route that helps create a more sustainable economy, Lee (2024). Promoting environmental sustainability requires a shift to green production and reducing the ecological impact of industrial processes (Ma, 2023). Enzyme technologies, biocatalysis, and biodegradation approaches have become increasingly popular because of the search for environment-friendly and sustainable biomanufacturing technologies. This highlights the significance of reducing environmental impact in industrial processes.

(e) Bio-construction industries: due to their sustainability and potential advantages for the environment, bio-based building materials have attracted a lot of attention. These goods produce materials like hempcrete, bio-based bricks, and bio-composite racing sailboats by using renewable resources like agro-wastes, bio-based plastics, and bio-cementation (Maraveas, 2020). In addition to lowering greenhouse gas emissions, the use of bio-based building materials promotes circular economy principles by making use of biomass from agricultural waste and bio-pozzolans (Hoxha et al., 2020; Duque-Acevedo et al., 2022; Aluga & Kambole, 2022).

Furthermore, studies have investigated the use of bio-based building materials for slope stabilization, erosion control, and asphalt mixtures, such as bio-grout and bio-ash (Omar, 2019; Šimun, 2023).

(f) Biopackaging industries: In response to the growing need for environment-friendly packaging options, the biopackaging industries are undergoing a noticeable shift toward the use of bio-based materials. The need to reduce waste generation and consumer preferences for eco-friendly products are the main driving forces behind the adoption of biopackaging materials (Mustaqim et al., 2021). In line with sustainability goals, these materials—which include biodegradable and biocompatible packaging—offer substitutes for conventional nonbiodegradable polymer packaging (Primožič et al., 2021). These substances, encompassing biodegradable and bio-based packaging are used in several industries, but it is most found in the food packaging industry, where sustainable and biodegradable packaging materials are made using edible films, coatings, and other processes derived from biological resources (Yuvaraj et al., 2021).

Biodegradable polymers are essential for packaging applications because they provide green alternatives for sectors such as food, agriculture, and medicine (Vroman & Tighzert, 2009). The main goals of creating cutting-edge sustainable packaging solutions within the context of a circular economy are fighting food waste, ensuring food safety, and protecting the environment (Guillard et al., 2018). Additionally, active and intelligent features are being integrated into biopackaging materials to improve performance and functionality. Antimicrobial and antioxidant properties, for instance, are being studied in bio-based packaging materials containing bacterial nanocellulose or silver nanoparticles (Kraśniewska et al., 2020; Ludwicka et al., 2020). Furthermore, natural biopolymer-based nanocomposite films are being developed to replace conventional plastic packaging materials with materials that have high barrier qualities and biodegradability (Venkateshaiah et al., 2022).

In the biopackaging industry, research and development efforts are focused on producing high-performance packaging materials that satisfy consumer needs for practicality, safety, and sustainability. The industry's commitment to addressing environmental issues related to traditional packaging materials is demonstrated using compostable and biodegradable alternatives to conventional plastics as well as the incorporation of extracts and nanomaterials in biodegradable composite-based food packaging (Alim et al., 2022; Song et al., 2009).

(g) Bio-surfactant Industries: The diverse properties and potential applications of bio-surfactants, which are amphiphilic compounds produced by microorganisms, have garnered attention (Cok et al., 2013). These molecules are promising for therapeutic applications, drug delivery, nanotechnology, and emulsions because of their distinctive structural features and adaptability (Gudiña et al., 2013). The bio-surfactant sector includes a variety of surfactant types with different compositions and origins, such as microbial bio-surfactants, synthetic surfactants, and bio-based surfactants (Moldes et al., 2021). When it comes to improving the bioavailability and biodegradation of contaminants like polycyclic aromatic hydrocarbons in soil remediation procedures, the use of bio-surfactants has demonstrated considerable promise in environmental applications (Volkering et al., 1995). Furthermore, because of their superior physicochemical properties, biodegradability, and ease of manufacturing, bio-surfactants derived from natural fatty acids have emerged as sustainable alternatives to synthetic surfactants (Bhadani et al., 2017).

According to Begum (2023), these bio-surfactants are generated when hydrocarbonoclastic microorganisms break down waste carbohydrate feedstocks a process known to support sustainable development practices. Utilizing substrates like neem oil to produce surfactants at different concentrations for better oil extraction processes, microbially derived bio-surfactants have also been investigated for applications in enhanced oil recovery (Ogunkunle et al., 2020).

Utilizing microbial cell factories, like Bacillus subtilis, to produce bio-surfactants has shown promise in producing surface-active agents from inexpensive, low-quality substrates (Koirala et al., 2021). Furthermore, a plethora of research has been conducted on the biodegradability and environmental safety of bio-surfactants, underscoring their potential as environmentally beneficial substitutes for conventional surfactants (Cowan-Ellsberry et al., 2014).

(h) Adhesives and Sealant Industries: Bioadhesives and sealants are indispensable in many fields, from tissue engineering to wound closure. These materials are biocompatible and biodegradable, and they are designed to stick to biological tissues well. The goal of research on bio-adhesives and sealants is to develop effective and safe techniques for tissue bonding. A tactic in the development of bio-adhesives is the application of tissue adhesives inspired by mussels, based on citrate, which are well-known for their high wet strength and appropriateness for sutureless wound closure (Mehdizadeh et al., 2012). Significant wet adhesion strength, regulated degradation, enhanced biocompatibility, and lower manufacturing costs are all provided by these materials. Another type of bio-adhesive that is frequently used in surgical procedures is fibrin sealant; this is because it is biocompatible and biodegradable (Lien et al., 2020). The development of elastic sealants for surgical applications highlights the significance of adhesion strength and physical properties in clinical settings by sealing wounds without obstructing tissue movement and function (Annabi et al., 2015). Furthermore, Assmann et al. (2017) have developed a naturally derived sealant that is based on gelatin methacryloyl and is highly adhesive, biocompatible, and biodegradable. This sealant offers a financially viable option for tissue bonding. Tissue sealants and cell delivery matrices are two uses for silk fibroin-based adhesives that have shown promise (Burke et al., 2015). These adhesives are modeled after the adhesive proteins found in mussels. These biodegradable and biocompatible materials provide flexible answers to tissue bonding needs. Furthermore, the mechanisms and applications of bioadhesive tissue-engineered scaffolds have been studied, highlighting the importance of promoting cellular activity for tissue regeneration (Chen et al., 2021).

The development of hydrogel-based materials with tunable properties for particular applications like wound healing and surgical procedures is the result of advancements in bioadhesives and sealants (Yu et al., 2013). These materials demonstrate their potential for clinical use by offering improved adhesion, flexibility, and hemostatic efficacy. Studies on the adhesion characteristics and biodegradation rates of biodegradable polymer blends have demonstrated the potential for fully and safely healing wounds (Milenkovic et al., 2018).

(i) Bio-plastic and bio-fiber Industries: The bioplastics and bio-fiber sectors are proactively involved in sustainable material development to mitigate the ecological footprint of conventional plastics and fibers. These sectors concentrate on using natural fibers and biodegradable polymers to produce environment-friendly substitutes for a range of uses.

The development of biodegradable polymers, such as poly(ε-caprolactone) (PCL) and poly(lactic acid) (PLA), for tissue engineering and packaging, has been facilitated by research on bio-plastics (Simões et al., 2008; Zhu, 2024). These polymers are biodegradable and flexible during processing, which makes them appropriate for a variety of uses. Chitosan and other natural polymers are also added to bio-plastics to improve their biodegradability and biocompatibility, which helps create sustainable material solutions (Zhu, 2024).

The use of natural fibers, such as ramie, in thermoplastic biodegradable composites has drawn interest in the bio-fiber industry for applications in a variety of fields (Lavania et al., 2023).

These composites are appropriate for ecologically friendly products because of their enhanced mechanical qualities and biodegradability. Furthermore, the creation of biodegradable polymer blends with adjustable qualities, like self-adhesiveness and viscoelasticity, provides flexible answers for various uses (Lin et al., 2022).

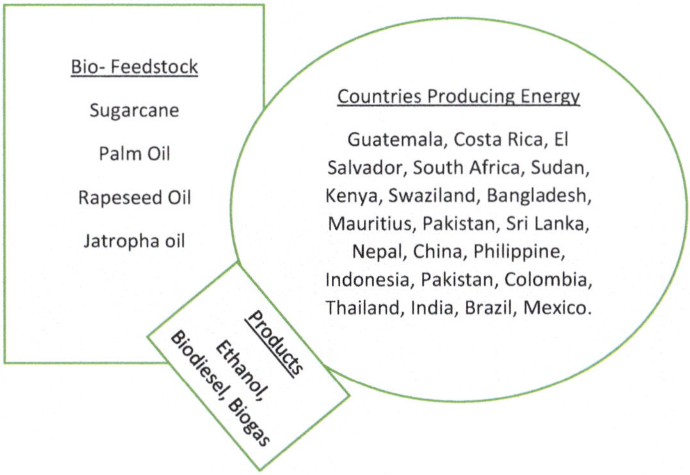

Fig. 2 Bio-energy producing countries and products

In the medical field, the incorporation of bioplastics and bio-fibers into tissue adhesives and wound dressings has demonstrated potential. Adhesive and biodegradable wound dressings with antimicrobial qualities have been made using biodegradable polymers such as poly(ethylene glycol) (PEG) and poly(lactic-co-glycolic acid) (PLGA) (Daristotle et al., 2019). These materials not only facilitate healing but also help to prevent infections and effectively close wounds. Furthermore, advancements in packaging materials are being fueled by the sustainable qualities of bio-fibers and bioplastics. To lessen plastic waste and environmental pollution, edible packaging solutions made of biodegradable polymers are being investigated (Nair et al., 2023). The potential for sustainable packaging solutions is demonstrated by the development of biodegradable copolyesters with tunable properties like viscoelasticity and self-adhesiveness (Lin et al., 2022). Figure 1 displays countries producing bio-product in the Global South (Figs. 2, 3 and 4).

4.3 Sustainability and Environmental Impact of Bio-Based Industries

Replacing fossil-based products with bio-based products has posed to have its shortfalls. The transition to bioeconomy of the Global South can be faced with challenges of sustainability and the consequential impact on the environment. The agricultural feedstock raises many issues whose severity can vary greatly between different crops as well as different geographical regions. These potential issues include competition of land allocation for plant-based feedstock generation against other purposes, use of water for irrigation, environmental impact of pesticides and fertilizer use, level of penetration of genetically modified seeds and natural habitat.

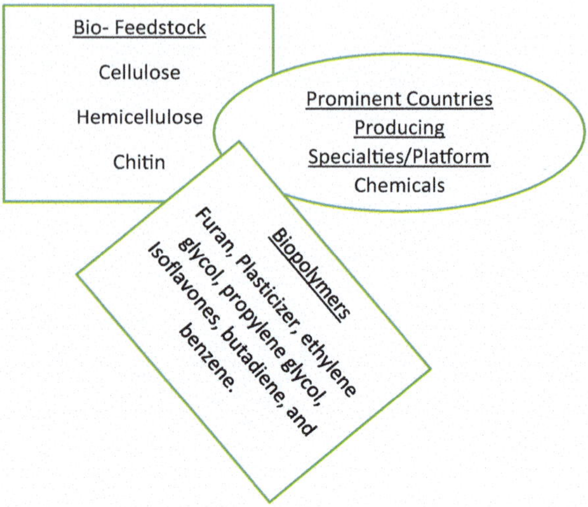

Fig. 3 Bio-chemical producing companies and products

Fig. 4 Biomaterial producing companies and their products

Limitation of ingredients from bio-based feedstock in the production of some useful polymers, energy consuming process for product derivation, variation in product properties from the desirable result. To support a more sustainable and environment-friendly bioeconomy, research on the sustainability and environmental effects of bio-based industries is essential for informing policy choices, technical developments, and industry practices. Figure 5 highlights the factors responsible for sustainable and eco-friendly industries.

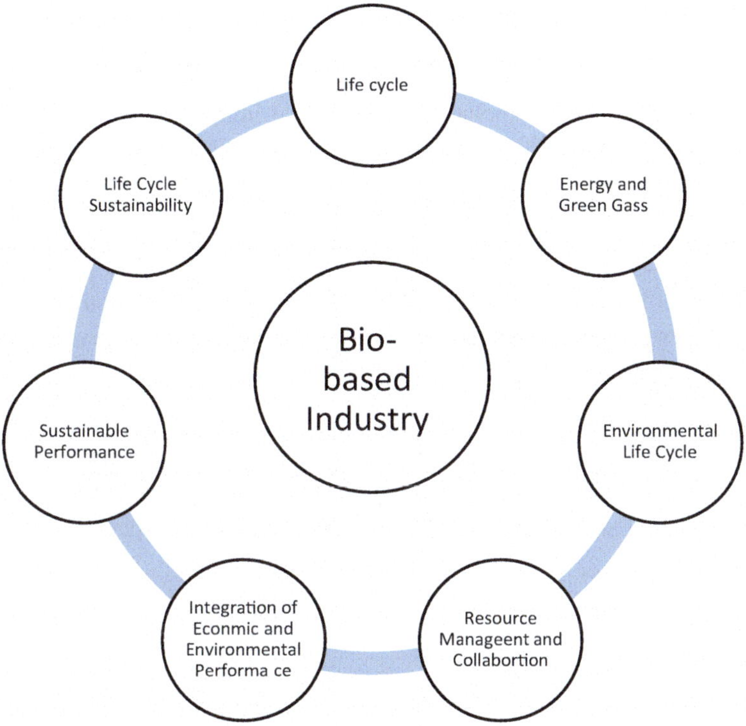

Fig. 5 Factors for assessment of bio-industry

5 Conclusion

The transition to a more sustainable future relies on the increasing competitiveness of bio-based products, driven by advancements in genetic engineering, industrial biotechnology, and biorefinery systems. The Global South can benefit significantly from its natural resources, abundant population potentially serving as a valuable human resource, and rich reserves of forest resources and agricultural waste, which form the foundation of the feedstock essential for global bio-industrialization.

The development of bio-based industries in the Global South is important to attain sustainable economic evolution and ecological protection. To end the fossil fuel dependency, these industries provide a feasible alternative that can utilize renewable biological resources to create eco-friendly products such as biodegradable polymers, biofuels, and bio-based chemicals. Such evolution not only addresses ecological concerns, it also paves ways for new economic opportunities and pathways for growth.

This chapter points out the position of various feedstocks and advanced technologies in the production processes, highlighting the need for sustainable practices and effective supply chain management (Obahiagbon & Ogwu, 2023a, b, 2024a, b). To reduce the greenhouse gas emissions and support for a circular economy, the potential of bio-based industries is promising in terms of ecological and economic benefits.

Nevertheless, there are challenges that bio-based industry sector, in the Global South, is facing. This often includes technological limitations, resources management issues, and competitive global trade dynamics, which must be tackled carefully. Then stride of these industries depends significantly on consumer acceptance and development of markets for bio-based products.

Despite these challenges, there is unique opportunity for the GS to harness the natural resources and favorable climate to become a front-runner in the bioeconomy. Countries in the GS can sustainable development, create employment opportunities, and foster economic resilience through strategic investment in research, innovation, and supportive policy frameworks in the region.

The GS has transformative potential with bio-based industries. With focus on sustainable resources utilization, technological innovation, and market growth, these industries can contribute immensely to a greener and more sustainable future. Inclusion of bio-based solutions into the economic fabric of the GS can address global environmental challenges and promoting long-term prosperity.

References

Agoh, E. C., Ogwu, M. C., Chukwuemeka, O. S., & Ekeledo, P. I. (2024). Environmental and human health effects of cassava processing and processing waste. In M. C. Ogwu, S. C. Izah, A. A. Cunha Alves, & S. C. Babu (Eds.), *Plant biology, sustainability and climate change, sustainable cassava* (pp. 203–219). Academic. https://doi.org/10.1016/B978-0-443-21747-0.00001-1

Akudugu, M. A., & Ogwu, M. C. (2024). Sustainable development policies and interventions: A bibliometric analysis of the contributions of the academic community. *Journal of Cleaner Production, 434*, Article # 139919. https://doi.org/10.1016/j.jclepro.2023.139919

Alim, A., Shirajuddin, S., & Anuar, F. (2022). A review of nonbiodegradable and biodegradable composites for food packaging application. *Journal of Chemistry, 2022*, 1–27. https://doi.org/10.1155/2022/7670819

Altoukhov, A., Kashkin, S., & Utkina, E. (2021). Bioeconomic development in Europe and in the world. *E3s Web of Conferences, 244*, 10061. https://doi.org/10.1051/e3sconf/202124410061

Annabi, N., Yue, K., Tamayol, A., & Khademhosseini, A. (2015). Elastic sealants for surgical applications. *European Journal of Pharmaceutics and Biopharmaceutics, 95*, 27–39. https://doi.org/10.1016/j.ejpb.2015.05.022

Arias, A., González-García, S., & Feijoo, G. (2021). Tannin-based bio-adhesives for the wood panel industry as sustainable alternatives to petrochemical resins. *Journal of Industrial Ecology, 26*(2), 627–642. https://doi.org/10.1111/jiec.13210

Assmann, A., Vegh, A., Ghasemi-Rad, M., Bagherifard, S., Cheng, G., Sani, E., et al. (2017). A highly adhesive and naturally derived sealant. *Biomaterials, 140*, 115–127. https://doi.org/10.1016/j.biomaterials.2017.06.004

Beaumont, M., Blanc, F., Cherbuy, C., Egidy, G., Giuffra, E., Lacroix-Lamandé, S., et al. (2021). Intestinal organoids in farm animals. *Veterinary Research, 52*(1), 33. https://doi.org/10.1186/s13567-021-00909-x

Bennich, T., & Belyazid, S. (2017). The route to sustainability—Prospects and challenges of the bio-based economy. *Sustainability, 9*(6), 887. https://doi.org/10.3390/su9060887

Berglund, L. A., & Burgert, I. (2018). Bioinspired wood nanotechnology for functional materials. *Advanced Materials, 20*, 1704285. https://doi.org/10.1002/adma.201704285

Besi, M., & McCormick, K. (2015). Towards a bioeconomy in Europe: National. *Regional and Industrial Strategies. Sustainability, 7*(8), 10461–10478. https://doi.org/10.3390/su70810461

Bhadani, A., Iwabata, K., Sakai, K., Koura, S., Sakai, H., & Abe, M. (2017). Sustainable oleic and stearic acid-based biodegradable surfactants. *RSC Advances, 7*(17), 10433–10442. https://doi.org/10.1039/c6ra27036k

Broeren, M., Mc, Z., Loop, S., Heugens, E., Posthuma, L., Worrell, E., et al. (2017). Environmental assessment of bio-based Chemicals in Early-Stage Development: A review of methods and indicators. *Biofuels Bioproducts and Biorefining, 11*(4), 701–718. https://doi.org/10.1002/bbb.1772

Burke, K., Roberts, D., & Kaplan, D. (2015). Silk fibroin aqueous-based adhesives inspired by mussel adhesive proteins. *Biomacromolecules, 17*(1), 237–245. https://doi.org/10.1021/acs.biomac.5b01330

Caferra, R., Falcone, P., Morone, A., & Morone, P. (2022). Is COVID-19 anticipating the future? Evidence from Investors' sustainable orientation. *Eurasian Economic Review, 12*(1), 177–196. https://doi.org/10.1007/s40821-022-00204-5

Chen, F. (2024). Pathways to ensuring food security in the context of the Chinese bio-economy landscape. *ACS Agricultural Science & Technology, 4*(1), 92–102. https://doi.org/10.1021/acsagscitech.3c00423

Chen, S., Gil, C., Ning, L., Jin, L., Perez, L., Kabboul, G., et al. (2021). Adhesive tissue engineered scaffolds: Mechanisms and applications. *Frontiers in Bioengineering and Biotechnology, 9*. https://doi.org/10.3389/fbioe.2021.683079

Choi, J., Yong, K., Choi, J., & Cowie, A. (2019). Recent advances in photo-cross-linkable hydrogels for biomedical applications. *BioTechniques, 66*(1), 40–53. https://doi.org/10.2144/btn-2018-0083

Cok, B., Tsiropoulos, I., Roes, A., & Patel, M. (2013). Succinic acid production derived from carbohydrates: An energy and greenhouse gas assessment of a platform chemical toward a bio-based economy. *Biofuels Bioproducts and Biorefining, 8*(1), 16–29. https://doi.org/10.1002/bbb.1427

Cowan-Ellsberry, C., Belanger, S., Dorn, P., Dyer, S., McAvoy, D., Sanderson, H., et al. (2014). Environmental safety of the use of major surfactant classes in North America. *Critical Reviews in Environmental Science and Technology, 44*(17), 1893–1993. https://doi.org/10.1080/10739149.2013.803777

Daristotle, J., Lau, L., Erdi, M., Hunter, J., Djoum, A., Srinivasan, P., et al. (2019). Sprayable and biodegradable, intrinsically adhesive wound dressing with antimicrobial properties. *Bioengineering & Translational Medicine, 5*(1), e10149. https://doi.org/10.1002/btm2.10149

Di, Y., Tkach, A., Yanamala, N., Stanley, S., Gao, S., Shurin, M., et al. (2013). Dual acute Proinflammatory and Antifibrotic pulmonary effects of short palate, lung, and nasal epithelium clone–1 after exposure to carbon nanotubes. *American Journal of Respiratory Cell and Molecular Biology, 49*(5), 759–767. https://doi.org/10.1165/rcmb.2012-0435oc

Djali, M., Kayaputri, I., Kurniati, D., Sukarminah, E., Mudjenan, I., & Utama, G. (2021). Degradation of lignocelluloses cocoa Shell (Theobroma Cacao l.) by various types of Mould treatments. *Journal of Food Quality, 2021*, 1–8. https://doi.org/10.1155/2021/6127029

Dunn, J. B., Adom, F., Sather, N., Han, J., Snyder, S., & He, C. (2015). *Life-cycle analysis of bioproducts and their conventional counterparts in greet*. Argonne National Laboratory. https://doi.org/10.2172/1250468

Duque-Acevedo, M., Lancellotti, I., Andreola, F., Barbieri, L., Ureña, L., & Camacho-Ferre, F. (2022). Management of Agricultural Waste Biomass as raw material for the construction sector: An analysis of sustainable and circular alternatives. *Environmental Sciences Europe, 34*(1). https://doi.org/10.1186/s12302-022-00655-7

Egboduku, W. O., Egboduku, T., Golohor, O. M., Imarhiagbe, O., & Ogwu, M. C. (2024). Cassava as raw material for sustainable bioeconomy development. In M. C. Ogwu, S. C. Izah, A. A. Cunha Alves, & S. C. Babu (Eds.), *Plant biology, sustainability and climate change, sustainable cassava* (pp. 57–73). Academic. https://doi.org/10.1016/B978-0-443-21747-0.00022-9

Feehan, J., & Petersen, J. (2003). *A framework for evaluating the environmental impact of biofuel use* (OECD Workshop on Biomass and agricultural). European Environmental Agency.
Fei, Q., O'Brien, M., Nelson, R., Chen, X., Lowell, A., & Dowe, N. (2016). Enhanced lipid production by Rhodosporidium Toruloides using different fed-batch feeding strategies with Lignocellulosic hydrolysate as the sole carbon source. *Biotechnology for Biofuels, 9*(1), 130. https://doi.org/10.1186/s13068-016-0542-x
Gelder, J., German, L., & Bailis, R. (2012). Biofuels investments in tropical forest-rich countries: Implications for responsible finance. *Sustainability Accounting Management and Policy Journal, 3*(2), 134–160. https://doi.org/10.1108/20408021211282296
Ghasemi, S., Sibi, M., Ulven, C., & Pourhashem, G. (2020). A preliminary environmental assessment of Epoxidized sucrose Soyate (ess)-based biocomposite. *Molecules, 25*(12), 2797. https://doi.org/10.3390/molecules25122797
Global Bioeconomy Assessment. (2024). *Coordinated efforts of policy, innovation, and sustainability for a greener future*. https://doi.org/10.59117/20.500.11822/45332
Gudiña, E., Rangarajan, V., Sen, R., & Rodrigues, L. (2013). Potential therapeutic applications of biosurfactants. *Trends in Pharmacological Sciences, 34*(12), 667–675. https://doi.org/10.1016/j.tips.2013.10.002
Guillard, V., Gaucel, S., Fornaciari, C., Angellier-Coussy, H., Buche, P., & Gontard, N. (2018). The next generation of sustainable food packaging to preserve our environment in a circular economy context. *Frontiers in Nutrition, 5*. https://doi.org/10.3389/fnut.2018.00121
Huang, C., Luo, M., Chen, X., Qi, G., Lin, X., Wang, C., et al. (2017). Combined "de novo" and "ex novo" lipid fermentation in a mix-medium of corncob acid hydrolysate and soybean oil by Trichosporon Dermatis. *Biotechnology for Biofuels, 10*(1), 147. https://doi.org/10.1186/s13068-017-0835-8
Huda, A. S. N., Mekhilef, S., & Ahsan, A. (2014). Biomass energy in Bangladesh: Current status and prospects. *Renewable and Sustainable Energy Reviews, 30*, 504–517. https://doi.org/10.1016/j.rser.2013.10.028. ISSN 1364-0321.
Isola, C., Sieverding, H., Raghunathan, R., Sibi, M., Webster, D., Sivaguru, J., et al. (2017). Life cycle assessment of photodegradable polymeric material derived from renewable bioresources. *Journal of Cleaner Production, 142*, 2935–2944. https://doi.org/10.1016/j.jclepro.2016.10.177
Jiang, F., Li, T., Li, Y., Zhang, Y., Gong, A., Dai, J., Hitz, E., Luo, W., & Hu, L. (2017). Wood-based nanotechnologies toward sustainability. *Advance Material*. https://doi.org/10.1002/adma.201703453. Wiley Library.
Jong, E., Higson, A., Walsh, P., & Wellisch, M. (2012). Product developments in the bio-based chemicals arena. *Biofuels Bioproducts and Biorefining, 6*(6), 606–624. https://doi.org/10.1002/bbb.1360
Kamusoko, R., Jingura, R., Parawira, W., & Chikwambi, Z. (2022). Characterization of lignocellulosic crop residues for potential biogas production in Zimbabwe. *Biofuels Bioproducts and Biorefining, 16*(5), 1165–1171. https://doi.org/10.1002/bbb.2396
Koirala, N., Khanal, S., Chaudhary, S., Gautam, S., Sah, S., Subba, P., et al. (2021). Potential surface active agent production using very low grade and cheap substrate by bacillus subtilis as microbial cell factory. Nepal. *Journal of Biotechnology, 9*(2), 21–28. https://doi.org/10.54796/njb.v9i2.41910
Kosoe, E. A., & Ogwu, M. C. (2024). Sustainable cassava processing: Processes, techniques, wastes, and waste streams. In M. C. Ogwu, S. C. Izah, A. A. Cunha Alves, & S. C. Babu (Eds.), *Plant biology, sustainability and climate change, sustainable cassava* (pp. 257–272). Academic. https://doi.org/10.1016/B978-0-443-21747-0.00011-4
Kraśniewska, K., Galus, S., & Gniewosz, M. (2020). Biopolymers-based materials containing silver nanoparticles as active packaging for food applications–a review. *International Journal of Molecular Sciences, 21*(3), 698. https://doi.org/10.3390/ijms21030698
Kumirska, J., Czerwicka, M., Kaczyński, Z., Bychowska, A., Brzozowski, K., Thöming, J., & Stepnowski, P. (2010). Application of spectroscopic methods for structural analysis of chitin and chitosan. *Marine Drugs, 8*(5), 1567–1636. https://doi.org/10.3390/md8051567

Lavania, S., Mehta, J., Bhardwaj, P., Tripathi, A., Gupta, N., & Gupta, P. (2023). Biocomposites: Prospects and manifold applications for human and environmental sustainability. *ECS Journal of Solid State Science and Technology, 12*(3), 037002. https://doi.org/10.1149/2162-8777/acbe17

Lee, S. (2024). Cell-free synthesis: Expediting biomanufacturing of chemical and biological molecules. *Molecules, 29*(8), 1878. https://doi.org/10.3390/molecules29081878

Lien, B., Brown, N., Ransom, S., Lehrich, B., Shahrestani, S., Tafreshi, A., et al. (2020). Enhancing peripheral nerve regeneration with neurotrophic factors and bioengineered scaffolds: A basic science and clinical perspective. *Journal of the Peripheral Nervous System, 25*(4), 320–334. https://doi.org/10.1111/jns.12414

Lin, Y., Ye, M., Zhang, X., Chen, Y., Chen, Y., Wu, J., et al. (2022). Biodegradable copolyesters based on a "soft" isohexide building block with tunable viscoelasticity and self-adhesiveness. *Polymer Chemistry, 13*(31), 4511–4523. https://doi.org/10.1039/d2py00586g

Lovett, J., de Bie, F., & Visser, D. (2022). *Sustainable sourcing of feedstocks for bioplastic: "Clarifyng sustainability aspects around feedstock use for the production of bioplastics"*. Whitepaper Sustainable Sourcing of Feedstocks for Bioplastics, Version 1.3.

Ludwicka, K., Kaczmarek, M., & Białkowska, A. (2020). Bacterial nanocellulose—A biobased polymer for active and intelligent food packaging applications: Recent advances and developments. *Polymers, 12*(10), 2209. https://doi.org/10.3390/polym12102209

Ma, L. (2023). Green production of a high-value mosquito insecticide of nootkatone from seaweed hydrolysates. *Journal of Agricultural and Food Chemistry, 71*(48), 18919–18927. https://doi.org/10.1021/acs.jafc.3c06708

Macrelli, S., Mogensen, J., & Zacchi, G. (2012). Techno-economic evaluation of 2nd generation bioethanol production from sugar cane bagasse and leaves integrated with the sugar-based ethanol process. *Biotechnology for Biofuels, 5*, 22. https://doi.org/10.1186/1754-6834-5-22

Majer, S., Wurster, S., Moosmann, D., Ladu, L., & Thrän, D. (2018). Gaps and research demand for sustainability certification and standardisation in a sustainable bio-based economy in the EU. *Sustainability, 10*(7), 2455. https://doi.org/10.3390/su10072455

Maraveas, C. (2020). Production of sustainable construction materials using agro-wastes. *Materials, 13*(2), 262. https://doi.org/10.3390/ma13020262

Mehdizadeh, M., Hong, W., Gyawali, D., Tang, L., & Yang, J. (2012). Injectable citrate-based mussel-inspired tissue bioadhesives with high wet strength for sutureless wound closure. *Biomaterials, 33*(32), 7972–7983. https://doi.org/10.1016/j.biomaterials.2012.07.055

Mellinas, C., Ramos, M., Jiménez, A., & Garrigós, M. (2020). Recent trends in the use of pectin from agro-waste residues as a natural-based biopolymer for food packaging applications. *Materials, 13*(3), 673. https://doi.org/10.3390/ma13030673

Meramo, S., Sukumara, S., & Fantke, P. (2022). Advances and opportunities in integrating economic and environmental performance of renewable products. *Biotechnology for Biofuels and Bioproducts, 15*(1), 144. https://doi.org/10.1186/s13068-022-02239-2

Merighi, S., Mazzocchetti, L., Benelli, T., & Giorgini, L. (2020). Evaluation of novel bio-based amino curing agent systems for epoxy resins: Effect of tryptophan and guanine. *Processes, 9*(1), 42. https://doi.org/10.3390/pr9010042

Milenkovic, B., Dimic-Janjic, S., & Popevic, S. (2018). Review of lung sealant technologies for lung volume reduction in pulmonary disease. *Medical Devices Evidence and Research, 11*, 225–231. https://doi.org/10.2147/mder.s127136

Moldes, A., Rincón-Fontán, M., López-Prieto, A., Vecino, X., & Cruz, J. (2021). Synthetic and bio-derived surfactants versus microbial biosurfactants in the cosmetic industry: An overview. *International Journal of Molecular Sciences, 22*(5), 2371. https://doi.org/10.3390/ijms22052371

Moreno, A. D., Susmozas, A., Oliva, J. M., & Negro, M. J. (2020). Chapter 1: Overview of bio-based industries. In *Biobased products and industries* (pp. 1–40). Elsevier. https://doi.org/10.1016/B978-0-12-818493-6.00001-4. ISBN 9780128184936.

Mouzouvi, C., Umerska, A., Bigot, A., & Saulnier, P. (2017). Surface active properties of lipid nanocapsules. *PLoS One, 12*(8), e0179211. https://doi.org/10.1371/journal.pone.0179211

Mustaqim, N. A., Sapuan, S. M., & Ilyas, R. A. (2021). Bio-based polymer for food packaging application: A review. In *Seminar on advanced bio-and mineral based natural fibre composites (SBMC 2021)*. ISBN: 978-983-44426-9-9.

Nair, S., Trafiałek, J., & Kolanowski, W. (2023). Edible packaging: A technological update for the sustainable future of the food industry. *Applied Sciences, 13*(14), 8234. https://doi.org/10.3390/app13148234

Nakhate, P., & Meer, Y. (2021). A systematic review on seaweed functionality: A sustainable bio-based material. *Sustainability, 13*(11), 6174. https://doi.org/10.3390/su13116174

National Research Council (US) Committee on Bio-based Industrial Products. (2000). *Bio-based industrial products: Priorities for research and commercialization*. National Research Council (US) Committee on Bio-based Industrial Products.

Neykov, N., Krišťáková, S., Hajdúchová, I., Sedliačiková, M., Antov, P., & Giertliová, B. (2021). Economic efficiency of forest enterprises—Empirical study based on data envelopment analysis. *Forests, 12*(4), 462. https://doi.org/10.3390/f12040462

Nguyễn, T., Kikuchi, Y., Noda, M., & Hirao, M. (2015). Impact of local conditions on bio-based chemical process design: Selection of input feedstock and production scale and scenario. *Environmental Progress & Sustainable Energy, 35*(1), 174–182. https://doi.org/10.1002/ep.12188

Obahiagbon, E. G., & Ogwu, M. C. (2023a). Sustainable supply chain management in the Herbal Medicine Industry. In S. C. Izah, M. C. Ogwu, & M. Akram (Eds.), *Herbal medicine phytochemistry* (Reference series in phytochemistry). Springer. https://doi.org/10.1007/978-3-031-21973-3_64-1

Obahiagbon, E. G., & Ogwu, M. C. (2023b). Consumer perception and demand for sustainable herbal medicine products and market. In S. C. Izah, M. C. Ogwu, & M. Akram (Eds.), *Herbal medicine phytochemistry* (Reference series in phytochemistry). Springer. https://doi.org/10.1007/978-3-031-21973-3_65-1

Obahiagbon, E. G., & Ogwu, M. C. (2024a). Organic food preservatives: The shift towards natural alternatives and sustainability in the global South's markets. In M. C. Ogwu, S. C. Izah, & N. R. Ntuli (Eds.), *Food safety and quality in the global south* (pp. 299–329). Springer. https://doi.org/10.1007/978-981-97-2428-4_10

Obahiagbon, E. G., & Ogwu, M. C. (2024b). The nexus of business, sustainability, and herbal medicine. In S. C. Izah, M. C. Ogwu, & M. Akram (Eds.), *Herbal medicine phytochemistry* (Reference series in phytochemistry). Springer. https://doi.org/10.1007/978-3-031-21973-3_67-1

Ögmundarson, Ó., Sukumara, S., Laurent, A., & Fantke, P. (2019). Environmental hotspots of lactic acid production systems. *GCB Bioenergy, 12*(1), 19–38. https://doi.org/10.1111/gcbb.12652

Ogunkunle, T., Fadairo, A., Rasouli, V., Ling, K., Oladepo, A., Chukwuma, O., et al. (2020). Microbial-derived bio-surfactant using neem oil as substrate and its suitability for enhanced oil recovery. *Journal of Petroleum Exploration and Production Technology, 11*(2), 627–638. https://doi.org/10.1007/s13202-020-01040-x

Ogwu, M. C., Iyiola, A. O., & Izah, S. C. (2023). Overview of African biological resources and environment. In S. C. Izah & M. C. Ogwu (Eds.), *Sustainable utilization and conservation of Africa's biological resources and environment* (Sustainable development and biodiversity) (Vol. 888, pp. 1–34). Springer. https://doi.org/10.1007/978-981-19-6974-4_1

Ogwu, M.C., Odozi, I.P., Ahonsi, O.C., Uleanya, K.O., and Odozi, E.B. (2024). Organic acid production from cassava. In: Ogwu, M.C., Izah, S.C., Cunha Alves, A.A. and Babu, S.C. (eds.). Plant biology, sustainability and climate change, sustainable cassava. Academic. 395–418. https://doi.org/10.1016/B978-0-443-21747-0.00009-6.

Omar, R. (2019). Geotextile from pineapple leaves and bio-grout for slope stabilization and erosion control. *International Journal of Geomate, 17*(60), 10.21660/2019.60.8319.

Patel, M., Béchu, A., Villegas, J., Bergez-Lacoste, M., Yeung, K., Murphy, R., et al. (2018). Second-generation bio-based plastics are becoming a reality—Non-renewable energy and greenhouse gas (ghg) balance of succinic acid-based plastic end products made from lignocellulosic biomass. *Biofuels Bioproducts and Biorefining, 12*(3), 426–441. https://doi.org/10.1002/bbb.1849

Popa, V. I. (2018). Biomass for fuels and materials. In *Biomass as renewable raw material to obtain bioproducts of high-tech value*. https://doi.org/10.1016/B978-0-444-63774-1.00001-6

Primožič, M., Knez, Ž., & Leitgeb, M. (2021). (bio)nanotechnology in food science—Food packaging. *Nanomaterials, 11*(2), 292. https://doi.org/10.3390/nano11020292
Ravichandran, R., Sundarrajan, S., Venugopal, J., Mukherjee, S., & Ramakrishna, S. (2012). Advances in polymeric systems for tissue engineering and biomedical applications. *Macromolecular Bioscience, 12*(3), 286–311. https://doi.org/10.1002/mabi.201100325
Reddy, L., Arias, J., Nicolas, J., & Couvreur, P. (2012). Magnetic nanoparticles: Design and characterization, toxicity and biocompatibility, pharmaceutical and biomedical applications. *Chemical Reviews, 112*(11), 5818–5878. https://doi.org/10.1021/cr300068p
Reynolds, T., & Martirosyan, D. (2016). Nutrition by design: A review of biotechnology in functional food. *Functional Foods in Health and Disease, 6*(2), 110. 10.31989/ffhd.v6i2.236.
Santhanam, N., Badri, D., Decker, S., Manter, D., Reardon, K., & Vivanco, J. (2011). Lignocellulose decomposition by microbial secretions. In *Secretions and exudates in biological systems* (pp. 125–153). https://doi.org/10.1007/978-3-642-23047-9_7
Seddiqi, H., Jin, J., Oliaei, E., Honarkar, H., Geonzon, L. C., Bacabac, R. G., & Klein-Nulend, J. (2021). *Cellulose and its derivatives: Towards biomedical applications*. Springer Nature. https://doi.org/10.1007/s10570-020-03674-w
Sikdar, P., Dip, T., Dhar, A., Bhattacharjee, M., Hoque, M., & Ali, S. (2022). Polyurethane (pu) based multifunctional materials: Emerging paradigm for functional textiles, smart, and biomedical applications. *Journal of Applied Polymer Science, 139*(38). https://doi.org/10.1002/app.52832
Simões, C., Viana, J., & Cunha, A. (2008). Mechanical properties of poly(ε-caprolactone) and poly(lactic acid) blends. *Journal of Applied Polymer Science, 112*(1), 345–352. https://doi.org/10.1002/app.29425
Sulik, G. (2011). 'Our diagnoses, our selves': The rise of the technoscientific illness identity. *Sociology Compass, 5*(6), 463–477. https://doi.org/10.1111/j.1751-9020.2011.00374.x
Sun, W., Darling, A., Starly, B., & Nam, J. (2004). Computer-aided tissue engineering: Overview, scope and challenges. *Biotechnology and Applied Biochemistry, 39*(1), 29–47. https://doi.org/10.1042/ba20030108
Szwarc, D., & Głowacka, K. (2021). Increasing the biogas potential of rapeseed straw using pulsed electric field pre-treatment. *Energies, 14*(24), 8307. https://doi.org/10.3390/en14248307
Tan, S., Chen, C., Chen, X., & Zhou, W. (2021). Crystalline micro- and nano-materials for medical and other biochemical applications. *Crystals, 11*(11), 1361. https://doi.org/10.3390/cryst11111361
Temesgen, S., Rennert, M., Tesfaye, T., & Nase, M. (2021). Review on spinning of biopolymer fibers from starch. *Polymers, 13*(7), 1121. https://doi.org/10.3390/polym13071121
Venkateshaiah, A., Timmins, R., Sehl, E., Černík, M., & Agarwal, S. (2022). High barrier, biodegradable nanocomposite films based on clay-coated and chemically modified gum kondagogu. *Macromolecular Materials and Engineering, 307*(8). https://doi.org/10.1002/mame.202200008
Verkerk, P. J., Hassegawa, M., Van Brusselen, J., Cramm, M., Chen, X., Maximo, Y. I., Koç, M., Lovrić, M., & Tegegne, Y. T. (2022). *Forest products in the global bioeconomy; "Enabling substitution by wood-based products and contributing to the sustainable development goals"*. FAO. https://doi.org/10.4060/cb7274en
Vinod, A., Sanjay, M. R., Suchart, S., & Jyotishkumar, P. (2020). Renewable and sustainable biobased materials: An assessment on biofibers, biofilms, biopolymers and biocomposites. *Journal of Cleaner Production, 258*(120978). https://doi.org/10.1016/j.jclepro.2020.120978. ISSN 0959-6526.
Volkering, F., Breure, A., Andel, J., & Rulkens, W. (1995). Influence of nonionic surfactants on bioavailability and biodegradation of polycyclic aromatic hydrocarbons. *Applied and Environmental Microbiology, 61*(5), 1699–1705. https://doi.org/10.1128/aem.61.5.1699-1705.1995
Vroman, I., & Tighzert, L. (2009). Biodegradable polymers. *Materials, 2*(2), 307–344. https://doi.org/10.3390/ma2020307
Wang, L. (2024). Advancing biomedical applications of polyoxometalate-based metal–organic frameworks: From design to therapeutic potential. *Inorganic Chemistry Frontiers, 11*(5), 1339–1365. https://doi.org/10.1039/d3qi02414h

Wang, Y., Lv, T., Yin, K., Feng, N., Sun, X., Zhou, J., et al. (2023). Carbon dot-based hydrogels: Preparations, properties, and applications. *Small, 19*(17), e2207048. https://doi.org/10.1002/smll.202207048

Xu, H., Fang, Z., Tian, W., Wang, Y., Ye, Q., Zhang, L., et al. (2018). Green fabrication of amphiphilic quaternized β-chitin derivatives with excellent biocompatibility and antibacterial activities for wound healing. *Advanced Materials, 30*(29), e1801100. https://doi.org/10.1002/adma.201801100

Yadav, M., & Agarwal, M. (2021). Biobased building materials for sustainable future: An overview. *Materials Today: Proceedings, 43*(Part 5), 2895–2902. https://doi.org/10.1016/j.matpr.2021.01.165. ISSN 2214-7853.

Yanovska, A., Artyukhov, A., Vakal, S., Vakal, V., & Shkola, V. (2021). Encapsulated organic–mineral fertilizers with nanoporous structure. *Applied Nanoscience, 12*(4), 1275–1283. https://doi.org/10.1007/s13204-021-01893-6

Yu, J., Kan, Y., Rapp, M., Danner, E., Wei, W., Das, S., et al. (2013). Adaptive hydrophobic and hydrophilic interactions of mussel foot proteins with organic thin films. *Proceedings of the National Academy of Sciences, 110*(39), 15680–15685. https://doi.org/10.1073/pnas.1315015110

Yuvaraj, D., Iyyappan, J., Gnanasekaran, R., Ishwarya, G., Harshini, R., Dhithya, V., et al. (2021). Advances in bio food packaging—An overview. *Heliyon, 7*(9), e07998. https://doi.org/10.1016/j.heliyon.2021.e07998

Zhang, Y., & Khademhosseini, A. (2017). Advances in engineering hydrogels. *Science, 356*(6337). https://doi.org/10.1126/science.aaf3627

Zhang, Y., Chen, M., Chen, Y., Hou, Y., & Hu, S. (2019). Characterization and exploration of recombinant wheat catalase for improvement of wheat-flour-processing quality. *Journal of Agricultural and Food Chemistry, 67*(9), 2660–2669. https://doi.org/10.1021/acs.jafc.8b06646

Zhao, C., Ma, Z., Huang, C., Wen, J., & Hassan, M. (2022). Editorial: From biomass to bioenergy and bio-chemicals: Pretreatment, thermochemical conversion, biochemical conversion and its bio-based applications. *Frontiers in Bioengineering and Biotechnology, 10*. https://doi.org/10.3389/fbioe.2022.975171

Zhu, S. (2024). Recent advances in the degradability and applications of tissue adhesives based on biodegradable polymers. *International Journal of Molecular Sciences, 25*(10), 5249. https://doi.org/10.3390/ijms25105249

Zhu, Y., Kong, B., Li, R., & Zhao, Y. (2022). Developing biomedical engineering technologies for reproductive medicine. *Smart Medicine, 1*(1), e20220006. https://doi.org/10.1002/smmd.20220006

Waste Streams of Biobased Products in the Global South

Adams Ovie Iyiola, Berchie Asiedu, Jacob Somorhire Ewutanure, Olamigoke Olawale Akingba, Sulaiman Kajogbola Alimi, and Christy Funmilayo Idowu

Abstract

The Global South is faced with challenges that are unique in terms of waste management streams of biobased products. These products are derived from renewable resources like agricultural wastes and plants which are more friendly to the environment when compared with fuel-based fossil products from traditional

A. O. Iyiola (✉)
Department of Fisheries and Aquatic Resources Management, Faculty of Renewable Natural Resources Management, College of Agriculture and Renewable Natural Resources, Ejigbo Campus, Osun State University, Ejigbo, Osun State, Nigeria
e-mail: adams.iyiola@uniosun.edu.ng

B. Asiedu
Department of Fisheries and Water Resources; School of Natural Resources, University of Energy and Natural Resources, Sunyani, Ghana
e-mail: berchie.asiedu@uenr.edu.gh

J. S. Ewutanure
Department of Fisheries and Aquaculture, Faculty of Environmental Management, Nigeria Maritime University, Okerenkoko, Delta State, Nigeria

O. O. Akingba
Nigerian Centre for Energy and Environment (NCEE) Administrative Block Complex, University of Benin, Benin, Edo State, Nigeria
e-mail: akingba.o@ncee.org.ng

S. K. Alimi
Department of Environmental Management, School of Computing, Engineering and Built Environment, Glasgow Caledonian University, Glasgow, Scotland, UK
e-mail: salimi300@caledonian.ac.uk

C. F. Idowu
National Biotechnology Research and Development Agency, Bioresources Development Centre, Ogbomoso, Oyo State, Nigeria
e-mail: Idowu.funmilayo@nabda.gov.ng

M. C. Ogwu et al. (eds.), *Sustainable Bioeconomy Development in the Global South*, https://doi.org/10.1007/978-981-96-0305-3_7

sources. However, waste streams are generated from the production and disposal of biobased products which requires strict management. Agricultural waste is a major waste stream in Global South countries (Africa, Asia, and Latin America). These countries rely majorly on agricultural activities as a principal economic activity, thereby resulting in increased generation of residue from crops like husks, straw, and stalks. Such residue can be utilized in the production of biobased products such as bioplastics, biofertilizers, and biofuels. The improper disposal of wastes from these activities is a major problem that can result in environmental pollution and health issues. Production processes also generate some form of waste generation associated with biobased products. For instance, biochar and glycerol are waste products from the production of biofuels that require proper disposal. Similarly, bioplastics may generate waste streams such as polymers and packaging wastes. These waste streams require waste management strategies that are effective in preventing contamination and resource recovery maximization. Also, the management of end-of-life biobased products is a great challenge as Global South countries lack the necessary infrastructure for biodegradation that is needed for composting, thereby affecting their disposal. With this, the biobased products are usually discarded indiscriminately in landfills where they decompose to cause pollution and the emission of greenhouse gases. There is a great need to ensure sustainable waste management practices in the Global South. This can be focussed on infrastructures for collection of wastes, recycling, composting, and segregation of biobased materials. Also, the government can formulate policies on disposal of biobased products and initiate recycling, and facilities for composting. The public may also be enlightened through campaigns and meetings on the need for waste management that is responsible for ensuring a positive behaviour change.

Keywords

Behavioural change · Global South · Biobased products · Resource recovery · Sustainability · Environmental management

Abbreviations

AU	African Union
BOD	Biological Oxygen Demand
COD	Chemical Oxygen Demand
CO_2	Carbon dioxide
CH_4	Methane
C & D	Construction and Demolition wastes
DFFE	Department of Forestry, Fisheries, and the Environment
EPA	Environmental Protection Agency
FAO	Food and Agricultural Organization

GHG	Greenhouse gas
IPCC	Intergovernmental Panel on Climate Change
ISWA	International Solid Waste Association
MSW	Municipal Solid Wastes
PBDE	Polybrominated diphenyl ethers
SDGs	Sustainable Development Goals
SWDS	Solid waste disposal systems
UNEP	United Nations Environmental Programme
WHO	World Health Organization

1 Introduction

Biobased product waste streams are an important resource that has significant benefits to the economy, the environment, and society. Reducing the dependence on limited fossil fuel resources and promoting sustainability, biobased products—which are sourced from renewable biological resources like biomass——offer viable substitutes for traditional petroleum-based products (Cherubini & Strømman, 2011). Nonetheless, the bioeconomy has obstacles as well as opportunities due to the diverse waste streams that are produced during the manufacture and use of biobased products. These include biomass leftovers, agricultural by-products, and organic wastes. Products that are biobased use renewable resources to provide hopeful answers to environmental problems. The manufacturing and elimination of these goods, especially in the Global South, creates several waste streams. According to Wojnowska-Baryła et al. (2020), comprehension of these waste streams is essential for waste management and sustainable development initiatives. These waste streams come from a variety of biobased industry sources, such as forestry operations, food processing facilities, biorefineries, and agricultural activities. A major supplier of biobased goods like biofuels and biomaterials is agriculture. However, management and disposal of agricultural waste, such as crop residues and animal dung, provide difficulties. Food waste streams are impacted by biobased goods made from food sources, such as biofuels and bioplastics. Concerns about food security in the Global South may worsen if these goods are disposed of improperly (Food and Agricultural Organization [FAO], 2019).

Biodegradability is a common selling point for items made from biomaterials. Ecosystems and human health may be harmed by the improper disposal of environmental wastes (Jönsson & Martín, 2016). A sizable amount of waste streams in the bioeconomy are composed of biomass residues, which include crop residues, forestry residues, and energy crops. These residues are produced during agricultural and forestry operations. Furthermore, the total amount of waste generated in biobased companies includes organic wastes from biorefining processes including lignin, wastewater, and leftover fermentation broth (Gnansounou & Dauriat, 2010). Moreover, additional waste streams that can be utilized in the bioeconomy include organic wastes from food processing, wastewater treatment facilities, and municipal solid waste (Ellen MacArthur Foundation, 2015; Nizzy & Ogwu, 2024). To maximize resource efficiency, minimize environmental effects, and promote circularity within the bioeconomy, waste streams from biobased products must be effectively managed and utilized (Geyer et al., 2017; Iyiola et al. 2022a, b, c). Through the establishment of new technologies and business models, these waste streams offer prospects for resource recovery, value creation, and innovation.

Additionally, waste can be converted into biofuels, biochemicals, biopolymers, and other value-added products through waste valorization techniques like anaerobic digestion, pyrolysis, and biochemical conversion (De Baere et al., 2016). This helps the bioeconomy move towards a more sustainable and circular form of production. To fully utilize waste streams in the bioeconomy, however, obstacles such as feedstock supply, quality fluctuation, technological limits, and regulatory limitations need to be overcome (Mata-Alvarez et al., 2014; Ogwu et al., 2024). The infrastructure for proper waste management is lacking in many Global South nations, making it difficult to handle trash from biobased products. To solve this issue, infrastructure development investments are crucial (Ouda, 2016). To encourage sustainable waste management techniques, governments must enact and enforce strict rules (Intergovernmental Panel on Climate Change [IPCC], 2019). Campaigns for public awareness and education are essential for encouraging the responsible use and disposal of biobased products. According to Ceballos et al. (2018), community involvement programs can enable people to take part in waste reduction efforts and make educated decisions. In the value chains of biobased products, implementing circular economy concepts can reduce waste production and increase resource efficiency. Waste streams can be reduced by using techniques like closed-loop systems and recyclable product design (Ellen MacArthur Foundation, 2015). Working together, stakeholders—policymakers, researchers, industry participants, and consumers—will be necessary to overcome these obstacles and capitalize on the social, economic, and environmental advantages linked to biobased product waste streams.

The complexity and growth of different wastes pose a threat to community health, cause social and economic problems, and negatively impact the environment (Mpofu et al., 2021; Capoor & Parida, 2021). Dealing with garbage presents several issues for developing and growing nations, such as ineffective waste management techniques that harm the environment and public health (Department of Forestry, Fisheries and the Environment [DFFE], 2018; International Solid Waste Association [ISWA], 2012). With insufficient infrastructure to handle the increasing amounts of trash (Ferronato & Torretta, 2019), garbage is primarily disposed in landfills, with large amounts of rubbish being burned and dumped in the open (Godfrey et al., 2014). When compared with developed countries, several waste streams pose challenges, such as organic waste, packaging materials, hazardous waste, and waste from construction and demolition (Greencape, 2017). Additionally, recycling levels are lower, with a significant share of recycling being done by the informal sector (Godfrey & Oelofse, 2017). Sufficient waste processing can help improve areas of sustainability like social fairness, environmental protection, and economic viability by reducing socioeconomic and environmental consequences (Iyiola et al. 2023a, b, c). This review presents an evaluation of the waste streams of biobased products in the Global South and their management. Literature was sought from reputable publishers such as Springer, Elsevier, Taylor and Francis, and Wiley, and information gotten were synthesized. The publications were peer-reviewed (journals and conference proceedings) related to biobased products, biological materials, and their significance.

2 Biobased Products and Global South

Biobased products are the ones produced from renewable biological resources such as animals, plants, and microorganisms (Ogwu et al., 2023; Egboduku et al., 2024). These resources are abundant in the Global South regions and the use of the bio-products from these resources has numerous benefits such as sustainable economic growth, improved food security, job creation and reduction of poverty, enhanced sustainability of the environment, and the reduced dependence on energy sources from fossil fuels. There are diverse types of biobased products found in the Global South, examples of such are: biofuels, which are produced from palm oils, corn, or sugarcane; bioplastics, which are produced from potatoes, sugarcane, or cassava; biocomposite,s which are produced from sisal, coconut, or bamboo; biopesticides, which are produced from Spinosad, neem; or pyrethrum; and biocosmetics, which are produced from shea butter, coconut oil, or argan oil.

2.1 Initiatives Supporting Biobased Products in the Global South

The Global South countries are very advantaged and can use their biomass resources and biodiversity to develop a biobased economy. This approach can cause sustainable development and reduce the Global South's dependence on the use of fossil fuels as a source of energy. Some of the initiatives that support biobased products are:

- The Bioeconomy Strategy of the African Union.
- The Asian Biomass and Biogas Alliance.
- The Latin American Biobased Economy Program.
- The Global Biobased Economy Forum.
- The United Nations Sustainable Development Goals (SDGs)—most importantly SDG 9 (Industry, Innovation, and Infrastructure) and SDG 12 (Responsible Consumption and Production).

These initiatives are aimed at promoting the use of biobased products, development, economic opportunities creation, and sustainable development in the Global South.

2.2 The Bioeconomy Strategy of the African Union

The strategy was created by the African Union (AU) in 2020 towards a responsive action at sustainable economic growth in the African Union regional countries. The strategy was developed based on the following objectives:

- The sustainable economic growth and creation of jobs.
- Improving the environmental condition and climate change mitigation.
- Using the resources to generate products useful in bioenergy, biobased industrial processes, food and nutrition and health (Stockholm Environment Institute, 2022).

The bioeconomy strategy of the African Union is focussed on the following goals as reported by Muyambi (2021):

- *Unity*: This aims at achieving unity between all African people and their countries.
- *Sovereignty*: This involves the territorial integrity, sovereignty defence and the independence of all member states in the AU.
- *Common positions*: This focusses on promoting and defending the positions of the African people on issues relating to the continent and interests of people.
- *Integration*: It focusses on the process of integrating political and issues that may occur within the AU.
- *Peace and security*: It focusses on the promotion of peace, security, and stability of countries within the continent.
- *International cooperation*: This goal encourages international engagements, with cognisance of the United Nations Charter and Universal Declaration of Human Rights.
- *Health*: The goal focusses on associations with international partners that are relevant in disease prevention and good health promotion in the continent.
- *Development*: The goal focuses on the advancement through research promotion, most especially in science and technology across the continent.
- *Regional economic communities*: It focuses on coordination and linkages of policies existing between the present and future and the attainment of the sustainable objective of the AU.
- *Sustainable development*: It aims at promoting the social, cultural, and economic aspects of the country and the integration of African economies.
- *Global economy*: It focuses on establishing the conditions needed for the continent to take its position in the global economy and negotiations in the international forum.
- *Democratic principles*: It focuses on the promotion of good governance, democratic principles and institutions, and participation.

3 Global South and Waste Streams

Global South countries are characterized by economies between the low and middle-income categories. The waste streams are very complex and challenging due to the nature of their economies and are faced with challenges such as inadequate waste disposal and management strategies, limited infrastructural facilities, and rapid urbanization and industrialization (Conlon et al., 2021; Kosoe & Ogwu, 2024; Agoh et al., 2024). There are different types of waste streams in the Global South as illustrated in Fig. 1, namely, municipal solid wastes (MSW), industrial wastes, hazardous wastes, electronic wastes (e-wastes), construction and demolition wastes (C &D), healthcare wastes, and agricultural wastes.

3.1 Municipal Solid Waste (MSW)

It describes the common materials that are disposed of by homes, businesses, and institutions (Environmental Protection Agency [EPA], 2021). Paper, plastics, glass, metals, and organic materials are all included in this trash category. The heterogeneous composition and vast amount of MSW present considerable issues for garbage management (Iyiola & Valbo-Jørgensen, 2023; Izah et al., 2023). Waste, or unneeded or undesired by-products, are produced during most processes and cannot be avoided. About 7–9 billion tonnes of garbage are produced annually worldwide (Wilson & Velis, 2015). MSW is a distinct type of home waste that, depending on the reporting standard, may also include commercial and industrial wastes (Wilson & Velis, 2015). Two billion tonnes of garbage were produced in 2016, of which two billion tonnes were municipal solid waste (MSW). Nonetheless, it merits consideration due to its effects on the environment at the local, regional, and global levels; its proximity to people and consequently possible health effects; and its potential for recovery through supply chains of the circular economy (Wilson & Velis, 2015; EC

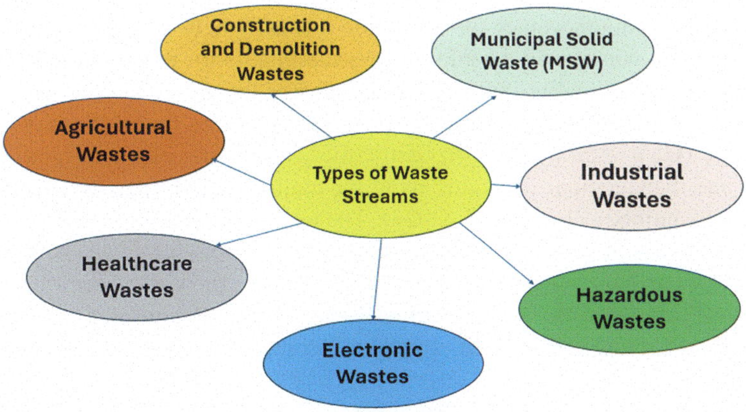

Fig. 1 Types of waste streams

(European Commission), 2015; Kaza et al., 2018). Depending on how wastes are disposed, different types of MSW can have different effects on the environment and human health. The long-term persistence of plastic wastes and their ingestion by creatures that may have an influence on human health as they move up the food chain are causes for growing worries on a worldwide scale.

A noteworthy long-term local impact of waste leachate is nitrogen pollution, which can lead to nutritional imbalances and sickness in adjacent water bodies (Burton & Watson-Craik, 1998). Moreover, it has been demonstrated that burning trash outdoors produces a considerable amount of dangerous air pollutants, which have a negative impact on human health, particularly in developing nations. Recycling, composting, or waste-to-energy incineration are ways in whereby waste materials can be recycled and re-incorporated into production flows as part of the shift to a circular economy (EC, 2015). Nevertheless, landfills and dumps, collectively referred to as solid waste disposal systems (SWDS), receive 70% of the garbage produced worldwide today (Kaza et al., 2018). As the most basic type of SWDS, dumps are a major source of waste pollution since they are large-scale waste storage facilities without any technical control. A landfill may use a range of technical techniques, such as impermeable layers and covers, to lessen and recover the quantity of leachate and gases generated (Iyiola et al., 2024a).

3.2 Industrial Wastes

Numerous waste streams are produced by industrial processes, including both hazardous and non-hazardous trash. Heavy metals, chemicals, solvents, and by-products of production operations are all considered forms of industrial waste. Hazards to human health and environmental contamination can result from improper industrial waste disposal. It is extremely concerning to think that hazardous pollutants from industrial wastes could seep into the ecosystem (He & Kappler, 2017). Industrial wastes containing radioactive elements and heavy metals have the potential to induce coastal erosion, pollution, and change in marine ecosystems. Additionally, after industrial pollutants enter surface water, it is extremely expensive and difficult to remove them, which leads to additional accumulation of the pollutants (Hairom et al., 2021). Additionally, industrial wastes harm the terrestrial ecosystem. For example, the composting and landfilling processes release harmful gases into the atmosphere (Gaur et al., 2020). Recently, the usage strategies, terminal disposal options, and management strategies about the control of industrial waste pollution have been examined (Zhang et al., 2021). The techniques for the subsequent treatment and disposal of industrial wastes from various sectors have been substantially enhanced by technological improvement. For example, the waste sector has undergone radical transformation due to various waste-treatment pathways such as incineration, gasification, combustion, pyrolysis, compaction, etc. (Bhattacharya & Kim, 2016).

Furthermore, in recent years, ideas such as waste-to-energy have gained prominence. However, the widespread adoption of these approaches across different nations has been severely limited by the associated high cost of operation, need for specialized infrastructure, and shortage of experienced labour (Das et al., 2019). Industrial wastes are not biodegradable, therefore releasing them into the environment has hastened the contamination of freshwater ecosystems, clean water sources, and terrestrial ecosystems (Azmi et al., 2018; Hairom et al., 2021). According to Haq et al. (2016), one of the main environmental pollutants is the waste produced by the pulp and paper industries, which is known for its high concentrations of hazardous substances such as heavy metals, phosphorus, sulphur, organic acids, and chlorophenols. About 20–100 m^3 of these hazardous wastes are dumped into lakes and rivers for every mg of the products (Lindholm-Lehto et al., 2015). Some sources of industrial wastes are wastes from cotton production, olive-oil production (Gaur et al., 2020), dairy industry (Sar et al., 2022), heavy metals (Ericson & Larsson, 2000; Iqbal et al., 2016; Soliman & Moustafa, 2020; Iyiola et al., 2024 b, c), mining (Sobihah et al., 2018; Covre et al., 2022), and textile wastes (Soto-Hernández et al., 2017).

3.3 Hazardous Wastes

A hazardous waste is considered to as any garbage that significantly endangers the environment or public health. Chemicals, insecticides, radioactive materials, and specific medical waste are found in this category (EPA, 2020, 2018). Hazardous waste must be handled, treated, and disposed of properly to reduce environmental concerns and avoid contamination.

3.4 Electronic Waste (E-Waste)

It is the term used to describe abandoned electronic equipment and devices, including computers, cell phones, and appliances. In addition to dangerous materials like lead and mercury, e-waste also contains valuable resources such as copper, silver, and gold. Recovering valuable resources and preventing environmental degradation require recycling and correct disposal of e-waste. The problem of e-waste has difficulties and effects in the Global South because the area has unique environmental and socioeconomic challenges that exacerbate the e-waste issue. Grant et al. (2013) highlight that a significant obstacle facing the Global South is the growing influx of e-waste from wealthy nations. E-waste from developed countries is frequently exported to the Global South, where recycling and disposal laws are either laxer or not implemented properly. By improperly processing and disposing of e-waste, this activity may cause pollution of the environment and put the health of the surrounding populations at danger. In the Global South, the unorganized sector is crucial to the management of e-waste. Many people, many of whom are impoverished and disenfranchised, make a living by participating in unofficial e-waste recycling operations. Yet, these unofficial recycling methods are frequently uncontrolled and may entail risky procedures that expose employees to harmful materials (Li et al., 2015).

The health and safety concerns that informal sector workers face is made worse by a lack of protective gear and a lack of awareness about the dangers involved in recycling e-waste. In addition, the Global South frequently lacks the resources and infrastructure needed to manage e-waste efficiently. Formal recycling systems are difficult to develop due to limited finance, poor technological skills, and restricted access to suitable recycling facilities. Therefore, when e-waste is inadequately treated, it ends up in landfills or unofficial recycling sites, releasing contaminants into the environment.

In the Global South, the effects of e-waste on the environment and human health are a major worry. Ecosystems and public health may be impacted by improper recycling and disposal methods that contaminate the air, water, and soil. Lead, mercury, and brominated flame retardants are examples of dangerous compounds that can cause exposure, which can result in several health concerns, including neurological abnormalities, reproductive problems, and lung problems. A multifaceted strategy is needed to address the e-waste issue in the Global South. It entails tightening laws and upholding them to stop the unauthorized importation of e-waste and to encourage appropriate recycling and disposal methods (Laha, 2015). According to Gollakota et al. (2020), governments and international organizations ought to back the establishment of formal recycling infrastructure by offering resources and training to guarantee secure and ecologically responsible e-waste disposal. In addition, it is critical to educate governments, workers, and communities on the dangers associated with improper management of e-waste. Initiatives aimed at educating and enhancing capacity can enable people to dispose of e-waste responsibly and promote the adoption of sustainable practices. To fully address the e-waste problem, cooperation between governments, businesses, and civil society organizations is crucial (Anuardo et al., 2023).

3.5 Construction and Demolition Wastes (C&D)

A variety of materials, including concrete, bricks, metals, wood, and plastics, are found in C&D waste in the Global South. These materials are impacted by regional building customs and resource availability (Medina, 2018). The development of C&D garbage is frequently greatly influenced by informal construction activities, which results in a diversified waste composition and more difficulties with waste management. Effective C&D waste management in the Global South is hampered by several issues. These include insufficient infrastructure for collecting trash, a lack of funding for waste management programs, and a lack of knowledge among stakeholders (Moyo & Iyiola, 2020). Unofficial disposal methods, such burning and open dumping, increase the threats to the public's health and the environment. The management of construction and demolition waste is governed by laws in many Global South nations; however, enforcement may be lax. To guarantee compliance and encourage sustainable waste management practices, rules must be strengthened, and strict monitoring systems must be put in place (UNEP, 2018; World Bank, 2019). Notwithstanding obstacles, creative approaches and programs are being developed in the Global South to deal with C&D waste management. To enhance garbage collection, recycling, and disposal procedures, public-private partnerships, technology-driven solutions, and community-based recycling programs are being put into place (Moyo & Iyiola, 2020; FAO, 2020).

3.6 Agricultural Wastes

The agricultural waste produced by farming, processing, and post-harvest operations encompasses a wide range of organic and inorganic components in the Global South. According to Nair et al. (2019), common agricultural waste types include crop leftovers, animal manure, agro-industrial by-products, and packaging materials. Area-specific farming methods, crop kinds, and socioeconomic variables all have an impact on the quantity and composition of agricultural waste. The Global South faces several obstacles in managing agricultural waste, including poor infrastructure, insufficient funding, and socioeconomic limitations. A large percentage of agricultural producers in the region are smallholder farmers, who frequently lack access to adequate waste management methods and facilities (Bhatia et al., 2020). Ineffective methods for gathering and disposing of trash lead to contamination of the environment, deterioration of soil, and hazards to public health. While some countries in the Global South have regulatory frameworks for agricultural waste management, enforcement may be weak, and implementation gaps exist. Strengthening policies, regulations, and institutional capacities is essential to promote sustainable waste management practices (FAO, 2019). International organizations, such as the Food and Agriculture Organization (FAO), provide technical assistance and capacity-building support to countries in the Global South to enhance their waste management systems.

Several strategies are employed in the Global South to manage agricultural waste, including composting, biogas production, and biochar utilization. Composting, for instance, converts organic waste into nutrient-rich compost, which can improve soil fertility and enhance crop productivity. Biogas production from animal manure and crop residues provides renewable energy sources for rural communities, reducing dependence on fossil fuels and mitigating greenhouse gas emissions. Sustainable solutions for agricultural waste management in the Global South include the adoption of circular economy principles, value addition, and waste valorization. By integrating circular economy principles, stakeholders can create closed-loop systems that minimize waste generation and maximize resource efficiency (Narayana et al., 2021). Value addition initiatives, such as the production of bio-based products and biofuels from agricultural waste, offer economic opportunities while reducing environmental impact.

3.7 Healthcare Wastes

Medical waste from healthcare institutions includes sharps, infectious materials, and medications. Disease transmission and environmental contamination can result from improper disposal of medical waste (World Health Organization [WHO], 2020). To protect the environment and the public health, strict laws and appropriate waste management practices are required.

4 Impacts of Waste Streams on Biobased Products

Waste streams are vital sources of feedstock for the manufacturing of biobased products. Biorefineries can use biomass residues, agricultural wastes, and organic by-products from a variety of sectors as raw materials to create biobased chemicals, fuels, and materials. The environmental impact of waste disposal is lessened, and resource efficiency is increased when waste streams are integrated into biobased production processes (Gnansounou & Dauriat, 2010). For instance, biochemical or thermochemical conversion processes can be used to transform lignocellulosic biomass leftovers, such as forestry and agricultural straws, into biofuels. Anaerobic digestion can be used to convert municipal solid waste (MSW) streams with organic parts into biogas, which can then be used to produce organic fertilizers and renewable energy. These illustrations show how waste streams can propel the environmentally friendly manufacturing of biobased goods, supporting the idea of a circular economy. The environmental sustainability of biobased products is contingent upon the efficient handling of waste streams. The environmental advantages of biobased manufacturing might be countered by improper waste treatment and disposal, which can lead to pollution, habitat damage, and greenhouse gas emissions (De Baere et al., 2016). Biobased companies can reduce these environmental effects through resource recovery and waste valorization by using waste streams as feedstock (Cherubini & Jungmeier, 2010). Waste stream-derived biobased products frequently have smaller carbon footprints than their fossil fuel-based equivalents. Throughout their lifecycle, bioplastics made from feedstocks obtained from organic waste, for example, have been demonstrated to lower energy consumption and greenhouse gas emissions (Behera & Ray, 2016). By substituting fossil fuels and lowering reliance on limited resources, the use of biofuels produced from trash can also help mitigate climate change.

Reduced raw material prices and new revenue sources are two ways that the incorporation of waste streams into biobased industrial processes can improve economic viability. In comparison to conventional feedstock sources, waste-derived feedstocks are frequently accessible at low or even negative cost (Mata-Alvarez et al., 2014). In addition, Nghiem et al. (2017) suggest that waste valorization has the potential to create supplementary sources of income by producing biobased goods with added value. However, to guarantee the economic competitiveness of waste-based biobased products, issues including feedstock supply, quality fluctuation, and processing costs need to be resolved (Dalgaard et al., 2008). To maximize the economic potential of waste streams in biobased enterprises and overcome these obstacles, technological breakthroughs in biomass conversion, process optimization, and waste management methods are crucial (Cherubini & Strømman, 2011).

4.1 Economic Importance of Waste Streams

During the production process, the biobased goods business in the global South produces a substantial amount of trash. A valuable resource may be produced from these waste streams, which include food waste, agricultural residues, and biomass conversion by-products. These waste streams can be converted into valuable products such as biofuels, biogas, and biobased chemicals using effective waste management systems and technology. The process of conversion generates new economic prospects that lead to the establishment of new sectors, revenue streams, and jobs (United Nations Environmental Programme [UNEP], 2018). Additionally, the global South's circular economy can benefit from the use of waste streams from biobased products. By closing the material and product cycle, the circular economy seeks to reduce waste production and increase resource efficiency. The circular economy can lessen reliance on virgin resources and encourage a more resilient and sustainable economy by repurposing waste streams into new goods or feedstocks. Businesses in the biobased goods industry may benefit financially from this shift to a circular economy as well as from better resource availability and enhanced competitiveness (Ellen MacArthur Foundation, 2019).

Furthermore, the value of biobased product waste streams to the economy goes beyond their immediate financial gains. Social cohesion and environmental sustainability can both be enhanced by appropriate waste management and utilization. For instance, turning waste biomass into bioenergy can lessen greenhouse gas emissions, slow down global warming, and enhance air quality. Furthermore, the construction of biorefineries or waste-to-energy facilities can offer localized energy options, particularly in rural areas with restricted access to power. This has the potential to improve local populations' quality of life, encourage rural development, and strengthen energy security. Encouraging regulations and incentives are essential if waste streams from biobased products are to be completely utilized economically in the global South. In addition to establishing clear waste management legislation and offering financial assistance for waste-to-value technology research and development, governments and regulatory agencies also need to foster a business-friendly atmosphere. Exchange of best practices and innovation in waste management and usage can also be facilitated by international cooperation and knowledge sharing, as well as collaboration between the public and commercial sectors (Amin et al., 2023).

5 Streams of Biobased Products from Organic Solid Waste

5.1 Biofertilizers

It is a biobased product that is manufactured from diverse technologies in bioconversion such as the process of preparing compost, vermicomposting, insect biorefinery and digestion in anaerobic means. This material reduces the dangerous effects fertilizers that are synthetic pose on the environment and this approach ensures prompt waste management (ChandiniKumar et al., 2019). The biofertilizer is economical and stimulates plant growth, health of the soil, and yield of crops. The use of compost has been used to boost several agricultural applications. Altieri and Esposito (2010) used waste from mill of olive as soil fertility organic compost and reduction of carbon dioxide emissions and the compost produced contained adequate nutrients for growth of plants. Kuligowski et al. (2023) reported the use of compost from wastes generated in the kitchen, and it gave higher growth in plants and nitrogen uptake. Black soldier fly frass was reported to promote the growth of plants when fed brewery waste (Beesigamukama et al., 2020). The bioconversion of organic solid wastes from municipal sources into compost with the use of larvae of black soldier fly was reported to enhance nitrogen, phosphorus, potassium contents, making it suitable for use as a fertilizer (Addo et al., 2022). Digestates, which are by-products of anaerobic digestion, can be used as biofertilizers. Owamah et al. (2014) obtained a digestate from anaerobic digestion of 12 kg wastes from food and 3 kg excreta from human and combined with water for 60 days at 22–31 °C. The biofertilizer was rich in nitrogen, total solids and reduced BOD, COD, ash content and organic carbon. The organisms prevalent were *Bacillus, Pseudomonas, Clostridium,* and *Penicillium.*

5.2 Biohydrogen

It is a renewable energy source that is clean, free of pollution, and is applied in pharmaceutical, food, and chemical industries. It is also used in oil refinery processes, generation of electricity and transportation processes (Zhang et al., 2022). The use of this type of fuel doesn't emit carbon dioxide and other gases when used as engine fuel in comparison with fossil fuel sources (Awogbemi & Kallon, 2022). Biohydrogen is produced from fermentation of materials organic in nature in the presence of microbes, electrolysis of microbes or water biophotolysis by microalgae (Chavan et al., 2022). The production techniques of biohydrogen can be dependent or independent of light and the production from biowaste is eco-friendly, economical, and sustainable (Xu et al., 2022). Aquaculture wastes, sewage sludge, residues from agro-industrial processes and wastes from pulp and paper production can be used for production of biohydrogen. The use of wastes from agro-industries produces easy conversion and high production of biohydrogen (Dong et al., 2018; Zhang et al. 2020a, b). Pawar et al. (2013) reported the wheat straw hydrolysate production using *Caldicellulosiruptor saccharolyticus* and produced 5.2 L hydrogen/L/day. Factors such as cost, availability, biodegradability, volatile solids, COD, BOD, presence of inhibitors, and nutrient contents determine the selection of organic waste that is suitable to produce biohydrogen (Chavan et al., 2022). Other factors such as pressure, temperature, pH, concentration of substrate, technology of bioconversion, and strain of microbes can also influence production of biohydrogen.

5.3 Bioethanol

It is a renewable biofuel that is in liquid and produced during starch, sugars or biomass fermentation from food wastes or agricultural materials (Awogbemi & Kallon, 2022). It the most used liquid biofuel in the world with the size in market at 40.8 million dollars in 2021 and forecast increase to 124.5 billion dollars in 2030 (Arya et al., 2022). It is commonly used as an alternative fuel and additives in personal care products, disinfectants, chemicals, and pharmaceuticals (Awogbemi & Kallon, 2022). The use of waste biomass from agriculture, municipal or industrial sources for production of bioethanol is eco-friendly, economical and promotes the management of wastes. Bioethanol is mostly produced from wastes such as sorghum straw, sugarcane bagasse, barley straw, sorghum straw, sweet sorghum bagasse, cassava, rice straw, etc. (Yogalakshmi et al., 2022). There are three main phases of process in production of bioethanol from organic wastes, which are the stage of pretreatment, cellulose hydrolysis, hemicellulose and starch stage to sugars that are fermentable and the stage of sugar fermentation in the presence of microbes into ethanol in the presence of *Saccharomyces cerevisiae, Trichoderma reesei, Zymonomas mobilis, Wickerhamomyces anomalus, Aspergillus niger,* etc. (Elsagan et al., 2023).

The most prominent culture microbes for production of bioethanol are *S. cerevisiae and Z. mobilis* as they have high tolerance to sugar and ethanol, have high productivity, and can grow in anaerobic or aerobic conditions (Bibra et al., 2023). Narisetty et al. (2022) produced ethanol from bread wastes that has been fermented and hydrolysed enzymatically by *S. cerevisiae* KL17. Similarly, Nunui et al. (2022) used *S. cerevisiae* on pretreated rubberwood taste and yielded substantial amount of bioethanol. Also pretreated lignocellulosic sawdust was pretreated and bioethanol was produced by *S. cerevisiae* at a speed of 300 rpm agitation, incubation time of 102 h, and temperature of 36.5 °C. Ma et al. (2016) used *Zymomonas mobilis* in fermenting pretreated kitchen and food wastes and yielded bioethanol. When culture mixtures are used for production of bioethanol, it promotes an interaction of organisms that is synergic in relationship and improves the metabolic activities, secretion of enzymes, hydrolysis of wastes and saccharification (Bibra et al., 2023). *S. cerevisiae* and *Fusarium oxysporum* were reported by Prasoulas et al. (2020) to be used in fermenting acid pretreated and hydrolysed food wastes. Potato waste was also reported to be fermented by *S. cerevisiae* and *Aspergillus niger* and the yield was bioethanol of 38 g/L (Izmirlioglu & Demirci, 2017). Bioethanol was reported to be produced from fermented food wastes by *Geobacillus* sp. and

Thermoanerobacter sp. at a temperature of 60 °C (Bibra et al., 2020). Likewise, bioethanol was produced from hydrolysed fermented palm kernel cake in the presence of recombinant *Geobacillus thermoglucosidaisus.*

5.4 Bioelectricity

It is a bioenergy produced from agricultural and forest residues, municipal and food wastes using methods such as gasification, direct combustion, thermochemical conversion, or Microbial Fuel Cell (MFC). A very effective and promising method is the use of MFC as it is an electrochemical technology that can convert organic contents in solid wastes into electricity under anaerobic conditions but electrogenic microorganisms must be present. The approach is a safe and clean form of producing electricity that is direct and eliminates pollution by removing all the components that be organic in nature and may be found in the solid waste. The FC has two chambers, the cathode and anode which are divided by a membrane made of proton. The solid wastes are usually oxidized with produce proton and electrons as by-products, which are transferred through the membrane and circuit that is external respectively (Sindhu et al., 2019). Factors such as pH, loading rate, hydraulic retention time, magnetic fields and concentration of substrate can determine the current and power MFC can generate. Miran et al. (2016) and Goud et al. (2011) converted orange peel wastes and canteen-based composite food material in the presence of *Enterococcus, Pseudomonas,* and *Paludibacter.*

5.5 Biogas

It is a gaseous fuel that is a by-product at the last stage in organic solid wastes digestion under anaerobic conditions. This conversion process is called methanogenesis (Rajendran et al., 2021). Crude biogas contains 50–70% methane, 30–49% carbon dioxide, trace amounts of water, nitrogen gas, ammonia, hydrogen gas, and siloxane, while biomethane is produced refined from biogas which contains methane (95–97%) and carbon dioxide (1–3%) (Chavan et al., 2022). Biogas is useful as fuel for vehicular movements, electricity, and heat production and operating power plants. The process eliminates emission of carbon, cost effective, sustainable and independence of energy. In some cases, obnoxious odour may be generated due to the presence of Hydrogen sulphide which may explode and can contribute to increasing greenhouse gases causing climate change (Awogbemi & Kallon, 2022). The use of food, municipal solid and agricultural wastes in the production of biogas is an economical, sustainable, and eco-friendly approach (Matsakas et al., 2017). Factors such as volatile solids, COD, BOD, carbon-to-nitrogen ratio, moisture content, and temperature can affect the biodigestibility of waste materials. Llano et al. (2021) investigated the production from municipal solid wastes, food wastes and lignocellulosic wastes and the municipal waste gave the highest optimum yield of biogas. Similarly, biogas was producer when municipal solid and domestic wastes were co-digested, and a yield of 360 mL methane/g was produced at an optimum rate of 2.9 g Volatile Solids/mL/d. The process of pre-heating, either physical, chemical, or biological, can improve the digestion of organic solid wastes and its efficacy in conversion (Dahunsi, 2019). A biomethane recovery of 518.58 mL/g/VS was produced because of the combination of thermophilic anaerobic co-digestion between corn stover and cattle dung after an initial pretreatment (Joseph et al., 2019).

6 Limitations in Organic Solid Waste Bioconversion into Biobased Products

The approach of organic solid waste biorefinery is seen as a technological approach that is sustainable, efficient, and eco-friendly. It can reduce the disposal of solid wastes and promotes reusing of materials termed wastes for product manufacture that are valuable. However, the approach is faced with some challenges that make them practicable on a large commercial scale. Some of the constraints are:

- Some feedstocks are complex and slow to biodegrade which is a major challenge on bioconversion into biobased products. Some pretreatment methods can be used to address these problems, such as thermal, chemical, biological, mechanical, and electrical techniques for improved efficiency bioconversion.
- The proper management of waste is affected by the inadequate methods of waste collection, transportation and separation of wastes, inadequate knowledges of techniques for bioconversion, financial constraints, and policies from the government. This can be addressed by proper and effective collection of wastes, public awareness on recent technologies, huge capital investments and enforcing strict penalties on indiscriminate dumping of refuse should be initiated by the government.
- The lack of design that is suitable for bioreactors, by-products accumulation, instability of processes hinders the production from organic solid wastes. A way out is the constant vacuum fermentation with liquid bleed and effective bioreactor construction that is by-product specific.
- The composting of organic solid waste usually has a long duration (< 100 days) before biobased products are formed. This great challenge can be addressed by using beds that are controlled for improving the activities of organisms.

7 Conclusion

Over the last decades, the rate at which solid wastes are generated and disposed indiscriminately have tremendously increased and caused serious environmental hazards and health threats. The wastes can be municipal solid wastes (MSW), industrial wastes, hazardous wastes, electronic wastes (e-wastes), construction and demolition wastes (C&D), healthcare wastes, and agricultural wastes. All these forms of wastes have varying effects on the environment. To address this, there is need for proper management of wastes to eliminate pollution and promote reuse of materials. The use of organic solid wastes for production of biobased products offers a lot of benefits, such as biohydrogen, biogas, bioethanol, biofertilizers, bioelectricity production. The use of this approach is a green and sustainable method for biobased products recovery from organic solid wastes. However, the release of obnoxious gases from indiscriminate burning and incineration can be addressed as global warming which is a major crisis most especially in the Global South are addressed. Gases released are clean, safe, and sustainable from biobased processes. For future enhancement of bioconversion techniques that can yield high bioproducts, the following are recommended:

- An eco-friendly approach such as the microbial fuel cell technology can be encouraged as it can produce high energy and gas yield from conversion of solid wastes.
- There should be more research into the use of cheap biocatalysts, industrial-scale plants and biomass mix for efficient processing and biobased product recovery.
- The in-depth understanding of organic solid wastes and their inoculants mechanisms of action is essential as the optimization strategies in bioprocesses are crucial in efficient formation of biobased products.
- Innovative technologies such as robotics, machine learning, artificial intelligence, modelling, and optimization tools should be encouraged towards enhanced bio efficiency and recover of products from organic solid wastes.

References

Addo, P., Oduro-Kwarteng, S., Gyasi, S. F., & Awuah, E. (2022). Bioconversion of municipal organic solid waste into compost using black soldier fly (*Hermetia illucens*). *International Journal of Recycling of Organic Waste in Agriculture, 11*, 515–526.

Agoh, E. C., Ogwu, M. C., Chukwuemeka, O. S., & Ekeledo, P. I. (2024). Environmental and human health effects of cassava processing and processing waste. In M. C. Ogwu, S. C. Izah, A. A. Cunha Alves, & babu, S.C. (Eds.), *Plant biology, sustainability and climate change, sustainable cassava* (pp. 203–219). Academic. https://doi.org/10.1016/B978-0-443-21747-0.00001-1

Altieri, R., & Esposito, A. (2010). Evaluation of the fertilizing effect of olive mill waste compost in short-term crops. *International Biodeterioration and Biodegradation, 64*, 124–128.

Amin, N., Aslam, M., Khan, Z., Yasin, M., Hossain, S., Shahid, M. K., Inayat, A., Samir, A., Ahmad, R., Murshed, M. N., Khurram, M. S., El Sayed, M. E., & Ghauri, M. (2023). Municipal solid waste treatment for bioenergy and resource production: Potential technologies, techno-economic-environmental aspects and implications of membrane-based recovery. *Chemosphere, 323*, 138196.

Anuardo, R. G., Espuny, M., Costa, A. C. F., Espuny, A. L. G., Kazançoğlu, Y., Kandsamy, J., & de Oliveira, O. J. (2023). Transforming E-waste into opportunities: Driving organizational actions to achieve sustainable development goals. *Sustainability, 15*(19), 14150.

Arya, P. S., Yagnik, S. M., Rajput, K. N., Panchal, R. R., & Raval, V. H. (2022). Valorization of agro-food wastes: Ease of concomitant-enzymes production with application in food and biofuel industries. *Bioresource Technology, 361*, 127738.

Awogbemi, O., & Kallon, D. V. V. (2022). Valorization of agricultural waste for biofuel applications. *Heliyon, 8*, e11117.

Azmi, A. A., Jai, J., Zamanhuri, N. A., & Yahya, A. (2018). Precious metals recovery from electroplating wastewater: A review. *IOP Conference Series: Materials Science and Engineering, 358*, 1757. https://doi.org/10.1088/1757-899X/358/1/012024

Beesigamukama, D., Mochoge, B., Korir, N. K., Fiaboe, K. K. M., Nakimbugwe, D., Khamis, F. M., Subramanian, S., Dubois, T., Musyoka, M. W., Ekesi, S., Kelemu, S., & Tanga, C. M. (2020). Exploring black soldier fly frass as novel fertilizer for improved growth, yield, and nitrogen use efficiency of maize under field conditions. *Frontiers in Plant Science, 11*, 574592.

Behera, S. S., & Ray, R. C. (2016). Solid state fermentation for production of microbial cellulases: Recent advances and improvement strategies. *International Journal of Biological Macromolecules, 86*, 656–669.

Bhatia, R., Bhatia, A., & Yadav, N. (2020). Agricultural waste management: A review. *Journal of Environmental Management, 269*, 110849.

Bhattacharya, S. S., & Kim, K.-H. (2016). Utilization of coal ash: Is vermitechnology a sustainable avenue? *Renewable and Sustainable Energy Reviews, 58*, 1376.

Bibra, M., Rathinam, N. K., Johnson, G. R., & Sani, R. K. (2020). Single pot biovalorization of food waste to ethanol by *Geobacillus* and *Thermoanaerobacter* spp. *Renewable Energy, 155*, 1032–1041.

Bibra, M., Samanta, D., Sharma, N. K., Singh, G., Johnson, G. R., & Sani, R. K. (2023). Food waste to bioethanol: Opportunities and challenges. *Fermentation, 9*, 8.

Burton, S. A. Q., & Watson-Craik, I. A. (1998). Ammonia and nitrogen fluxes in landfill sites: Applicability to sustainable landfilling. *Waste Management & Research, 16*, 41–53.

Capoor, M. R., & Parida, A. (2021). Current perspectives of biomedical waste management in context of COVID 19. *Indian Journal of Medical Microbiology, 39*, 171–178.

Ceballos, F., Hernández-Acosta, E., & Fonseca, G. (2018). Public awareness, education and participation for sustainable development: Regional insights from Mexico. *Sustainability, 10*(2), 531.

ChandiniKumar, R., Kumar, R., & Prakash, O. (2019). The impact of chemical fertilizers on our environment and ecosystem. In P. Sharma (Ed.), *Research trends in environmental sciences* (pp. 69–86). AkiNik Publications.

Chavan, S., Yadav, B., Atmakuri, A., Tyagi, R. D., Wong, J. W. C., & Drogui, P. (2022). Bioconversion of organic wastes into value-added products: A review. *Bioresource Technology, 344*, 126398.

Cherubini, F., & Jungmeier, G. (2010). LCA of a biorefinery concept producing bioethanol, bioenergy, and chemicals from switchgrass. *International Journal of Life Cycle Assessment, 15*(1), 53–66.

Cherubini, F., & Strømman, A. H. (2011). Life cycle assessment of bioenergy systems: State of the art and future challenges. *Bioresource Technology, 102*(2), 437–451.

Conlon, K., Jayasinghe, R., & Dasanayake, R. (2021). *Circular economy: Waste-to-wealth, jobs the global south*. Urban Studies and Planning Faculty Publications and Presentations. 308. https://pdxscholar.library.pdx.edu/usp_fac/308

Covre, W. P., Ramos, S. J., da Silveira Pereira, W. V., de Souza, E. S., Martins, G. C., Teixeira, O. M. M., et al. (2022). Impact of copper mining wastes in the Amazon: Properties and risks to environment and human health. *Journal of Hazardous Materials, 421*, 126688.

Dahunsi, S. (2019). Mechanical pretreatment of lignocelluloses for enhanced biogas production: Methane yield prediction from biomass structural components. *Bioresource Technology, 280*, 18–26.

Dalgaard, R., Schmidt, J., & Halberg, N. (2008). Life cycle assessment of margarine. *Journal of Cleaner Production, 16*(7), 786–792.

Das, S., Lee, S. H., Kumar, P., Kim, K. H., Lee, S. S., & Bhattacharya, S. S. (2019). Solid waste management: Scope and the challenge of sustainability. *Journal of Cleaner Production, 228*, 658–678.

De Baere, L., Mattheuws, B., De Vos, J., Vandecasteele, C., & Dewulf, J. (2016). Sustainability assessment of chemical processes: A tool for early-stage design? *Sustainable Production and Consumption, 8*, 82–99.

Department of Forestry, Fisheries and the Environment [DFFE]. (2018). *South Africa state of waste report* (p. 2018). Department of Forestry, Fisheries and the Environment.

Dong, L., Cao, G., Zhao, L., Liu, B., & Ren, N. (2018). Alkali/urea pretreatment of rice straw at low temperature for enhanced biological hydrogen production. *Bioresource Technology, 267*, 71–76.

EC (European Commission). (2015). *Directive of the European Parliament and of the council amending directive 2008/98/EC on waste* (p. 4). European Commission.

Egboduku, W. O., Egboduku, T., Golohor, O. M., Imarhiagbe, O., & Ogwu, M. C. (2024). Cassava as raw material for sustainable bioeconomy development. In M. C. Ogwu, S. C. Izah, A. A. Cunha Alves, & babu, S.C. (Eds.), *Plant biology, sustainability and climate change, sustainable cassava* (pp. 57–73). Academic. https://doi.org/10.1016/B978-0-443-21747-0.00022-9

Ellen MacArthur Foundation. (2015). *Towards the circular economy: Accelerating the scale-up across global supply chains*. Ellen MacArthur Foundation.

Ellen MacArthur Foundation. (2019). *Towards the circular economy: Accelerating the scale-up across global supply chains*.

Elsagan, Z. A., Ali, R. M., El-Naggar, M. A., El-Ashtoukhy, E. S. Z., & AbdElhafez, S. E. (2023). New perspectives for maximizing sustainable bioethanol production from corn Stover. *Renewable Energy, 209*, 608–618.

EPA. (2018). *Construction and demolition materials*. United States Environmental Protection Agency.

EPA. (2020). *Hazardous waste*. United States Environmental Protection Agency.

EPA. (2021). *Municipal Solid Waste*. United States Environmental Protection Agency.

Ericson, G., & Larsson, Å. (2000). DNA adducts in perch (*Perca fluviatilis*) living in coastal water polluted with bleached pulp mill effluents. *Ecotoxicology and Environmental Safety, 46*(2), 167–173.

FAO. (2019). *The state of food and agriculture 2019. Moving forward on food loss and waste reduction*. Food and Agriculture Organization of the United Nations.

FAO (2020) Sustainable agricultural practices for mitigating environmental impacts. Food and Agriculture Organization of the United Nations.

Ferronato, N., & Torretta, V. (2019). Waste mismanagement in developing countries: A review of global issues. *International Journal of Environmental Research and Public Health, 16*, 1060.

Gaur, V. K., Sharma, P., Sirohi, R., Awasthi, M. K., Dussap, C. G., & Pandey, A. (2020). Assessing the impact of industrial waste on environment and mitigation strategies: A comprehensive review. *Journal of Hazardous Materials, 398*, 123019.

Geyer, R., Jambeck, J. R., & Law, K. L. (2017). Production, use, and fate of all plastics ever made. *Science Advances, 3*(7), e1700782.

Gnansounou, E., & Dauriat, A. (2010). Techno-economic analysis of lignocellulosic ethanol: A review. *Bioresource Technology, 101*(13), 4980–4991.

Godfrey, L., & Oelofse, S. (2017). Historical review of waste management and recycling in South Africa. *Resources, 6*, 57.

Godfrey, L., Rivers, M., & Jindal, N. (2014). *Trends in waste management and priority streams for the waste RDI roadmap*. Pretoria, South Africa.

Gollakota, A. R., Gautam, S., & Shu, C. M. (2020). Inconsistencies of e-waste management in developing nations–Facts and plausible solutions. *Journal of Environmental Management, 261*, 110234.

Goud, R. K., Babu, P. S., & Mohan, A. V. (2011). Canteen based composite food waste as potential anodic fuel for bioelectricity generation in single chambered microbial fuel cell (MFC): Bio-electrochemical evaluation under increasing substrate loading condition. *International Journal of Hydrogen Energy, 36*, 6210–6218.

Grant, K., Oteng-Ababio, M., & Yeboah, D. (2013). The global north's export of e-waste to Africa: Environmental and economic implications. *Environmental Science and Pollution Research, 20*(5), 3224–3235.

Greencape. (2017). Waste economy: Market intelligence report 2017. In L. Basson, S. Bronkhorst, M. Mulcahy, S. O'Carroll, & C. Pineo (Eds.). Greencape.

Hairom, N. H. H., Soon, C. F., Mohamed, R. M. S. R., et al. (2021). A review of nanotechnological applications to detect and control surface water pollution. *Environmental Technology and Innovation, 24*, 102032. https://doi.org/10.1016/j.eti.2021.102032

Haq, I., Kumar, S., Kumari, V., Singh, S. K., & Raj, A. (2016). Evaluation of bioremediation potentiality of ligninolytic Serratia liquefaciens for detoxification of pulp and paper mill effluent. *Journal of Hazardous Materials, 305*, 190–199.

He, J., & Kappler, A. (2017). Recovery of precious metals from waste streams. *Microbial Biotechnology, 10*(5), 1194.

IPCC. (2019). *Climate change and land: An IPCC special report on climate change, desertification, land degradation, sustainable land management, food security, and greenhouse gas fluxes in terrestrial ecosystems*. IPCC.

Iqbal, M., Iqbal, N., Bhatti, I. A., Ahmad, N., & Zahid, M. (2016). Response surface methodology application in optimization of cadmium adsorption by shoe waste: A good option of waste mitigation by waste. *Ecological Engineering, 88*, 265–275.

ISWA. (2012). *Globalization and waste management*. Rotterdam, The Netherlands.

Iyiola, A. O., & Valbo-Jørgensen, J. (2023). Intrinsic values of the African environment: A sustainable perspective. In S. C. Izah & M. C. Ogwu (Eds.), *Sustainable utilization and conservation of Africa's biological resources and environment. Sustainable development and biodiversity* (Vol. 888). Springer. https://doi.org/10.1007/978-981-19-6974-4_23

Iyiola, A. O., Ipinmoroti, M. O., Kolawole, A. S., Akinsorotan, A. M., Oyewole, E. O., & Akinrinade, A. J. (2022a). Weather factors and their implications for fish species in Owalla reservoir, Osun State, Nigeria. *Journal of Global Affairs and Sustainable Development (JoGASD), 1*, 93–106.

Iyiola, A. O., Akinrinade, A. J., & Ajayi, F. O. (2022b). Effects of water pollution on biodiversity across coastal regions. In S. C. Izah (Ed.), *Biodiversity in Africa: Potentials, threats and conservation, sustainable development and biodiversity 29* (pp. 345–367). Springer. https://doi.org/10.1007/978-981-19-3326-4_13

Iyiola, A. O., Berchie, A., Oyewole, O. O., & Akinrinade, A. J. (2022c). Impacts of climate change on biodiversity in Africa. In S. C. Izah (Ed.), *Biodiversity in Africa: Potentials, threats and conservation, sustainable development and biodiversity 29* (pp. 369–394). Springer. https://doi.org/10.1007/978-981-19-3326-4_14

Iyiola, A. O., Akinsorotan, O. A., Ojeleye, A. E., & Fajimolu, A. O. (2023a). An overview of environmental resources in Africa: Emerging issues and sustainable exploitation. In S. C. Izah & M. C. Ogwu (Eds.), *Sustainable utilization and conservation of Africa's biological resources and environment. Sustainable development and biodiversity* (Vol. 888). Springer. https://doi.org/10.1007/978-981-19-6974-4_20

Iyiola, A. O., Kolawole, A. S., & Oyewole, E. O. (2023b). Sustainable alternatives to agrochemicals and their socio-economic and ecological values. In M. C. Ogwu & S. Chibueze Izah (Eds.), *One health implications of agrochemicals and their sustainable alternatives, sustainable development and biodiversity*. Springer. https://doi.org/10.1007/978-981-99-3439-3_25

Iyiola, A. O., Akinsorotan, A. M., & Ewutanure, J. S. (2023c). Impacts of agrochemicals on fish composition in natural waters: A sustainable management approach. In M. C. Ogwu & S. Chibueze Izah (Eds.), *One health implications of agrochemicals and their sustainable alternatives, sustainable development and biodiversity*. Springer. https://doi.org/10.1007/978-981-99-3439-3_24

Iyiola, A. O., Kolawole, A. S., Ajayi, F. O., Ogidi, O. I., & Ogwu, M. C. (2024a). Sustainable water use and management for agricultural transformation in Africa. In S. Madhav, A. L. Srivastav, S. C. Izah, & E. van Hullebusch (Eds.), *Water resources management for rural development:*

Challenges and mitigation (pp. 287–300). Elsevier. https://doi.org/10.1016/B978-0-443-18778-0.00001-5
Iyiola, A. O., Setufe, S. B., Ofori, E., Bilikoni, J., & Ogwu, M. C. (2024b). Biomarkers for the detection of pollutants from the water environment. In S. C. Izah, M. C. Ogwu, & H. Hamidifar (Eds.), *Biomonitoring of pollutants in the global south*. Springer. https://doi.org/10.1007/978-981-97-1658-6_16
Iyiola, A. O., Kolawole, A. S., Setufe, S. B., Bilikoni, J., Ofori, E., & Ogwu, M. C. (2024c). Fish as a sustainable biomonitoring tool in aquatic environments. In S. C. Izah, M. C. Ogwu, & H. Hamidifar (Eds.), *Biomonitoring of pollutants in the global south*. Springer. https://doi.org/10.1007/978-981-97-1658-6_12
Izah, S. C., Iyiola, A. O., Yarkwan, B., & Richard, G. (2023). Impact of air quality as a component of climate change on biodiversity-based ecosystem services. In A. L. Srivastav, A. K. Dubey, A. Kumar, S. K. Narang, & M. A. Khan (Eds.), *Visualization techniques for climate change with machine learning and artificial intelligence* (pp. 123–147). Elsevier. https://doi.org/10.1016/B978-0-323-99714-0.00005-4
Izmirlioglu, G., & Demirci, A. (2017). Simultaneous saccharification and fermentation of ethanol from potato waste by co-cultures of *Aspergillus niger* and *Saccharomyces cerevisiae* in biofilm reactors. *Fuel, 202*, 260–270.
Jönsson, L. J., & Martín, C. (2016). Pretreatment of lignocellulose: Formation of inhibitory by-products and strategies for minimizing their effects. *Bioresource Technology, 199*, 103–112.
Joseph, G., Zhang, B., Mahzabin, R. Q., Wang, L., & Shahbazi, A. (2019). Two-stage thermophilic anaerobic co-digestion of corn Stover and cattle manure to enhance biomethane production. *Journal of Environmental Science and Health, Part A, 54*, 452–460.
Kaza, S., Yao, L., Bhada-Tata, P., & Woerden, F. V. (2018). *What a waste 2.0: A global snapshot of solid waste management to 2050*. International Bank for Reconstruction and Development/The World Bank.
Kosoe, E. A., & Ogwu, M. C. (2024). Sustainable cassava processing: Processes, techniques, wastes, and waste streams. In M. C. Ogwu, S. C. Izah, A. A. Cunha Alves, & babu, S.C. (Eds.), *Plant biology, sustainability and climate change, sustainable cassava* (pp. 257–272). Academic. https://doi.org/10.1016/B978-0-443-21747-0.00011-4
Kuligowski, K., Konkol, I., Świerczek, L., Chojnacka, K., Cenian, A., & Szufa, S. (2023). Evaluation of kitchen waste recycling as organic N-fertiliser for sustainable agriculture under cool and warm seasons. *Sustainability, 15*, 7997.
Laha, S. (2015). *Formality in E-waste movement & management in the global economy*. The University of Manchester.
Li, J., Zeng, X., Li, X., & Huang, J. (2015). E-waste dismantling processes and their environmental impacts: An overview. *Waste Management, 45*, 11–20.
Lindholm-Lehto, P. C., Knuutinen, J. S., Ahkola, H. S., & Herve, S. H. (2015). Refractory organic pollutants and toxicity in pulp and paper mill wastewaters. *Environmental Science and Pollution Research, 22*, 6473–6499.
Llano, T., Arce, C., & Finger, D. C. (2021). Optimization of biogas production through anaerobic digestion of municipal solid waste: A case study in the capital area of Reykjavik, Iceland. *Journal of Chemical Technology & Biotechnology, 96*, 1333–1344.
Ma, K., Ruan, Z., Shui, Z., Wang, Y., Hu, G., & He, M. (2016). Open fermentative production of fuel ethanol from wood waste by an acid-tolerant mutant strain of *Zymomonas mobilis*. *Bioresource Technology, 203*, 295–302.
Mata-Alvarez, J., Dosta, J., Romero-Güiza, M. S., Fonoll, X., & Peces, M. (2014). Quantitative characterization of food waste in Spanish households. *Waste Management, 34*(11), 1977–1984.
Matsakas, L., Gao, Q., Jansson, S., Rova, U., & Christakopoulos, P. (2017). Green conversion of municipal solid wastes into fuels and chemicals. *Electronic Journal of Biotechnology, 26*, 69–83.
Medina, M. (2018). Construction and demolition waste management in developing countries: A review. *Waste Management and Research, 36*(7), 578–592.

Miran, W., Nawaz, M., Jang, J., & Lee, D. S. (2016). Conversion of orange peel waste biomass to bioelectricity using mediator-less microbial fuel cell. *Science of The Total Environment, 547*, 197–205.

Moyo, N. A., & Iyiola, O. A. (2020). Construction and demolition waste management in sub-Saharan Africa: Challenges and opportunities. *Sustainable Cities and Society, 62*, 102402.

Mpofu, A. B., Oyekola, O. O., & Welz, P. J. (2021). Anaerobic treatment of tannery wastewater in the context of a circular bioeconomy for developing countries. *Journal of Cleaner Production, 296*, 126490.

Muyambi, F. (2021). *The East African regional bioeconomy strategy – A summary*. https://atpsnet.orh/wp-content/uploads/2021/10/Final-Summary-EAC-BIOECONOMY-STRATEGY.pdf

Nair, R. B., Lee, J., & Kook, J. W. (2019). Potential bioenergy production from agricultural waste in Asia: Current status and future trends. *Renewable Energy, 134*, 38–48.

Narayana, S., Ramakrishna, P. V., & Prasad, J. V. N. S. (2021). Managing agricultural waste: A review. *Journal of Cleaner Production, 290*, 125732.

Narisetty, V., Nagarajan, S., Gadkari, S., Ranade, V. V., Zhang, J., Patchigolla, K., Bhatnagar, A., Awasthi, M. K., Pandey, A., & Kumar, V. (2022). Process optimization for recycling of bread waste into bioethanol and biomethane: A circular economy approach. *Energy Conversion and Management, 266*, 115784.

Nghiem, L. D., Davarzani, H., & Hai, F. I. (2017). Anaerobic membrane bioreactors for sustainable wastewater treatment and energy generation. *Current Opinion in Biotechnology, 45*, 51–56.

Nizzy, A. M., & Ogwu, M. C. (2024). Valorization of cassava processing by-products into biofuel for a sustainable environment. In M. C. Ogwu, S. C. Izah, A. A. Cunha Alves, & babu, S.C. (Eds.), *Plant biology, sustainability and climate change, sustainable cassava* (pp. 291–309). Academic. https://doi.org/10.1016/B978-0-443-21747-0.00010-2

Nunui, K., Boonsawang, P., Chaiprapat, S., & Charnnok, B. (2022). Using organosolv pretreatment with acid wastewater for enhanced fermentable sugar and ethanol production from rubberwood waste. *Renewable Energy, 198*, 723–732.

Ogwu, M. C., Iyiola, A. O., & Izah, S. C. (2023). Overview of African biological resources and environment. In S. C. Izah & M. C. Ogwu (Eds.), *Sustainable utilization and conservation of Africa's biological resources and environment* (Sustainable Development and Biodiversity, volume 888) (pp. 1–34). Springer. https://doi.org/10.1007/978-981-19-6974-4_1

Ogwu, M. C., Odozi, I. P., Ahonsi, O. C., Uleanya, K. O., & Odozi, E. B. (2024). Organic acid production from cassava. In M. C. Ogwu, S. C. Izah, A. A. Cunha Alves, & babu, S.C. (Eds.), *Plant biology, sustainability and climate change, sustainable cassava* (pp. 395–418). Academic. https://doi.org/10.1016/B978-0-443-21747-0.00009-6

Ouda, O. K. (2016). Treated wastewater use in Saudi Arabia: Challenges and initiatives. *International Journal of Water Resources Development, 32*(5), 799–809.

Owamah, H. I., Dahunsi, S. O., Oranusi, U. S., & Alfa, M. I. (2014). Fertilizer and sanitary quality of digestate biofertilizer from co-digestion of food waste and human excreta. *Waste Management, 34*, 747–752.

Pawar, S. S., Nkemka, V. N., Zeidan, A. A., Murto, M., & van Niel, E. W. (2013). Biohydrogen production from wheat straw hydrolysate using *Caldicellulosiruptor saccharolyticus* followed by biogas production in a two-step uncoupled process. *International Journal of Hydrogen Energy, 38*, 9121–9130.

Prasoulas, G., Gentikis, A., Konti, A., Kalantzi, S., Kekos, D., & Mamma, D. J. F. (2020). Bioethanol production from food waste applying the multienzyme system produced on-site by *Fusarium oxysporium* F3 and mixed microbial cultures. *Fermentation, 6*, 39.

Rajendran, N., Gurunathan, B., Han, J., Krishna, S., Ananth, A., Venugopal, K., & Sherly, P. R. B. (2021). Recent advances in valorization of organic municipal waste into energy using biorefinery approach, environment and economic analysis. *Bioresource Technology, 337*, 125498.

Sar, T., Harirchi, S., Ramezani, M., Bulkan, G., Akbas, M. Y., Pandey, A., & Taherzadeh, M. J. (2022). Potential utilization of dairy industries by-products and wastes through microbial processes: A critical review. *Science of the Total Environment, 810*, 152253.

Sindhu, R., Gnansounou, E., Rebello, S., Binod, P., Varjani, S., Thakur, I. S., Nair, R. B., & Pandey, A. (2019). Conversion of food and kitchen waste to value-added products. *Journal of Environmental Management, 241*, 619–630.
Sobihah, N. N., Zaharin, A. A., Nizam, M. K., Juen, L. L., & Kyoung-Woong, K. (2018). Bioaccumulation of heavy metals in maricultured fish, *Lates calcarifer* (Barramudi), *Lutjanus campechanus* (red snapper) and *Lutjanus griseus* (grey snapper). *Chemosphere, 197*, 318–324.
Soliman, N. K., & Moustafa, A. F. (2020). Industrial solid waste for heavy metals adsorption features and challenges; a review. *Journal of Materials Research and Technology, 9*(5), 10235–10253.
Soto-Hernández, M., Tenango, M. P., & García-Mateos, R. (Eds.). (2017). *Phenolic compounds: Natural sources, importance and applications*. BoD–Books on Demand.
Stockholm Environment Institute [SEI]. (2022). *New strategy to kickstart the bioeconomy in East Africa*. www.sei.org/features/strategy-bioeconomy-east-africa/
United Nations Environment Programme (UNEP). (2018). *Global waste management outlook*. https://www.unenvironment.org/resources/report/global-waste-management-outlook-2018
WHO. (2020). *Safe management of wastes from health-care activities*. World Health Organization.
Wilson, D. C., & Velis, C. A. (2015). Waste management–still a global challenge in the 21st century: An evidence-based call for action. *Waste Management & Research, 33*(12), 1049–1051.
Wojnowska-Baryła, I., Kulikowska, D., & Bernat, K. (2020). Effect of bio-based products on waste management. *Sustainability, 12*(5), 2088.
World Bank. (2019). *What a waste 2.0: A global snapshot of solid waste management to 2050*. Retrieved from https://openknowledge.worldbank.org/handle/10986/30317
Xu, M., Yang, M., Sun, H., Gao, M., Wang, Q., & Wu, C. (2022). Bioconversion of biowaste into renewable energy and resources: A sustainable strategy. *Environmental Research, 214*, 113929.
Yogalakshmi, K. N., Sivashanmugam, P., Kavitha, S., Kannah, Y., Varjani, S., AdishKumar, S., & Kumar, G. (2022). Lignocellulosic biomass-based pyrolysis: A comprehensive review. *Chemosphere, 286*, 131824.
Zhang, Y., Qiu, X., Chen, C., Yu, Z., & Hong, H. (2020a). Recent progress in microbial production of succinic acid. *CIESC Journal, 71*, 1964–1975.
Zhang, Y., Yuan, J., & Guo, L. (2020b). Enhanced bio-hydrogen production from cornstalk hydrolysate pretreated by alkaline-enzymolysis with orthogonal design method. *International Journal of Hydrogen Energy, 45*, 3750–3759.
Zhang, X., Zhou, M., Li, J., Wei, L., Dong, Y., Hou, H., Chen, C., & Wang, Z. (2021). Analysis of driving factors on China's industrial solid waste generation: Insights from critical supply chains. *Science of the Total Environment, 775*, 145185.
Zhang, Z., Jiang, D., Zhang, H., & Zhang, Q. (2022). Integrated technologies for biohydrogen production. In S. Varjani, A. Pandey, T. Bhaskar, S. V. Mohan, & D. C. W. Tsang (Eds.), *Biomass, biofuels, biochemicals* (pp. 141–159). Elsevier.

Waste-to-Value Technologies in the Global South

Ojeleye Abiola Elizabeth, Badmus Mubarak Olalekan, and Adeniji Adedeji Adeola

Abstract

Waste management challenges in the Global South are due to various reasons related to rapid urbanization, massive waste stream, improper waste collection and sorting, inadequate policies, and inappropriate treatment and recycling systems which contribute to environmental pollution, public health risks, and hindered sustainable development. Waste in the Global South is from plastics, electronics, organic, industries, clinics, construction, and demolition sites. The GS bioeconomy strategy for waste management continually involves an expansive and sustainable approach, incorporating various technologies to support biological resources efficiently. The strategies adopt biological treatments, bioenergy production, bioremediation, circular economy practices, sustainable agriculture, bioplastics, and biobased materials, with research and innovations. Noticeably, electronic waste is significant in the Global South. Improper disposal of electronic devices leads to environmental contamination and health hazards due to the presence of harmful components and heavy metals, making it a pressing concern for public health and the environment. Mechanical shredding and sorting, pyrometallurgical and hydrometallurgical processes, biotechnological methods, recycling, and refining technologies have been adopted to convert e-waste to valuable materials. Similarly, mechanical, pyrolysis, and chemical recycling,

O. A. Elizabeth (✉)
Department of Agronomy, College of Agriculture, Ejigbo Campus, Osun State University, Osogbo, Nigeria
e-mail: abiola.ojeleye@uniosun.edu.ng

B. M. Olalekan
Centre of Agriculture for Sustainable Development, University of Greenwich, London, UK

A. A. Adeola
Centre for Geography and Environmental Science, Penryn Campus, University of Exeter, Exeter, UK

M. C. Ogwu et al. (eds.), *Sustainable Bioeconomy Development in the Global South*, https://doi.org/10.1007/978-981-96-0305-3_8

among others, have been adopted in converting plastic wastes. Autoclaving, microwave treatment, waste-to-energy, chemical disinfection, shredding, and compaction are adopted to ensure safe and responsible disposal of medical waste. Composting methods, anaerobic digestion, bioconversion, and gasification are adopted in converting organic waste to valuable products. Technologies convert various types of waste into valuable resources energy or recycled materials. Others contribute to value in terms of resource recovery and environmental impact reduction. Implementing such solutions can contribute to sustainable development, reduce pollution, and foster innovation in the regions facing waste management issues. A predicted future gap in the waste-to-value technology in the GS can simply be bridged with improved infrastructure and accessible technology. Other key elements to tackle are regulatory frameworks, awareness and education, financial barriers, and e-waste complexity. Collaborative efforts with governments, private sectors, and communities are mandatory to promote research, investment, and development of sustainable waste-to-value solutions in the Global South.

Keywords

Global South · Technology · Public health · Sustainable development · Value · Waste · e-waste

Abbreviations

AD	Anaerobic digestion
BC	Bokashi Composting
CBC	Community-Based Composting
EEAW	Electrical and Electronic Appliances Wastes
WtE	Waste-to-Energy
WC	Window Composting
MBT	Mechanical Biological Treatment
HTC	Hydrothermal Carbonization
PET	Polyethylene Terephthalate
THC	Traditional Heap Composting

1 Introduction

To create a sustainable environmental system, it is necessary to share waste management strategies, monitoring tools, possibilities due to urban sprawl and industrialization as well as aggressive human population, setbacks, and opportunities to build a sustainable ecosystem (Srivastava et al., 2015). According to Hoornweg et al. (2015), waste also referred to as garbage accumulates quicker than other pollutants, such as GHGs. Between 2001 and 2011, the population of India increased from 27.81% to 31.16%: thus, it is referred to as the second largest and important Global South country with high urbanization. Just like in every country, urbanization is placing immense strain on food, shelter, and other natural resources (Cointreau, 2006a, b; Kathiravale & Muhd Yunus, 2008; Ogwu, 2019a, b). The world urban centers generate around 1300 million tons/ear (1.2 kg/capita/day) waste volume which may reach 2200 million tons/year by 2025 (World Bank, 2012).

Increasing waste has spurred scientists, NGOs, and companies around the Global South and developing countries to ruminate on ideas and invent technologies to reduce the building process. Wastes are sourced from different systems, Industrial waste, municipal solid waste, medical waste, etc. Waste disposal means the management of trash from its inception to its ultimate disposal (UNSD, 2017; Agoh et al., 2024). The waste management processes encompass different stages, including waste collection, transport, treatment, and disposal. Additionally, waste-related laws, technologies, and economic mechanisms are monitored and regulated. Wastes are either categorized as solid, liquid, or gas; each category has varied methods of managing them. Waste management handles waste categories from different systems. Oftentimes, wastes pose threats to humans and the environment (Giusti, 2009). Municipal solid waste from commercial, industrial, and residential activities makes up a significant portion of waste/trash management (US, EPA, 2023a, b). There are diverse waste management approaches depending on countries, regions, and sectors (Davidson, 2011). To build a safe and sustainable environment, proper management of waste is important, but it's a challenge affecting developing countries and cities in the Global South.

Waste made of plastic and electronics doesn't biodegrade and therefore will only pile up wherever it is disposed of. Any greenhouse gas that is present in plastic and electronic wastes will gradually leak into the atmosphere. E-waste releases harmful substances and toxic materials including mercury. These heavy metals hurt the environment and the health of people and animals exposed to them, but when properly disposed of, these metals also have adverse effects on the environment and the health of people and animals that come in contact to them.

Waste management is a pressing issue in many Global South regions due to rapid urbanization, population growth, and limited infrastructure. However, this challenge also presents an opportunity to adopt innovative waste-to-value (WtV) technologies that can transform waste into valuable resources while addressing environmental and socio-economic concerns. Waste-to-value technologies encompass a wide range of processes aimed at converting various types of waste into useful products, energy, or materials (Zheng et al., 2017; Tony, 2022).

Bioeconomy is expressed as a system that banks on the use *biological*, such as crops, forests, and microorganisms, to manufacture a range of products and services that support economic growth while preserving the environment. Government and policymakers are investing in research and development and renewable resources use (Kosoe et al., 2023; Akudugu & Ogwu, 2024). For a more sustainable, circular economy with reduced waste and more effective resource use can be achieved through the bioeconomy. Contemporary waste management methods often take sociological factors into account in addition to technological ones. The methods of managing waste have advanced, and more countries are gradually moving from using landfills to the waste-to-energy options listed below:

Biogas production: For GS regions with massive organic waste, biogas production represents a sustainable solution (UNEP, 2016). The process is carried out in the absence of oxygen *Anaerobic* digestion breaks down the organic materials to produce biogas, which majorly comprises gases like methane and carbon dioxide, which is a renewable energy source. The gases are used for cooking, heating, or generating electricity. *Waste-to-energy (WtE) plants* method embraces thermal processes like incineration or gasification to convert solid waste into heat or electricity energy. The method releases emissions that serve as and potential environmental threat but the modern WtE technologies incorporate advanced filtration systems to minimize pollution. In the Global South, small-scale decentralized WtE facilities can provide localized energy solutions and reduce reliance on fossil fuels (IRENA, 2019). Organic waste is converted to high-quality compost for agricultural use, increasing soil fertility and lowering demand for chemical fertilizers. *Composting* is a natural breakdown of organic waste into nutrient-rich soil amendments (FAO, 2017). *Plastic trash* is transformed into raw materials for new products using mechanical and chemical recycling procedures which minimizes incineration and lowers the demand for pure plastics (WEF, 2020).

Innovative technologies are emerging to recover valuable resources from various waste streams. From electronic waste (e-waste), metals and other valuable materials can be extracted while from sewage sludge nutrients like phosphorus and nitrogen can be recovered. In many developing countries in the GS, informal waste pickers aid in waste management. Community-based approaches empower these individuals by providing training, infrastructure, and support for waste segregation, recycling, and composting initiatives, creating opportunities for economic empowerment and environmental stewardship (UNEP, IETC 2016–2020). Implementing waste-to-value technologies in the Global South requires a multi-dimensional approach that addresses technical, economic, social, and policy challenges. Governments, NGOs, private sector actors, and local communities must collaborate to invest in infrastructure, promote innovation, raise awareness, and enact supportive policies to foster the adoption and scale-up of these technologies. By leveraging waste as a valuable resource, countries in the Global South can achieve sustainable development goals while creating a cleaner, healthier, and more prosperous future for all.

This study examines data gathered from over a hundred published studies on waste management techniques that span various regions in the global south. The studies presented are gathered from conferences, technical papers, reports, thesis, and peer-reviewed journals. Using specific words like "waste management," "bio-economy," "Global South," and "technology," pertinent works were found in the body of knowledge. These works were examined to comprehend the slow but steady change in waste management in the Global South. The chapter highlights the following primary goals while presenting a broad range of strategic advancements in WtE technologies: (i) to demonstrate the current global South Bio-economy Strategies for waste management; (ii) to give a summary of significant waste on prevailing scenario of critical and hazardous waste across selected countries in the global south; and (iii) to talk about the waste-to-value projects in the Global South. Additionally, the chapter provides a thorough description of the technologies used for waste conversion in the Global South for waste conversion.

2 Global South Bioeconomy Strategies on Waste Management

The bioeconomy in the Global South presents significant opportunities for sustainable waste management practices, leveraging organic materials to create value-added products, generate renewable energy, and mitigate environmental pollution. Bioeconomy strategies are in line with three important waste hierarchies: reduce, reuse, and recycle. Most waste reduction strategies rely on the waste hierarchy because they maximize product benefits and generate minimal end waste (Albert, 2011).

2.1 Waste of Importance in the Global South

2.1.1 Electricity Wastes

According to the UN, e-waste is any abandoned object that has a battery or plug and contains dangerous and poisonous materials like mercury that can severely endanger the health of people and the environment. Elevated levels of discretional income, growing urbanization, and industrialization, led the society to increased demand for electrical and electronic appliances (EEA). A high percentage of the EEA are short-lived, difficult, and or expensive to repair depending on the brand. Brands create products on consumers' demand while downgrading the old technology even when they are serviceable. Disposal of these appliances after they are damaged results in e-waste (electronic waste). Most e-waste encompasses valuable recyclable materials and nonrecyclable toxic materials like lead, mercury, and cadmium, hazardous to health and the environment (Arabella, 2024). There are two major types of substances in e-waste: hazardous *substances,* for example, cadmium, chromium, lead, mercury, chlorofluorocarbon, polyaromatic hydrocarbons ((PAHs), polybrominated diphenyl ethers (PBDEs), and (PCDD/Fs); and nonhazardous substances, for example, copper, selenium, zinc, Ag, gold, and platinum.

Both kinds of compounds have the potential to harm the ecosystem. When e-waste is processed improperly, the majority of them are released into the environment. Its effects have been significant for ecological risk assessments demonstrating that the soil extensively contaminated by PAHs is concentrated in areas with high population density, thus subjecting the residents to high health risk (Wierckx et al., 2015).

According to the *e-waste monitor*, there will be over 347 Mt. of un-recycled e-waste on earth in 2024 with China, the USA, and India producing the most e-waste (Landrigan et al., 2018). Table 1 shows the top five e-waste-producing countries in the Global South in terms of the total volume produced (million metric Tons).

2.1.2 Medical Wastes

Medical waste refers to any waste generated as a result of medical, dental, and veterinary activities, including diagnosis, treatment, immunization, or research (World Health Organization, 2017). Medical wastes are specifically differentiated according to their potential hazard and the risk they pose to human health and the environment. Medical wastes can be categorized into; infectious (wastes from surgeries and autopsies on patients with infectious diseases), pathological (human tissues, organs, body parts, and animal carcasses used in research or contaminated with human pathogens), chemical (expired, unused, or contaminated drugs and vaccines, and waste from the production of pharmaceuticals), pharmaceutical waste (solvents, disinfectants, and heavy metals contained in medical devices), genotoxic (substances that are (carcinogenic, mutagenic, or teratogenic used in cancer treatment), radioactive (contaminated with radioactive substances), and nonhazardous wastes, for example, paper and cardboards.

2.2 Waste-to-Value Projects in the Global South

"*Waste-to-value*" projects in the global south aim to convert various types of waste into valuable products, resources, or energy, thereby addressing environmental challenges, promoting sustainable development, and creating economic opportunities. These projects leverage innovative technologies, business models, and partnerships to extract value from waste streams while minimizing environmental impacts (Obahiagbon and Ogwu, Obahiagbon & Ogwu, 2023a, b, Obahiagbon & Ogwu, 2024a, b).

Organic waste to biogas production explains community-scale biogas digesters in rural areas, industrial-scale anaerobic digestion facilities for municipal organic waste, and decentralized biogas plants at agro-industrial sites, recycling plastics, paper, metals, and glass, for reuse, remanufacturing, or recycling into new products. Nonrecyclable wastes are converted into heat, electricity, or biofuels through incineration, gasification, or pyrolysis processes to convert waste to energy. Circular economy Initiatives involve closed-loop systems for waste management and resource utilization, aiming to minimize waste generation, maximize resource efficiency, and create value from waste streams. Innovative technologies and processes convert waste into high-value products, such as biobased materials, biochemicals, biofuels, and bioproducts. These projects often involve collaboration between

Table 1 The top 5 e-waste producing countries in the Global South

Country	Rank	E-waste generated (million metric tons)
China	1	10.1
India	2	3.2
Brazil	3	2.1
Indonesia	4	1.6
Mexico	5	1.2

research institutions, startups, and industry partners to develop and commercialize novel waste valorization technologies. Social enterprise and community-based projects that empower local communities, create livelihood opportunities, and address social and environmental challenges through waste management and value-added activities, for example, community recycling cooperatives, waste pickers' organizations, and entrepreneurship programs focused on waste collection, sorting, recycling, and up-cycling. Public-private partnerships and sustainable financing mechanisms that mobilize investment, technical expertise, and resources for waste-to-value projects, including blended finance, impact investing, and development assistance (Table 2).

3 Technologies Engaged in the Global South for Waste Conversion

Several countries and organizations have upgraded their national constitution to improve the reuse, recycling, and other forms of recovering waste to reduce disposal and handling errors. Recycling is an important method not only from the point of waste treatment view but also from the point of recovering valuable materials (Jirand & Eric, 2003). Generally, technologies adopted in the GS wrap around anaerobic digestion (Khanal, 2008), composting (Makan & Tabatabaei, 2019), pyrolysis (Bridgwater, 2012), gasification (Rezaei & Brown, 2014), incineration (Themelis & Ulloa, 2007), waste-to-energy (WtE) plants (Williams, 2013), bioremediation (Singh et al., 2019), mechanical biological treatment (MBT), and hydrothermal carbonization (HTC) (Funke & Ziegler, 2010). These technologies offer a range of options for converting waste into valuable products, energy, or resources, contributing to sustainable waste management practices in the Global South.

3.1 E-waste Technology

3.1.1 Mechanical Shredding and Sorting

Electrical and Electronic Appliances Wastes (EEAW) is a diverse and complex system, concerning the materials, components makeup, and brand. The particle size of waste is significant in the development of a cost-effective, sustainable, and friendly recycling system for the massive electrical electronic appliances waste stream. Consequently, in the process of recycling EEAW, selective disassembly—that is, separating potentially dangerous and/or valuable components—is essential. Mechanical methods that have been widely used in the recycling industry include screening, shape separation, magnetic separation, electrostatic separation, and jigging. Recycling EEAW is just the first step, though. The best shredded EEAW for separating criterion is less than 5 or 10 mm in size, with uniformly tiny particles. Disassembly (selective, focusing on removing valuable or hazardous materials), upgrading (mechanical/metallurgical processing to upgrade desirable material

Table 2 Waste-to-value projects/program in the Global South

Country	Project	Description	Reference
Nigeria	Waste to Wealth	Recycling and recovery of waste	Federal Ministry of Environment, Nigeria. "National Policy on Environment," 2013.
India	Plastics for Change	Plastic waste recycling and fair-trade principles for plastic waste pickers	Plastics for Change. https://plasticsforchange.org/
Brazil	Cataki	Recycling opportunities, social inclusion, and environmental sustainability	Cataki. https://cataki.org/
Indonesia	Waste4Change	Waste management services, waste reduction, recycling, and community empowerment	Waste4Change. https://waste4change.com/
South Africa	Waste to Energy Projects	Converts municipal solid waste into energy through incineration or gasification technologies	Department of Environmental Affairs, South Africa. "National Waste Management Strategy," 2011.
Kenya	Takataka Solutions	Waste collection, recycling, and composting services	Takataka Solutions. https://takatakasolutions.com/
Bangladesh	Waste concerns	Community-based composting projects, and Composting	Waste Concern. http://www.wasteconcern.org/
Philippines	The Plastic Flamingo Project	Recycles plastic waste into eco-bricks, providing affordable building materials	The Plastic Flamingo Project. Https://plasticflamingoproject.com/
Mexico	EcoCasa	Constructs affordable housing using recycled materials, including plastic bottles, tires, and glass	EcoCasa. https://ecocasa.mx/

content in preparation for the third step), and refining are currently the three main processes of EEAW recycling.

3.1.2 Pyrometallurgical and Hydrometallurgical Processes

Pyrometallurgical Processes

This involves the use of high temperatures to thermally decompose e-waste and recover metals through smelting, roasting, and refining. The end products from pyrometallurgical processes are metals like copper, aluminum, gold, silver, and palladium. The process of pyrometallurgical generates hazardous emissions and residues, requiring proper pollution control measures and waste management practices to minimize environmental and health impacts (Awasthi & Li, 2020). Despite the economic advantage the release of toxic fumes into the environment are the major limitations (Karthikeyan et al., 2021).

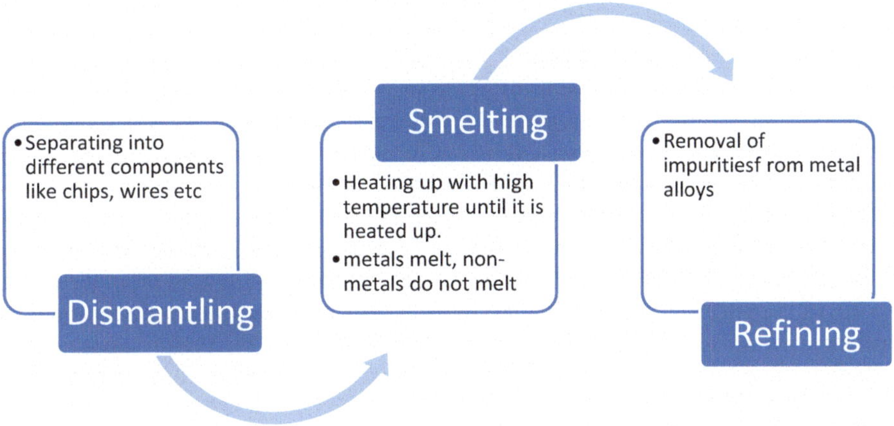

Hydrometallurgical Process

This process engages the use of aqueous solutions to dissolve metals from e-waste, followed by precipitation, purification, and recovery steps. *Leaching*: E-waste is crushed and treated with acidic or alkaline solutions to dissolve metals into the solution. *Precipitation*: Metal ions in solution are precipitated as insoluble compounds through chemical reactions or electrolysis. *Purification*: The precipitated metal compounds are further purified through filtration, solvent extraction, or other separation techniques. *Recovery*: Pure metal salts or metals are obtained from the purified solutions through crystallization, electrolysis, or other recovery methods. *Metals Recovered*: Hydrometallurgical processes can recover a wide range of metals from e-waste, including gold, silver, copper, palladium, and rare earth elements (Awasthi & Li, 2020).

Hydrometallurgical processes require careful control of process parameters, such as pH, temperature, and leaching agents, to optimize metal recovery efficiency and minimize chemical consumption and waste generation. These processes are often employed in recycling facilities and specialized e-waste treatment plants in the global south to recover valuable resources from discarded electronic devices.

3.1.3 Biotechnological Methods

Biotechnological methods for electronic waste (e-waste) treatment involve the use of biological processes or organisms to recover valuable materials or detoxify hazardous components; while biotechnological approaches are less common than physical or chemical methods, they offer potential advantages in terms of environmental sustainability and resource recovery. Biotechnological methods offer innovative approaches to address the environmental and health challenges associated with e-waste while simultaneously recovering valuable resources and promoting sustainable waste management practices in the Global South. Some biotechnological methods used for e-waste treatment are listed and explained below:

A. *Bioleaching*

Bioleaching involves the use of microorganisms to solubilize metals from e-waste materials by oxidizing them. This process can be used to extract valuable metals such as copper, gold, and silver from printed circuit boards (PCBs) and other electronic components. Microorganisms such as acidophilic bacteria and archaea produce organic acids and enzymes that break down metal sulfides in e-waste, releasing metal ions into solution (Ogwu et al., 2024). Bioleaching is relatively eco-friendly compared to traditional chemical leaching methods, as it reduces the need for harsh chemicals and generates less hazardous waste. Bioleaching may require optimization of process conditions (e.g., pH, temperature, oxygen supply) and selection of suitable microbial strains to maximize metal recovery efficiency.

B. *Bioremediation*

Bioremediation employs microorganisms to degrade or detoxify hazardous substances in e-waste, such as polychlorinated biphenyls (PCBs), brominated flame retardants, and heavy metals. Certain bacteria and fungi possess enzymes capable of breaking down organic pollutants or transforming them into less toxic forms through metabolic pathways. Bioremediation can be applied to treat e-waste-contaminated soil, water, or sediment, as well as to remove contaminants from wastewater generated during e-waste recycling processes. Bioremediation offers a cost-effective and sustainable approach to e-waste detoxification, particularly in cases where conventional methods are impractical or costly.

C. *Phytomining*

Phytomining involves the use of hyper-accumulating plants to extract and accumulate metals from e-waste-contaminated soils through their roots. Hyper-accumulating plants possess the ability to absorb and concentrate high levels of metals in their tissues without being adversely affected. After harvesting, the metal-rich plant biomass can be processed to recover metals.

Applications: Phytomining can be applied to remediate soils contaminated with heavy metals from e-waste disposal or recycling activities, while simultaneously recovering valuable metals for recycling. Phytomining offers a sustainable and environmentally friendly method for metal recovery and soil remediation, with minimal disruption to ecosystems.

D. *Microbial Fuel Cells (MFCs)*

Microbial fuel cells utilize microorganisms to catalyze oxidation process of organic matter in e-waste, generating electricity as a byproduct. Microorganisms, typically bacteria or archaea, oxidize organic substrates in the anode chamber of the MFC, releasing electrons that flow through an external circuit to the cathode, where they reduce an electron acceptor (e.g., oxygen). MFCs can be integrated into e-waste treatment systems to generate renewable electricity from organic waste streams, such as leachate from landfills or wastewater from recycling processes. MFCs offer a sustainable and decentralized energy generation solution, with potential applications in off-grid or remote e-waste recycling facilities. MFC performance may be affected by factors such as substrate availability, microbial activity, electrode materials, and system design. Optimization of MFC operation and scale-up for practical applications are ongoing challenges.

3.1.4 Recycling and Refining Technologies

Recycling and refining technologies for electronic waste (e-waste) are essential for recovering valuable materials and minimizing environmental pollution. These technologies involve various processes aimed at separating, recovering, and refining different components of e-waste. Cucchiella, et al. 2015) Common recycling and refining technologies used for e-waste are highlighted below.

Mechanical Separation

Mechanical separation processes involve shredding, crushing, grinding, and sorting e-waste components based on size, density, and magnetic properties. Mechanical separation is used to disassemble and segregate e-waste into individual components, such as plastics, metals, and printed circuit boards (PCBs), which can then be further processed using other recycling technologies. Mechanical separation is a cost-effective and scalable method for e-waste preprocessing, enabling the recovery of valuable materials for recycling. Mechanical separation may generate dust, noise, and mechanical wear, requiring proper ventilation, noise control, and maintenance measures to ensure worker safety and equipment longevity.

Electrostatic Separation

Electrostatic separation utilizes differences in electrical conductivity or surface charge to separate conductive and nonconductive materials in e-waste. E-waste materials are charged electrostatically and subjected to an electric field, causing particles with different conductivity or charge properties to be attracted to or repelled from electrodes, facilitating their separation. Electrostatic separation is particularly effective for separating plastics from metals in e-waste, allowing for the recovery of high-purity plastic fractions suitable for recycling. Electrostatic separation offers a nondestructive and environmentally friendly method for e-waste sorting, with high separation efficiency and minimal water or chemical consumption. Electrostatic separation performance may be influenced by factors such as particle size, moisture content, and surface characteristics, requiring optimization of process parameters for optimal separation outcomes.

Hydrometallurgical Processing

Hydrometallurgical processes involve aqueous solutions to dissolve and extract metals from e-waste materials, followed by precipitation, purification, and refining steps. Hydrometallurgical processing includes leaching, solvent extraction, precipitation, and electrochemical refining to recover metals (gold, silver, copper, and palladium) from e-waste. Hydrometallurgical processing is widely used for recovering precious metals from e-waste, particularly from PCBs and electronic components with high metal content. Hydrometallurgical processes offer high metal recovery rates, flexibility in processing different types of e-waste, and relatively low energy consumption compared to pyrometallurgical methods. Hydrometallurgical processing requires careful management of chemical reagents, wastewater treatment, and byproduct disposal to minimize environmental impacts and ensure process sustainability.

Pyrometallurgical Processing

Pyrometallurgical processes involve high-temperature treatment of e-waste materials to recover metals through smelting, roasting, and refining.

Steps in pyrometallurgical processing include thermal decomposition, metal melting, slag formation, and gas purification to extract metals from e-waste and produce metal alloys or ingots. Pyrometallurgical processing is commonly used for recovering base metals such as copper, aluminum, and iron from e-waste, as well as for treating residues from hydrometallurgical processes. Pyrometallurgical processing may generate emissions of air pollutants, hazardous residues, and energy-intensive operations, requiring pollution control measures and waste management strategies to mitigate environmental and health risks.

Plasma Arc Recycling

Plasma arc recycling utilizes high-temperature plasma torches to dissociate and vaporize e-waste materials into their constituent elements, which can then be recovered and refined. Plasma arc recycling is suitable for treating complex e-waste streams containing diverse materials, including refractory metals, ceramics, and composites, allowing for efficient metal recovery and slag valorization. Plasma arc recycling offers rapid and complete decomposition of e-waste materials, with minimal emissions, energy consumption, and waste generation compared to conventional pyrometallurgical methods.

Refining and Purification

Refining and purification processes involve further processing of recovered metals to remove impurities, increase purity levels, and meet quality specifications for reuse or resale.

Refining techniques such as electrolysis, distillation, and chemical precipitation are used to purify metals obtained from e-waste recycling processes, ensuring compliance with industry standards and market requirements. Refining and purification are essential steps in the recycling value chain for e-waste materials, enabling the production of high-quality recycled metals and alloys for manufacturing new electronic products. Refining processes improve the quality and market value of recycled metals, enhancing their competitiveness and sustainability as substitutes for virgin materials in electronics and other industries. Refining operations may consume additional energy, chemicals, and resources, increasing production costs and environmental footprint, and necessitating optimization of process efficiency and resource utilization.

3.2 Plastic Wastes

3.2.1 Mechanical

Mechanical technologies for plastic waste management in the global south involve various processes for sorting, shredding, washing, and pelletizing plastic waste to produce recyclable materials or products. These technologies play a crucial role in diverting plastic waste from landfills, reducing environmental pollution, and promoting circular economy practices.

A. *Sorting and Segregation*

Sorting and segregation technologies involve the separation of different types of plastic waste based on their polymer type, color, size, and density. Automated or manual sorting systems use techniques such as conveyor belts, air classifiers, optical sensors, and manual picking to separate plastics into distinct streams for further processing. Sorting and segregation are essential initial steps in plastic recycling facilities, enabling the separation of recyclable plastics (e.g., PET, HDPE, PP) from nonrecyclable materials and contaminants (e.g., PVC, mixed plastics, nonplastic items). Efficient sorting and segregation enhance the quality and value of recycled plastics, minimize contamination, and facilitate downstream processing and recycling operations. Challenges may include the complexity of plastic waste streams, labor-intensive sorting processes, and the need for advanced sorting technologies to handle diverse plastic materials effectively.

B. *Shredding and Size Reduction*

Shredding and size reduction technologies reduce PW into smaller fragments or granules, facilitating subsequent processing and recycling steps. Shredders, granulators, and crushers mechanically break down plastic waste into smaller particles through cutting, grinding, or shearing actions. Shredding and size reduction are critical for preparing plastic waste for recycling processes such as washing, extrusion, and pelletizing, as well as for volume reduction and storage optimization. Shredding and size reduction improve the handling, transportation, and processing efficiency of plastic waste while increasing surface area for cleaning and drying operations. Challenges may include the selection of appropriate shredding equipment, optimization of shredding parameters (e.g., speed, feed rate), and maintenance of equipment performance and durability.

C. *Washing and Cleaning*

Washing and cleaning technologies remove contaminants, residues, and impurities from shredded plastic waste through mechanical, chemical, and/or thermal treatment. Washing systems utilize water, detergents, agitation, and filtration to remove dirt, labels, adhesives, oils, and other contaminants from plastic flakes or pellets. Washing and cleaning are essential for improving the quality and purity of recycled plastics, particularly for food-grade and high-value applications requiring stringent cleanliness standards. Washing and cleaning enhance the marketability and usability of recycled plastics by reducing odor, color, and foreign matter content, while meeting regulatory requirements for product safety and quality. Challenges may include water consumption, wastewater treatment, energy usage, and operational costs associated with washing and cleaning processes, as well as the selection of appropriate cleaning agents and equipment.

D. *Pelletizing and Extrusion*

Pelletizing and extrusion technologies transform cleaned and dried plastic flakes or granules into uniform pellets or filaments suitable for manufacturing new plastic products. Pelletizers and extruders melt, compress, and shape plastic feedstock into cylindrical pellets or continuous profiles through heating, melting, and forming processes. Pelletizing and extrusion are key steps in the production of recycled plastic raw materials for injection molding, blow molding, and extrusion processes used in various industries. Pelletizing and extrusion convert plastic waste into value-added products with consistent quality, mechanical properties, and process ability, enabling closed-loop recycling and circular economy practices. Challenges may include the need for equipment maintenance, process control, and material characterization to ensure product quality and performance, as well as market demand and pricing fluctuations for recycled plastic pellets.

3.2.2 Pyrolysis

Pyrolysis of plastic waste in the GS is a thermal conversion process that decomposes plastic materials into smaller molecules in the absence of oxygen, typically at elevated temperatures. This process can help address plastic pollution and waste management challenges by converting plastic waste into valuable products such as fuel, chemicals, or solid residues (Al-Salem et al., 2009; UNEP, 2015).

Pyrolysis is a thermochemical decomposition process that breaks down complex organic compounds in PW into simpler molecules through the application of heat in the absence of oxygen. Plastic waste is fed into a reactor or pyrolysis chamber where it undergoes heating at temperatures typically ranging from 300 to 800 °C, depending on the type of plastic and desired products. In the absence of oxygen, plastic polymers decompose into gases (e.g., hydrocarbons), liquids (e.g., pyrolysis oil or bio-oil), and solid residues (e.g., char or carbon black).

Plastic waste is sorted, shredded, and dried to remove contaminants and moisture, preparing it for the pyrolysis process. The prepared PW is fed into a reactor or kiln, where it is heated to high temperatures in the absence of oxygen, initiating the pyrolysis reaction. The pyrolysis products, including gases, liquids, and solids, are recovered and separated through condensation, cooling, and filtration processes. Pyrolysis oil and gas can be further processed or refined to remove impurities and obtain high-quality fuel or chemical products. The product yield and composition of pyrolysis depend on various factors such as the type of plastic feedstock, pyrolysis conditions (temperature, residence time), and reactor design. Pyrolysis typically yields a mixture of products, including pyrolysis oil (liquid fuel), pyrolysis gas (syngas or combustible gases), and solid char or carbon black. The composition of pyrolysis products may vary, with pyrolysis oil containing a mixture of hydrocarbons suitable for fuel applications, while pyrolysis gas can be used for heat or electricity generation. Pyrolysis of plastic waste offers several potential applications in the GS, including:

(a) Production of liquid fuels: Pyrolysis oil can be used as a substitute for diesel fuel or as feedstock for refining into transportation fuels.
(b) Generation of syngas: Pyrolysis gas can be utilized for heat and power generation through combustion or gasification processes.
(c) Recovery of carbon black: Solid residues from pyrolysis can be processed and used as filler or reinforcement in rubber, plastics, or construction materials.

Pyrolysis of PW produces valuable products, reducing reliance on fossil fuels and mitigating environmental pollution; Energy recovery from plastic waste, contributing to sustainable waste management and resource conservation; Potential for decentralized and community-based pyrolysis systems, providing economic opportunities and energy access in underserved areas.

Despite its potential benefits, pyrolysis of PW faces challenges such as Technological and infrastructure constraints, including the need for appropriate pyrolysis equipment, operational expertise, and environmental controls; Economic viability and market competitiveness of pyrolysis products compared to conventional fuels or raw materials; Environmental and health risks associated with emissions, residues, and byproducts from pyrolysis operations, requiring appropriate mitigation measures and regulatory oversight.

Pyrolysis of plastic waste presents a promising technology for converting PW into valuable resources, addressing environmental challenges, and promoting sustainable development in the GS. However, successful implementation requires careful consideration of technical, economic, environmental, and social factors to ensure its effectiveness and sustainability.

3.2.3 Chemical Recycling

Chemical recycling is a process that involves breaking down plastic waste into its molecular building blocks or converting it into new chemicals or fuels through chemical reactions. It is also termed feedstock recycling. Chemical recycling aims to chemically degrade or transform plastic polymers into simpler molecules that can be used as raw materials for the production of plastics, fuels, or other chemical products.

Chemical recycling typically involves one of the following processes:

1. Depolymerization: This process breaks down polymer chains into monomers or smaller molecules using chemical catalysts, solvents, or other agents. The resulting monomers can then be purified and used to produce new plastic materials with properties similar to virgin plastics. Depolymerization is particularly suitable for plastics like polyethylene terephthalate (PET) and polystyrene (PS) (Geyer et al., 2015; Kühnert et al., 2021).
2. Solvolysis: This involves dissolving plastics in a solvent and subjecting them to chemical reactions to break down polymer chains into smaller molecules or monomers. Solvolysis can be performed using various solvents and reaction conditions, and it offers the advantage of being able to handle mixed or contaminated plastic waste streams (Arrieta et al., 2020; Chakrabarti et al., 2021).
3. Hydrolysis: It is a chemical reaction in which polymers are broken down into monomers or oligomers by reacting with water or steam. This process is commonly used for plastics like PET, polyamides (e.g., nylon), and polycarbonates. The resulting monomers can be purified and reused to produce new plastics (Silva et al., 2018; Sonne et al., 2019).
4. Glycolysis: This involves reacting plastics with glycols, such as ethylene glycol, to break down polymer chains and obtain monomers or oligomers. Glycolysis is particularly effective for recycling PET and polyester-based plastics, yielding high-purity monomers that can be used to produce new PET or polyester materials (Geyer et al., 2015; Ciacci et al., 2021).

Chemical recycling offers several potential advantages over mechanical recycling, including the ability to handle mixed or contaminated plastic waste streams, produce high-quality recycled materials, and close the loop on plastic waste by returning it to the production of new plastics.

3.3 Medical Wastes

Medical wastes (MW), as the name implies, are generated in healthcare facilities and hospitals. Mishandling of medical wastes exposes humans to environmental and occupational health risks. Most GS countries are constrained by resource especially when it comes to appropriate management of MS. The absence of knowledge and awareness on proper handling and inadequate safety equipment during collection and gathering of medical waste by hospital sanitary workers and scavengers calls for attention. Adopting any or integrated practices of these technologies in management practices can bring about magnanimous reduction of toxic effects of hospital wastes (Ali et al., 2017). Waste minimization, labeling, separation, and segregation are examples of MW management approaches. Additionally, hospital waste can be efficiently handled by technology using landfilling, autoclaving, chemical disinfection, microwave disinfection, plasma gasification, and pyrolysis (Magda & El-Salam, 2010; Shareefdeen, 2022) (Figs. 1 and 2).

3.3.1 Autoclaving

Autoclaving of medical waste in the GS follows a process similar to that in more developed regions, but it often faces unique challenges due to resource constraints, infrastructure limitations, and varying regulatory frameworks. Autoclaving follows the trend below:

Segregation and Collection: Medical waste is initially segregated at the source into different categories such as infectious waste, sharps, pathological waste, pharmaceutical waste, and nonhazardous waste. This segregation is crucial for efficient and safe disposal. In some cases, this segregation might be lacking due to limited awareness or infrastructure. ***Transportation***: Once segregated, the medical waste is collected and transported to a central facility equipped with an autoclave. In the

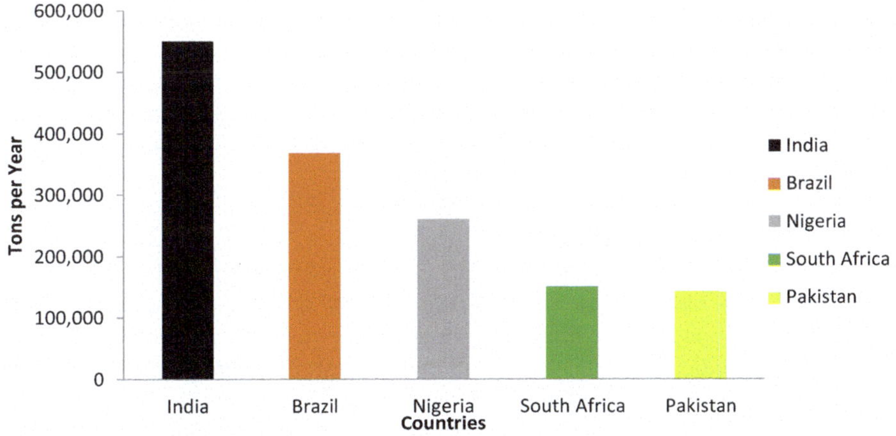

Fig. 1 Estimated medical waste generation (tons/year)

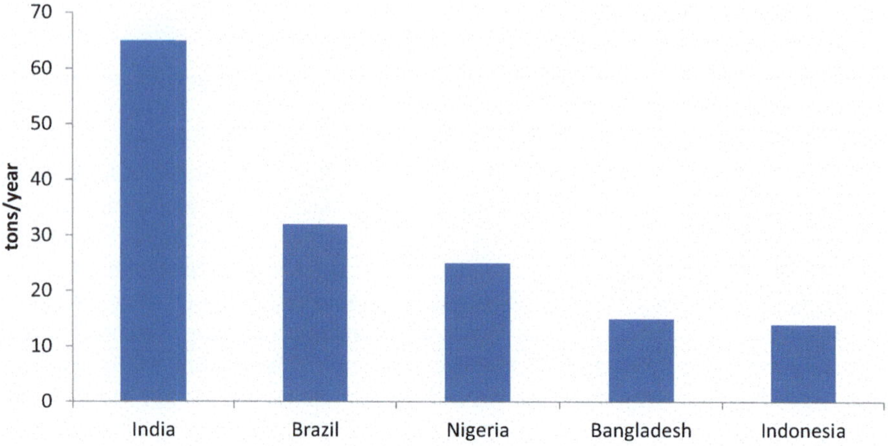

Fig. 2 Estimated organic waste generation in selected countries in the Global South. (Source; Ministry of Environment (Brazil, Nigeria, and Indonesia); Bangladesh Planning Commission and The Energy and Resource Institute (TERI))

global south, transportation logistics might be challenging due to inadequate road infrastructure, limited access to vehicles, or long distances between healthcare facilities and disposal sites. ***Pretreatment***: Before autoclaving, certain types of MW may undergo pretreatment processes such as shredding or compacting to reduce volume and facilitate more efficient autoclaving. Again, the availability of such pretreatment facilities may vary based on resource availability. ***Autoclaving***: The MW is loaded into the autoclave, which is a sealed chamber that uses high pressure and steam to sterilize the waste. Typically, the waste is subjected to temperatures of around 121–134 °C (250–273 °F) for a specified period (usually around 15–45 minutes), effectively killing pathogens and rendering the waste noninfectious.

After sterilization, the waste is allowed to cool and dry before further handling to prevent microbial growth and to ensure the waste is safe for subsequent handling and disposal. Regular monitoring of autoclave parameters such as temperature, pressure, and cycle duration must be carried out for quality control check. Once sterilized, the MW is typically disposed of through landfilling, incineration, or other approved methods. In some cases, treated waste may also be recycled or repurposed if feasible and safe.

3.3.2 Microwave Treatment

Microwaving of MW in the Global South is a less common method compared to autoclaving or incineration. However, it offers several advantages, including lower energy consumption, shorter processing times, and potentially reduced emissions. As with other medical waste management methods, proper segregation and collection of waste streams are essential. Collected medical waste is transported to a facility equipped with a microwave treatment system. Before microwaving, medical waste may undergo preprocessing steps such as shredding or size reduction to facilitate more efficient treatment.

In the microwave treatment process, MW is subjected to high-frequency electromagnetic radiation, typically at a frequency of 915 MHz or 2.45 GHz. This radiation heats the waste rapidly and uniformly, effectively sterilizing it by killing pathogens. After microwave treatment, the sterilized waste is allowed to cool before further handling. Proper cooling is essential to ensure the safety of workers and prevent microbial recontamination. Quality control measures, such as temperature monitoring and validation of sterilization efficacy, are implemented to ensure that the microwave treatment process is effective in rendering the waste noninfectious. Once sterilized, the treated MW can be disposed of through landfilling or other approved methods. In some cases, depending on local regulations and waste characteristics, the treated waste may also be suitable for recycling or repurposing.

3.3.3 Waste-to-Energy

The conversion of different waste types to energy (electricity, heat, or fuel) is termed waste-to-energy (WtE) technologies In the Global South, where energy demand often outstrips supply and waste management infrastructures may be inadequate, WtE technologies offer potential solutions for both waste management and energy generation. The first stage is the collection of health facility wastes infectious materials, sharps, and pharmaceuticals and segregating them to ensure proper handling. Segregation is crucial to prevent contamination and ensure the safe processing of waste. MW may undergo *pretreatment* steps such as shredding, sterilization, or drying to reduce volume, eliminate pathogens, and improve combustion efficiency. Technologies suitable for processing medical waste include incineration, pyrolysis, and gasification. *Incineration* is the method of combusting waste at high temperatures to generate heat that can be used to produce steam to drive turbines and generate electricity. The process of pyrolysis and gasification involve heating waste in the absence of oxygen to produce syngas used for power generation or other forms of energy such as liquid fuels or heat.

The energy produced from WtE processes can be used to meet the electricity and heat demands of local communities or healthcare facilities, reducing reliance on fossil fuels and mitigating greenhouse gas emissions. In some cases, excess energy generated from WtE plants can be fed into the grid for wider distribution while WtE technologies offer the potential for sustainable waste management and energy generation, they also raise environmental concerns such as air pollution, ash disposal, and greenhouse gas emissions. Proper emission control systems, ash management practices, and regulatory oversight are essential to mitigate these impacts and ensure the environmental sustainability of WtE facilities.

3.4 Organic Wastes

Organic wastes are generated from different sources including households, agriculture, food processing/manufacturing industries, and markets. Organic wastes can be categorized as materials ranging from food scraps, garden waste, agricultural residues, and animal manure. Organic wastes are generated from households (kitchen waste, including fruit and vegetable peels, leftover food, and food packaging materials); agricultural activities (crop residues, pruning waste, and animal manure); food processing industries (by-products, trimmings, and expired or unsold products); and markets/vendors/hawkers (unsold fruits, vegetables, and peels, other perishable items). In the GS, managing organic waste is a significant challenge due to increasing population growth, rapid urbanization, change in consumption patterns, and inadequate waste management infrastructure.

The major challenge of organic waste management is lack of infrastructure for organic waste collection, transportation, and treatment, leading to open disposal and incineration of waste, which contributes to environmental pollution and other health hazards. The role of informal Waste pickers in collecting and recycling organic waste is challenged with marginalization, health risks, and lack of adequate training and equipment. In addition, public awareness and behavior Change must be emphasized among the populace. Sorting of waste cannot be overemphasized; inappropriate segregation obstructs recycling.

Weak policy implementation, inadequate enforcement of regulations, and limited funding for waste management projects pose significant challenges.

Sustainable organic waste management practices are: *Composting*: a natural process that converts organic waste into nutrient-rich compost, which can be used to improve soil fertility and enhance agricultural productivity. *Biogas Generation*: Anaerobic digestion of organic waste produces biogas, a renewable energy source that can be used for cooking, heating, or electricity generation. Biogas plants are increasingly being implemented in rural and peri-urban areas to manage organic waste and provide clean energy. Also, *Waste-to-Energy*: Converts some organic waste, such as agricultural residues and animal manure, to feedstock for biomass and energy generation through processes such as gasification or combustion, providing an alternative to fossil fuels and reducing greenhouse gas emissions.

3.4.1 Composting Methods

Composting is a widely practiced method of managing organic waste in the Global South due to its simplicity, low cost, and environmental benefits. Various composting methods are employed, ranging from traditional heap composting to more advanced techniques tailored to local conditions and available resources (Paritosh et al., 2017). Some common composting methods used for organic waste management in the GS:

Traditional heap composting (THC): This involves piling organic waste in layers, alternating between green (nitrogen-rich) and brown (carbon-rich) materials. The compost pile is periodically turned or aerated to promote decomposition and facilitate the breakdown of organic matter by microorganisms. THC is simple, requires minimal investment in equipment, and can be implemented at the household or community level. It produces nutrient-rich compost that can be used to improve soil fertility and enhance crop yields. THC may take several months to produce mature compost, and the process can be labor-intensive, especially for large-scale operations.

Vermicomposting: Earthworms are employed to decompose organic waste and convert into nutrient-rich vermicompost, also known as worm castings. Organic waste is placed in a vermicomposting bin or bed along with bedding material (e.g., shredded paper, coconut coir), and earthworms are introduced to the system. The worms feed on the organic matter, digesting it and excreting nutrient-rich castings. Vermicomposting is well-suited for household or small-scale operations, as it requires minimal space and can be done indoors or outdoors. It produces high-quality compost with a balanced nutrient content and promotes the growth of beneficial soil microorganisms. Vermicomposting may require careful management of environmental conditions such as moisture, temperature, and pH to ensure optimal worm activity and composting efficiency.

Windrow composting (WC): It involves arranging organic waste in long, narrow piles or windrows, which are periodically turned to aerate the composting material and accelerate decomposition. Organic waste is placed in windrows typically on a prepared composting pad or surface. The windrows are periodically turned using specialized equipment or manually to ensure proper aeration and mixing of the composting material. WC is suitable for large-scale operations and can process significant quantities of organic waste efficiently. It produces compost relatively quickly compared to heap composting and can be adapted to different types of organic waste, including agricultural residues and municipal green waste. This method requires space, infrastructure, and equipment for turning the compost piles, which may not be readily available in resource-constrained settings.

Bokashi composting (BC): This is a fermentation-based method that utilizes a special mix of microorganisms to break down organic waste anaerobically. Organic waste is mixed with a Bokashi bran inoculated with beneficial microorganisms and packed tightly into an airtight container or bucket. The waste undergoes fermentation, producing a precompost material that can be further composted or buried in soil. BC is fast, odor-free, and can handle a wide range of organic waste types, including meat, dairy, and cooked food scraps. It is well-suited for urban and household composting applications. BC requires the procurement of Bokashi bran or microbial inoculants, which may not be readily available in some regions. It also requires careful management of fermentation conditions to prevent mold growth and ensure effective decomposition of waste.

Community-based composting (CBC): involves the collective participation of community members in composting activities. Organic waste is collected from households, markets, or institutions and processed at centralized composting facilities or decentralized composting sites within the community. Community members may be involved in waste segregation, composting operations, and distribution or use of compost products. CBC promotes social cohesion, environmental awareness, and sustainable waste management practices.

3.4.2 Anaerobic Digestion

In the Global South, organic waste is increasingly being treated via anaerobic digestion (AD). Anaerobic digestion can simultaneously treat organic waste and produce biogas for use in energy generation (Khanal, 2008). In anaerobic digestion, organic matter is broken down by microorganisms into digestate and biogas while there is no oxygen present. Organic waste, such as food scraps, agricultural residues, animal manure, and wastewater sludge, is placed in a sealed anaerobic digester, where it undergoes decomposition by anaerobic bacteria. In anaerobic digesters, organic waste is mixed with water and heated to an optimal temperature range (typically between 35 and 55 °C) to create conditions conducive to microbial activity. Anaerobic bacteria break down complex organic molecules into simpler compounds through a series of biochemical reactions, producing biogas as a byproduct. Biogas primarily consists of methane (CH_4) and carbon dioxide (CO_2), with smaller amounts of other gases such as hydrogen sulfide (H_2S) and trace contaminants. The produced biogas can be collected and used as a renewable energy source for various applications.

Anaerobic digestion of organic waste is applicable to a wide range of waste types and can be implemented at different scales, from household or community-level digesters to large-scale industrial facilities. In the global south, small-scale anaerobic digesters are often used in rural areas for the treatment of agricultural residues, animal manure, and household organic waste. These digesters provide renewable energy for cooking and lighting, as well as nutrient-rich digestate for agriculture. Larger-scale anaerobic digestion facilities are also being developed in urban areas to treat municipal organic waste, food waste from markets and restaurants, and wastewater sludge.

Anaerobic digestion of organic waste offers several environmental, economic, and social benefits, including reduction of methane emissions from organic waste decomposition in landfills, mitigating climate change; treatment and valorization of organic waste, converting it into valuable products such as biogas and digestate; creation of job opportunities in waste management, renewable energy, and agriculture sectors; improvement of sanitation and public health through the safe and hygienic treatment of organic waste, reducing odor and disease transmission; and challenges associated with anaerobic digestion of organic waste in the GS (Table 3).

3.4.3 Gasification

Gasification is regarded as a clean and effective process of converting low-quality biomass (organic waste) to higher-valued gas for solving various waste utilization challenges (Basu, 2013; Caminati et al., 2022). The process converts biomass and other organic materials into synthesis gas (syngas), a mixture of carbon monoxide, hydrogen, and other gases, through a thermochemical process in a controlled environment. Syngas can be used as a fuel to generate heat and power. It can also serve as feedstock for the production of biofuels and chemicals. Gasification offers a promising technology for organic waste management in regions with abundant biomass resources, providing an environmentally sustainable alternative to landfilling and open burning (Bridgwater, 2012).

Gasification is a thermochemical conversion process that converts organic materials into syngas by reacting them with a controlled amount of oxygen or steam at elevated temperatures (typically 700–1500 °C) in a gasifier. Organic waste

Table 3 Bioconversion methods for organic waste management in the Global South

Bioconversion method	Description	Applications	Advantages	Challenges	References
Composting (humus-rich soil conditioner)	Biological decomposition of organic waste by microorganisms under controlled conditions	Soil amendment; organic fertilizer; waste diversion from landfills; carbon sequestration	Produces nutrient-rich fertilizer; reduces greenhouse gas emissions; low-cost and scalable technology	Require space, time, and labor; odor and pest management; monitoring and quality control	Lo et al. (2020)
Vermicomposting	Biological decomposition of organic waste using earthworms (e.g., red wigglers), which consume and digest organic matter, producing vermicompost	Organic fertilizer production; soil health improvement; waste reduction and recycling	Accelerates composting process; enhances soil structure and fertility; suitable for small-scale and decentralized operations	Requires suitable earthworm species and habitat; moisture and temperature management; odor and pest control	Domínguez et al. (2005)
Anaerobic digestion	Microbial breakdown of organic waste in the absence of oxygen, producing biogas (methane and carbon dioxide) and digestate (nutrient-rich fertilizer)	Biogas production for energy; organic waste treatment; nutrient recovery for agriculture	Renewable energy generation; waste valorization and resource recovery; reduction of methane emissions from landfills	Requires suitable feedstock composition and pretreatment process optimization for biogas yield; digestate management and utilization	Mata-Alvarez et al. (2000)
Fermentation	Microbial conversion of organic waste into useful products (e.g., ethanol, organic acids) through Anaerobic or aerobic fermentation processes	Biofuel production (e.g., ethanol), chemicals and biochemical Waste valorization and resource recovery	Production of renewable fuels and chemicals; utilization of diverse feedstocks (e.g., food waste, agricultural residues); reduction of waste and environmental pollution	Process optimization for product yield and purity; selection of Suitable microbial strains and fermentation conditions; economic viability and market competitiveness	Mohd and Abdullah (2021)
Aerobic degradation	Microbial degradation of organic waste in the presence of oxygen, resulting in the breakdown of organic compounds and production of carbon dioxide, water, and heat	Waste treatment and stabilization; soil and environmental remediation; bioremediation of organic pollutants	Efficient organic waste management; soil health improvement and restoration; mitigation of greenhouse gas emissions	Requires aeration and mixing for oxygen supply; control of moisture and temperature; monitoring and management of process	Xiao and Wan (2018)

feedstocks for gasification can include agricultural residues, forestry residues, municipal solid waste (MSW), organic sludge, and energy crops. The gasification process occurs in several stages, including drying, pyrolysis (thermal decomposition), gasification, and combustion or oxidation of remaining char (Mohan et al., 2006).

Process steps Organic waste feedstock is sorted, shredded, and dried to remove moisture and contaminants, preparing it for gasification. The prepared feedstock is fed into the gasifier, where it undergoes thermochemical reactions at high temperatures in the presence of a controlled amount of oxygen, steam, or a combination of both. Gasification produces syngas, which consists mainly of carbon monoxide (CO), hydrogen (H_2), carbon dioxide (CO_2), and traces of other gases such as methane (CH_4). The raw syngas may undergo cleaning and conditioning processes to remove impurities such as tar, sulfur compounds, and particulates, improving its quality for downstream applications. The cleaned syngas can be used for various purposes, including heat and power generation through combustion in boilers, engines, or gas turbines, or as a feedstock for the production of biofuels, chemicals, or hydrogen.

Applications Several applications work with the gasification process of organic waste; *Electricity and heat generation*: Syngas or combined heat and power (CHP) systems can be used as fuel for power plants providing renewable energy for local communities. *Biofuels and chemicals*: Syngas can be converted into liquid fuels (e.g., ethanol, methanol) or chemicals (e.g., ammonia, synthetic natural gas) through catalytic processes such as Fischer–Tropsch synthesis. *Waste-to-energy projects*: Gasification plants can be integrated with waste management systems to divert organic waste from landfills, reduce greenhouse gas emissions, and generate revenue from energy sales or carbon credits. Gasification of organic waste presents a promising technology for sustainable waste management and renewable energy production in the global south, offering environmental, social, and economic benefits.

4 Future Gaps

Due to a lack of legislation and suitable treatment facilities, improper waste handling is one of the major causes of environmental pollution and the degradation of many cities, especially in developing nations. Numerous researchers claim that it is a known fact that India's e-waste-related heavy metal pollution has moved from unofficial activities into the neighboring environment including plants (Fujimori et al., 2012; Ejiogu, 2013).

A predicted future gap in the waste-to-value technology in the Global South can simply be bridged with improved infrastructure and accessible technology. Other key elements to tackle are regulatory frameworks, awareness and education, financial barrier, and e-waste complexity. Addressing these gaps requires collaborative efforts with the governments, private sectors and communities to promote research, investment, and development of sustainable waste-to-value solutions in the Global South.

The administrators of health facilities waste care services are required to apply succinct attention and diligence to necessary practices to avoid negative outcomes. Improper handling can lead to diverse exposure to toxic and infectious agents. To achieve effective and improved management of health-related waste; practices that promote the reduction of generating and accumulating waste should be encouraged. Also, waste segregation should be emphasized; strategies and systems with vision in view should be established by each organization; technologies like autoclaving, microwaving, and chemical treatment should be adopted over incineration; erecting a comprehensive system coupled with allocating responsibilities to handle, sort, and dispose should be sustained and improved; in addition, safety measure should be put in place for waste managers within the organization to avoid being contaminated with toxic and infectious substances. The issue of electronic waste (e-waste) is significant in the Global South due to various factors: increased consumption of EEA, inadequate disposal structure, and the alarming increased import of e-waste.

5 Conclusion

The global south faces complicated, multifaceted, and interconnected waste issues. While some organizations have emphasized the significance of participatory procedures in waste care, these techniques need to be customized to the specific conditions and players in the local community. Lack of government involvement, general knowledge, and inadequate GS infrastructure are a few of the challenges and restrictions. Consequently, it is important to combine participatory formats with appropriate frameworks and equitable participation from all pertinent parties. To have a better knowledge of the efficacy of involvement for initiatives addressing environmental contamination, among other things, more empirical study is required. In urban or metropolitan locations in the Global South, such research might highlight the (hidden) potentials of participatory formats for sustainable waste management.

The Global South and the rest of the world have both given thought to waste from electric and electronic appliances, or EEAW. Concerns about climate change notwithstanding, mechanical and physical processing is crucial to EEAW upgrading. The characterization of EEAW offers a reliable and strong starting point for creating efficient separation methods. But when it comes to the kind, dimensions, and configuration of the materials and components, EEAW is remarkably complicated and heterogeneous. As a result, it was noted that 78 nations had national e-waste policies, laws, or regulations in 2019. This accounted for over 71% of the global populace, a notable increase from just 44% 5 years earlier. Nonetheless, some nations—particularly developing nations—have inadequate or nonexistent legal enforcement. Moreover, certain nations in the Group of States continue to lack laws mandating the recycling of electronic trash. Because some e-waste contains various valuable raw materials that are not recovered because of environmental or economic difficulties, the global e-waste market is to improve. Expectedly, the market should nearly triple in value to about $143,870 million by 2028 from its 2020 evaluation of $49,880 million. One of the most frequent problems with waste management is a lack of knowledge about the risks involved with improper handling of garbage, particularly medical waste. The Global South should embrace emerging technologies that turn garbage into profit and enforce policies around waste management.

References

Agoh, E. C., Ogwu, M. C., Chukwuemeka, O. S., & Ekeledo, P. I. (2024). Environmental and human health effects of cassava processing and processing waste. In M. C. Ogwu, S. C. Izah, A. A. Cunha Alves, & babu, S.C. (Eds.), *Plant biology, sustainability and climate change, sustainable cassava* (pp. 203–219). Academic. https://doi.org/10.1016/B978-0-443-21747-0.00001-1

Akudugu, M. A., & Ogwu, M. C. (2024). Sustainable development policies and interventions: A bibliometric Analysis of the contributions of the academic community. *Journal of Cleaner Production*, Article # 139919. https://doi.org/10.1016/j.jclepro.2023.139919

Albert, R. (2011). The Proper Care and Use of a Garbage Disposal. Disposal Mag. Retrieved 3 March 2017

Al-Salem, S. M., Lettieri, P., & Baeyens, J. (2009). Recycling and Recovery Routes of Plastic Solid Waste (PSW): A Review. *Waste Management, 29*, 2625–2643.

Arrieta, M. P., Samper, M. D., Aldas, M., López, J., & Ferri, J. M. (2020). Valorization of plastic waste to obtain chemicals and fuels: A review. *Journal of Cleaner Production, 277*, 123800. https://doi.org/10.1016/j.jclepro.2020.123800

Awasthi, A. K., & Li, J. (2020). A review on e-waste management and recycling practices. *Environmental Science and Technology, 54*(19), 11692–11706. https://doi.org/10.1021/acs.est.0c01901

Basu, P. (2013). *Biomass gasification and pyrolysis: Practical design and theory*. Academic.

Bridgwater, A. V. (2012). Review of fast pyrolysis of biomass and product upgrading. *Biomass and Bioenergy, 38*, 68–94.

Caminati, V., Ragazzi, M., Silvestri, S., Berghi, F., Guirreri, D., Castellucci, A., & Rama, I. (2022). Collection and composting of organic waste: A case study in the City of Beira, Mozambique. *International Journal of Environmental Impacts, 5*(3), 259–270. https://doi.org/10.2495/EI-V5-N3-259-270

Chakrabarti, H., Maity, R., Baishya, S., & Maity, N. P. (2021). An accurate model for threshold voltage analysis of dual material double gate metal oxide semiconductor field effect transistor. *Silicon, 13*, 1851–1861.

Ciacci, L., Angellotti, M., Soccio, M., Guan, J., Gambarotti, C., Lotti, N., & Munari, A. (2021). Chemical recycling of poly(ethylene terephthalate): Glycolysis and hydrolysis. *Polymers, 13*(2), 272. https://doi.org/10.3390/polym13020272

Cointreau, S. (2006a). *Occupational and environmental health issues of solid waste management: Special emphasis on middle and lower-income countries*. Report to the Waste Management Unit of the World Health Organization, Regional Office in Europe.

Cointreau, S. (2006b). *Occupational and environmental health issues of solid waste management*. World Bank Urban Sector Board. Urban paper series no. UP-2, p. 48.

Cucchiella, F., Idiano, A. S., Lenny, K., & Paolo, R. (2015). Recycling of WEEEs: An economic assessment of present and future e-waste streams. *Renewable and Sustainable Energy Reviews, 51*, 263–272. ISSN 1364-0321. https://doi.org/10.1016/j.rser.2015.06.010

Davidson, G. (2011). *Waste management practices: Literature review* (pp. 1–59). Dalhousie University.

Domínguez, J., Edwards, C. A., & Edwards, C. A. (2005). Vermicomposting of sewage sludge: Effect of bulking materials on the growth and reproduction of the earthworm Eisenia andrei. *Waste Management, 25*(2), 195–200. https://doi.org/10.1016/j.wasman.2004.10.006

Ejiogu, R. A. (2013). E-waste economics: A Nigerian perspective. *Management of Environmental Quality, 24*(2), 199–213. https://doi.org/10.1108/14777831311303083

FAO. (2017). Food and Agriculture Organization of the United Nations, Rome, ItalyAgency, Abu Dhabi.

Fujimori, T., Takigami, H., Agusa, T., et al. (2012). Impact of metals in surface matrices from formal and informal electronic-waste recycling around Metro Manila, the Philippines, and intra-Asian comparison. *Journal of Hazardous Materials, 221*(222), 139–149.

Funke, A., & Ziegler, F. (2010). Hydrothermal carbonization of biomass: A summary and discussion of chemical mechanisms for process engineering. *Biofuels, Bioproducts and Biorefining, 4*(2), 160–177.

Geyer, R., Jambeck, J. R., & Law, K. L. (2015). Production, use, and fate of all plastics ever made. *Science Advances, 3*(7), e1700782. https://doi.org/10.1126/sciadv.1700782

Giusti, L. (2009). A review of waste management practices and their impact on human health. *Waste Management, 29*, 2227–2239. https://doi.org/10.1016/j.wasman.2009.03.028

Hoornweg, D., Bhada-Tata, P., & Kennedy, C. (2015). Peak waste: When is it likely to occur? *Journal of Industrial Ecology, 19*(1), 117–128.

IRENA. (2019). *Renewable energy statistics 2019*. The International Renewable Energy.

Karthikeyan, B., Kathyayini, R., Kumar, V. A., Uthra, V., & Kumaran, S. S. (2021). Effect of dumped iron ore tailing waste as fine aggregate with steel and basalt fibre in improving the performance of concrete. *Materials Today: Proceedings, 46*, 7624–7632.

Kathiravale, S., & Muhd Yunus, M. N. (2008). Waste to wealth. *Asia Europe Journal, 6*, 359–371. https://doi.org/10.1007/s10308-008-0179-x

Khanal, S. K. (2008). *Anaerobic biotechnology for bioenergy production: Principles and applications*. Wiley.

Kosoe, E. A., Achana, G. T. W., & Ogwu, M. C. (2023). Regulations and policies for herbal medicine and practitioners. In S. C. Izah, M. C. Ogwu, & M. Akram (Eds.), *Herbal medicine phytochemistry* (Reference series in phytochemistry). Springer. https://doi.org/10.1007/978-3-031-21973-3_33-1

Kühnert, I., Papadaki, E., Wang, C., Boukis, N., & Bikiaris, D. (2021). Depolymerization of polyethylene terephthalate waste: a review. *Green Chemistry, 23*(3), 1094–1135. https://doi.org/10.1039/d0gc03386d

Landrigan, P. J., Fuller, R., Hu, H., Caravanos, J., Cropper, M. L., Hanrahan, D., Sandilya, K., Chiles, C., Kumar, P., & Suk, W. A. (2018). Pollution and global health – An agenda for prevention. *Environmental Health Perspectives, 126*(8). https://doi.org/10.1289/EHP3141

Lo, K. V., Xing, B., & Tsang, D. C. W. (2020). Composting and compost utilization for remediation of contaminated soils and water. *Reviews in Environmental Science and Bio/Technology, 19*(3), 433–471. https://doi.org/10.1007/s11157-020-09523-7

Magda, M., & El-Salam (2010). Hospital waste management in El-Beheira Governorate, Egypt. *Journal of Environmental Management, 91*, 618–629.

Makan, A., & Tabatabaei, M. (2019). Composting as a waste management technique in developing countries. In *Sustainable waste management* (pp. 165–190). Springer. https://doi.org/10.1007/978-3-030-04684-1_8

Mata-Alvarez, J., Macé, S., & Llabrés, P. (2000). Anaerobic digestion of organic solid wastes. An overview of research achievements and perspectives. *Bioresource Technology, 74*(1), 3–16. https://doi.org/10.1016/S0960-8524(99)00160-5

Mohan, D., & Pittman, J. C. U., & Steele P. H. (2006). Pyrolysis of wood/biomass for bio-oil: A critical review. *Energy & Fuels, 20*(3), 848–889.

Mohd, A. K. A., & Abdullah, N. (2021). Solid-state fermentation (SSF) of food waste for ethanol production: A review. *Journal of Environmental Management, 278*, 111542. https://doi.org/10.1016/j.jenvman.2020.111542

Obahiagbon, E. G., & Ogwu, M. C. (2023a). Sustainable supply chain management in the herbal medicine industry. In S. C. Izah, M. C. Ogwu, & M. Akram (Eds.), *Herbal medicine phytochemistry* (Reference series in phytochemistry). Springer. https://doi.org/10.1007/978-3-031-21973-3_64-1

Obahiagbon, E. G., & Ogwu, M. C. (2023b). Consumer perception and demand for sustainable herbal medicine products and market. In S. C. Izah, M. C. Ogwu, & M. Akram (Eds.), *Herbal medicine phytochemistry* (Reference series in phytochemistry). Springer. https://doi.org/10.1007/978-3-031-21973-3_65-1

Obahiagbon, E. G., & Ogwu, M. C. (2024a). Organic food preservatives: The shift towards natural alternatives and sustainability in the global south's markets. In M. C. Ogwu, S. C. Izah, & N. R. Ntuli (Eds.), *Food safety and quality in the global south* (pp. 299–329). Springer. https://doi.org/10.1007/978-981-97-2428-4_10

Obahiagbon, E. G., & Ogwu, M. C. (2024b). The nexus of business, sustainability, and herbal medicine. In S. C. Izah, M. C. Ogwu, & M. Akram (Eds.), *Herbal medicine phytochemistry* (Reference series in phytochemistry). Springer. https://doi.org/10.1007/978-3-031-21973-3_67-1

Ogwu, M. C. (2019a). Towards sustainable development in Africa: The challenge of urbanization and climate change adaptation. In P. B. Cobbinah & M. Addaney (Eds.), *The geography of climate change adaptation in Urban Africa* (pp. 29–55). Springer Nature. https://doi.org/10.1007/978-3-030-04873-0_2

Ogwu, M. C. (2019b). Understanding the composition of food waste: An "-omics" approach to food waste management. In A. P. Gunjal, M. S. Waghmode, N. N. Patil, & P. Bhatt (Eds.), *Global initiatives for waste reduction and cutting food loss* (pp. 212–236). IGI Global. https://doi.org/10.4018/978-1-5225-7706-5.ch011

Ogwu, M. C., Odozi, I. P., Ahonsi, O. C., Uleanya, K. O., & Odozi, E. B. (2024). Organic acid production from cassava. In M. C. Ogwu, S. C. Izah, A. A. Cunha Alves, & babu, S.C. (Eds.), *Plant biology, sustainability and climate change, sustainable cassava* (pp. 395–418). Academic. https://doi.org/10.1016/B978-0-443-21747-0.00009-6

Paritosh, K., Kushwaha, S. K., Yadav, M., Pareek, N., Chawade, A., Vivekanand, V., & Singh, A. (2017). Food waste to energy: An overview of sustainable approaches for food waste management and nutrient recycling. *Bioresource Technology, 248*, 2–12. https://doi.org/10.1016/j.biortech.2017.06.120

Shareefdeen, Z. (2022). Hazardous Waste Management: Advances in Chemical and Industrial Waste Treatment and Technologies.

Silva, M. T., Morais, D. A. L., & Pinto, J. C. (2018). Hydrolysis of Polyethylene Terephthalate (PET): A Review. *Journal of Environmental Treatment Techniques, 6*(1), 100–106.

Singh, J. A., Guyatt, G., Ogdie, A., Gladman, D. D., Deal, C., Deodhar, A., ... & Reston, J. (2019). 2018 American College of Rheumatology/National Psoriasis Foundation guideline for the treatment of psoriatic arthritis. *Journal of Psoriasis and Psoriatic arthritis, 4*(1), 31–58.

Sonne, C., Carlsson, M. V., Fernández-Rodríguez, C., & Astrup, T. (2019). Hydrolysis of post-consumer PET for monomer recycling: A review. *Waste Management, 85*, 301–315. https://doi.org/10.1016/j.wasman.2018.12.006

Srivastava, A., Avan, B. I., & Rajbangshi, P. (2015). Determinants of women's satisfaction with maternal health care: a review of literature from developing countries. *BMC Pregnancy and Childbirth, 15*, 97. (2015). https://doi.org/10.1186/s12884-015-0525-0

Themelis, N. J., & Ulloa, P. A. (2007). Methane Generation in Landfills, Earth Engineering Center and Department of Earth and Engineering. *Renewable Energy, 32*, 1243–1257. https://doi.org/10.1016/j.renene.2006.04.020

Tony, M. A. (2022). Valorization of undervalued aluminum-based waterworks sludge waste for the science of "The 5 Rs' criteria". *Applied Water Science, 12*(2), 20.

United Nations Environment Programme. (2016). *Waste to energy: Opportunities and challenges for developing and transition economies*. Available at: https://www.unenvironment.org/resources/report/waste-energy-opportunities-and-challenges-developing-and-transition-economies

United Nations Environment Programme, UNEP. (2015). *Waste plastic – Conversion technologies*. United Nations Environment Programme. https://wedocs.unep.org/handle/20.500.11822/8740. World Economic Forum Geneva, Switzerland 2000 The Global Competitiveness Report 2000.

UNSD. (2017). *System of Environmental- Economic Accounting— SEEA Revision. Internet site* https://unstats.un.org/unsd/envaccounting/seearev/(open in a new window). United Nations.

U.S. EPA. (2023a). Wasted Food Report: Estimates of Generation and Management of Wasted Food in the United States in 2019. United States Environmental Protection Agency.

U.S. EPA. (2023b). Anaerobic digestion facilities processing food waste in the United States (2019). EPA 530-R-23-003. United States Environmental Protection Agency.

UNEP IETC (2016–2020) UNEP International Environmental Technology Center (IETC) resources. Provides access to all IETC publications, including the Global and Regional Waste Management Outlooks. Available at: https://www.unep.org/ietc/resources. Accessed 3 December 2024.

Wierckx, N., Prieto, M. A., Pomposiello, P., De Lorenzo, V., O'Connor, K., & Blank, L. M. (2015). Plastic waste as a novel substrate for industrial biotechnology. *Microbial Biotechnology, 8*(6), 900–903. https://doi.org/10.1111/1751-7915.12312

Williams, P. T. (2013). *Waste treatment and disposal*. Wiley.

World Bank. (2012). What a Waste: A Global Review of Solid Waste Management. https://www.ghanabusinessnews.com/2012/06/08/world-bank-predicts-annual-cost-of-treating-solid-waste-rising-to-375b-by-2025/

Xiao, B., & Wan, Y. (2018). Aerobic composting of organic solid wastes: A review. *Journal of Environmental Management, 217*, 644–658. https://doi.org/10.1016/j.jenvman.2018.03.055

Zheng, L., Wu, H., Zhang, H., Duan, H., Wang, J., Jiang, W., & Song, Q. (2017). Characterizing the generation and flows of construction and demolition waste in China. *Construction and Building Materials, 136*, 405–413.

Sustainable Bioenergy Generation from Biological Resources in the Global South

T. F. Oyewusi, A. R. Olaniyan, O. E. Ogundahunsi, O. O. Ilori, and A. R. Soji-Adekunle

Abstract

The whole world without the exception of the Global South nations is concentrated on reducing and replacing non-renewable energy sources (fossil fuels) due to the key global priority of ensuring energy security and achieving net-zero carbon emissions. Based on this information, alternative energy produced from bioresources has been found to be essential for achieving net-zero carbon emissions and energy security. This book chapter, therefore, examined sustainable bioenergy generation from bioresources in the Global South. This is with a view to discovering the bioresources that are available in the regions (Africa, Asia, Latin America, and the Caribbean) for the generation of bioenergy that can replace fossil fuels. Based on this, the chapter established four general categories of biological resources that are available in the Global South. Also, major forms of bioenergy development that are found in some countries across the three regions of the Global South were discussed. Before now, the Global South was known for using traditional bioenergy (fuel wood and charcoal) that is consumed through open-air fires and other inefficient conversion methods. But modern forms of bioenergy (bioethanol, biodiesel, biogas, etc.) have been discovered to have the potential to be an alternative energy, displacing a significant amount of fossil fuels in the Global South regions. With this discovery, bioenergy has become increasingly important in the Global South as it offers opportunities and

T. F. Oyewusi (✉) · A. R. Olaniyan
Department of Agricultural Engineering, Osun State University, Osogbo, Nigeria
e-mail: toyese.oyewusi@uniosun.edu.ng

O. E. Ogundahunsi
Department of Agricultural Engineering, First Technical University, Ibadan, Nigeria

O. O. Ilori · A. R. Soji-Adekunle
Department of Mechanical Engineering, Adeleke University, Ede, Nigeria

M. C. Ogwu et al. (eds.), *Sustainable Bioeconomy Development in the Global South*, https://doi.org/10.1007/978-981-96-0305-3_9

benefits that are either directly or indirectly related to achieving the United Nations Sustainable Development Goals. In this regard, modern bioenergy technologies and policies are required to achieve more sustainable energy and balance in the bioeconomy to create more prospects that will be beneficial to the environment and people living in it. Therefore, it is suggested that a Global South bioenergy forum be established to foster communication, education, and collaboration as well as to guarantee that the transformative, beneficial aspects of bioenergy are realized and the harmful ones are avoided. Based on this, it is expected that the information given in this chapter will provide essential guidance to decision-makers or policymakers in the energy sector to achieve more sustainability in the economy, society, and environment in the Global South.

Keywords

Biofuels · Biomass · Bioresources · Energy production energy security · Renewable energy · Sustainable development

Abbreviations

ANP	Anuário Estatístico Brasileiro
BCM	Billion Cubic Metres
CO_2	Carbon dioxide
FAO	Food and Agriculture Organization
GBEP	Global Bioenergy Partnership
GHG	Greenhouse Gas
H_2O	Water
IEA	International Energy Agency
LAC	Latin America and the Caribbean
MCM	Million Cubic Metres
NDC	Nationally Determined Contribution
RSB	Roundtable on Sustainable Biomaterials
SDGs	Sustainable Development Goals
IRENA	The International Renewable Energy Agency
UN	United Nation
UNDP	United Nations Development Programme
UNEP	United Nations Environment Programme

1 Introduction

Energy is the foundation of the modern industrial economy. Almost all human endeavours require energy as a basic element. Among other things, it supports industrial development, food production and storage, health, education, and transportation (Østergaard et al., 2020). This implies that energy services are a major driver of both social and economic development. In view of this, energy consumption has increased along with economic expansion in the Global South (Dirisu et al., 2024). Between 2000 and 2015, the energy consumption of the Global South regions increased by almost 50%, from about 3750 million metric tonnes to 7000 million metric tonnes of oil equivalent (Oyedepo, 2012). According to Dirisu et al. (2024), the majority of the world's energy is currently consumed by nearly six billion people in developing nations, and over the next 15 years, it is projected that their demand will increase by an additional 30%. Presently, in the Global South, the consumer class has continued to increase and become wealthier, driving up energy consumption as they buy more goods and services that raise living standards (such as housing, transportation, healthcare, and education) and are driven in huge part by quickly growing economies. Therefore, the per capita energy consumption is a measure of the per capita income as well as a measure of the prosperity of a nation (Oyedepo, 2012).

Fossil fuels that are both non-renewable and high in carbon dioxide (CO_2) have been the primary supply of energy for years right from the emergence of the Industrial Revolution and the pattern has remained consistent in recent years (Dirisu et al., 2024). Fossil fuels are the major source of energy for electricity and transportation worldwide, constituting 64% of all gas emissions, which consists of 38 gigatonnes of CO_2 in 2019 (UNEP, 2020). Around 9.25 GTOE of fossil and renewable energy are now consumed globally, 3.8 of them in developing nations (Dirisu et al., 2024). Between 2016 and 2017, fossil fuels market increases more than that of renewable energy. Renewable energy is trailing (0.7%), whereas coal, oil, and natural gas contributed to the 1.6% rise in the primary power source between 2016 and 2017 (IEA, 2020). However, using fossil fuels for energy has taken a huge toll on humans and the environment (Baz et al., 2021). This is because they contribute to air and water pollution along with a rise in global temperature that has caused depletion of the ozone layer resulting in global warming (Anyaoha & Zhang, 2022). In addition to all these downsides, relying solely on fossil fuel is insecure because of the depletion of its resources. Hence, the whole world without the exception of the Global South nations is concentrated on reducing and replacing non-renewable energy sources (fossil fuels) due to the key global priority of achieving net-zero carbon emissions to ensure energy security (Field et al., 2020).

Energy from biological resources, or bioenergy, has been found to play a critical role in achieving net-zero carbon emissions and serves as a substitute for fossil fuels (Oyewusi et al., 2018, 2023, 2024; Aransiola et al., 2019; Adeboye et al., 2022; Ogundahunsi et al., 2024; Soji-Adekunle et al., 2024). The term "bioenergy" refers to all energy produced by substances with a biological origin known as biomass (Li et al., 2023). According to Chong et al. (2021), bioenergy represents the major renewable energy source in the world, providing for over two-thirds of green energy sources, which contribute to approximately 55% of all renewable energy and 6% of energy supply worldwide. The primary energy supply from biomass as a whole fell by 0.9 EJ in 2017 to reach 55.6 EJ, practically matching the previous year's level (IEA, 2020). The biomass supply decreased for the first time since 2010 during this time. Despite reducing the supply, biomass continues to play a noteworthy part in providing energy from sustainable sources, concerning 70% of the overall renewable stream. The IRENA (2017) reported the several advantages and numerous benefits of the transition from fossil fuels to bioenergy, which include lowering reliance on imported energy sources and improving energy security through source diversification in addition to preventing emissions from direct fuel combustion. In their report, Dirisu et al. (2024) presented that the development of bioenergy has enormous economic, societal, and environmental benefits. In addition, under the Paris Climate Agreement to reduce GHG emissions is the development and use of bioenergy, which is one of the main approaches put forth by many nations to assist in meeting Nationally Determined Contributions (NDC) (Torvanger, 2019). Hence, it can be deduced that energy derived from sustainable bioresources can prevent adverse effects and provide benefits when combined with extensive environmental protections and social needs.

Many Global North nations have established bioenergy sustainability frameworks as well as certification programmes to guarantee social, economic, and environmental protections that prevent negative effects at every stage of the energy life cycle from the production of feedstock to the end-use of energy (GBEF, 2011; RSB, 2016). Environmental, social, and economic protections are intended by these frameworks. On the other hand, many countries in the Global South, particularly countries in Africa and Asia since time immemorial have been using biomass in a traditional way for cooking, heating, and lighting (Dirisu et al., 2024). Such uses of biomass impose considerable costs on health, local ecology, and productivity even as they use up considerable biomass resources only to obtain low-quality energy services (Johnson, 2017; Ogwu et al., 2024). In several identified hotspots, they are also contributing significantly to GHG emissions (Bailis et al., 2015). A key aim in the transition to a sustainable bioeconomy must be to replace traditional biomass uses with more efficient and effective alternatives, including modern end-use systems coupled with modern bioenergy in the form of biogas, bioethanol, or briquettes. However, in many of the Global South nations, both modern and large-scale bioenergy systems are challenging, owing to high capital costs and sustainable supply of biomass. But with about half of the biomass used for traditional heating and cooking in the Global South, particularly in African and Asian countries, there is a real opportunity for effective and efficient, small-scale, biomass-based power generation that can help to achieve Sustainable Development Goal 7 (SDG 7), which is access to affordable, reliable, sustainable, and modern energy for all by 2030 (Jagger et al., 2019).

Globally, a growing number of industrialized and developing nations are implementing national bioeconomy policies and supporting advancements in renewable energy (Stephenson & Damerell, 2022). Based on this, sustainability in bioenergy is a requirement for further growth and it has the potential to be a key tactic in support of low-carbon economic growth as well as the expansion of a bioeconomy in the Global South. Hence, sustainable bioenergy generation could help the Global South to capitalize on the economic potential of the sustainable and renewable bioresources that are available throughout the regions. In addition, the production of energy from biological resources can offer the Global South a number of

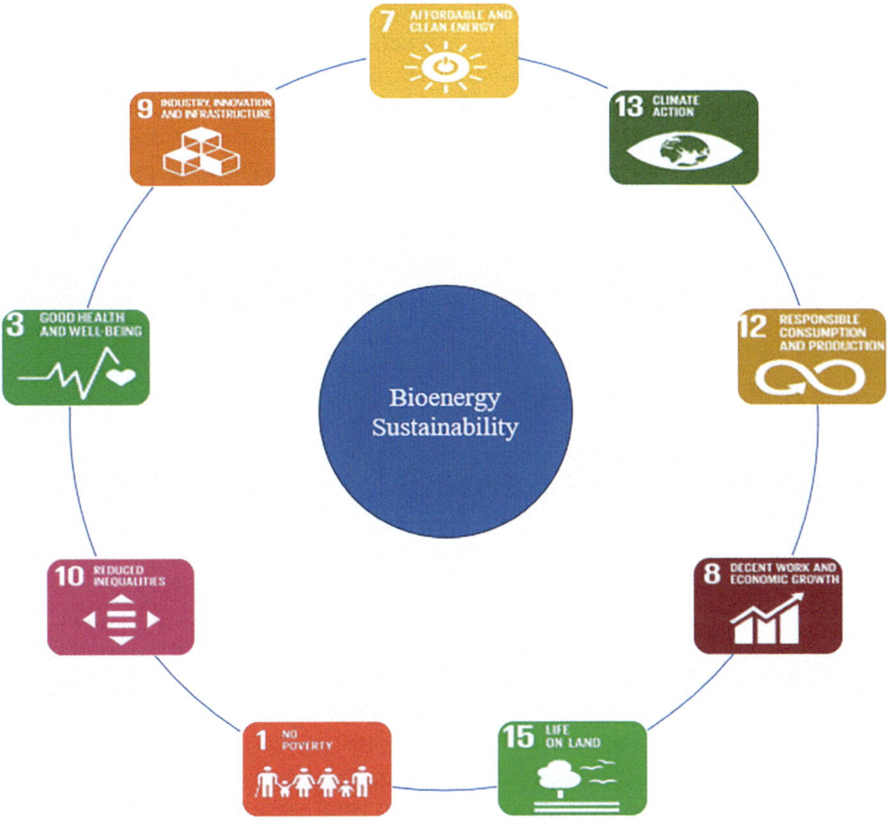

Fig. 1 Direct and indirect contributions of bioenergy to sustainable development goals

opportunities and co-benefits that are either directly or indirectly related to the United Nations SDGs (Fig. 1).

In view of this, Destek et al. (2021) identified five areas in which bioenergy directly contributes to the SDGs: good health and well-being (SDG 3), affordable and clean energy (SDG 7), decent work and economic growth (SDG 8), climate change (SDG 13) and life on land (SDG 15). However, the economic advantages of bioenergy also indirectly address four other SDGs: SDG 1 (no poverty), SDG 9 (industry, innovation, and infrastructure), SDG 10 (reducing inequality), and SDG 12 (responsible consumption and production). The advantages—both direct and indirect—should serve as a guide or road map for the Global South to better understand the potential and sustainability of bioenergy from biological resources and resolve to fully embrace the production for the betterment of all the regions (Kosoe et al., 2023; Akudugu & Ogwu, 2024).

This chapter, therefore, examined sustainable bioenergy generation from biological resources in the Global South regions of the world. To accomplish this book chapter, an overarching literature review of published books and online articles was thoroughly done in a systematic way to gather all reports presented in this work. Therefore, the book chapter focused on bioenergy production (from different categories of bioresources) that served as invention, advancement, and delivery of indispensable services required for a sustainable economy in the Global South regions—Africa, Asia, Latin America, and the Caribbean. It is imperative that the Global South fully shift to a sustainable bioeconomy considering the increasing population, environmental challenges, and the diminishing availability of fossil-based resources. It is expected that the information given in this chapter will provide essential guidance to decision-makers or policymakers in the energy sector to achieve a more sustainable economy, society, and environment in the Global South.

2 Biological Resources

2.1 General Overview

Biological resources are broadly referred to as any resources that come from nature or biological origin; hence, they comprise renewable resources derived from living or, recently living, organisms and are sometimes called bioresources or biomass (Ingle et al., 2020). In this regard, biological resources are defined as naturally occurring, sustainably renewable, and biodegradable materials resulting from living things (Narodoslawsky, 2014). Bioresources have always been intimately linked to the history of human development. Primitive human ancestors have lived in deep forests since the dawn of human civilization, where they were warmly supported by an abundance of natural resources. It is suggested that this close connection between humans and nature denotes the transition from rural society to an urban setting. Thus, biological resources are considered an important part of life and play a major role in the economy, environment, and society, and by comparing global bioresources to other resources—especially fossil fuels—we find a greater diversity in bioresources (Nimachow et al., 2008).

Presently, in the Global South regions (Africa, Asia, and Latin America and the Caribbean), bioresources are very vital to meet human needs, which primarily include those for fuel, food, feed, shelter, fibre, with other various bio-products; hence, they have been discovered as an essential component of the bioeconomy (Ingle et al., 2020). The word "bioeconomy" generally relates to the use of bioresources for primary production through broad divisions, including forestry, agriculture, fisheries, and aquaculture, as well as all other industries that use and process bioresources, such as feed, food, paper, and pulp, chemical and energy sectors. In addition, bioresources are necessary for research on health and the production of suitable quantities of pharmaceuticals. The results of many studies using bioresources, from basic research to the development of new drugs, have been helping our lives. In general, Asia, Africa, and the South America have the most extensive domestic biomass supply of all the continent but Asia and Africa make up a significant amount of the world's biomass due to the widespread use of fuel wood and charcoal for heating and cooking (Olabi et al., 2022). Therefore, in the Global

South, bioresources are thought to be among the elements that either directly or indirectly cause the socioeconomic developments (Ingle et al., 2020).

2.2 Categories of Biological Resources

The major bioresources available in the Global South include agricultural residues and wastes, forestry wood, organic industrial residues, residues from society, common and aquatic weeds, marine resources, and so on (Ingle et al., 2020; Egboduku et al., 2024). The four general categories into which all of these bioresources fall are explained as follows:

1. Primary Bioresources

   In the Global South, bioresources are produced in forestry, agriculture, and aquaculture with a specific application in mind to enable the production of energy, food, and valuable bio-products (Ingle et al., 2020). Examples include wood, bamboo, algae, etc. Due to efficiency concerns, the use of primary bioresources in the production of energy is questionable. Meanwhile, primary bioresources can be separated even further into the following categories:

   (i) Crops: Crops are one of the most important categories of bioresources that are utilized to meet a variety of basic requirements for both humans and animals (Oyewusi et al., 2018; Ogundahunsi et al., 2022a). Any plants can be directly grown or harvested in large quantities for food or financial gain. These crops are group into seven broad categories—food, feed, fibre, oil, ornamental, industrial, and energy crops. Although, large scale production of bioenergy from crops may lead to competition with food but data from various sources indicate that the Global South regions have abundant and underutilized arable land which can be effectively used for mass production of energy crops for production of Energy (Ogbonna et al., 2013).

   (ii) Forest wood: Forests are thought to be a vital component of the bioeconomy with great potential, encompassing forestry, the production of furniture, wood processing, and numerous other high-value goods (Gravelsins et al., 2017). Oyewusi et al. (2023) reported that wood timbers can be processed or converted into briquettes, sawdust, chips, logs, pellets, and sawdust (M'hamdi et al., 2017). For the poor in the Global South nations, both urban as well as rural, wood is usually the principal source of energy for cooking and heating. But to avoid deforestation, forest wood are not encouraged to be used for energy production, especially modern energy production in large scale. However, forest residues can be used for energy production due to their composition, which consists of lignocellulosic materials (Oyewusi et al., 2023).

   (iii) Marine bioresources: Due to their immense potential for use, marine bioresources have been investigated by humans over time; however, uncontrolled use of these resources has raised many concerns worldwide (Fu et al., 2018). In terms of human use, marine resources are primarily utilized for three purposes: aquaculture farming, fishing, and tourism (Fabinyi, 2018). Due to its variety and distinctive processes, resources from marine have drawn a lot of attention over the years (Qin, 2018). Marine algae is a useful raw material for biorefineries because of its versatility and capacity to antioxidants, fatty acids, produce oils, pigments, and other value-added bio-based goods (Balina et al., 2017). However, marine bioresources are comparatively new and emerge as a promising frontier for energy production in the Global South.

2. Secondary Bioresources.

Bioresources of this class primarily consist of wastes or by-products from industrial, societal, or agricultural processes. This category of biological resources allows land to be used for other purposes and, generally, their use does not compete with it. Following is a brief discussion of some of the major secondary bioresources.

(i) Agricultural wastes: These are primarily of two categories: "field residues" which consist of items such as seedpods, leaves, stalks, and plant stems; and "process residues," which consist of important materials like roots, seeds, husks, molasses, bagasse, etc. The primary source of these residues is the cultivation and processing of crops. According to Obi et al. (2016), the wastes could be in the form of solids, liquids, or slurries and animal waste like dung and carcasses. Kataki et al. (2016) stated the majority of these agro wastes are typically starchy and lignocellulosic, consisting of about 25% lignin, 30% hemicellulose, and 40% cellulose. Many of the Global South countries today produce huge amounts of agricultural wastes that can be used for energy generation (Ogwu, 2019).

(ii) Industrial wastes: These are produced by a range of small- and large-scale businesses. They typically consist of a wide range of wastes, such as those from abattoirs and food and fruit industries. Usually, these kinds of wastes are composed of a range of substances that are high in proteins, sugar, minerals, and moisture. Industrial waste in the Global South has a fast-growing trend shortly, particularly for the production of energy.

(iii) Municipal or urban waste: frequently called trash or garbage, this comprises waste that is produced by society and municipal organizations. Examples of this type of waste include yard waste, market waste, and food waste (Gupta & Gupta, 2015). However, only biowastes—common household waste items such as kitchen wastes (vegetables, fruits, etc.), food waste, paper waste, wastewater, spent vegetable oil, and organic municipal waste, etc.—are covered here. There is large potential for energy generation from these wastes in the Global South (Adeboye et al., 2022).

(iv) Common weeds: Weed control is essential to agriculture since weeds are one of the most detrimental biotic factors that affect crop yield, severely reducing crop yields (Gharde et al., 2018). In the Global South regions, weeds present a promising source of energy for the generation of biofuels because of their high calorific values, availability, and lower cost (Ciesielczuk et al., 2016; Nizzy & Ogwu, 2024).

3. Tertiary Bioresources.

   They are by-products or leftovers from virgin materials that have been separated during the production process. They are residues, though, are smaller in quantities and need to have valuable materials as compared to secondary bioresources. They might have happened as a result of uncontrollable changes like deterioration during storage. In general, tertiary bioresources are less valuable than secondary bioresources. However, their energetic potential is becoming increasingly acknowledged among the Global South nations.
4. Quaternary Bioresources

   These happen following product usage. They fall into three categories: short-, mid-, and long-term depending on how long they have been generated after the beginning of product use. They are produced in all situations involving the consumption of food and feed and sometimes could be in form of animal excrement as well as human faeces and urine; thus, they have potential to be used for energy generation (biogas) in the Global South.

3 Potential of Biological Resources in the Global South

Global South is often used for nations with lower income or less economic development. It appears that the majority of the countries that make up the Global South are located in the Northern Hemisphere; hence, the term "Global South" does not refer to the southern region of the globe or the world. Therefore, the phrase refers to groupings of nations in Africa, Latin America, and Asia that fall short of developed nations' standards in terms of industry and economy (Kowalski, 2021) as indicated in Table 1. Consequently, the Global South describes all countries that are newly industrialized or in the process of industrializing as such, it is mostly referred to as developing or third world. According to Lal et al. (2021), the Global South is defined by millions of people with diverse social backgrounds who live in less developed regions and who also have minor differences in norms, values, behaviours, religion, and languages. In addition, Shimeles et al. (2018) presented the Global South as a very poor region with a low standard of living. Thus, the region requires earnest action to contribute to the creation, development, and provision of essential services needed for a sustainable economy. One way to achieve this is through the use of bioresources for sustainable production which can benefit rural communities and developing nations by creating new jobs and economic opportunities, as well as environmental benefits.

Carbohydrates are the basic building blocks of bioresources that are created by a process called photosynthesis, which is the outcome of the interaction between

Table 1 An index of the Global South's member nations arranged according to region

Countries	Region
Algeria, Angola, Bahrain, Benin, Botswana, Burkina Faso, Burundi, Cabo Verde, Cameroon, Central African Republic, Chad, Comoros, Congo Dem. Rep., Congo Rep., Côte d'Ivoire, Djibouti, Egypt, Ethiopia, Eritrea, Equatorial Guinea, Gabon, Gambia, Ghana, Guinea, Guinea-Bissau, Kenya, Lesotho, Liberia, Libya, Madagascar, Malawi, Mali, Mauritania, Mauritius, Morocco, Mozambique, Namibia, Niger, Nigeria, Rwanda, São Tomé and Principe, Seychelles, Senegal, Sierra Leone, Somalia, South Africa, South Sudan, Sudan, Eswatini, Tanzania, Togo, Tunisia, Uganda, Zambia, Zimbabwe	Africa
Afghanistan, American Samoa, Azerbaijan, Bangladesh, Bhutan, Brunei, Darussalam, Cambodia, China, Fiji, French Polynesia, Guam, China, India, Indonesia, Kazakhstan, Kiribati, Kyrgyzstan, Iran, Iraq, Jordan, Kuwait, Lao PDR, Lebanon, Malaysia, Maldives, Malta, Marshall Islands, Mongolia, Myanmar, Nepal, Nauru, New Caledonia, North Korea, Northern Mariana Islands, Oman, Palau, Pakistan, Papua New Guinea, Philippines, Qatar, Samoa, Saudi Arabia, Solomon Islands, Sri Lanka, Syrian, Thailand, Timor-Leste, Tajikistan, Tonga, Turkmenistan, Tuvalu, Uzbekistan, Vanuatu, Vietnam, Yemen	Asia
Antigua and Barbuda, Aruba, Argentina, Bahamas, The Barbados, Belize, Bolivia, Brazil, Cayman Islands, Chile, Costa Rica, Colombia, Cuba, Curacao, Dominica, Dominican Republic, Ecuador, El, Salvador, Grenada, Guatemala, Guyana, Haiti, Honduras, Jamaica, Mexico, Nicaragua, Panama, Paraguay, Peru, Puerto Rico, St. Kitts and Nevis, St. Martin, St. Lucia, St. Vincent and the Grenadines, Suriname, Trinidad and Tobago, Turks and Caicos Islands, Uruguay, Venezuela, Virgin Islands	Latin America and the Caribbean

water, sunlight, and CO_2 in the atmosphere (Anastassiadis, 2016). Typically, this process turns 1% of sun energy that is available into energy and stores it in the structural elements of the bioresources. To effectively extract the stored energy, the bioresources are processed to oxidize carbon to produce CO_2 and H_2O. In this way, the chemical energy is released as heat after the bioresources are burnt; hence, they are considered as renewable resources with the potential for energy generation as shown in Fig. 2 (Anastassiadis, 2016). Although bioresources are widely distributed throughout the world and abundant in nature, the distribution of these resources depends on geographic location. According to ETC (2010), the Global South is home to 86% of the world's biomass; hence, it has potential for bioenergy generation. Conversely, the World Bank report that this same Global South has the largest "yield gaps" of productive capacity that are underutilized (World Bank, 2010). Nevertheless, if bioresources are well utilized as a source of bioenergy production, they may meet the demand for sustainable energy, particularly in the Global South (Debez et al., 2017). Based on this potential of biological resources, the Global South can rise to wealthy nations by taking advantage of benefits derived from bioresources and using them for energy generation to advance the region.

Based on the aforementioned, bioresources can produce various types of energy as illustrated in Fig. 3. It can be either burnt directly to generate heat (biofuel), or the flue gases generated during the burning can be used as steam to power turbines

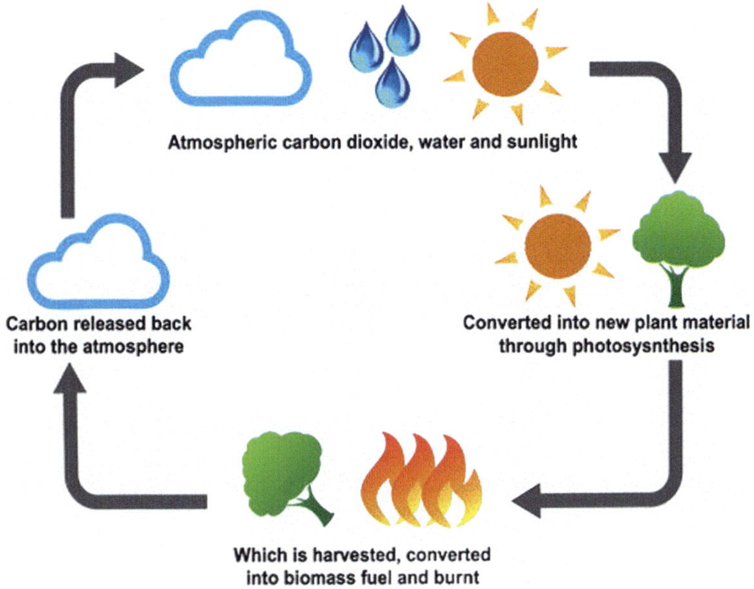

Fig. 2 Renewable nature of bioresources conversion into bioenergy

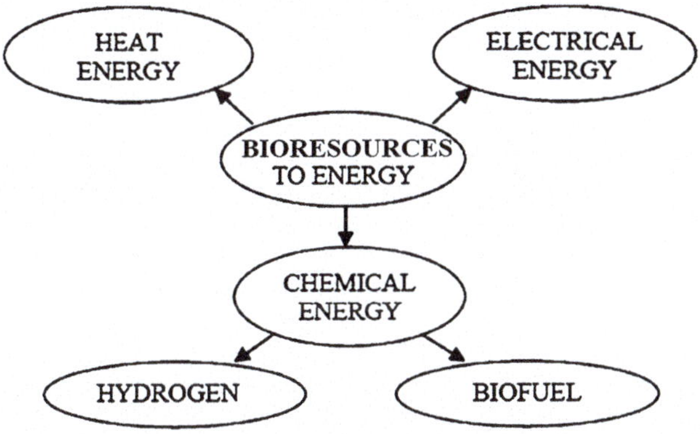

Fig. 3 Bioresources conversion into different forms of energy

to generate electricity (biopower or bioelectricity). However, biofuel represents perhaps the most common form of bioenergy around the world (Lebaka, 2013). In the Global South, biofuel has been the primary supply of energy for years, which has been used in forms of fuel wood and charcoal. These forms are by far the most heavily used and are consumed through open-air fires and other inefficient conversion methods. With global policy and technological developments, it is anticipated that the disadvantages of biofuels will be significantly reduced and it will gain the

Table 2 Classification of biofuels

Classification	Description	Examples
Physical state	Solid biofuels	Fuel wood, charcoal, briquette, etc.
	Liquid biofuels	Bioethanol, biodiesel, bio-oil, etc.
	Gaseous biofuels	Biogas, bio-syngas, biohydrogen, etc.
Technology maturity	Conventional biofuels	Biogas, bioethanol, and biodiesel are generated via anaerobic digestion, fermentation, and transesterification, respectively
	Advanced biofuels	Examples are microalgae biodiesel, biomass-generated esters, biohydrogen, hydrotreated vegetable oil, and lignocellulosic bioethanol
Generation of feedstock	First-generation feedstock (1GF)	Corn, sugar cane, sugar beetroot, wheat, soybean oil, sunflower oil, palm oil, rapeseed oil, etc.
	Second-generation feedstock (2GF)	Cellulose, hemicellulose, and lignin bioresources are feedstock that are economically viable to produce biofuels
	Third-generation feedstock (3GF)	Algae (macroalgae and microalgae), which originates from non-food origin
	Fourth-generation feedstock (4GF)	Genetically or metabolically engineered algae
Generation of products	Primary biofuels	Fuel wood, wood chips, pellets, manure, and leftovers from farms and forests are all used exactly as they are found—no processing, pre-treatment, or changes of any kind
	First-generation biofuels (1GB)	Biodiesel, biogas, bioethanol, bio-syngas, etc. are biofuels produced from edible crops such as edible vegetable oil, soybeans, corn, palm oil, wheat, etc.
	Second-generation biofuels (2GB)	Biodiesel, biogas, bioethanol, bio-syngas, briquette, etc. are generated from non-edible feedstocks such as waste animal fats, waste cooking oil, lignocellulosic biomass, etc.
	Third-generation biofuels (3GB)	Hydrogenated biodiesel, bio-gasoline, green aviation fuel, bio-oil, etc. originate from non-food sources like algae, yeast, etc.
	Fourth-generation biofuels (4GB)	Green aviation fuel, hydrogenated diesel, biodiesel, etc. are biofuels generated from genetically modified algae

potential to emerge as the primary substitute fuel, displacing a significant amount of fossil fuels (Zabed et al., 2016). Meanwhile, there are four criteria that can be used to broadly classify biofuels: physical state, technological maturity, feedstock generation, and product generation (Table 2). According to Zabed et al. (2016), biofuels are considered the most promising alternative to fossil fuels due to their huge benefits in the aspects of energy security, economic concerns, and environmental sustainability but may also raise some drawbacks in the aspects of feedstocks, technology, and policy as discussed in Table 3.

Table 3 Some indicators and descriptions of benefit and drawback aspects of biofuel production

Aspects	Indicators	Descriptions
Benefits	Energy security	Higher reliability Domestic energy source Well-connected supply–demand chain Locally distributed Reduced use of fossil fuels Renewability Domestic distribution Ready availability
	Economic impacts	Sustainability Price stability Fuel diversity Employment generation Rural development Agricultural development International competitiveness Increased income tax New industrial dimension
	Environmental impact	Improved land and waste utilization Higher combustion efficiency Reduced greenhouse gas emissions Carbon sequestration Reduction in landfill sites Reduced air pollution
Drawbacks	Technology	Efficiency improvement Pretreatment Technology cost Enzyme production Production of value-added products
	Policy	Pilot scale demonstration Fund for research and development Commercial scale deployment Land use change Tax credits on production and utilization of biofuels Policy for biofuels Procurement of subsidies on biofuel production
	Feedstock	Food fuel competition Collection network Storage facilities

4 Bioenergy Generation in the Global South

4.1 Bioenergy Development in Africa

Nowadays, bioenergy is becoming increasingly popular among the widely accepted energy sources, and more forms of bioenergy are being produced from different existing bioresources (Gupta et al., 2014). Of course, more developed nations have achieved impressive levels of research and development in the production and use of bioenergy over the past few decades (Guo et al., 2015). Presently, a growing number of African nations are embracing a variety of renewable energy, primarily due to the pressing need to find viable substitutes for fossil fuels, which are the energy source that is currently most commonly used (Ullah et al., 2015). This is mostly predicated on the knowledge that fossil fuels will eventually become extinct. According to Romeu-Dalmau et al. (2018), it will be appropriate for all African countries to be well involved in the mainstream of developed nations that produce bioenergy. This will lead to qualitative societal developments, as Africa is especially blessed with a favourable climate, rich soils, a vast land area, and other ecologically friendly resources that are critical to technological advancements. Also, Africa has one more noteworthy benefit to grow its economy and develop its nations through bioenergy generation as the second most populous continent in the world after Asia and representing 10% of the global population, according to Amigun et al. (2008). Ohimain and Izah (2017) with Russo and von Blottnitz (2017) reported that it will be the best option for the continental energy sector to start considering bioenergy production more specifically, bioethanol, biodiesel, and biogas as practical substitutes for fossil fuels. These various forms of bioenergy have been and will continue to find uses in various applications, such as driving, cooking, heating, and jet engine power.

Although, a few decades ago, numerous African nations jumped at the chance to sign significant agreements involving biofuels since the programmes started in the 1980s in nations like Malawi and Zimbabwe (Benti et al., 2023). However, policymakers, business, and development professionals now feel a sense of urgency due to the interplay of climate change, volatile fuel prices, the persistent food crisis, and the worldwide economic downturn. This urgency is demonstrated by the recent and sharp rise in global biofuel markets, production, and policy (Amigun et al., 2011; Obahiagbon & Ogwu, 2023a, b, 2024a, b). As of right now, Sudan is making good progress toward producing liquid biofuel. In 2009, approximately 5 million litres of bioethanol were exported to the European Union by one of its sugar companies called Kenana, with plans to produce 200 million litres annually (Miftah & Mutta, 2024). The year 2016 saw the production of roughly 480 million litres of bioethanol by Swaziland and Ghana produced 50 million litres of biodiesel for the international market from jatropha oil and sugar cane molasses, respectively (Benti et al., 2023). The production of biodiesel is frequently fed from jatropha oil seeds in many African countries, and the production of bioethanol is commonly fed from sugarcane molasses (Birhanu & Ayalew, 2017). In every nation, bioethanol is produced from sugarcane molasses, except for Benin, which uses cassava, and Guinea, which uses cashew. Jatropha was the primary crop used to produce biodiesel in every nation (Miftah & Mutta, 2024). Africa has thus experienced a portion of the paradigm shift in bioenergy in the Global South.

Large-scale bioenergy processing facilities can be found all over Africa these days and adequate steps have been implemented to guarantee the security of energy production and marketing (Dahunsi et al., 2020). However, studies on bioenergy production have been growing slowly throughout Africa, and the few studies that have been done have only made use of a small portion of the continent's resources. Whereas most of these bioresources are available year-round the essential responsibility of evaluating the fundamental chemical, physical, and structural elements of the numerous available biological resources has not received the attention it deserves. According to Nigam and Singh (2011), Austin and Morris (2012), Betiku et al. (2016), and Ogundahunsi et al. (2022b), the most commonly available biomaterials that can be exploited for bioenergy research include fruit peels, plant residues, animal and human excreta, and household wastes. Only a small portion of the vast array of these bioresources found on the African continent have been used to produce bioenergy (Saladini et al., 2016). Therefore, the generation and sustainable distribution of bioenergy, mostly derived from biomass, that is environmentally benign and renewable must be welcomed wholeheartedly by Africa as proved by the sharp rise in energy consumption in each of its countries as a result of periodic population booms (Maqhuzu et al., 2017; Dahunsi et al., 2020).

4.2 Bioenergy Development in Asia

Asia is the largest continent in the world, making up around 30% of the planet's surface (Roy et al., 2021). According to Prange (2019), it is also the continent with the greatest population in the world, making up about 60% of all people. The eastern and northern hemispheres contain the majority of the continent. Additionally, the Asian continent in its population has about 55% of the global population living on it, even with the average population density worldwide of 0.54 persons per hectare; the Asia continent has 1.87 people per hectare (Roy et al., 2021). On the Asian continent, the Global South comprises nations like Bangladesh, Bhutan, Nepal, Sri Lanka, Malaysia, Indonesia, Vietnam, Pakistan, China, and India. This expanding population has a daily rising energy demand (IEA, 2020). Therefore, sustainable energy production has become imperative, and the UN SDGs state that bioenergy is essential to this region's support in the context of energy security and climate change (UNDP, 2019). Also, IRENA (2021) stated that the generation of sustainable energy, such as bioenergy, really concerns the region when it comes to environmental sustainability and planet protection. In line with international efforts, the governments of the majority of Asian countries are supporting bioenergy programmes using these bioresources to address issues of energy security, the environment, and national income. These countries are also recognized for having an abundance of bioresources with high calorific values. In light of the security of energy and climate change, sustainable energy production has become imperative, and the UN SDGs state that bioenergy is essential to supporting this region.

Many Asian nations have mandated or set targets for the use of bioenergy in recent years, along with related policies for promotion (Elder & Hayashi, 2018). This is mostly motivated by the government's efforts to reduce poverty, promote economic development (especially through strengthening trade balances and growing the agriculture sector), and ensure energy security. These goals have varied from being low to being high. Asia's Global South countries have been producing a sizable quantity of biofuels, including bioethanol from China and India and biodiesel from Indonesia and Malaysia (Zhou & Thomson, 2009). East Asian nations like

Thailand, the Philippines, and Vietnam are known for bioethanol, biodiesel, and bioethanol, respectively, and have produced more biofuels, albeit on a much smaller scale. In light of this, most of the bioenergy producer nations in Asia have their bioenergy strategies focused around the nation's main bioresources. Presently, the countries that produce the most biofuel in the region are Malaysia, the Philippines, Thailand, Indonesia, China, and India while Indonesia and Malaysia lead Asia in the production of biodiesel, China, India, and Thailand concentrate primarily on the production of bioethanol (Börjesson, 2009).

According to Mahmud et al. (2022), China and Indonesia are the two largest countries in the region and are well-known for producing bioethanol and biodiesel, respectively. With four initial bioethanol plants that were subsidized by the government, bioethanol is currently produced on an industrial scale in several provinces, making China the third-largest national producer of biofuels in the world, behind the United States and Brazil. As far as renewable diesel production goes, Indonesia leads the world in biodiesel production with an output of 13.65 billion litres in 2022 (Statistical Research Department SRD, 2024). This was more than the combined total of the USA and Brazil which are in second and third positions, correspondingly. Taking the above into account, the availability of biofuel feedstock and the size of the biofuel market vary widely among Asian nations. For instance, there is a lot of potential to produce biodiesel in Indonesia and Malaysia because they are the two countries that produce the most palm oil globally, which is used to make biodiesel (Yamaguchi et al., 2013). However, the amount of bioethanol produced in these two nations is not very high. On the other hand, Thailand has a sizable domestic market and a wealth of potential for producing bioethanol; however, the nation's capacity to produce biodiesel is constrained because there are only a few palm plantations in its southern region (Ngammuangtueng et al., 2020). To maximize the economic conditions of biofuels in the region, an integrated market for trade in biofuels across national borders is anticipated.

4.3 Bioenergy Development in Latin America and the Caribbean

The developing economies of Latin America and the Caribbean (LAC) have a lot of potential for biofuels because they have an abundance of local resources, a growing demand for sustainable energy, and land that can be used to produce biofuels (Ordoñez Olivo & Lakner, 2023). Cantarella et al. (2023) presented an estimated 360 million hectares of spare, usable marginal land available for LAC. Canabarro et al. (2023) reported the potential for biofuel expansion in LAC by taking into account policies, local infrastructure, crop yields, land availability, techno-economics, technology readiness level, and climate impacts. According to Trindade et al. (2022), the application of biofuels is widespread and well-established throughout South America (apart from Chile and Venezuela), as it is in nations like Thailand, Indonesia, Malaysia, the Philippines, and Malaysia, which have decades of experience using liquid biofuels. Due to this, several Latin American and Caribbean

(LAC) nations have implemented legal blending mandates to encourage the generation and application of biodiesel and bioethanol. These mandates aim to address environmental concerns, enhance energy security, foster rural development and job creation, and investigate export market opportunities (Antar et al., 2021). To lessen their reliance on oil for energy production, several LAC countries are currently putting programmes in place to encourage the conversion of bioresources to provide bioenergy in the form of liquid, gaseous, and solid biofuels.

Some nations are leaders in the use of biomass in LAC research advancements. One of them is Argentina, which started producing bioethanol from biomass in the 1970s and 1980s as a way to produce energy (Vega et al., 2024). However, compared to biodiesel, bioethanol production has not developed as much in Argentina. First-generation feedstocks are used to produce biodiesel because they are made from soybean oil. The quantity of biodiesel generated in Argentina for 2019, 2020, 2021 and 2022 were 2.15, 1.16, 1.72, and 1.91 Mton, respectively, being used domestically and primarily exported to the Netherlands (IRENA, 2023). There are currently 62 public and private biogas plants in sixteen provinces in Argentina. In 2022, they produced 418 GW-h (Vega et al., 2024). A percentage of the biogas produced is used by the governmental and private domains, respectively, for the production of 44% of electricity and 38% of thermal energy (FAO, 2020). In this nation, the steps taken in biofuel generation convert a sizable portion of the energy derived from biomass into electrical energy. Because of the wide range of substrates available for this process, the production of biofuels in this nation offers several benefits and aids in the development of the nation's circular economy.

According to Vidal (2019), Brazil is the leader in the generation of biofuels in LAC. It was the first country to generate energy from solids and then liquid biofuels. Soy is the main feedstock used for biodiesel production in this nation, though castor, sesame, sunflower, and oil palm have also been utilized. At the moment, the country's central and southern regions are the primary locations for the production of biodiesel (Vidal, 2019). Brazil has quadrupled its production of B100 biodiesel over the past 10 years, producing 5.9 million cubic metres (MCM) in 2019, 6.4 MCM in 2020, 6.8 MCM in 2021, and 6.3 MCM in 2022 (Bhatia et al., 2018). Currently, around the world, Brazil generates 85% of bioethanol primarily from sugar cane. They produce both first- and second-generation bioethanol. Brazil generated 35.3 MCM of bioethanol in 2019, 10.4 MCM were anhydride and 24.9 MCM were hydrated (Mączyńska et al., 2019). When anhydride is mixed with petrol, automobiles with compatible engines can use it immediately (CIBIOGÁS, 2020). The Anuário Estatístico Brasileiro (ANP) (2020) stated that its combustible fuel contains at least 27.5% bioethanol. In their study, czapp and Aranha (czapp & Aranha, 2023) reported that 30.75 billion litres of bioethanol were produced in 2022 while its policy aims to support the regional bioethanol industry by reducing the import of petrol and crude oil. The same authors also reported that 2022 saw 2.8 BCM of biogas produced from 810 plants, though it is one of the least biofuels generated in the nation.

The generation of biofuels is significant in Mexico, where the most common fuels are solid fuels and then biogas (Paredes-Cervantes et al., 2020). Biodiesel and other liquid biofuels together do not constitute a significant production. Jatropha is the most significant oil crop used to produce biodiesel; in 2017, 2.4 million litres were produced from it (Vega et al., 2024). In the same year, 49,200 litres of oil palm were harvested as the second one. Together, jatropha and oil palm represent almost 46% of the country's total usage of biodiesel raw materials (SENER, 2018). According to Vega et al. (2024), Mexico has the potential for bioethanol generation from high-sugar crops, which include sugar cane, beets, rice, and sweet sorghum. In addition to forest residues, wheat bran and its residues produced in this country offer a promising source of feedstock for biofuel production though the high cost of conversion has caused a barrier to harness the full potential. However, the production of biogas is less than 1% of that of solid biofuels. One of the most significant biogas producers in LAC is Mexico, nevertheless (Vega et al., 2024).

Other than Argentina, Brazil, and Mexico, biofuel-producing countries in the LAC also include Chile, Colombia, Costa Rica, etc. According to data from Vega et al. (2024), the rate of adoption of biomass-based technology is low in Chile to produce bioenergy, but it encourages industries to use the bioresources because of the positive impacts the biofuels have on the economy, environment, and society. For this reason, the nation produces a greater amount of solid biofuel than liquid biofuel due to the absence of sophisticated technologies in the production of liquid biofuels (biodiesel and bioethanol) and the lack of soil suitable for growing crops with high production potential for liquid biofuels (Vega et al., 2024). Although Colombia is behind Brazil and Mexico in biofuel production the nation's biofuel production is very significant. Even as of 2022, Colombia produced 779 million litres of biodiesel generated from palm oil with other used cooking oils (in smaller quantities), which is an increase of about 8.2% from the year before (Vega et al., 2024). However, just like Chile, most of the bioenergy generated in Colombia is in the form of solid biofuels while liquid biofuels come in second position. For Costa Rica, the capacity to generate energy from biomass is close to 600 MW, and projects that are currently underway have an estimation of 122 MW output (Vega et al., 2024). In conclusion, Cremonez et al. (2015) reported that the generation of liquid biofuels represented mostly by biodiesel and bioethanol stands out among renewable energy production in LAC. In LAC, Brazil and Argentina set the standard for the situation and future of biofuel generation with Peru, Uruguay, and a host of other nations still searching for substitutes to meet these demands while fighting for policies and laws that encourage biofuel applications in the energy grid (Cremonez et al., 2015).

5 Bioenergy as a Sustainable Development in the Global South

To create development that takes into account the complex interplay of energy, land, economic, and climatic factors, it is essential to understand the concept of sustainable energy in terms of bioenergy generation. Bioenergy is essential as it ensures the reduction of poverty and the promotion of sustainable development, according to UNDP (2019), on sustainable energy. It has an impact on all facets of development, including population growth, gender-related issues, access to water, agricultural productivity, livelihoods, and health and education (UNDP, 2019). Without an appreciable rise in the quantity and quality of energy services in the Global South, none of the SDGs can be achieved in their entirety. Also, Wiek et al. (2015) reported that in some rural parts of the Global South where access to energy is limited and living standards are low, bioenergy can help ensure energy and livelihood stability through transdisciplinary collaboration that focuses on workable solution options beyond sectorial constraints. Furthermore, bioenergy plays a key role in energy self-sufficiency in rural areas where demand and prices for cooking fuels are much higher (Brewer et al., 2018). Also, it has been discovered that there are significant technical, social, and financial barriers to bioenergy usage in the Global South (Karanja et al., 2020); however, cleaner bioenergy for cooking is still possible, and it is an essential part of the sustainability passage in developing nations.

According to Lynd and Woods (2011), there are several ways that bioenergy could aid in development. These include creating jobs, helping rural populations gain skills that are transferable and marketable, introducing agricultural infrastructure and expertise, and enhancing currency valuation and balance of payments. Others include accessible energy that enables self-reliance and makes agricultural machinery and processes more attainable and provision of an economically viable means of rehabilitating Africa's vast degraded land areas. Consequently, for bioenergy projects in the Global South to be stable and sustainable, the following priority areas need immediate attention:

1. Formulation of favourable policies on bioenergy production as well as alignment regulations for stakeholders.
2. Development of competitive business environments for bioenergy and their allied products/derivatives.
3. Establishment of appropriate expertise in the selection of suitable feedstock.
4. Introduction of high-level technical know-how into the bioenergy sector.
5. Prevention and eradication of unethical and predatory behaviours.
6. Internalization of environmental sustainability mentality in stakeholders in the bioenergy sector.
7. Instituting appropriate land tenure systems and protection in the bioenergy sector.
8. Minimization/reduction in food-fuel competition.
9. Eradication of harmful environmental practices.
10. Promotion of fuels end uses that promote social and environmental benefits.
11. Establishment of appropriate health schemes or policies for bioenergy workers, operators, and users.

6 Conclusion

The global priority of ensuring energy security and reaching net-zero carbon emissions has focused attention on replacing and reducing non-renewable energy sources, or fossil fuels. Therefore, this book chapter examined sustainable bioenergy generation from biological resources in the Global South regions of the world. This chapter has established that the Global South regions (Africa, Asia, Latin America, and the Caribbean) have a lot of potential for bioenergy generation because they have a lot natural resources that support abundant production of biological resources in different categories. As a result of abundant bioresources, from time immemorial, the Global South is known for using traditional bioenergy sources like fuel wood and charcoal but now, it has been found that more and more modern forms of bioenergy, such as bioethanol, biodiesel, and biogas, have the potential to replace a sizable portion of fossil fuels in the three regions. Consequently, the development of modern bioenergy in these regions will open up opportunities, chief among them sustainable energy security for the environment, society, and economy, thereby providing benefits that are either directly or indirectly related to achieving the United Nations Sustainable Development Goals.

Despite the fact that there are many advantages and benefits attached to generation of bioenergy from bioresources, the transition from the well-established fossil-based economy to the growing bioeconomy is undoubtedly difficult. Some of the basic challenges are the lack of large-scale production, technological obstacles, decentralization of the biomass supply chain, and high initial costs. The creation of effective and optimal procedures for bioenergy production should be the main objective in resolving these challenges. Furthermore, more sustainable energy and economic balance must be achieved through the use of modern bioenergy technologies and policies. Therefore, it is recommended that a Global South bioenergy forum is needed to support dialogue, learning and cooperation, and to help ensure that the positive, transformative aspects of bioenergy are realized and deleterious ones avoided. Also, it is expected that the information given in this chapter will provide essential guidance to decision-makers or policymakers in the energy sector to achieve more sustainability in the economy, society, and environment in the Global South.

References

Adeboye, B. S., Idris, M. O., Adedeji, W. O., Adefajo, A. A., Oyewusi, T. F., & Adelekun, A. (2022). Characterization and energy potential of municipal solid waste in Osogbo metropolis. *Cleaner Waste Systems, 2*, 100020.

Akudugu, M. A., & Ogwu, M. C. (2024). Sustainable development policies and interventions: A bibliometric analysis of the contributions of the academic community. *Journal of Cleaner Production, 434*, Article # 139919. https://doi.org/10.1016/j.jclepro.2023.139919

Amigun, B., Sigamoney, R., & von Blottnitz, H. (2008). Commercialisation of biofuel industry in Africa: a review. *Renewable and Sustainable Energy Reviews, 12*(3), 690–711.

Amigun, B., Musango, J. K., & Stafford, W. (2011). Biofuels and sustainability in Africa. *Renewable and Sustainable Energy Reviews, 15*(2), 1360–1372.

Anastassiadis, S. G. (2016). Carbon sources for biomass, food, fossils, biofuels and biotechnology-review article. *World Journal of Biology and Biotechnology, 1*(1), 1–32.

Antar, M., Lyu, D., Nazari, M., Shah, A., Zhou, X., & Smith, D. L. (2021). Biomass for a sustainable bioeconomy: An overview of world biomass production and utilization. *Renewable and Sustainable Energy Reviews, 139*, 110691.

Anuário Estatístico Brasileiro (ANP). (2020). *Anuário estatístico brasileiro do petroleo, gás natural e biocombustíveis*. [Online]. Available: http://www.anp.gov.br/images/publicacoes/anuario-estatistico/2018/anuario_2018.pdf. Accessed on 6 Apr 2024.

Anyaoha, K. E., & Zhang, L. (2022). Transition from fossil-fuel to renewable-energy-based smallholder bioeconomy: Techno-economic analyses of two oil palm production systems. *Chemical Engineering Journal Advances, 10*, 100270.

Aransiola, E. F., Oyewusi, T. F., Osunbitan, J. A., & Ogunjimi, L. A. (2019). Effect of binder type, binder concentration and compacting pressure on some physical properties of carbonized corncob briquette. *Energy Reports, 5*, 909–918.

Austin, G., & Morris, G. (2012). Biogas production in Africa. In *Bioenergy for sustainable development in Africa* (pp. 103–115).

Bailis, R., Drigo, R., Ghilardi, A., & Masera, O. (2015). The carbon footprint of traditional woodfuels. *Nature Climate Change, 5*(3), 266–272.

Balina, K., Romagnoli, F., & Blumberga, D. (2017). Seaweed biorefinery concept for sustainable use of marine resources. *Energy Procedia, 128*, 504–511.

Baz, K., Cheng, J., Xu, D., Abbas, K., Ali, I., Ali, H., & Fang, C. (2021). Asymmetric impact of fossil fuel and renewable energy consumption on economic growth: A nonlinear technique. *Energy, 226*, 120357.

Benti, N. E., Aneseyee, A. B., Geffe, C. A., Woldegiyorgis, T. A., Gurmesa, G. S., Bibiso, M., Asfaw, A. A., Milki, A. W., & Mekonnen, Y. S. (2023). Biodiesel production in Ethiopia: Current status and future prospects. *Scientific African, 19*, e01531.

Betiku, E., Akintunde, A. M., & Ojumu, T. V. (2016). Banana peels as a biobase catalyst for fatty acid methyl esters production using Napoleon's plume (Bauhinia monandra) seed oil: A process parameters optimization study. *Energy, 103*, 797–806.

Bhatia, S. K., Joo, H. S., & Yang, Y. H. (2018). Biowaste-to-bioenergy using biological methods—a mini-review (Dec.). *Energy Conversion and Management, 177*, 640–660.

Birhanu, A., & Ayalew, S. (2017). A review on potential and status of biofuel production in Ethiopia. *Journal of Plant Sciences, 5*(2), 82–89.

Börjesson, P. (2009). Good or bad bioethanol from a greenhouse gas perspective–what determines this? *Applied Energy, 86*(5), 589–594.

Brewer, J. P., Vandever, S., & Johnson, J. T. (2018). Towards energy sovereignty: biomass as sustainability in interior Alaska. *Sustainability Science, 13*, 417–429.

czapp, & Aranha, C. (2023). *Biogas Production Increases 21.3% in Brazil in 2022*. [Online]. Available: Biogas Production Increases 21.3% in Brazil in 2022 | Czapp. Accessed on 06 Apr 2024.

Canabarro, N. I., Silva-Ortiz, P., Nogueira, L. A., Cantarella, H., Maciel-Filho, R., & Souza, G. M. (2023). Sustainability assessment of ethanol and biodiesel production in Argentina, Brazil, Colombia, and Guatemala. *Renewable and Sustainable Energy Reviews, 171*, 113019.

Cantarella, H., Leal Silva, J. F., Nogueira, L. A. H., Maciel Filho, R., Rossetto, R., Ekbom, T., Souza, G. M., & Mueller-Langer, F. (2023). Biofuel technologies: Lessons learned and pathways to decarbonization. *GCB Bioenergy, 15*(10), 1190–1203.

Chong, C. T., Loe, T. Y., Wong, K. Y., Ashokkumar, V., Lam, S. S., Chong, W. T., Borrion, A., Tian, B., & Ng, J. H. (2021). Biodiesel sustainability: The global impact of potential biodiesel production on the energy–water–food (EWF) nexus. *Environmental Technology & Innovation, 22*, 101408.

CIBIOGÁS. (2020). Panorama do Biogás no Brasil em 2019. *Foz do Iguaçu, 5*(3), 248–253.

Ciesielczuk, T., Poluszyńska, J., Rosik-Dulewska, C., Sporek, M., & Lenkiewicz, M. (2016). Uses of weeds as an economical alternative to processed wood biomass and fossil fuels. *Ecological Engineering, 95*, 485–491.

Cremonez, P. A., Feroldi, M., Feiden, A., Teleken, J. G., Gris, D. J., Dieter, J., de Rossi, E., & Antonelli, J. (2015). Current scenario and prospects of use of liquid biofuels in South America. *Renewable and Sustainable Energy Reviews, 43*, 352–362.

Dahunsi, S. O., Fagbiele, O. O., & Yusuf, E. O. (2020). Bioenergy technologies adoption in Africa: A review of past and current status. *Journal of Cleaner Production, 264*, 121683.

Debez, A., Belghith, I., Friesen, J., Montzka, C., & Elleuche, S. (2017). Facing the challenge of sustainable bioenergy production: Could halophytes be part of the solution? *Journal of Biological Engineering, 11*(1), 27.

Destek, M. A., Sarkodie, S. A., & Asamoah, E. F. (2021). Does biomass energy drive environmental sustainability? An SDG perspective for top five biomass consuming countries. *Biomass and Bioenergy, 149*, 106076.

Dirisu, J. O., Salawu, E. Y., Ekpe, I. C., Udoye, N. E., Falodun, O. E., Oyedepo, S. O., Ajayi, O. O., & Kale, S. A. (2024). Promoting the use of bioenergy in developing nations: A CDM route to sustainable development. *Frontiers in Energy Research, 11*, 1184348.

Egboduku, W. O., Egboduku, T., Golohor, O. M., Imarhiagbe, O., & Ogwu, M. C. (2024). Cassava as raw material for sustainable bioeconomy development. In M. C. Ogwu, S. C. Izah, A. A. Cunha Alves, & babu, S.C. (Eds.), *Plant biology, Sustainability and climate change, sustainable cassava* (pp. 57–73). Academic Press. https://doi.org/10.1016/B978-0-443-21747-0.00022-9

Elder, M., & Hayashi, S. (2018). A regional perspective on biofuels in Asia. In *Biofuels and sustainability: Holistic perspectives for policy-making* (pp. 223–246).

ETC Group. (2010). The new biomassters: Synthetic biology and the next assault on biodiversity and livelihoods. ETC Group Communiqué 104.

Fabinyi, M. (2018). Environmental fixes and historical trajectories of marine resource use in Southeast Asia. *Geoforum, 91*, 87–96.

FAO. (2020). Actual. Del. Balance De. biomasa con fines energ'eticos En. la Argent. https://doi.org/10.4060/ca8764es

Field, J. L., Richard, T. L., Smithwickc, E. A. H., Caie, H., Laserf, M. S., LeBauerg, D. S., Longh, S. P., Paustian, K., Zhangcai Qinc, Z., & Shcchann, J. J. (2020). Robust paths to net greenhouse gas mitigation and negative emissions via advanced biofuels. *PNAS, 117*(36), 21968–21977. https://www.pnas.org/doi/10.1073/pnas.1920877117

Fu, X. M., Zhang, M. Q., Liu, Y., Shao, C. L., Hu, Y., Wang, X. Y., Su, L. R., Wang, N., & Wang, C. Y. (2018). Protective exploitation of marine bioresources in China. *Ocean and Coastal Management, 163*, 192–204.

GBEP, Global Bioenergy Partnership. (2011). *The global bioenergy partnership sustainability indicators for bioenergy* (1st edn.). Global Bioenergy Partnership. http://www.globalbioenergy.org/fileadmin/user_upload/gbep/docs/Indicators/The_GBEP_Sustainability_Indicators_for_Bioenergy_FINAL.pdf. Accessed on 06 Apr 2024

Gharde, Y., Singh, P. K., Dubey, R. P., & Gupta, P. K. (2018). Assessment of yield and economic losses in agriculture due to weeds in India. *Crop Protection, 107*, 12–18.

Gravelsins, A., Blumberga, A., Blumberga, D., & Muizniece, I. (2017). Economic analysis of wood products: System dynamics approach. *Energy Procedia, 128*, 431–436.

Guo, M., Song, W., & Buhain, J. (2015). Bioenergy and biofuels: History, status, and perspective. *Renewable and Sustainable Energy Reviews, 42*, 712–725.

Gupta, N., & Gupta, R. (2015). Solid waste management and sustainable cities in India: the case of Chandigarh. *Environment and Urbanization, 27*(2), 573–588.

Gupta, V. K., Potumarthi, R., O'Donovan, A., Kubicek, C. P., Sharma, G. D., & Tuohy, M. G. (2014). Bioenergy research: An overview on technological developments and bioresources. In *Bioenergy research: Advances and applications* (pp. 23–47).

IEA, International Energy Agency. (2020). *Introduction to System Integration of Renewables*. IEA. https://www.iea.org/reports/introduction-to-system-integration-of-renewables

Ingle, A. P., Philippini, R. R., Martiniano, S., Marcelino, P. R. F., Gupta, I., Prasad, S., & da Silva, S. S. (2020). Bioresources and their significance: Prospects and obstacles. In *Current developments in biotechnology and bioengineering* (pp. 3–40).

IRENA, International Renewable Energy Agency. (2017). *Biofuel potential in Southeast Asia: Raising food yields, reducing food waste and utilising residues*. International Renewable Energy Agency. https://www.irena.org/publications/2017/Jun/Biofuel-potential-in-Southeast-Asia-Raisingfood-yields-reducing-food-waste-and-utilisingresidues. Accessed on 06 Apr 2024.

IRENA, International Renewable Energy Agency. (2021). *World energy transitions outlook: 1.5°C pathway*. International Renewable Energy Agency, Abu Dhabi. Accessed on 06 Apr 2024.

IRENA, Renewable Energy Statistics. (2023). *International Renewable Energy Agency*. 2023. [Online]. https://mc-cd8320d4–36a1–40ac-83cc-3389-cdnendpoint.azureedge.net/-/media/Files/IRENA/Agency/Publication/2023/Mar/IRENA_RE_Capacity_Statistics_2023.pdf? Accessed on 06 Apr 2024.

Jagger, P., Bailis, R., Dermawan, A., Kittner, N., & McCord, R. (2019). SDG 7: Affordable and clean energy–How access to affordable and clean energy affects forests and forest-based livelihoods. In *Sustainable development goals: Their impacts on forests and people* (pp. 206–236). Cambridge University Press.

Johnson, F. X. (2017). Biofuels, bioenergy and the bioeconomy in North and South. *Industrial Biotechnology, 13*(6), 289. https://doi.org/10.1089/ind.2017.29106.fxj

Karanja, A., Mburu, F., & Gasparatos, A. (2020). A multi-stakeholder perception analysis about the adoption, impacts and priority areas in the Kenyan clean cooking sector. *Sustainability Science, 15*, 333–351.

Kataki, S., West, H., Clarke, M., & Baruah, D. C. (2016). Phosphorus recovery as struvite from farm, municipal and industrial waste: Feedstock suitability, methods and pre-treatments. *Waste Management, 49*, 437–454.

Kosoe, E. A., Achana, G. T. W., & Ogwu, M. C. (2023). Regulations and policies for herbal medicine and practitioners. In S. C. Izah, M. C. Ogwu, & M. Akram (Eds.), *Herbal medicine phytochemistry. Reference series in phytochemistry*. Springer. https://doi.org/10.1007/978-3-031-21973-3_33-1

Kowalski, A. M. (2021). Global south-global north differences. In *No poverty* (pp. 389–400). Springer International Publishing.

Lal, R., Bouma, J., Brevik, E., Dawson, L., Field, D. J., Glaser, B., Hatano, R., Hartemink, A. E., Kosaki, T., Lascelles, B., & Monger, C. (2021). Soils and sustainable development goals of the United Nations: An International Union of Soil Sciences perspective. *Geoderma Regional, 25*, 00398.

Lebaka, V. (2013). Potential bioresources as future sources of biofuels production: An overview. In V. Gupta & M. Tuohy (Eds.), *Biofuel technologies*. Springer.

Li, F., Li, Y., Novoselov, K. S., Liang, F., Meng, J., Ho, S. H., Zhao, T., Zhou, H., Ahmad, A., Zhu, Y., Hu, L., Ji, D., Jia, L., Liu, R., Ramakrishna, S., & Zhang, X. (2023). Bioresource upgrade for sustainable energy, environment, and biomedicine. *Nano-Micro Letters, 15*(1), 35. https://doi.org/10.1007/s40820-022-00993-4

Lynd, L. R., & Woods, J. (2011). Perspective: A new hope for Africa. *Nature, 474*(7352), S20–S21.

M'hamdi, A. I., Kandri, N. I., Zerouale, A., Blumberga, D., & Gusca, J. (2017). Life cycle assessment of paper production from treated wood. *Energy Procedia, 128*, 461–468.

Mączyńska, J., Krzywonos, M., Kupczyk, A., Tucki, K., Sikora, M., Pińkowska, H., Bączyk, A., & Wielewska, I. (2019). Production and use of biofuels for transport in Poland and Brazil–The case of bioethanol. *Fuel, 241*, 989–996.

Mahmud, S., Haider, A. R., Shahriar, S. T., Salehin, S., Hasan, A. M., & Johansson, M. T. (2022). Bioethanol and biodiesel blended fuels—feasibility analysis of biofuel feedstocks in Bangladesh. *Energy Reports, 8*, 1741–1756.

Maqhuzu, A. B., Yoshikawa, K., & Takahashi, F. (2017). Biofuels from agricultural biomass in Zimbabwe: feedstock availability and energy potential. *Energy Procedia, 142*, 111–116.

Miftah, F. K., & Mutta, D. (2024). Potential markets and policies for sustainable liquid biofuel production with emphasis to Eastern Africa countries: a review. *Energy, Sustainability and Society, 14*(1), 1.

Narodoslawsky M (2014) *Utilizing bio-resources: Rational strategies for a sustainable bio-economy*. ITA-manu14-02. Available at: http://epub.oeaw.ac.at/ita/ita-manuscript/ita_14_02.pdf. Accessed on 18 Apr 2024.

Ngammuangtueng, P., Jakrawatana, N., & Gheewala, S. H. (2020). Nexus resources efficiency assessment and management towards transition to sustainable bioeconomy in Thailand. *Resources, Conservation and Recycling, 160*, 104945.

Nigam, P. S., & Singh, A. (2011). Production of liquid biofuels from renewable resources. *Progress in Energy and Combustion Science, 37*(1), 52–68.

Nimachow, G., Taga, T., Tag, H., & Dai, O. (2008). Linkages between bio-resources and human livelihood: A case study of Adi tribes of Mirem Village, Arunachal Pradesh (India). *The Initiation, 2*(1), 183–198.

Nizzy, A. M., & Ogwu, M. C. (2024). Valorization of cassava processing by-products into biofuel for a sustainable environment. In M. C. Ogwu, S. C. Izah, A. A. Cunha Alves, & babu, S.C. (Eds.), *Plant biology, sustainability and climate change, sustainable cassava* (pp. 291–309). Academic. https://doi.org/10.1016/B978-0-443-21747-0.00010-2

Obahiagbon, E. G., & Ogwu, M. C. (2023a). Sustainable supply chain management in the herbal medicine industry. In S. C. Izah, M. C. Ogwu, & M. Akram (Eds.), *Herbal medicine phytochemistry. Reference series in phytochemistry*. Springer. https://doi.org/10.1007/978-3-031-21973-3_64-1

Obahiagbon, E. G., & Ogwu, M. C. (2023b). Consumer perception and demand for sustainable herbal medicine products and market. In S. C. Izah, M. C. Ogwu, & M. Akram (Eds.), *Herbal medicine phytochemistry. Reference series in phytochemistry*. Springer. https://doi.org/10.1007/978-3-031-21973-3_65-1

Obahiagbon, E. G., & Ogwu, M. C. (2024a). Organic food preservatives: The shift towards natural alternatives and sustainability in the global south's markets. In M. C. Ogwu, S. C. Izah, & N. R. Ntuli (Eds.), *Food safety and quality in the global south* (pp. 299–329). Springer. https://doi.org/10.1007/978-981-97-2428-4_10

Obahiagbon, E. G., & Ogwu, M. C. (2024b). The nexus of business, sustainability, and herbal medicine. In S. C. Izah, M. C. Ogwu, & M. Akram (Eds.), *Herbal medicine phytochemistry. Reference series in phytochemistry*. Springer. https://doi.org/10.1007/978-3-031-21973-3_67-1

Obi, F. O., Ugwuishiwu, B. O., & Nwakaire, J. N. (2016). Agricultural waste concept, generation, utilization and management. *Nigerian Journal of Technology, 35*(4), 957–964.

Ogbonna, J. C., Nomura, N., & Aoyagi, H. (2013). Bioenergy production and food security in Africa. *African Journal of Biotechnology, 12*(52), 7147–7157.

Ogundahunsi, O. E., Akpan, G. E., Ayorinde, T. A., Babafemi, O. P., & Ayodele, A. A. (2022a). Industrial applications of biodiversity potentials in Africa. In S. Chibueze Izah (Ed.), *Biodiversity in Africa: Potentials, threats and conservation. Sustainable development and biodiversity* (Vol. 29, pp. 23–49). Springer. https://doi.org/10.1007/978-981-19-3326-4_2

Ogundahunsi, O. E., Ogunsina, B. S., & Aransiola, E. F. (2022b). Kariya Pod-husks: A novel biocatalyst in biodiesel production. *Journal of Energy Research and Reviews., 11*(4), 48–54.

Ogundahunsi, O. E., Akpan, G. E., Agbaje, C. O., Oyeniyi, K., Olaoye, I. O. & Oyewusi, T. F. (2024). Energy Audit and Alternative Energy Source for Sustainable Processing of Cassava Products in Oyo State, Nigeria. *Journal of Engineering Research and Reports, 26*(7), 462–471.

Ogwu, M. C. (2019). Understanding the composition of food waste: An "-Omics" approach to food waste management. In A. P. Gunjal, M. S. Waghmode, N. N. Patil, & P. Bhatt (Eds.), *Global initiatives for waste reduction and cutting food loss* (pp. 212–236). IGI Global. https://doi.org/10.4018/978-1-5225-7706-5.ch011

Ogwu, M. C., Odozi, I. P., Ahonsi, O. C., Uleanya, K. O., & Odozi, E. B. (2024). Organic acid production from cassava. In M. C. Ogwu, S. C. Izah, A. A. Cunha Alves, & babu, S.C. (Eds.), *Plant biology, sustainability and climate change, sustainable cassava* (pp. 395–418). Academic. https://doi.org/10.1016/B978-0-443-21747-0.00009-6

Ohimain, E. I., & Izah, S. C. (2017). A review of biogas production from palm oil mill effluents using different configurations of bioreactors. *Renewable and Sustainable Energy Reviews, 70*, 242–253.

Olabi, A. G., Obaideen, K., Elsaid, K., Wilberforce, T., Sayed, E. T., Maghrabie, H. M., et al. (2022). Assessment of the pre-combustion carbon capture contribution into sustainable development goals SDGs using novel indicators. *Renewable and Sustainable Energy Reviews, 153*, 111710.

Ordoñez Olivo, M. L., & Lakner, Z. (2023). Food security and biofuels in Latin America and the Caribbean Region: A data panel analysis on eight countries. *Energies, 16*(23), 7799.

Oyedepo, S. O. (2012). Energy and sustainable development in Nigeria: The way forward. *Energy, Sustainability and Society, 2*, 1–17.

Oyewusi, T. F., Aransiola, E. F., Oyinlola, T. A., & Adesigbin, J. A. (2018). Comparative study of the calorific values and gas emissions of briquettes produced from carbonized and uncarbonized process of corncob. *Journal of the Nigerian Institution of Mechanical Engineers, 8*(2), 20–36.

Oyewusi, T. F., Alebiowu, G., Aransiola, E. F., Soji-Adekunle, A. R., & Adeboye, B. S. (2023). Effect of pretreatment on the physical properties and heating values of briquettes. *Journal of Renewable Energy and Environment, 10*(1), 26–33.

Oyewusi, T. F., Soji-Adekunle, A. R., Oguntunji, M. A., Ayo, V. O., Omotosho, O. A., & Ogundahunsi, O. E. (2024). Biopotential property of maggots as inoculum in anaerobic digestion of substrates for biogas production. *Journal of Engineering in Agriculture and the Environment, 10*(2), 43–54.

Østergaard, P. A., Duic, N., Noorollahi, Y., Mikulcic, H., & Kalogirou, S. (2020). Sustainable development using renewable energy technology. *Renewable Energy, 146*, 2430–2437. https://doi.org/10.1016/j.renene.2019.08.094

Paredes-Cervantes, S. A., Barahona-Pérez, L. F., Barroso-Tanoira, F. G., & Ponce-Marbán, D. V. (2020). Biofuels and their potential in the mexican energy market. *Revista de economía, 37*(94), 35–56.

Prange, C. (2019). Asia: The continent of superlatives and contradictions. In *Management practices in Asia: Case studies on market entry, CSR, and coaching* (pp. 3–9).

Qin, Y. (2018). Seaweed bioresources. In *Bioactive seaweeds for food applications* (pp. 3–24). Academic.

Romeu-Dalmau, C., Gasparatos, A., von Maltitz, G., Graham, A., Almagro-Garcia, J., Wilebore, B., & Willis, K. J. (2018). Impacts of land use change due to biofuel crops on climate regulation services: five case studies in Malawi, Mozambique and Swaziland. *Biomass and Bioenergy, 114*, 30–40.

Roy, N., Kumar, A., Majumder, S., & Mondal, P. (2021). Consequences of anthropogenic disturbance on variation of soil properties and food security: An Asian story. In *Soil science: Fundamentals to recent advances* (pp. 693–727).

RSB, Roundtable on Sustainable Biomaterials. (2016). *RSB principles and criteria*. RSBSTD-01-001 (Version 3.0). Roundtable on Sustainable Biomaterials. https://rsb.org/wp-content/uploads/2020/06/RSB-STD-01-001_Principles_and_Criteria-DIGITAL.pdf

Russo, V., & Von Blottnitz, H. (2017). Potentialities of biogas installation in South African meat value chain for environmental impacts reduction. *Journal of Cleaner Production, 153*, 465–473.

Saladini, F., Patrizi, N., Pulselli, F. M., Marchettini, N., & Bastianoni, S. (2016). Guidelines for energy evaluation of first, second and third generation biofuels. *Renewable and Sustainable Energy Reviews, 66*, 221–227.

SENER. (2018). *Reporte de Inteligencia tecnológica Biodiésel*.

Shimeles, A., Verdier-Chouchane, A., & Boly, A. (2018). Introduction: Understanding the challenges of the agricultural sector in Sub-Saharan Africa. In A. Shimles, A. Verdier-Chouchane, & A. Boly (Eds.), *Building a resilient and sustainable agriculture in sub-Saharan Africa* (pp. 1–12). Palgrave Macmillan.

Soji-Adekunle, A. R., Betiku, E., Asere, A. A., Falowo, O. A., Oyewusi, T. F., & Adeboye, B. S. (2024). Optimization and modeling of the performance of compression ignition engine fired on biofuel from non-edible vegetable oils. *Cleaner Energy Systems, 7*, 100105.

Statista Research Department. (2024). *Global biodiesel production 2022, by country*. https://www.statista.com/statistics/271472/biodiesel-production-in-selected-countries/#:~:text=Indonesia%20is%20the%20world's%20largest,output%20of%2013.65%20billion%20liters. Accessed on 8 Apr 2024.
Stephenson, P. J., & Damerell, A. (2022). Bioeconomy and circular economy approaches need to enhance the focus on biodiversity to achieve sustainability. *Sustainability, 14*(17), 10643. https://doi.org/10.3390/su141710643
Torvanger, A. (2019). Governance of bioenergy with carbon capture and storage (BECCS): accounting, rewarding, and the Paris agreement. *Climate Policy, 19*(3), 329–341.
Trindade, S. C., Nogueira, L. A. H., & Souza, G. M. (2022). Relevance of Latin America and Sub-Saharan Africa (LACAf) biofuels for global sustainability. *Biofuels, 13*(3), 279–289.
Ullah, K., Sharma, V. K., Dhingra, S., Braccio, G., Ahmad, M., & Sofia, S. (2015). Assessing the lignocellulosic biomass resources potential in developing countries: A critical review. *Renewable and Sustainable Energy Reviews, 51*, 682–698.
UNDP. (2019). The Sustainable Development Goals Report 2019, https://unstats.un.org/sdgs/report/2019/The-Sustainable-Development-Goals-Report-2019.pdf. Accessed 20 May 2024.
UNEP, United Nations Environment Programme. (2020). Emissions gap report 2020 executive summary. United Nations Environment Programme. https://www.unep.org/emissions-gap-report-2020. Accessed on 06 Apr 2024.
Vega, L. P., Bautista, K. T., Campos, H., Daza, S., & Vargas, G. (2024). Biofuel production in Latin America: A review for Argentina, Brazil, Mexico, Chile, Costa Rica and Colombia. *Energy Reports, 11*, 28–38.
Vidal, M. D. E. F. (2019). Produção e uso de biocombustíveis no Brasil. *Cad. Set. ETENE*, 1–13.
Wiek, A., Harlow, J., Melnick, R., van der Leeuw, S., Fukushi, K., Takeuchi, K., Farioli, F., Yamba, F., Blake, A., Geiger, C., & Kutter, R. (2015). Sustainability science in action: A review of the state of the field through case studies on disaster recovery, bioenergy, and precautionary purchasing. *Sustainability Science, 10*, 17–31.
World Bank. (2010). *Rising global interest in farmland: Can it yield sustainable and equitable benefits?* World Bank.
Yamaguchi, K., Matsumoto, M., Kusdiana, D., Ikeda, T., Hoshi, H., & Kan, S. (2013). *Study on ASIA potential of biofuel market*. ERIA Research Project Report, (25).
Zabed, H., Sahu, J. N., Boyce, A. N., & Faruq, G. (2016). Fuel ethanol production from lignocellulosic biomass: an overview on feedstocks and technological approaches. *Renewable and Sustainable Energy Reviews, 66*, 751–774.
Zhou, A., & Thomson, E. (2009). The development of biofuels in Asia. *Applied Energy, 86*, S11–S20.

Biohybrid Processes for the Valorization of Wastewater as a Circular Supply in a Microalgae Biorefinery

Beatriz Bonola, Ángel Eduardo Yáñez Ríos,
Yanet Lissette Antonio Lopéz,
and S. Montserrat García-Solares

Abstract

The growing demand to meet the needs of the population has generated various environmental impacts that are sometimes irreversible, specifically in the area of water, and to minimize impacts, conventional treatments have been implemented, which are divided into physical, chemical, and biological; however, these are not designed for the removal of emerging compounds (EC) such as: pesticides, flame retardants, microplastics, surfactants, analgesics, and antibiotics, among others. Therefore, biohybrid treatments have currently been proposed, which consist of the combination of a conventional process with a biological treatment, coupled with high efficiency, and value-added bioproducts can also be obtained. Some examples of biohybrid treatments, but not limited to them, are: (i) membrane biological processes (MBR), (ii) biological treatment with a moving bed reactor, (iii) disinfection processes with biological treatment, and (iv) advanced oxida-

B. Bonola
Unidad Profesional Interdisciplinaria en Ingeniería y Tecnologías Avanzadas, Instituto Politécnico Nacional, Gustavo A. Madero, Ciudad de México, Mexico

Á. E. Y. Ríos
Laboratorio de electroquímica y corrosión, Escuela Superior de Ingeniería Química e Industrias Extractivas (ESIQIE), Ciudad de México, Mexico

Y. L. A. Lopéz
Escuela Superior de Ingeniería Química e Industrias Extractivas (ESIQIE), Ciudad de México, Mexico

S. M. García-Solares (✉)
Centro Mexicano para la Producción más Limpia, Instituto Politécnico Nacional, Ciudad de México, Mexico

Laboratorio Nacional de Desarrollo y Aseguramiento de la Calidad de Biocombustibles (LaNDACBio), Ciudad de México, Mexico
e-mail: sgarcias@ipn.mx

M. C. Ogwu et al. (eds.), *Sustainable Bioeconomy Development in the Global South*, https://doi.org/10.1007/978-981-96-0305-3_10

tion processes (AOPs) with physicochemical and biological treatments. These biohybrid treatments offer new opportunities to revolutionize environmental remediation, up to biomedical applications directly impacting the economy. Given this concern, microalgae turns out to be an excellent renewable raw material for the refining of different types of wastewater combined with conventional treatment. In addition, they are commonly found in various ecosystems, and value-added products are obtained that can be used directly or as a precursor of bioenergetics, considering it an attractive process for the circular economy. Therefore, this chapter addresses the issue of wastewater treatment with AOPs to subsequently refine them with microalgae as an excellent alternative for water reuse, generation of value-added bioproducts, and production of bioenergetics, to be incorporated into the bioeconomy cycle and contribute to the fulfillment of the objectives of the 2030 agenda.

Keywords

Biorefinery · Emerging compounds · Biohybrid processes · Wastewater and microalgae

Abbreviations

AOP	Advanced oxidation processes
BIEX	Biological ion exchange
EC	Emerging compounds
EPA	United States Environmental Protection Agency
FAME	Fatty acid ester
IEA	International Energy Agency
IGES	Institute for Global Environmental Strategies
MBR	Membrane biological processes
MPL	Maximum permissible limits
NMC	Native microalgal consortia
PHA	Polyhydroxyalkanoate
SDG	Global Sustainable Development Goals
USEPA	United States Environmental Protection Agency
WHO	The World Health Organization
WWTP	Wastewater treatment plants

1 Introduction

The growing demand to meet the needs of the population has generated various environmental impacts that are sometimes irreversible; thus, it is necessary to implement sustainable solutions to mitigate those impacts. One of the most affected natural resources is water since it is required for daily life activities and all processes, generating around 380 billion m^3 per year (380 trillion liters) of wastewater worldwide, of which only 24% is treated before being discharged into rivers and lakes, due to this, treated wastewater should increase the percentage to conserve water resources. Wastewater is classified according to its source of generation into: (a) *domestic*: are those produced by daily household activities, such as cooking, laundry, and toilet, generally, contain some of the different types of pollutants, such as human excrement, liquid chemicals, organic matter, pathogens; (b) *industrial:* it is effluent of variable quality, composition, and volume from manufacturing processes, and the cleaning activities that are carried out along with these processes, typically contain: organic matter, dyes, metal ions, fats, and emerging pollutants that put health and environment at risk; (c) agricultural wastewater from crop irrigation can contain heavy metals, pesticides, and insecticides that easily filter into the soil and groundwater; and (d) commercial from establishments such as restaurants and shopping centers. Wastewater may contain a variety of contaminating materials that make it difficult to make it drinkable and cause damage to health.

To minimize the impacts generated by wastewater, various treatment technologies have been implemented, physical, chemical, and biological, allowing the removal of contaminating material; however, these treatments are not designed to eliminate emerging compounds (EC) such as pesticides, flame retardants, microplastics, surfactants, analgesics and antibiotics, among others; they have even been alarmingly found in treated water. In general, the presence of EC in wastewater is due to the increase in population, the lack of responsible consumption, and the recent SARS-CoV-2 pandemic, as an example. Eheneden et al. (2023) reported that in 8 wastewater treatment plants (WWTP), they found 14 different types of antibiotics, in concentrations of 3.1 and 1.2 μg/L in influents and effluents, respectively; likewise, they state that the low probability that antibiotic residues damage the microorganisms and organisms involved in the treatment of these effluents, due to their adaptive capacity, so that biological treatment with organisms and/or microorganisms native to a WWTP increases the efficiency of conventional treatments and is considered a plausible, sustainable, and low-cost solution for the removal of emerging contaminating material and other persistent types such as heavy metals. Dubey et al. (2023) studied the behavior of antibiotics in 10 WWTPs, obtaining influent and effluent oscillations between 11.94 and 0.70 kg/day respectively, showing a limited elimination by conventional treatment methods, identifying that the sorption process with sludge does not influence the removal of emerging contaminating material. On the other hand, microalgae contributed to the circular economy in approximately $3.4 billion in 2020, particularly, within the global market, and it is expected to increase to approximately $4.6 billion by 2027 (Wang et al., 2022).

Conventional treatments have a poor performance in the elimination of emerging contaminating material and are influenced by geographic location, kind of contaminating material, daily plant influent, and type of wastewater, making it necessary to implement new and advanced treatments for their elimination. One option is the use of hybrid treatments which combine two or more technologies to increase the diversity of the kind of contaminating material that can be treated and the efficiency of elimination.

Some examples of hybrid treatments for different types of wastewaters, but not limited to: (i) membrane biological processes (MBR), (ii) biological treatment with a moving bed reactor, (iii) disinfection processes with biological treatment, and (iv) advanced oxidation processes (AOPs) with physicochemical treatments. All those conventional processes that are combined with a biological treatment are called biohybrid, these systems together with the high removal efficiency of emerging contaminating material and such kinds as nitrogen and phosphorus in wastewater have the characteristic of being sustainable providing new opportunities to revolutionize various fields, from environmental remediation to biomedical application impacting directly on the economy (Fu et al., 2023). Some biohybrid treatment systems also focus on resource recovery, such as the generation of value-added bioproducts. Given this concern, in the last year, 24% of the studies carried out have focused on biological wastewater treatments for the removal of emerging contaminating material, mainly because of their simple operation, relatively low process cost, and availability of various types of microorganisms; however, due to the persistence of contaminating material and high removal efficiencies, AOPs and their combinations have been studied more frequently (28% and 39% respectively) (Pariente et al., 2022).

In this sense, microalgae are an excellent renewable raw material for the refining of different types of wastewaters, they are also commonly found in WWTPs, they have a low operating cost, and value-added products are obtained, being considered an attractive process for the circular economy.

Furthermore, the subject of "water" is addressed in the 2030 Agenda, which is a global action plan adopted by the Member States of the United Nations in September 2015. The overarching objective of the Agenda is to eradicate poverty, protect the planet, and ensure prosperity for all. This objective is to be achieved through a balanced approach that integrates the economic, social, and environmental dimensions of sustainable development. The Agenda is structured around 17 Sustainable Development Goals (SDGs), each of which is comprised of a set of targets. In total, 169 specific actions are required to achieve the goals. The specific targets are designed to be achieved by 2030 and are measured by indicators that allow progress to be monitored. These goals are universally applicable, regardless of a country's level of economic development. The SDGs address a multitude of global challenges, including poverty, hunger, health, education, gender equality, clean water and sanitation, affordable and clean energy, decent work and economic growth, reducing inequalities, sustainable cities, responsible consumption and production, climate action, underwater life, life on land, peace and justice, and partnerships to achieve these goals.

The responsibility for implementing and monitoring the SDGs falls upon governments. These entities must adapt global targets to local contexts and develop policies, programs, and monitoring frameworks to ensure progress. In addition, society, the productive and academic-scientific sector, and other entities contribute to this effort through initiatives, financing, innovation, and the promotion of sustainable practices. The advancement of the SDGs is gauged through a network of global, regional, and national indicators. Countries are required to submit regular reports to the United Nations on their progress in implementing the SDGs. Periodic assessments are conducted to facilitate the necessary adjustments to ensure that countries remain on track to achieve the goals by 2030.

In particular, Sustainable Development Goal 6 (SDG 6) is concerned with ensuring the availability and sustainable management of water and sanitation for all. This goal acknowledges the critical role of access to safe drinking water and sanitation in human health, well-being, and sustainable development. This includes the eradication of poverty, food security, health, education, gender equity, and economic growth. The specific targets set out in SDG 6 are as follows:

1. Universal and equitable access to safe and affordable drinking water for all
2. Access to adequate and equitable sanitation and hygiene services for all and an end to open defecation, with special attention to the needs of women and girls and people in vulnerable situations
3. Water quality by reducing pollution, eliminating dumping and minimizing the release of chemicals and hazardous materials, halving the proportion of untreated wastewater, and substantially increasing recycling and safe reuse globally
4. Efficient use of water in all sectors and ensuring the sustainability of freshwater withdrawal and supply to address water scarcity and substantially reduce the number of people suffering from water scarcity
5. Integrated water resources management at all levels, including through transboundary cooperation as appropriate
6. Protection and restoration of water-related ecosystems
7. Involvement of local communities in improving water management and sanitation
8. International cooperation and support to developing countries in activities and programs related to water and sanitation, including water harvesting, desalination, efficient use of water resources, wastewater treatment, recycling, and reuse technologies
9. Training and education development in activities and programs related to water and sanitation, including water harvesting, desalination, efficient use of water resources, wastewater treatment, recycling, and reuse technologies

Therefore, this chapter addresses the issue of wastewater with other physicochemical treatments to be subsequently refined with microalgae and to be incorporated into the cycle of the economy, being an excellent alternative for water reuse and a mechanism for minimizing the environmental impacts caused by water extraction and its poor disposal.

2 Importance of Water in the Economy and Its Use as a Circular Input

The planet Earth is covered by a third of water, although not all of it corresponds to freshwater bodies, the latter have been overexploited due to the growing demands to be met by anthropogenic activities, since almost every aspect of daily life requires water, poor management, low treatment efficiency, the presence of recalcitrant contaminating material that are difficult to eliminate, increase the lack of drinking water and water insecurity in society, together with the imbalance of the biogeochemical water cycle, the scarce recharge of aquifers and climate change; This makes innovations for water purification and monitoring.

The United Nations, since the adoption of the Global Sustainable Development Goals (SDGs) in 2015, has as a principle to minimize and mitigate the factors that potentiate climate change through sustainable cities and jobs that contribute to the economy of society, additionally promotes universal and equitable access to safe and affordable drinking water for all by 2030.

Globally, various alternatives have been proposed for water purification, reuse, recycling and water technologies that contribute to the fulfillment of the SDGs, specifically in SDG 6 "clean water and sanitation," which is focused on the proper management of water through 13 goals, it is estimated that these goals exceed 100 billion dollars annually with a total cost of 3.3 trillion dollars from 2015 to 2030, in addition to these costs may decrease or increase depending on the country, without representing a total compliance in less time (Dilekli & Cazcarro, 2019).

Leong (2021), reported a bibliometric analysis concerning water issues, finding a total of 1849 documents divided into scientific articles, books, book chapters and conferences, concluding that the challenges in water issues are linked to human activities, the understanding of published documents and the link with social behavior change. On the other hand, Obaideen et al. (2022) cite that although there are several global efforts for the fulfillment of SDG 6, this has been diminished due to all the industrialization activities that generate impact on the biogeochemical water cycle, as well as the various and new contaminating material that are found, which have hindered the treatment of wastewater affecting the availability of drinking water considering the scarcity of this resource; This has led to the need for wastewater to be collected, treated and reused in any country, as well as having an impact on the economy, making the application of sustainable development through the interrelationships of the SDGs a challenge today and a topic of vital importance, moving from an isolated approach to an integrated approach for quantifying compliance and identifying the synergies of the SDGs, for this purpose the Institute for Global Environmental Strategies (IGES) proposed a four-step statistical analysis methodology for "SDG Interlinkages and Indicators of SDG Compliance" for 27 countries in Asia and Africa, using the SDG Interlinkages Analysis & Visualization (V4.0) web tool, which are described below:

(i) Identification of causal links through the literature

(ii) Selection of indicators with traceable data for SDG targets based on the SDG Global Indicators and other proxy indicators (e.g., World Bank Development Indicators)
(iii) Data collection in different periods for 27 countries in Asia and Africa
(iv) Quantification of identified causal linkages between Step I targets

This methodology allows users to know with greater precision the compliance and synergies between the SDGs and their targets, although for the moment it is only designed for 27 countries on two continents.

Since all components of an ecosystem are interrelated, water as a basic requirement for humans, animals, and plants is required for the development of a society; therefore, all SDGs are related to the SDG 6 goals (Fig. 1) directly to SDGs 2, 3, 8, 9, 11, 12, 13, 14, and 15 and indirectly to the other SDGs (Obaideen et al., 2022). Water must comply with the maximum permissible limits (MPL) of several parameters which indicate the quality that allows it to be assigned for a specific activity. MPLs are issued by different international and national governmental agencies such as World Health Organization (WHO), the United States Environmental Protection Agency (EPA), and the national standards that each country issues, to comply with a legislated quality that can be used, which is why if a society does not have

Fig. 1 Relationship of sustainable development goals

adequate management of water resources, it would be difficult to move toward a more solid economy and reduce poverty (SDG 1), if the MPLs are not met the quality of water is at risk as well as the health of consumers, this way is that the goal of SDG 6; 6. 1 "provision of safe drinking water and sanitation," is related to SDG 3 "good health and well-being" because if the population is exposed to poor water quality it can affect their health.

One of the targets of Goal 13 "climate action" is to strengthen resilience and adaptive capacity to climate-related risks, which is complemented by the targets of Goal 14, which in summary indicate: reducing ocean acidification to minimize climate change, management of marine resources involving sustainable and artisanal fisheries, and one of the targets of SDG 15 is to promote the conservation, restoration and sustainable use of terrestrial and aquatic ecosystems as they provide shelter and food for people (Kosoe & Ogwu, 2024). Therefore, these goals have direct synergistic effects with the goals; 2 "zero hunger" (goal 2.4 to ensure sustainability of food production systems and apply resilient agricultural practices that increase productivity and production and contribute to the maintenance of ecosystems) and 6 "clean water and sanitation" specifically in goal 6.3 improve water quality, reducing untreated wastewater, significantly increasing recycling and safe reuse globally, goal 6.4 to make efficient use of water resources to ensure the sustainability of freshwater extraction and supply, goal 6.5 to manage water resources in an integrated manner, and goal 6.6 protect and restore water-related ecosystems, because water quality and quantity are essential elements for food production, i.e., water use for irrigation and livestock intake increases proportionally to population growth, demanding overexploitation of water resources and total compliance with MPLs when treated water is used to prevent crops irrigated with this type of water and the fauna involved in the food chain from ingesting contaminated water, in addition, most humans worldwide consume fish and other aquatic animals, it is therefore of utmost importance to prevent sewage and water that is not properly treated from being discharged into rivers, as they eventually lead to join the sea, thus greatly affecting aquatic life causing the death of living organisms, plants, and animals that are part of oceans and seas having a long term impact not only on animals and plants (Akudugu & Ogwu, 2024). According to a United Nations report, in developing countries, 23% of the world's population still receives contaminated water that causes diarrheal diseases and hepatitis, which compromises the fulfillment of SDG 3 "good health and well-being" (Sherif et al., 2023; Egboduku et al., 2024). The sustainable management of terrestrial and aquatic ecosystems through the adoption of clean technologies and processes will generate fair employment opportunities by promoting economic growth (SDGs 8 and 9), reducing the negative environmental impact of cities (SDG 11) and contributing to responsible consumption by avoiding food loss and waste, significantly reducing the release of contaminating material into the atmosphere, water and soil and promoting the philosophy of prevention, reduction, recycling and reuse (SDG 12), which will prevent environmental degradation. Therefore, the goals and indicators of the Sustainable Development Goals have synergistic effects on each other, which is relevant to study for each country.

The overall goal of meeting the SDGs is to preserve natural resources, which is why the unsustainable linear "take-make-use-waste-dispose" global economy must be adapted to sustainable management that involves reducing, recovering, reusing, and recycling waste, that is, a circular approach. To minimize the use of the unsustainable linear model, the circular economy has emerged, which has made significant progress in its implementation in some industries such as automotive, electronics, and construction (Delgado et al., 2024; Vinayagam et al., 2024). It is of utmost importance to study the circular economy of water because it is one of the most used natural resources in most activities; industrial, daily life, and habitat, this involves an approach to proper water management that includes minimizing its use, reuse, and reduction of contaminating material that are discharged into it, mitigating the imbalance in water resources, promoting equitable access to water and promoting the water-economy-ecosystem nexus (Fig. 1) (Sharma et al., 2021).

Water is one of the most important elements in the world economy since most industrial activities and daily life depend on it; it is, therefore, necessary to be aware of the potential risks of untreated wastewater for public health and the economy, due to its high environmental impacts on receiving water bodies and its overall operating costs, it is estimated that in recent years the costs for wastewater treatment that complies with the MPL and can be discharged are high amounting to billions annually.

Recently, the circular economy has become a global trend, and in the last 5 years, there has been a growing awareness and interest in the circularity of materials used in production processes. Implied by the global urgency to minimize the extraction of virgin materials and the incorporation of inputs into the global productive circularity, the Circularity Gap reports have provided an annual analysis and theory of the gap between the current state of the global economy and a fully circular economy since 2018 by Circle Economy, a non-profit organization based in the Netherlands (https://www.circularity-gap.world/global). This first report revealed that the global economy is only 9.1% circular, a data point that highlights the continued reliance on a linear economic model in which most materials are extracted, used, and then discarded. The Circularity Gap has had a significant impact on the global conversation around sustainability and the circular economy, highlighting the urgent need to close the circularity gap to ensure a more sustainable future through economic model policies and practices.

The circular economy is an economic model that focuses on reusing, recycling, and renewing products and materials to minimize waste and dependence on finite natural resources. The Circularity Gap quantifies the extent to which resources are used in a circular manner (i.e., recycled, reused, or restored) rather than extracted, consumed, and discarded through global circularity metrics and mapping of global material flows. The circularity metric is also known as the socioeconomic cycle, which indicates the relationship between the consumption of extracted materials and secondary materials, representing the amount of materials remaining in the production cycle. All biological materials that are returned to the environment in any form (i.e., solid, liquid, or gaseous) are considered in another indicator called the ecological cycle potential.

With the purpose to obtain the circularity metric, data selection is done through reports from international organizations and feedback from a scientific committee consisting of the World Resources Institute (WRI), the Netherlands Organization for Applied Scientific Research (TNO), the University of Natural Resources and Applied Life Sciences, Vienna (BOKU), the International Resource Panel (IRP), the Institute of Environmental Sciences at the University of Leiden (CML), the Doughnut Economics Action Lab (DEAL), the Netherlands Environmental Assessment Agency (PBL), and Statistics Netherlands (CBS).

According to the GAP reports, the global economy is significantly "linear," with a high percentage of materials ending up as waste after their initial use. For example, the 2023 report found that only 7.2% of materials used globally are recycled or reused, meaning that 92.8% of materials still follow a linear economic model, i.e., they are extracted.

The importance of measuring circularity is to reduce greenhouse gas emissions, conserve natural resources and create a sustainable economic model, using indicators such as assessing the flow of materials in the economy, from extraction and use to disposal and recycling. It analyzes what percentages of materials are recycled, reused or recovered, and what percentage are lost or discarded. It helps to understand how circularity can be improved in different sectors and regions, which requires public policies that promote the circular economy, such as incentives for recycling, waste reduction and the design of durable and repairable products. Businesses play a key role in the transition to circularity by adopting circular business models such as eco-design, the service economy (renting, leasing, sharing) and waste recovery. Changes in consumer and business behavior are essential to reduce the demand for virgin resources and increase the demand for recycled and reusable products. The Circularity GAP reflects the gap between current resource use and the potential of a global circular economy. Closing this gap is essential to move toward a more sustainable and resilient future, where resources are managed more efficiently and environmental impacts are significantly reduced, and this Circularity Report is linked to achieving the goals of the 2030 Agenda.

A report from *Circularity GAP* 2024 indicated that the circular economy has reached "megatrend" status, with the volume of discussions, debates, and articles on the concept nearly tripling over the past 5 years. However, this growth has not yet resulted in concrete actions or measurable impacts, suggesting that global circularity continues to decline. In the absence of decisive and expeditious action to transition to a circular economy, industries and governments may inadvertently resort to practices that could be perceived as "circular washing," despite their efforts to mitigate the impacts associated with the extraction of virgin materials and the disposal of waste in landfills. This is particularly relevant in the context of a linear economy, where the utilization of waste in circularity is not a predominant practice.

In the latest GAP 2024 report, it is stated that the proportion of secondary materials consumed by the global economy has declined from 9.1% in 2018 to 7.2% in 2023. This represents a 21% decline over the course of 5 years, with consumption continuing to accelerate. During this same period, humanity has consumed over 500 gigatons of materials, which represents 28% of all materials consumed since 1900.

The GAP 2024 report indicates that six of the nine key planetary boundaries that measure environmental health in land, water, and air have been exceeded, primarily due to the impacts of the linear "take-make-dispose" economy. It proposes the adoption of 16 circular economy-based solutions. To reverse the excess planetary boundaries, which are essentially the result of material extraction, it is necessary to reduce the global need for extraction by one-third. This can be achieved by implementing solutions in industries that are responsible for 15% of the planetary boundaries. The three key systems that have been identified as having the greatest impact on planetary boundaries are:

(i) The food system is a fundamental component of the global population and represents a significant employer, with approximately half of the global workforce engaged in its production and distribution. However, it is responsible for a quarter of the excess in the planetary limit of climate change due to the production of greenhouse gases and is the largest contributor to biodiversity loss. Furthermore, livestock operations utilize land areas equivalent to the size of the Americas, and approximately a quarter of freshwater resources are lost due to food waste.

(ii) The built environment, encompassing housing, commercial buildings, and infrastructure, is indispensable to human society. However, the extraction of minerals for construction materials has a profound environmental impact, accounting for approximately a quarter of global land use change. Approximately 40% of global greenhouse gas emissions are attributable to the construction, use, and demolition of buildings, with the associated processes accounting for nearly a third of total materials consumption.

(iii) The manufacture of goods represents a significant source of employment, yet the production of these goods is dependent on fossil fuels, which contribute considerably to environmental degradation and account for approximately one-third of the excess in the climate change limit due to greenhouse gas emissions. Furthermore, material- and energy-intensive industrial activities are linked to deforestation and account for 15% of the land and freshwater use limits. Furthermore, the production of these goods results in the generation of considerable quantities of hazardous industrial waste and chemical contamination.

The GAP 2024 initiative prioritizes the implementation of transformative circular solutions for the key systems. These solutions encompass the establishment of policies that prioritize circular practices, the securing of land rights and tenure policies to protect smallholder farmers, the setting of concrete policy targets for soil, water and biodiversity, the unlocking of investment in climate mitigation and adaptation, the provision of credit to farmers and landowners engaged in regenerative agriculture, the facilitation of the transition from the informal to the formal economy for both workers and economic units, the promotion of knowledge transfer, training and comprehensive skills development programs involving all relevant stakeholders. It is recommended that traditional circular construction techniques be supported, that the dominant regional waste streams be assessed, and that guidelines on waste integration in construction be provided.

The transition from a linear to a circular economy is based on changing legislation and policies that incentivize sustainable and circular practices while penalizing harmful ones in all sectors. These policies must be integrated into society so that circular solutions provide greater benefits than traditional linear ones. Furthermore, the implementation of the circular economy in any social sector must be a systematic and team-based practice. No single actor can drive change alone, and it is essential to ensure that green and circular jobs are of quality, with better representation, decent wages, and better working conditions. This will ensure a dignified life for all. Of all the inputs, wastewater has the greatest potential to be incorporated into a circular system.

In the circular water economy (Fig. 2); the contribution of wastewater treatment plants comes from their ability to recover nutrients, reuse water and recover energy, and they have been identified as one of the main ways to preserve the world's water resources. However, WWTPs are in charge of the effluent generated within a specific geographic area and commonly have conventional processes for the elimination of contaminating material, it is important to implement new treatments and/or

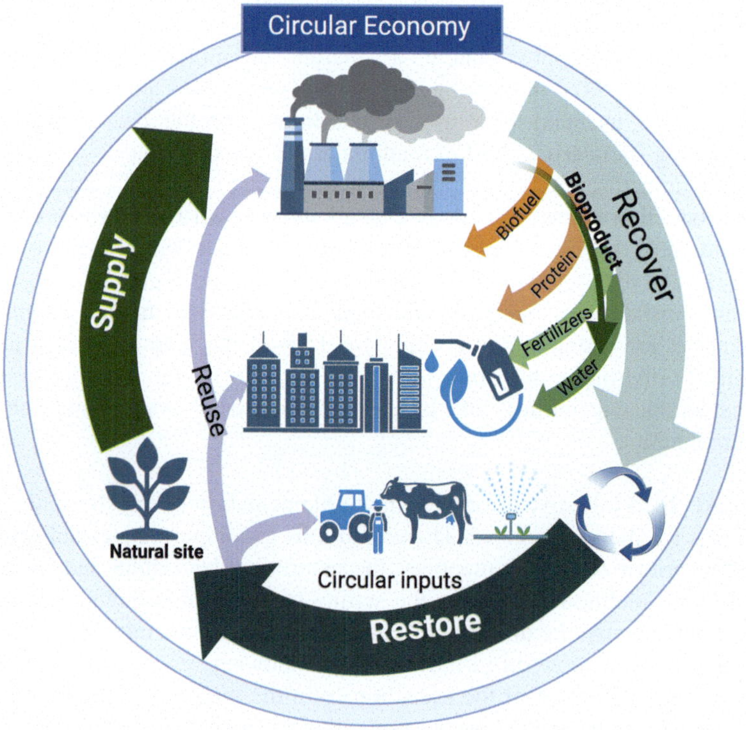

Fig. 2 Water in the circular economy. (Adapted from https://www.wicer-tool.com; and created in Biorender)

hybrid treatments, since several contaminating material are generated daily, such as emerging ones derived from the use of drugs that make the potabilization by traditional methods difficult and when the water is deposited in the receiving bodies, they cause imbalance, causing health hazards. Nishshanka et al. (2023) reported that wastewater treatment plants with capacities ranging from 8000 to 200,000 tons/day have been installed worldwide to discharge effluent with improved quality parameters as water scarcity affects between 1.69 and 2.37 billion urban residents.

Wastewater treatment and reuse are of significant economic importance due to the following factors: (i) *Resource recovery*: Wastewater contains nutrients such as nitrogen and phosphorus that can be recovered and utilized as fertilizers, thereby reducing the need for external inputs. Furthermore, valuable materials such as metals and chemicals can be extracted, thereby contributing to a circular economy and reducing dependence on natural resources. (ii) During wastewater treatment, anaerobic digestion processes can generate biogas, a renewable energy source. The biogas can be utilized to generate electricity and heat, thereby reducing the energy costs associated with treatment plants while simultaneously generating additional revenue. Furthermore, the reuse of treated water can alleviate the pressure on water resources, particularly in regions where water is scarce. (iii) The reuse of treated water can be employed for a variety of purposes, including agricultural irrigation, industrial processes, aquifer recharge, and recreational uses. This approach contributes to the sustainability of water resources and reduces the costs associated with the extraction and transportation of freshwater. (iv) The implementation of efficient wastewater treatment methods can also mitigate the contamination of water bodies, thereby enhancing the quality of public health. The health of ecosystems is of critical importance to some economic sectors, including fisheries, tourism, and agriculture, which depend on the quality of the water.

The management and treatment of wastewater have a substantial economic impact, as they facilitate the recovery of valuable resources, the generation of renewable energy, the reuse of water, a reduction in treatment costs, the creation of employment opportunities, compliance with environmental regulations, and the protection of ecosystems. These benefits contribute to the development of a more sustainable and resilient economy.

Recently, the practice of wastewater refining has been adopted as a strategy to be a circular input and obtain value-added products, mainly biofuels and treated water that can be incorporated back into the water cycle, which generates a decrease in environmental impacts, conservation of resources, and incorporation of water into the economy, and improves the quality of life for all.

Wastewater biorefining emerges as a sustainable tool of the circular economy due to the growing interest in sustainable and alternative energy sources, bio-based products and removal of persistent contaminating material, which implies a lower cost compared to the use of non-renewable primary materials, driving the paradigm shift of economic models that captures the full value of water: as a service and input and making the sector more financially and environmentally sustainable while ameliorating climate change and mitigating emissions (Delgado et al., 2024).The use of microorganisms and organisms, as well as the synergies between them, are essential elements in a biorefinery (Table 1) (Nishshanka et al., 2023).

Hereby, microalgae with the potential to effectively remove nutrients from wastewater streams have been exploited in wastewater treatment at the primary, secondary and tertiary treatment stages. Microalgae-based bioremediation generates valuable biomass with metabolites, namely lipids, proteins and carbohydrates, which could be used in the value-added production of biofuels, biofertilizers, etc. In addition, integrated wastewater treatment systems with microalgae would substantially remove residual contaminating material, nutrients, and pathogens with high removal efficiency. Therefore, the integration of microalgae into the conventional wastewater treatment process improves the sustainability of the process and at the same time contributes to the concept of circular bioeconomy. However, there are limited studies available on the potential of integrating microalgae into conventional wastewater treatment plants for real-world applications, although there are several reviews available in the literature that focus on microalgae-based wastewater treatment in a general context. Therefore, the current review aims to address this gap in the literature by comprehensively assessing the prospects of integrating phytoremediation as secondary and tertiary/advanced wastewater treatment processes, while discussing the challenges and prospects in the research domain.

Microalgae have a substantial economic impact, offering sustainable solutions in a range of industrial sectors. These include the production of biofuels such as biodiesel, bioethanol, and biogas, which represent a promising source of renewable energy. Such a shift could diminish reliance on fossil fuels and bolster the sustainability of our energy sources. They are employed as food supplements and in the production of functional foods, and are also utilized as animal feed for fish, mollusks, and crustaceans in aquaculture. They provide essential nutrients and enhance water quality in culture systems, which can augment the productivity and sustainability of aquaculture. The bioactive compounds present in microalgae, including antioxidants, fatty acids, and natural pigments, are of value to the pharmaceutical and cosmetic industries. They are employed in the formulation of pharmaceuticals, nutritional supplements, and personal care products. The capture of carbon dioxide

Table 1 Types of circular inputs in a biorefinery

Circular inputs	Microorganism/organism	Bioproduct	References
Tannery wastewater	Native microalgae consortium	Lipids, proteins, and water treats	Moreno-García et al. (2021)
Agricultural runoff wastewater	*Chlorella* sp. *S5*	The biomass was used as a potential source for the production of biofuels	Rana et al. (2024)
Waste-activated sludge (WAS)	Wet oxidation and *Yarrowia lipolytica*	Lipids and a slow-release fertilizer	Moreno et al. (2024)
CO_2 and dairy wastewater	*Chlorella vulgaris*	Water treats	Pasquarelli et al. (2024)
Primary treatment sludge	Dark fermentation	Short-chain fatty acids	Dubey et al. (2023)

(CO_2) and the production of bioplastics, in addition to wastewater treatment, are among the applications of microalgae that enable the absorption of nutrients and water pollutants, including nitrogen and phosphorus. This process represents a sustainable and efficient alternative to conventional water treatment methods.

3 Biohybrid Wastewater Treatments

Numerous methods have been applied in WWTPs, the efficiency of which is usually linked to the physicochemical composition of the effluent to be treated and the components of each technology. The treatments typically used to remove contaminating material from water are applied in the following order: (i) *pretreatments*: all those that retain floating solids to prevent damage to downstream treatment equipment; (ii) *primary treatments*: which consist of sedimentation of solids; (iii) *secondary treatments*: which include all aerobic and anaerobic biological treatments; and (iv) *tertiary treatments*: which include advanced filtration, disinfection, adsorption, chemical precipitation, and advanced oxidation processes (AOP) (Table 2).

When these treatments are implemented in combinations different from the traditional ones and also incorporate new technologies, they are known as hybrid treatments, which involve at least two different methods that can be primary, secondary and tertiary, since they have the potential to overcome the barriers of a single process and a conventional treatment train; achieving a higher efficiency of removal of

Table 2 Conventional treatments of wastewater

Treatments	Technology	Types of pollutants	References
Primary	Adsorption	Alcohols and acids are organic	Xiao et al. (2025)
Tertiary	Photocatalysis	Tetracycline	Latif et al. (2024)
Tertiary	Catalytic ozonation	Quinoline	Wu et al. (2024)
Primary	Adsorption	Denitrification and recovery of phosphate in water	Ding et al. (2024)
Tertiary	Electro-Fenton	Carbamazepine	Cruz del Álamo et al. (2024)
Secondary	*Aspergillus flavus* biomass	Tetracycline	Kiani Kori et al. (2024)
Primary	Wetlands	Emerging organic contaminants	Wolff et al. (2024)
Primary	Adsorption with granular activated carbon	Antibiotics	Sompornpailin et al. (2024)
Primary	Adsorption with granular activated carbon and surfactants	Pharmaceutical wastewater	Verdini et al. (2024)
Primary	*Bacillus flexus*	Decolorization of azo dyes and textile wastewater treatment	Batool et al. (2024)
Tertiary	Advanced oxidation processes (AOPs)	SARS-CoV-2 viral	Pramanik et al. (2023)

various contaminating material for a relatively low cost and time, in addition to the hybrid treatments that include biological treatment with enzymes, microorganisms, algae, microalgae and fungi, can be used as a bioprocess for obtaining value-added products, incorporating wastewater into the cycle of the economy, such systems are known as biohybrid treatments (Fig. 3) (Mumtaz et al., 2024).

Hybrid and biohybrid systems are normally applied in most wastewater because they usually consist of two or more contaminating material, even in ng/mL concentrations, and are promising sustainable solutions to meet the need to comply with MPLs for treated wastewater so that discharges do not affect the ecosystem and can be incorporated more quickly into the biogeochemical water cycle with the intention of recharging aquifers and minimizing water scarcity.

The choice of treatment depends mainly on the time required to bring the quality of the wastewater to a level of compliance with the MPLs stipulated in the geographical area where it is discharged and the type of contaminating material, for example, physical unit operations are commonly used for the removal of solids.

There are several reports of the application of hybrid treatments using physical-chemical systems, some of which are grouped in Table 3.

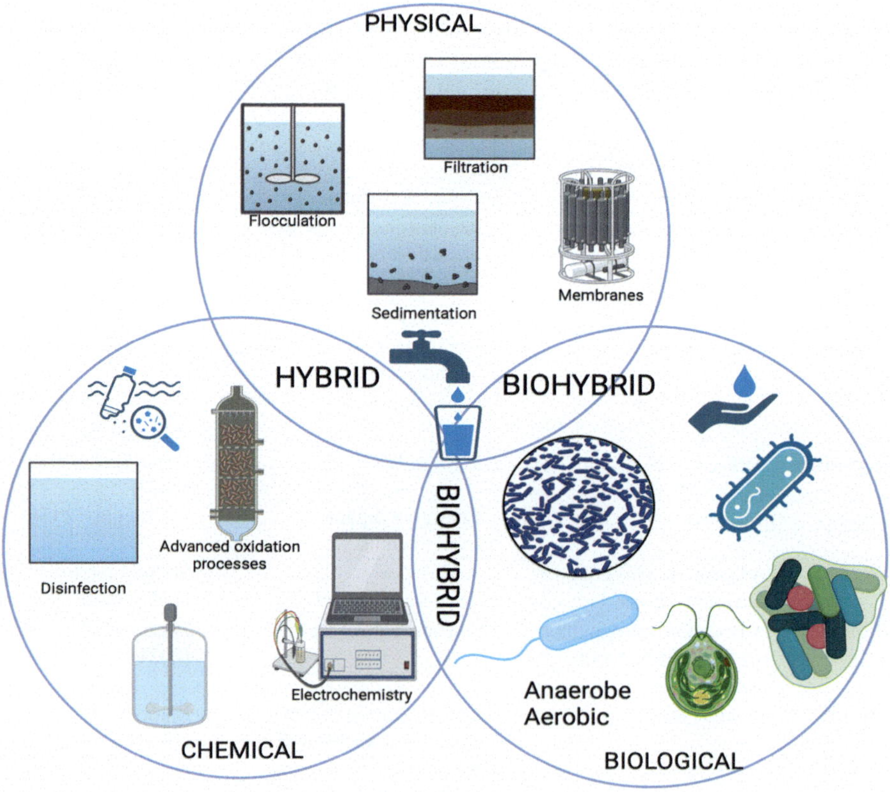

Fig. 3 Water treatments hybrid. (Created in Biorender)

Table 3 Hybrid systems wastewater

Hybrid systems	Pollution	References
Electrocoagulation and electrooxidation	Waste water	Asfaha et al. (2021)
Physical and biotechnologies	Phenolic	Mumtaz et al. (2024)
Advanced oxidation processes *(AOP)-biological*	Pharmaceutical industries	Babu Ponnusami et al. (2023)
Microbial fuel cell coupled hybrid systems	Dye wastewater	Singh et al. (2021)
Wetland system with three different plant *Canna indica*, *Agave americana*, *Pistia stratiotes*, and *Tagetes erecta*	Waste water	Savio et al. (2023)
The double-stage process using nano hybrid PES/SiO_2-PES membrane and PES/ZnO-PES membranes	Oily wastewater	Kusworo et al. (2017)
Flocculation with PolyDADMAC and polyacrylamide	Pulp and paper mills waste water	Ariffin et al. (2012)
Wetlands hybrid with zeolite and normal sand	Pharmaceutical industry wastewater	Al-Mashaqbeh et al. (2024)
Membrane filtration–electrolysis–ultraviolet	Nutrient media in microalgal cultivation	Jwa et al. (2023)
Filtration hybrid process with biological ion exchange (BIEX) resin and gravity-driven ceramic membrane	River water treatment for drinking water	Rasouli et al. (2024)
Phytoremediation and photocatalysis	Landfill leachate	Aibuedefe and Tina (2024)
Synergy of bioelectrochemical system and anaerobic digestion	Nitrogen and phosphorus	Jadhav et al. (2024)

Biohybrid treatments have been most frequently studied and implemented for the removal of contaminating material in wastewater due to the benefits they provide such as cost-effective, improved efficiency, quality of treated water, costs and times being relatively lower compared to a single treatment, various contaminating material that are integrated into the metabolisms of organisms and microorganisms used as sources of nutrients and substrates for products such as sugars, proteins, biofuels, and soil improvers, which achieves environmental benefits mainly by decreasing the impacts of greenhouse gases, in the extraction of natural raw materials and generation of solid waste (Nizzy & Ogwu, 2024; Kosoe & Ogwu, 2024).

The organisms and microorganisms commonly used in wastewater treatment using various technologies and growth conditions are shown in Table 4. These biological systems combined with physicochemical processes present a sustainable approach. Some biological methods have even been combined to enhance the removal of contaminating material from wastewater, such as the use of microalgae, which have proven to be an ally in the implementation of wastewater treatment systems, since they are unicellular photosynthetic microorganisms found in aquatic ecosystems that convert sunlight, water, and carbon dioxide into bioproducts such as carbohydrates, proteins, treated water and bi-fuels such as biodiesel, bioethanol and biogas.

4 Microalgae Biorefinery Compounded with Conventional Wastewater Treatments

In previous sections the hybrid and biohybrid treatments for wastewater were discussed, in this section the application of effluent from conventional wastewater treatment as raw material in the biorefinery with microalgae will be discussed, since these innovative and current treatment trains have the main advantage compared to conventional treatment trains of not only treating the water but also obtaining diverse bioproducts.

The term "biorefinery" was first used in 1981 and to date, a Scopus analysis found 15,785 documents referring to this concept, which has become popular, especially in the last decade, reflecting the growing interest in this research topic (Fig. 4).

A biorefinery is defined as the use of various processes and/or bioprocesses using liquid, solid, and gaseous renewable circular inputs as feedstock for the generation of value-added products contributing to a more sustainable industrial system by reducing greenhouse gas emissions, preserving non-renewable resources, promoting waste minimization, mitigation and that the systems used for the characterization of biorefinery products are by sustainability. It can refer to a concept, a facility, a process, a plant, or even a group of facilities (García-Solares et al., 2024; Olabi et al., 2023).

The biorefinery system is not rigid, and highly heterogeneous compared to oil refineries, these multifunctional systems have the main characteristic of diversifying the integration of their components, technologies, and facilities, which is why they can be classified according to the raw materials they use as solid, liquid and gaseous circular inputs, the type of microorganisms and organisms used for the bioprocess protagonist, the chemical, physical and thermal processes involved, by the products generated as: biofuels, acids, treated wastewater, PHA and by their

Table 4 Microorganisms and organisms used in wastewater treatment

Organism	Contaminant removed	Technology	Reference
Canna indica, Agave americana, Pistia stratiotes, and *Tagetes erecta*	Wastewater	Wetland system	Savio et al. (2023)
Ceratophyllum demersum	Tetracycline	Phytoremediation	Ali et al. (2023)
Lemna minor	Ionic dyes	Phytoremediation	Wibowo et al. (2023)
Aspergillus sydowii NCSL-XY1 pellets	Hypersaline industrial wastewater	Aerobic granular sludge	Han et al. (2024)
Scedosporium dehoogii Saitozyma and *Candida intermedia*	Total nitrogen removal and organic matter	Aerobic reactor	Ma et al. (2023)
Anaerobic bacteria	Nitrogen and phosphorus	Anaerobic digestion	Jadhav et al. (2024)
Anaerobic bacteria	Wastewater from petroleum production	Anaerobic digestion	Huang et al. (2021)

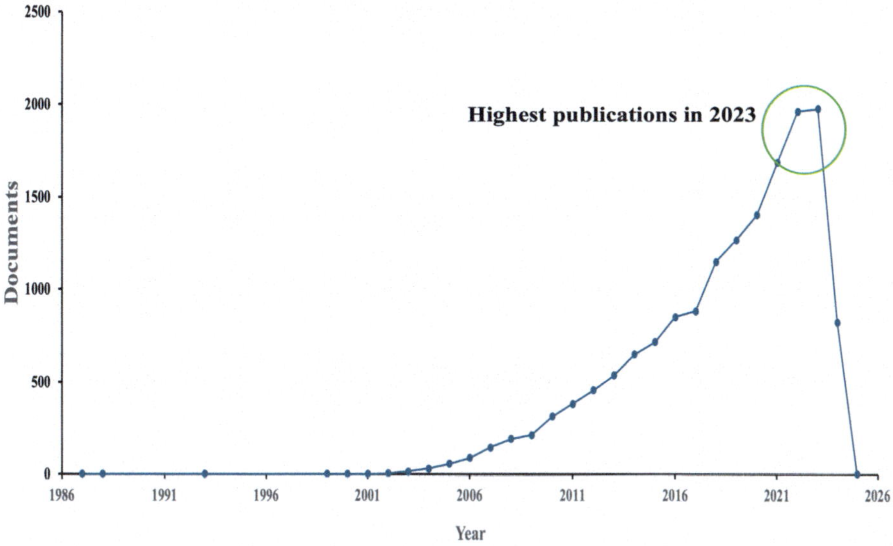

Fig. 4 Trends in the number of articles identified by the Scopus database related to the word biorefineries from 1970 to May 2024

potential for sustainability and flexibility promoting the circular economy by reducing greenhouse gas emissions and contributing to climate change mitigation and planetary boundary management (Agoh et al., 2024; Ogwu et al., 2024). Such diversification allows for the creation of products with greater resilience to fluctuations in fossil fuel prices, thereby increasing economic resilience (Fig. 5). The principal challenges associated with biorefineries pertain to the costs of operation, the advancement of technology, and the processes involved in converting biomass into monomeric sugars.

In general terms, biomass is defined as a non-fossil product with complex chemical structures that originate mainly from the process of photosynthesis. The general classification of biomass varieties can be divided into several groups according to their predominant biological diversity. These include lignocellulosic, herbaceous, and agricultural, aquatic, animal and human biomass residues, and contaminated biomass residues and industrial biomass. Further subdivision into subgroups is also possible (Vassilev et al., 2010). Biomass conversion processes are classified into three main categories: physical, chemical, and biological. These categories include shredding, densification, pyrolysis, gasification, and digestion (Yuan-Chung et al., 2014). However, the utilization of wastewater as a feedstock in a biorefinery results in cost savings, as it obviates the necessity for conversion and/or pretreatment processes, which are otherwise required for biomass.

First-generation biorefineries utilize food crops, including corn, sugar cane, wheat, soybeans, palm, and others, as the primary feedstock to produce biofuels and other derived products. These have constituted the initial basis for the development of the bioindustry, although they are the subject of debate due to the potential for

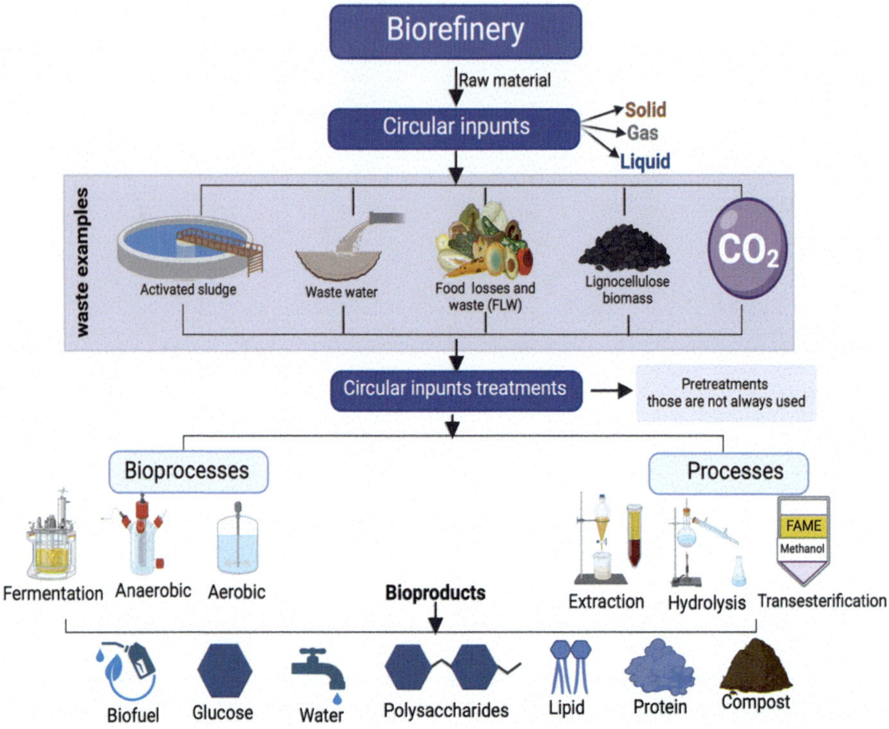

Fig. 5 Circular inputs and their transformation processes applied in biorefineries. (Created in Biorender)

competition with food production, which could result in higher food prices and affect food security. Considering the challenges and limitations associated with food crops, other types of biomasses, including agricultural, lignocellulosic, and algal biomass, have been increasingly employed to enhance sustainability and reduce competition with food production.

According to the International Energy Agency (IEA) in its Task 42: Biorefining in a Circular Economy, its work program has the following objectives: to provide quantitative, scientifically sound, and comprehensible data on the technical-economic-environmental assessment of biorefineries, to create a global biorefinery atlas portal, the current status of biorefineries, their potential contribution to a circular economy and to use the agency's international platform for the dissemination and exchange of information through conferences, webinars, workshops and national events. The IEA also indicates four categories that form the basis for the classification of biorefineries: platforms, products, feedstocks, and conversion processes, and each of these categories has subgroups.

Because of the heterogeneity of biorefineries, they are becoming increasingly relevant in their study and exploration with diverse technologies and stakeholders, including industry, policy makers, and researchers, as well as being a strategy for the implementation of a circular economy that focuses on sustainability and efficiency. On the other hand, microalgae have contributed exponentially to the circular economy within the global market in 2020 was approximately $3.4 billion, and is expected to increase to approximately $4.6 billion by 2027 (Wan Mahari et al., 2022).

Due to their contribution to the global circular economy, microalgae in a biorefinery are no exception. This is generally due to the fact that microalgae have a simple unicellular or multicellular structure, they are photosynthetic prokaryotic or eukaryotic microorganisms, their replication time is fast compared to other microorganisms, they can live in different environments including wastewater because they have a high potential to remove contaminating material and assimilate C, N, P from effluents which they then transform into bioproducts contributing to the biogeochemical carbon cycle and prompted the reorganization of traditional wastewater treatment processes (Kosoe & Ogwu, 2024).

Microalgae are a varied group of species and genera that have been used individually in the food, pharmaceutical, and environmental industries due to their virtues. The cultivation of microalgae monocultures must have the environmental conditions of light, salt concentration, pH and nutrients controlled and innocuous to avoid contamination by other species of microorganisms, including microalgae that are not metabolically compatible, to coexist in symbiosis with the culture, decreasing the efficiency of the bioprocess. The use of microalgae consortia with yeasts and/or bacteria is an alternative to potentiate the virtues of microalgae utilizing metabolic synergies, these systems present greater resistance to changes in environmental conditions and external contaminating agents, the obtaining of the consortia is by different strategies such as genetic modifications, evolutionary adaptation and the isolation of contaminated ecosystems which have presented better bioprocess efficiency compared to the two previous strategies (Moreno-García et al., 2021; García-Solares et al., 2024).

The use of microalgae or microalgae consortia in a biorefinery for wastewater treatment is a promising, low-cost, easily accessible, and maintainable biohybrid system that simultaneously treats different water-contaminating material in a single stage through metabolic symbiosis, recycling, and recovering nutrients (Parsy et al., 2024), and Sun et al. (2022) reported that the formation of granules with microalgae and bacteria in different types of reactors increases the efficiency in the removal of contaminating material from industrial and domestic wastewater. The application of consortia between microalgae and bacteria are emerging processes in wastewater treatment that emit low carbon emissions and require little energy in the biological removal of contaminating material including nitrogen and phosphorus in the long term (Chen et al., 2024; Garieri et al., 2024; Moreno-García et al., 2021). The use of native microalgal consortia in wastewater refining is a relatively new topic, which can be explored in depth in terms of mechanisms of resistance promoted by metabolic synergies for contaminant removal, mechanisms of adaptation and metabolic stress, optimal growth conditions, lethal dose of contaminating material, among others, which would allow expanding the applications of native microalgal consortia (NMC).

The physicochemical wastewater treatments coupled to a biorefinery with microalgae, either in single stage or in consortium, will have a better effect, since as discussed in previous sections, the composition of wastewater is varied and contains from two to more contaminating material of all types, which can sometimes produce toxicity to microalgae and decrease the income from the commercialization of bioproducts as well as their efficiency. For this reason, the study of physicochemical processes combined with microalgae for wastewater treatment has been initiated as a strategy that contributes to the circular economy using renewable sources, and circular inputs to pay back a percentage of the resources extracted from the environment and the impacts produced in it.

The main objective of the physicochemical processes that are combined with microalgae is to provide water quality favorable for the development of microalgae, that is, to convert it into a more bioavailable culture medium, for example, in the physicochemical treatment, suspended solids, volatile solids and organic matter can be effectively removed, likewise, the concentration of emerging contaminating material is decreased so that traces of these are used by microalgae in the generation of bioproducts, they are also able to incorporate in their metabolism ammonia nitrogen into nitrate and phosphorus which are a preferred source of nutrients for microalgae growth, both nitrogen and phosphorus promote eutrophication at concentration ranges of 0.5 to 1.2 mgL^{-1} and 0.03 to 0.1 mgL^{-1} respectively, their removal by low-carbon and energy-saving mechanisms is indispensable and microalgae have demonstrated good performance on nitrogen and phosphorus removal (Chen et al., 2024; Ren et al., 2023; Yu et al., 2023). On the other hand, the United States Environmental Protection Agency (USEPA) report suggested that the concentrations of nitrogen and phosphorus with which wastewater can be discharged and/or treated to rivers and lakes should be less than 2 and 0.2 mgL^{-1} respectively (Yu et al., 2023). In addition to the benefit of combining physicochemical treatments with microalgae biorefinery for nitrogen and phosphorus removal, carbon supply can be regulated to improve wastewater refining and increase the production of lipids that are precursors for clean, renewable biofuels which are of research and market interest to achieve a circular economy (Wang et al., 2022). In Table 5, several reports of biohybrid treatments with microalgae are compiled, which were implemented to treat wastewater with various contaminating material and physicochemical processes.

5 Bibliometric Analysis of Hybrid and Biohybrid Processes for Wastewater Treatment

In recent years, efforts have been made to meet the objectives of the 2030 agenda on water, focusing research on the innovation of conventional wastewater treatment technologies to enhance their efficiency and direct them toward sustainability, according to reports from the Scopus database, since the creation of the 2030 agenda in 2015 and with the use of VOSviewer, a bibliometric analysis was made, which is an effective tool to quantitatively investigate the current situation and the trend of

Table 5 Physicochemical treatments combined with a biorefinery

Type of wastewater	Microalgae	Bioproduct	Reference
Cigarette butt leachate wastewater	*Chlorella protothecoides*	Biomass, protein and carbohydrate	Fu et al. (2024)
Wastewater with microplastics	Native microalgae consortia	Removal microplastic	Afonso et al. (2024)
Simulated wastewater containing volatile fatty acids	*Chlorella sorokiniana*	Lipids, removal nitrogen and phosphorus	Lu et al. (2024)
Food industry wastewater	*Chlorella vulgaris* CCAP 211/11B, *Nannochloropsis oculata* CCAP 849/1, and *Scenedesmus* sp. CCAP 276/41	Treated water	Anagnostopoulou et al. (2024)
Mariculture wastewater with inorganic phosphate	*Chlorella pyrenoidosa*	Carbohydrates, lipids, proteins, and biomass	Cao et al. (2024)
BG11 medium	*Chlamydomonas* sp.	Methyl ester and malic acid	Sivaramakrishnan et al. (2024)
Tannery waste water	Native microalgae consortium	Lipids, proteins and water treats	Moreno-García et al. (2021)
Synthetic wastewater	*Dunaliella tertiolecta, Dunaliella salina, Chlorella minutissima,* and *Desmodesmus* sp. MCC34	FAMES for biofuel	Nagappan and Kumar (2021)
Domestic wastewater	*Scenedesmus* sp.	Biodiesel	Thangam et al. (2021)
Wastewater	Consortium with *Chlorella minutissima, Scendesmus* spp., and *Nostoc muscorum*	Biodiesel and biofertilizers	Khan et al. (2019)
Wastewater	*Chlorella minutissima*	Lipids	Khan et al. (2018)
Municipal wastewater from two different (primary and secondary) levels	*Chlorella vulgaris*	Lipids	Otondo et al. (2018)

scientific research, based on most of the publications that exist on a specific topic using statistics.

A total of 493 original research articles were published on the topic of wastewater treatment. Original research articles written in English and the words: wastewater, sustainable, removal, treatment and conventional, were the inclusion criteria considered for this analysis, later refined with the inclusion of the words hybrid and biohybrid, excluding publications referring to review, conference proceedings and book chapters (Figs. 6, 7, and 8), obtaining that in the last decade the number of publications on wastewater treatment has increased by approximately 33% and according to the 4 clusters the words included are related to: life cycle analysis, chemical oxygen demand and microbial community (Figs. 6a and 7). By including the word hybrid in the analysis, Scopus detected only 36 papers that have been published with these restrictions since 2016, and to date there has been an almost two-fold increase in publications (Fig. 6b), which indicates a steady increase in research on the subject, additionally the three clusters are associated with the words biological treatments, bioreactor and hybrid systems (Fig. 8a). Finally, the search was further restricted to specific research topics of wastewater, sustainable, removal, treatment, hybrids, and microalgae obtaining a total of 13 papers in a time interval from 2017 to 2023 (Fig. 6c), the words included in the two clusters are displayed in Fig. 8b. The objective of this analysis was to identify that these topics have become relevant in research with significant contributions to the field; however, this analysis indicates that biohybrid water treatment with microalgae is obviously weak and is a field to be explored.

6 Conclusion

In the current era with the accelerated pace of water consumption to meet basic daily needs, it is crucial to study and understand the biochemical cycle of water for an equitable supply, as well as a period of recharge of aquifers, to minimize the socioeconomic and environmental impacts of the acquisition of this resource, which involves the implementation of innovative and sustainable wastewater treatments that provide a water quality corresponding to the MPL, in addition to incorporating this liquid circular input to the bioeconomy through the participation of all consumers, so as to increase an adequate management of water resources, through quantitative and qualitative balances of consumption and supply, following up with indicators that contribute to the circular economy to recognize and capture the full value of water, and generation of bioproducts, mainly energy, making consumption more sustainable from the financial and environmental point of view, as well as improving resilience to climate change and waste generation.

It is essential to study and apply circular water systems that are integrated by sustainable processes, awareness programs in the use of water, energy, and reduction of greenhouse gas emissions, for the supply of this resource that allows a better quality of life and can be an ally to achieve several objectives of the 2030 agenda.

Although conventional treatments have been diversified to comply with the MPLs of each geographic area, it is essential to continue with the studies of those biohybrid treatments that correspond to sustainability and contribute to the circular economy since, according to bibliometric analyses, research is still scarce.

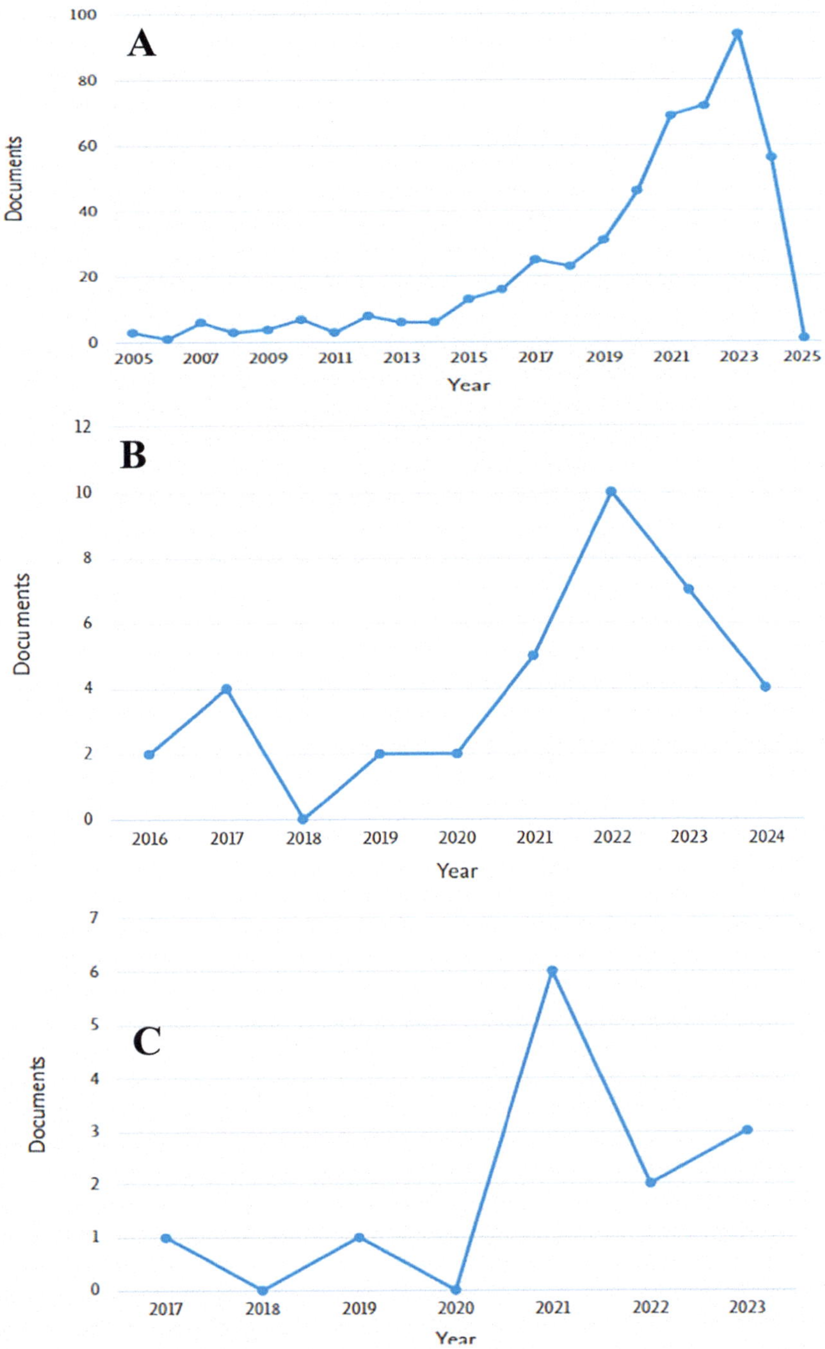

Fig. 6 Relevance of the study of wastewater treatments and their couplings in the last 10 years. (**a**) Physicochemical–biological treatments, (**b**) Hybrid treatments, and (**c**) Hybrid treatments with microalgae

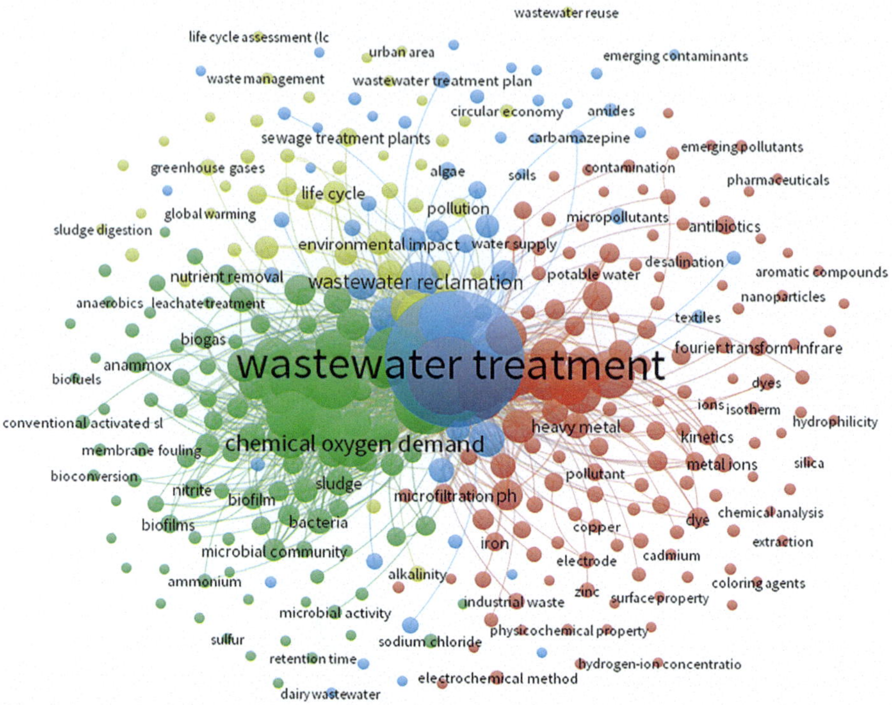

Fig. 7 Themes analysis by keywords related to water treatment

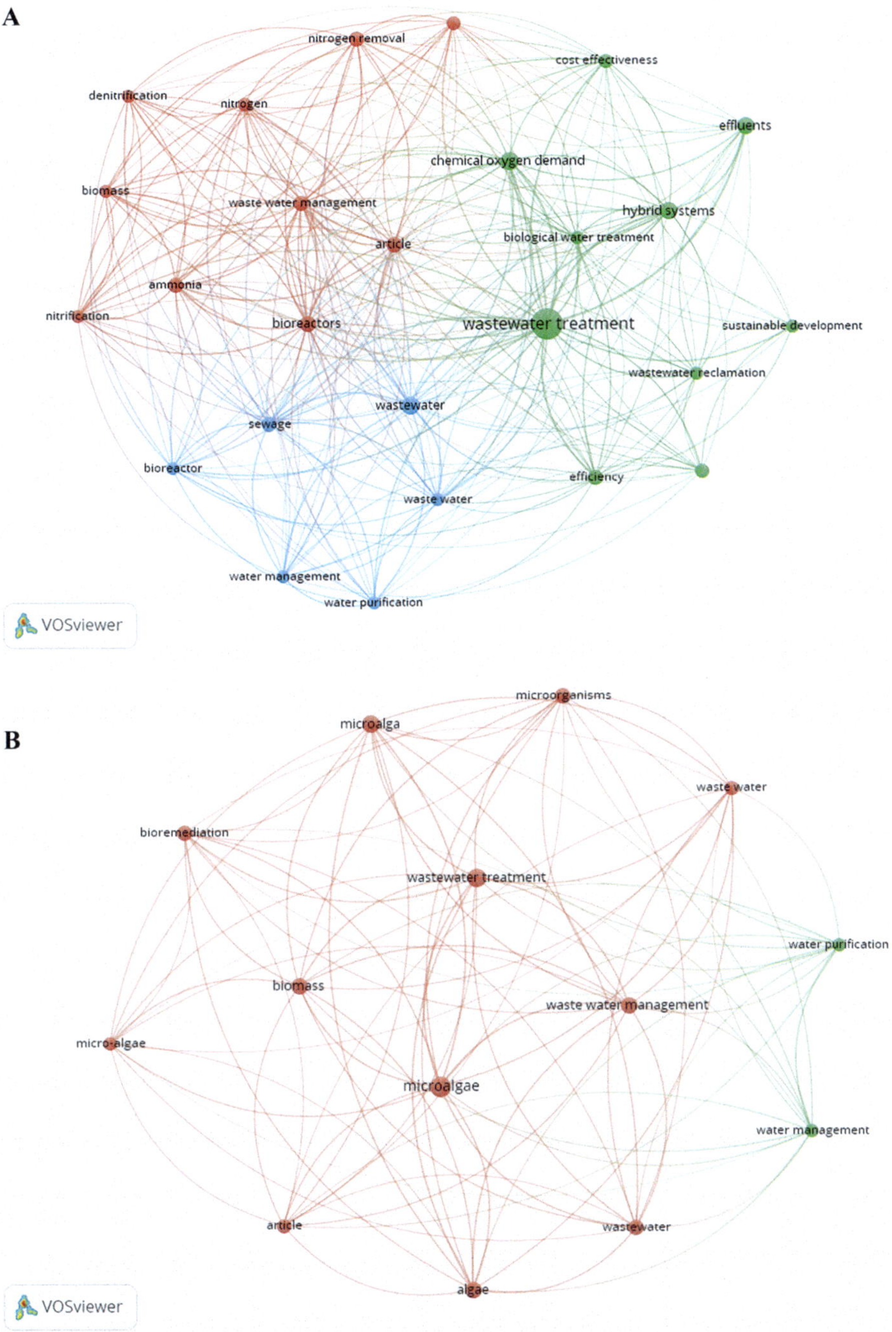

Fig. 8 Mapping of networks on wastewater treatments for the last 10 years (2015 to May 2024). (**a**) Hybrid wastewater treatments. (**b**) Wastewater treatments coupled with microalgae

References

Afonso, V., Borges, R., Rodrigues, B., Barros, R., João Bebianno, M., & Raposo, S. (2024). Are native microalgae consortia able to remove microplastics from wastewater effluents? *Environmental Pollution, 349*(April), 123931. https://doi.org/10.1016/j.envpol.2024.123931

Agoh, E. C., Ogwu, M. C., Chukwuemeka, O. S., & Ekeledo, P. I. (2024). Environmental and human health effects of cassava processing and processing waste. In M. C. Ogwu, S. C. Izah, A. A. Cunha Alves, & S. C. Babu (Eds.), *Plant biology, sustainability and climate change, sustainable cassava* (pp. 203–219). Academic Press. https://doi.org/10.1016/B978-0-443-21747-0.00001-1

Aibuedefe, A. F., & Tina, A. E. (2024). Optimization of landfill leachate through hybrid treatment of phytoremediation and photocatalysis using response surface methodology. *Sustainable Chemistry for the Environment, 5*(February), 100076. https://doi.org/10.1016/j.scenv.2024.100076

Akudugu, M. A., & Ogwu, M. C. (2024). Sustainable development policies and interventions: A bibliometric analysis of the contributions of the academic community. *Journal of Cleaner Production, 434*, 139919. https://doi.org/10.1016/j.jclepro.2023.139919

Ali, M. F., Ali, Q. A., M-Ridha, M. J., Mohammed, S. J., & Bohan, H. R. (2023). Phytoremediation of tetracycline via the coontail Ceratophyllum demersum in antibiotics-contaminated water. *Biocatalysis and Agricultural Biotechnology, 53*(October), 102887. https://doi.org/10.1016/j.bcab.2023.102887

Al-Mashaqbeh, O., Alsalhi, L., Salaymeh, L., Dotro, G., & Lyu, T. (2024). Treatment of pharmaceutical industry wastewater for water reuse in Jordan using hybrid constructed wetlands. *Science of the Total Environment, 939*(May), 173634. https://doi.org/10.1016/j.scitotenv.2024.173634

Anagnostopoulou, C., Papachristou, I., Kontogiannopoulos, K. N., Mourtzinos, I., & Kougias, P. G. (2024). Optimization of microalgae cultivation in food industry wastewater using microplates. *Sustainable Chemistry and Pharmacy, 39*(January), 101510. https://doi.org/10.1016/j.scp.2024.101510

Ariffin, A., Razali, M. A. A., & Ahmad, Z. (2012). PolyDADMAC and polyacrylamide as a hybrid flocculation system in the treatment of pulp and paper mills waste water. *Chemical Engineering Journal, 179*, 107–111. https://doi.org/10.1016/j.cej.2011.10.067

Asfaha, Y. G., Tekile, A. K., & Zewge, F. (2021). Hybrid process of electrocoagulation and electrooxidation system for wastewater treatment: A review. *Cleaner Engineering and Technology, 4*, 100261. https://doi.org/10.1016/j.clet.2021.100261

Babu Ponnusami, A., Sinha, S., Ashokan, H., Paul, M. V., Hariharan, S. P., Arun, J., Gopinath, K. P., Hoang Le, Q., & Pugazhendhi, A. (2023). Advanced oxidation process (AOP) combined biological process for wastewater treatment: A review on advancements, feasibility and practicability of combined techniques. *Environmental Research, 237*(P1), 116944. https://doi.org/10.1016/j.envres.2023.116944

Batool, F., Shahid, M., Mahmood, F., Shahzad, T., Azeem, F., Hussain, S., Algarni, T. S., Elshikh, M. S., Onazi, W. A., & Mustafa, S. (2024). Biosynthesis of copper nanoparticles using Bacillus flexus and estimation of their potential for decolorization of azo dyes and textile wastewater treatment. *Journal of King Saud University – Science*, 103309. https://doi.org/10.1016/j.jksus.2024.103309

Cao, W., Xing, Y., Xing, D., Zhao, Y., Gao, M., Jin, C., & Guo, L. (2024). The response to inorganic phosphates forms for mariculture wastewater treatment with microalgae: Biochemical component synthesis, cellular morphology and phosphorus storage. *Process Safety and Environmental Protection, 188*(May), 480–491. https://doi.org/10.1016/j.psep.2024.05.114

Chen, J., Liu, X., Lu, T., Liu, W., Zheng, Z., Chen, W., Yang, C., & Qin, Y. (2024). The coupling of anammox with microalgae-bacteria symbiosis: Nitrogen removal performance and microbial community. *Water Research, 252*(November), 121214. https://doi.org/10.1016/j.watres.2024.121214

Cruz del Álamo, A., Puga, A., Dias Soares, C. M., Pariente, M. I., Pazos, M., Molina, R., Sanromán, M. A., Martínez, F., & Delerue-Matos, C. (2024). Novel 3D electro-Fenton reactor based on a catalytic packed bed reactor of perovskite/carbon microelectrodes for the removal of carbamazepine in wastewater. *Journal of Environmental Chemical Engineering, 12*(4), 113154. https://doi.org/10.1016/j.jece.2024.113154

Delgado, A., Rodriguez, D. J., Amadei, C. A., & Makino, M. (2024). Water in circular economy and resilience (WICER) framework. *Utilities Policy, 87*(June 2023), 101604. https://doi.org/10.1016/j.jup.2023.101604

Dilekli, N., & Cazcarro, I. (2019). Testing the SDG targets on water and sanitation using the world trade model with a waste, wastewater, and recycling framework. *Ecological Economics, 165*(February), 106376. https://doi.org/10.1016/j.ecolecon.2019.106376

Dubey, M., Vellanki, B. P., & Kazmi, A. A. (2023). Emerging contaminants in conventional and advanced biological nutrient removal based wastewater treatment plants in India- insights into the removal processes. *Science of the Total Environment, 894*(June). https://doi.org/10.1016/j.scitotenv.2023.165094

Ding, Y., Bai, R., Yang, Y., Zhu, H., Song, J., Yang, H., & Zhao, S. (2024). Modified rare earth waste composite material for simultaneous denitrification and recovery of phosphate in water. *Journal of Environmental Chemical Engineering, 12*(4). https://doi.org/10.1016/j.jece.2024.113138

Egboduku, W. O., Egboduku, T., Golohor, O. M., Imarhiagbe, O., & Ogwu, M. C. (2024). Cassava as raw material for sustainable bioeconomy development. In M. C. Ogwu, S. C. Izah, A. A. Cunha Alves, & S. C. Babu (Eds.), *Plant biology, sustainability and climate change, sustainable cassava* (pp. 57–73). Academic Press. https://doi.org/10.1016/B978-0-443-21747-0.00022-9

Eheneden, I., Wang, R., & Zhao, J. (2023). Antibiotic removal by microalgae-bacteria consortium: Metabolic pathways and microbial responses. *Science of the Total Environment, 891*(March), 164489. https://doi.org/10.1016/j.scitotenv.2023.164489

Fu, X. Z., Wu, J., Li, J., Ding, J., Cui, S., Wang, X. M., Wang, Y. J., Liu, H. Q., Deng, X., Liu, D. F., & Li, W. W. (2023). Heavy-metal resistant bio-hybrid with biogenic ferrous sulfide nanoparticles: pH-regulated self-assembly and wastewater treatment application. *Journal of Hazardous Materials, 446*(October 2022), 130667. https://doi.org/10.1016/j.jhazmat.2022.130667

Fu, Y., Jiang, Y., Chen, H., Sivakumar, E., Li, H., Zhu, K., & Wang, S. (2024). Cultivation of Chlorella protothecoides in the medium amended with cigarette butt leachate wastewater for enhanced biomass and macromolecular productions. *Biomass and Bioenergy, 184*(November 2023), 107216. https://doi.org/10.1016/j.biombioe.2024.107216

García-Solares, M., Mena-Cervantes, V. Y., Arreola-Valerio, E., & Romero, C. P. (2024). Chapter 24: Native microalgae consortia: A potential contribution to a biorefinery based on liquid circular inputs. In *Book biorefinery of industrial effluents for a sustainable circular economy*. Paperback ISBN: 9780443218019, eBook ISBN: 9780443218026.

Garieri, T., Allen, D. G., Gao, W., & Liao, B. (2024). A review of emerging membrane-based microalgal-bacterial processes for wastewater treatment: Process configurations, biological and membrane performance, and perspectives. *Science of the Total Environment, 927*(April), 172141. https://doi.org/10.1016/j.scitotenv.2024.172141

Han, X., Niu, X., Jin, Y., & Yu, J. (2024). Rapid cultivation of aerobic granular sludge for shale gas flowback water treatment by bioaugmentation with inoculating multifunctional fungal pellets. *Journal of Cleaner Production, 457*(April), 142483. https://doi.org/10.1016/j.jclepro.2024.142483

https://www.wicer-tool.com

Huang, Z., He, X., Nye, C., Bagley, D., Urynowicz, M., & Fan, M. (2021). Effective anaerobic treatment of produced water from petroleum production using an anaerobic digestion inoculum from a brewery wastewater treatment facility. *Journal of Hazardous Materials, 407*(October 2020), 124348. https://doi.org/10.1016/j.jhazmat.2020.124348

Jadhav, D. A., Yu, Z., Hussien, M., Kim, J. H., Liu, W., Eisa, T., Sharma, M., Vinayak, V., Jang, J. K., Wilberforce Awotwe, T., Wang, A., & Chae, K. J. (2024). Paradigm shift in nutrient-energy-water centered sustainable wastewater treatment system through synergy of bioelectrochemical system and anaerobic digestion. *Bioresource Technology, 396*(January), 130404. https://doi.org/10.1016/j.biortech.2024.130404

Jwa, E., Na, O. S., Jeung, Y. C., Jeong, N., Nam, J. Y., & Lee, S. (2023). Recycling of nutrient medium to improve productivity in large-scale microalgal culture using a hybrid electrochemical water treatment system. *Water Research, 246*(May), 120683. https://doi.org/10.1016/j.watres.2023.120683

Khan, S. A., Malla, F. A., Malav, L. C., Gupta, N., & Kumar, A. (2018). Potential of wastewater treating Chlorella minutissima for methane enrichment and CO_2 sequestration of biogas and producing lipids. *Energy, 150*, 153–163. https://doi.org/10.1016/j.energy.2018.02.126

Khan, S. A., Sharma, G. K., Malla, F. A., Kumar, A., & Rashmi, & Gupta, N. (2019). Microalgae based biofertilizers: A biorefinery approach to phycoremediate wastewater and harvest biodiesel and manure. *Journal of Cleaner Production, 211*, 1412–1419. https://doi.org/10.1016/j.jclepro.2018.11.281

Kiani Kori, A., Ramavandi, B., Mahmoodi, S. M. M., & Javanmardi, F. (2024). Magnetization and ZIF-67 modification of Aspergillus flavus biomass for tetracycline removal from aqueous solutions: A stable and efficient composite. *Environmental Research, 252*(P2), 118931. https://doi.org/10.1016/j.envres.2024.118931

Kosoe, E. A., & Ogwu, M. C. (2024). Sustainable cassava processing: Processes, techniques, wastes, and waste streams. In M. C. Ogwu, S. C. Izah, A. A. Cunha Alves, & S. C. Babu (Eds.), *Plant biology, sustainability and climate change, sustainable cassava* (pp. 257–272). Academic Press. https://doi.org/10.1016/B978-0-443-21747-0.00011-4

Kusworo, T. D., Qudratun, & Utomo, D. P. (2017). Performance evaluation of double stage process using nano hybrid PES/SiO2-PES membrane and PES/ZnO-PES membranes for oily waste water treatment to clean water. *Journal of Environmental Chemical Engineering, 5*(6), 6077–6086. https://doi.org/10.1016/j.jece.2017.11.044

Latif, M., Nawaz, R., Hammad Aziz, M., Asif, M., Noor, F., Aligayev, A., Mansoor Ali, S., Alam, M., Papanikolaou, S., & Huang, Q. (2024). Evaluation of tetracycline photocatalytic degradation using NiFe2O4/CeO2/GO nanocomposite for environmental remediation: In silico molecular docking, antibacterial performance, degradation pathways, and DFT calculations. *Separation and Purification Technology, 351*(May), 128074. https://doi.org/10.1016/j.seppur.2024.128074

Leong, C. (2021). Narratives and water: A bibliometric review. *Global Environmental Change, 68*(January), 102267. https://doi.org/10.1016/j.gloenvcha.2021.102267

Lu, T., Su, K., Ma, G., Jia, C., Li, J., Zhao, Q., Song, M., Xu, C., & Song, X. (2024). The growth and nutrient removal properties of heterotrophic microalgae Chlorella sorokiniana in simulated wastewater containing volatile fatty acids. *Chemosphere, 358*(April), 142270. https://doi.org/10.1016/j.chemosphere.2024.142270

Ma, B., Niu, L., Li, N., Pan, S., Li, A., Chu, M., Liu, H., Kosolapov, D. B., Xin, X., Zhi, W., Hou, L., Chen, Z., Zhang, Y., Cao, S., Huang, T., & Zhang, H. (2023). Promoted aerobic denitrification through denitrifying fungal communities: Co-occurrence patterns and treatment of low C/N micro-polluted water. *Science of the Total Environment, 884*(February), 163859. https://doi.org/10.1016/j.scitotenv.2023.163859

Moreno, J. F., Oulego, P., Collado, S., & Díaz, M. (2024). Biorefinery of waste activated sludge: Nutrient recovery and microbial lipid production by Yarrowia lipolytica. *Journal of Environmental Management, 363*(May), 121315. https://doi.org/10.1016/j.jenvman.2024.121315

Moreno-García, A. F., Neri-Torres, E. E., Mena-Cervantes, V. Y., Altamirano, R. H., Pineda-Flores, G., Luna-Sánchez, R., García-Solares, M., Vazquez-Arenas, J., & Suastes-Rivas, J. K. (2021). Sustainable biorefinery associated with wastewater treatment of Cr (III) using a native microalgae consortium. *Fuel, 290*(January). https://doi.org/10.1016/j.fuel.2020.119040

Mumtaz, F., Li, B., Al Shehhia, M. R., Feng, X., & Wang, K. (2024). Treatment of phenolic-wastewater by hybrid technologies: A review. *Journal of Water Process Engineering, 57*(October 2023), 104695. https://doi.org/10.1016/j.jwpe.2023.104695

Nagappan, S., & Kumar, G. (2021). Investigation of four microalgae in nitrogen deficient synthetic wastewater for biorefinery based biofuel production. *Environmental Technology and Innovation, 23*, 101572. https://doi.org/10.1016/j.eti.2021.101572

Nishshanka, G. K. S. H., Thevarajah, B., Nimarshana, P. H. V., Prajapati, S. K., & Ariyadasa, T. U. (2023). Real-time integration of microalgae-based bioremediation in conventional wastewater treatment plants: Current status and prospects. *Journal of Water Process Engineering, 56*(June), 104248. https://doi.org/10.1016/j.jwpe.2023.104248

Nizzy, A. M., & Ogwu, M. C. (2024). Valorization of cassava processing by-products into biofuel for a sustainable environment. In M. C. Ogwu, S. C. Izah, A. A. Cunha Alves, & S. C. Babu (Eds.), *Plant biology, sustainability and climate change, sustainable cassava* (pp. 291–309). Academic Press. https://doi.org/10.1016/B978-0-443-21747-0.00010-2

Obaideen, K., Shehata, N., Sayed, E. T., Abdelkareem, M. A., Mahmoud, M. S., & Olabi, A. G. (2022). The role of wastewater treatment in achieving sustainable development goals (SDGs) and sustainability guideline. *Energy Nexus, 7*(July), 100112. https://doi.org/10.1016/j.nexus.2022.100112

Ogwu, M. C., Odozi, I. P., Ahonsi, O. C., Uleanya, K. O., & Odozi, E. B. (2024). Organic acid production from cassava. In M. C. Ogwu, S. C. Izah, A. A. Cunha Alves, & S. C. Babu (Eds.), *Plant biology, sustainability and climate change, sustainable cassava* (pp. 395–418). Academic Press. https://doi.org/10.1016/B978-0-443-21747-0.00009-6

Olabi, A. G., Shehata, N., Sayed, E. T., Rodriguez, C., Anyanwu, R. C., Russell, C., & Abdelkareem, M. A. (2023). Role of microalgae in achieving sustainable development goals and circular economy. *Science of the Total Environment, 854*(September 2022), 158689. https://doi.org/10.1016/j.scitotenv.2022.158689

Otondo, A., Kokabian, B., Stuart-Dahl, S., & Gude, V. G. (2018). Energetic evaluation of wastewater treatment using microalgae, Chlorella vulgaris. *Journal of Environmental Chemical Engineering, 6*(2), 3213–3222. https://doi.org/10.1016/j.jece.2018.04.064

Pariente, M. I., Segura, Y., Álvarez-Torrellas, S., Casas, J. A., de Pedro, Z. M., Diaz, E., García, J., López-Muñoz, M. J., Marugán, J., Mohedano, A. F., Molina, R., Munoz, M., Pablos, C., Perdigón-Melón, J. A., Petre, A. L., Rodríguez, J. J., Tobajas, M., & Martínez, F. (2022). Critical review of technologies for the on-site treatment of hospital wastewater: From conventional to combined advanced processes. *Journal of Environmental Management, 320*(February), 115769. https://doi.org/10.1016/j.jenvman.2022.115769

Parsy, A., Monlau, F., Guyoneaud, R., & Sambusiti, C. (2024). Nutrient recovery in effluents from the energy sectors for microalgae and cyanobacteria biomass production: A review. *Renewable and Sustainable Energy Reviews, 191*(December 2023), 114207. https://doi.org/10.1016/j.rser.2023.114207

Pasquarelli, F., Oliva, G., Mariniello, A., Buonerba, A., Li, C. W., Belgiorno, V., Naddeo, V., & Zarra, T. (2024). Carbon neutrality in wastewater treatment plants: An integrated biotechnological-based solution for nutrients recovery, odour abatement and CO2 conversion in alternative energy drivers. *Chemosphere, 354*(December 2023), 141700. https://doi.org/10.1016/j.chemosphere.2024.141700

Pramanik, R., Bodawar, N., Brahme, A., Kamble, S., & Dharne, M. (2023). Comparative evaluation of advanced oxidation processes (AOPs) for reducing SARS-CoV-2 viral load from campus sewage water. *Journal of Environmental Chemical Engineering, 11*(3), 1–9. https://doi.org/10.1016/j.jece.2023.109673

Rana, Q. U., Latif, S., Perveen, S., Haq, A., Ali, S., Irfan, M., Gauttam, R., Shah, T. A., Dawoud, T. M., Wondmie, G. F., Bourhia, M., & Badshah, M. (2024). Utilization of microalgae for agricultural runoff remediation and sustainable biofuel production through an integrated biorefinery approach. *Bioresources and Bioprocessing, 11*(1), 8. https://doi.org/10.1186/s40643-023-00720-w

Rasouli, Y., Maltais-Tariant, R., Barbeau, B., Peldszus, S., Boudoux, C., & Claveau-Mallet, D. (2024). Performance of biological ion exchange resin and gravity-driven ceramic membrane hybrid process for surface water treatment. *Separation and Purification Technology, 332*(November 2023), 125769. https://doi.org/10.1016/j.seppur.2023.125769

Ren, H., Ni, J., Shen, M., Zhou, D., Sun, F., & Loke Show, P. (2023). Enhanced carbon dioxide fixation of Chlorella vulgaris in microalgae reactor loaded with nanofiber membrane carried iron oxide nanoparticles. *Bioresource Technology, 382*(April). https://doi.org/10.1016/j.biortech.2023.129176

Savio, N., Pandey, D., & Srivastava, R. K. (2023). Potentialities of plant based hybrid wetland systems for the treatment of household waste water using Canna indica, Agave americana, Pistia stratiotes and Tagetes erecta. *Materials Today: Proceedings, 77*, 217–222. https://doi.org/10.1016/j.matpr.2022.11.264

Sharma, A. K., Sharma, A., Singh, Y., & Chen, W. H. (2021). Production of a sustainable fuel from microalgae Chlorella minutissima grown in a 1500 L open raceway ponds. *Biomass and Bioenergy, 149*(March). https://doi.org/10.1016/j.biombioe.2021.106073

Sherif, M., Abrar, M., Baig, F., & Kabeer, S. (2023). Gulf Cooperation Council countries' water and climate research to strengthen UN's SDGs 6 and 13. *Heliyon, 9*(3), e14584. https://doi.org/10.1016/j.heliyon.2023.e14584

Singh, A., Dahiya, S., & Mishra, B. K. (2021). Microbial fuel cell coupled hybrid systems for the treatment of dye wastewater: A review on synergistic mechanism and performance. *Journal of Environmental Chemical Engineering, 9*(6), 106765. https://doi.org/10.1016/j.jece.2021.106765

Sivaramakrishnan, R., Suresh, S., & Incharoensakdi, A. (2024). Development of Chlamydomonas sp. biorefinery for sustainable methyl ester and malic acid production. *Fuel, 371*(PB), 132017. https://doi.org/10.1016/j.fuel.2024.132017

Sompornpailin, D., Pulgerd, P., Sangsanont, J., Thayanukul, P., & Punyapalakul, P. (2024). Removal of antibiotics, bacterial toxicity, and occurrence of antibiotic resistance genes in secondary hospital effluents treated with granular activated carbon and the impact of preceding chlorination. *Science of the Total Environment, 927*(April), 172095. https://doi.org/10.1016/j.scitotenv.2024.172095

Sun, Y., Chang, H., Zhang, C., Xie, Y., & Ho, S. H. (2022). Emerging biological wastewater treatment using microalgal-bacterial granules: A review. *Bioresource Technology, 351*(February), 127089. https://doi.org/10.1016/j.biortech.2022.127089

Thangam, K. R., Santhiya, A., Sri, S. R. A., MubarakAli, D., Karthikumar, S., Kumar, R. S., Thajuddin, N., Soosai, M. R., Varalakshmi, P., Moorthy, I. G., & Pugazhendhi, A. (2021). Bio-refinery approaches based concomitant microalgal biofuel production and wastewater treatment. *Science of the Total Environment, 785*, 147267. https://doi.org/10.1016/j.scitotenv.2021.147267

Vassilev, S. V., Baxter, D., Andersen, L. K., & Vassileva, C. G. (2010). An overview of the chemicalcomposition of biomass. *Fuel, 89*(5), 913–933. https://doi.org/10.1016/j.fuel.2009.10.022

Verdini, F., Canova, E., Solarino, R., Calcio Gaudino, E., & Cravotto, G. (2024). Integrated physicochemical processes to tackle high-COD wastewater from pharmaceutical industry. *Environmental Pollution, 342*(November 2023), 123041. https://doi.org/10.1016/j.envpol.2023.123041

Vinayagam, V., Sikarwar, D., Das, S., & Pugazhendhi, A. (2024). Envisioning the innovative approaches to achieve circular economy in the water and wastewater sector. *Environmental Research, 241*(November 2023), 117663. https://doi.org/10.1016/j.envres.2023.117663

Wan Mahari, W. A., Wan Razali, W. A., Manan, H., Hersi, M. A., Ishak, S. D., Cheah, W., Chan, D. J. C., Sonne, C., Show, P. L., & Lam, S. S. (2022). Recent advances on microalgae cultivation for simultaneous biomass production and removal of wastewater pollutants to achieve circular economy. *Bioresource Technology, 364*(October), 128085. https://doi.org/10.1016/j.biortech.2022.128085

Wang, H., Deng, L., Qi, Z., & Wang, W. (2022). Constructed microalgal-bacterial symbiotic (MBS) system: Classification, performance, partnerships and perspectives. *Science of the Total Environment, 803*, 150082.

Wibowo, Y. G., Syahnur, M. T., Al-Azizah, P. S., Arantha Gintha, D., Lululangi, B. R. G., & Sudibyo. (2023). Phytoremediation of high concentration of ionic dyes using aquatic plant (Lemna minor): A potential eco-friendly solution for wastewater treatment. *Environmental*

Nanotechnology, Monitoring and Management, 20(June), 100849. https://doi.org/10.1016/j.enmm.2023.100849

Wolff, D., Abou-Kandil, A., Azaizeh, H., Wick, A., & Jadoun, J. (2024). Influence of vegetation and substrate type on removal of emerging organic contaminants and microbial dynamics in horizontal subsurface constructed wetlands. *Science of the Total Environment, 927*(November 2023), 172346. https://doi.org/10.1016/j.scitotenv.2024.172346

Wu, Z., Ge, S., Han, Y., Zong, M., He, S., & Zhang, J. (2024). Effect of inorganic anions on quinoline removal by catalytic ozonation: Performance and mechanism. *Journal of Environmental Chemical Engineering, 12*(4), 113289. https://doi.org/10.1016/j.jece.2024.113289

Xiao, L., Xiaolong, Z., Chunyu, G., Qinglan, H., Jie C., Xin, H., Yongwang, L., & Botao T. (2025). Effect of surface species on the alcohol-acid adsorption from FTS wastewater on graphene: A structure-capacity study. *Journal of Environmental Sciences, 149*, 651–662. https://doi.org/10.1016/j.jes.2024.02.031

Yu, Q., Yin, M., Chen, Y., Liu, S., Wang, S., Li, Y., Cui, H., Yu, D., Ge, B., & Huang, F. (2023). Simultaneous carbon dioxide sequestration and nitrate removal by Chlorella vulgaris and Pseudomonas sp. consortium. *Journal of Environmental Management, 333*(December 2022). https://doi.org/10.1016/j.jenvman.2023.117389

Yuan-Chung, L., Tzi-Yi, W., Wan-Yu, L., & Yi-Hsing, H. (2014). Production of hydrogen from rice straw using microwave-induced pyrolysis. *Fuel, 119*, 21–26. https://doi.org/10.1016/j.fuel.2013.11.046

Environmental Management Using Renewable Biological Resources in the Global South

Adams Ovie Iyiola and Matthew Chidozie Ogwu

Abstract

The Global South faces a multitude of challenges stemming from environmental, economic, and social factors. Increasing demands for food, energy, resources, and services have intensified pressure on biological resources in these regions. Renewable biological resources, such as plants, animals, and aquatic systems, are critical to the Global South due to their ability to regenerate naturally. However, the reliance on these resources is pronounced because many of these countries are economically disadvantaged, underdeveloped, and have a lower standard of living. Consequently, issues such as pollution, human-induced pressures, climate change, habitat fragmentation, and the over-exploitation of renewable resources arise, threatening the sustainability of these regions. Effective environmental management, heavily reliant on human activities, is essential for the sustainable use of renewable biological resources. Key areas of focus include biodiversity conservation, ozone layer protection, and climate change mitigation. By employing strategies such as agroforestry, sustainable agriculture, deforestation control, and the use of renewable energy sources like solar, wind, and biomass, environmental management can help reduce climate change impacts, improve biodiversity conservation, build resilience to climate disasters, and promote sustainable development in the Global South. This chapter will highlight

A. O. Iyiola (✉)
Department of Fisheries and Aquatic Resources Management, Faculty of Renewable Natural Resources Management, College of Agriculture and Renewable Natural Resources, Ejigbo Campus, Osun State University, Ejigbo, Osun State, Nigeria
e-mail: adams.iyiola@uniosun.edu.ng

M. C. Ogwu
Goodnight Family Department of Sustainable Development, Appalachian State University, Boone, NC, USA
e-mail: ogwumc@appstate.edu

M. C. Ogwu et al. (eds.), *Sustainable Bioeconomy Development in the Global South*, https://doi.org/10.1007/978-981-96-0305-3_11

the various types of renewable biological resources, their characteristics, and sensitivities, as well as the forms of pollution and their environmental effects. It will also outline management strategies aimed at achieving environmental sustainability in the Global South, fostering healthier and more resilient biological resources for the future.

Keywords

Global South · Biological resources · Renewable resources · Sustainability · Anthropogenic activities

Abbreviations

EIA	Environmental Impact Assessment
EM	Environmental Management
EMS	Environmental Management Systems
EP	Environmental Policy
EPI	Environmental Performance Indicators
ERA	Environmental Risk Assessment
IPM	Integrated Pest Management
ISO	International Standard Organization
LCA	Life Cycle Assessment
PEM	Preventive Environmental Management
RES	Renewable Energy Sources
SDG	Sustainable Development Goals
UN	United Nations
UNFCCC	United Nations Framework Convention on Climate Change
UNSDG	United Nations Sustainable Development Goals
UNEP	United Nations Environment Program
USEPA	United States Environmental Protection Agency

1 Introduction

The contamination and pollution of the environment occurs when contaminants are released either purposely or unpurposely into the soil, air, and water causing a drastic change in the environment and health of humans (Hill, 2020). This pollution may occur naturally from volcanic eruptions and wildfires, but majorly by anthropogenic activities such as agriculture, transportation, and industrialization (Ukaogo et al., 2020). These can cause various health issues in humans such as cancer, respiratory problems, and chronic diseases (Kim et al., 2018). A major contaminant in aquatic ecosystems is the chemical contaminants which have increased the level of toxicity in water, soil, and air (Izah et al., 2022, 2023). The economic consequences of the pollutants and contaminants are very significant as the cost of cleanup of sites that are contaminated and chronic human health issues that may arise are huge (López-Pacheco et al., 2019). The environment plays an important role in the sustaining of life processes and biodiversity conservation but the rising of contaminants and pollutants in the Global South has become a major issue over the last decades (Ukaogo et al., 2020; Singh, 2015). Environmental pollutants are of various forms such as water, air, soil, and noise pollution (Iyiola & Valbo-Jørgensen, 2023; Nnaemeka, 2020). The pollutants released by industries can be reduced by using cleaner production techniques and the government can release policies or penalties for industries that violate the environmental standards (Song & Zhou, 2021). Waste management strategies such as composting and recycling can be enforced by the government for proper environmental management (Kolawole & Iyiola, 2023; Haywood et al., 2021). It is essential to prevent pollution before it occurs because of the difficulty in the management of some pollutants as explained by the Preventive Environmental Management (PEM) approach. The history of Environmental Policy (EP) is very rich and has been explored and reported by authors (Ji et al., 2022; Smith & Jacques, 2022; Caldwell & Caldwell, 2020; Hill, 2020); however, it has not been used to adequately address the issue of pollution in the Global South as little attention is placed on EP because of the different constraints it has on its implementation.

Addressing environmental pollution and management involves a process of changing the human and industrial processes into one that is friendly to the environment and has less polluting effects. This can be done by first locating the point of pollution and carrying out measures that are preventive to reduce the impact of the pollutants in the environment. A key approach is the adoption of clean production methods, reducing chemical usage, and reducing the generation of waste from industries (Prince et al., 2020). These may transit to an environment that is sustainable and compliant with the standards of the environment set aside for industrial activities (Ahmed et al., 2021). In the long run, a sustainable environment is created and the planet is healthier for all and future generations. With the peculiarity of Global South countries, in which they lack institutional framework, nonresilience to climate change, low economic power, and institutional framework, they are exposed to issues that affect their biodiversity and natural resources in the environment. This review therefore employs the use of peer-reviewed literature search to derive updated information on the pollutants and the strategies of environmental management in the Global South (Pullin & Stewart, 2007; Xiao & Watson, 2017; Mengist et al., 2020). Likewise, the PEM, Environmental Risk Assessment (EIA), EP, and Environmental Performance Indicators (EPI) will discussed as possible approaches to sustainable environmental management.

1.1 United Nations Sustainable Development Goals (UNSDG) and the Environment

The UNSDG, along with its indicators and targets, provides information that are strategic for the natural resources and environment toward sustainability (Kanter et al., 2016). These Sustainable Development Goals (SDGs) focus on the harmonization between the economic, environmental, cultural, and societal agendas and policies. The approach is built on three main pillars: indivisibility, which focuses on the implementation of all the goals; inclusion—which all shall benefit from the actions; and acceleration—which focuses on actions that can generate several developmental dividends (Mann et al., 2018). The implementation of SDGs in the environment is based on ecological knowledge that is fostered by local communities, owners of lands and businesses, managers, and civil societies (Bieling & Plieninger, 2017). Therefore, the desired SDG output only occurs when there is a close relationship between the scientific community, land-use actors, and public and private sectors. The outputs desired are the alleviation of poverty, livelihood security, and improvement, management of ecosystem services, and better coordination between the system of governance (Bridgewater et al., 2015). The UNSDG addresses the issues of the environment and biological resources through specific targets that is aimed at conserving and ensuring the well-being of the environment. Life under Water (Goal 14) targets the conservation and sustainable use of the water bodies, reducing pollution, regulation of fishing activities, and restoration of fish stocks toward sustainable fisheries. Life on land (Goal 15) is aimed at restoring and protecting the terrestrial ecosystem, forests, wetlands, and biodiversity by reducing the rate of land degradation, deforestation, and biodiversity loss. Zero hunger (Goal 2) is focused on food security and improved nutrition by ensuring agriculture is sustained. Clean water and sanitation (Goal 6) ensure that water in the environment is clean and accessible to all. It fights water pollution, water wastage, and water pollution reduction and improves water-related ecosystems. Affordable and clean energy (Goal 7) ensures production of renewable energy and reduces pressure on fossil fuels which can degrade the environment due to the release of greenhouse gases. Sustainable cities and communities (Goal 11) ensure the urban areas are well planned and ensure that human activity is reduced which may have negative impacts on the environment and biodiversity. Responsible consumption and production (Goal 12) is achieved by minimization of human impacts on biological resources.

The implementation of SDGs through environmental management builds on the land-use systems which are multifunctional and can contribute to the production processes and the aim of environmental conservation (O'Farrell & Anderson, 2010). Agroforestry which comprises agriculture and forestry is an example of agroecological land use systems (Van Noordwijk, 2019) that play a crucial function in mitigation and adaptation to climate change (Waldron et al., 2017). About 43% of the agricultural lands are covered by trees of at least 10% (Zomer et al., 2016), and this illustrates the use of agroforestry for sustainable environmental management (Carmenta et al., 2020; Milder et al., 2014). The study of agroforestry systems is

Table 1 Renewable biological resources and their importance

Resources	Importance	References
Wild animals	As food or biomaterials, e.g., fish, ducks, seals, and deer	Arneth et al. (2020)
Forest biomass	Harvested for fiber, energy, or lumber	Gasparatos et al. (2017)
Wild plants	Gathered as food sources	Pullin and Stewart (2007)
Plants	Cultivated as food, medicine, materials, or energy source	Galparsoro et al. (2022)

numerous and relates to policy, practice, and land-use science which demonstrates the multifunctional use of trees and crops (Van Noordwijk, 2019; Kanter et al., 2016).

1.2 Renewable Biological Resources and Energy Production

Renewable resources are a category of natural resources found in the environment. Renewable resources may be seen as products from agriculture and its by-products that are utilized in industrial processes. Generally, renewable resources are any resource that can replace itself or by human activities and they can be utilized forever. Most resources categorized as renewable are mostly biological and they have various functions in the natural environment (Table 1). The resources can play a significant role in the livelihood and sustenance of Global South communities and can be sources of food, economic development, security, and biodiversity conservation. The biological resources must be sustained to ensure their availability on a long term and benefit generations to come.

It should be noted that the production growth in renewable energy should not cause harm to compromise the objectives of the environment. Projects to be embarked upon in the environment must be compatible with the protection of biodiversity and conservative objectives such as the Sustainable Development Goals or the Convention on Biological Diversity post-2020 targets (Arneth et al., 2020). Environmental activities must apply an ecosystem-based approach which ensures activities carried out in the environment do not negatively affect the health, resilience, and sustainability of the environment.

2 Preventive Environmental Management (PEM) in the Global South

PEM is a concept that is built around strategies such as the prevention of pollution, reduction of wastes, and conservation of resources in identifying and determining the potential risks to the environment (Gibbs, 2009). It is a tool that is cheap in terms of cost of application and helps organizations to reduce risks and liability when they comply (Van der Heijden et al., 2010). PEM is implemented through strategies such as Life Cycle Assessment (LCA), environmental management, green procurement, environmental performance indicators, Environmental Management Systems (EMS), and Environmental Risk Assessment (ERA) (Bansal & Bogner, 2002; Hauschild & Huijbregts, 2015; Kazancoglu, 2017).

2.1 Environmental Management (EM)

It is the process by which planning, monitoring, and implementing strategies are employed to manage natural resources, the ecosystem, and pollution to ensure their sustainable use and protection (Wahab et al., 2023). This approach ensures a reduction in the negative impacts of anthropogenic activities on the environment (Das, 2020). Faced with a lot of challenges, the benefits of EM are significant as it reduces the cost incurred from control of pollution and economic benefits (Matuszak-Flejszman, 2009), improvement in resource efficiency, and increased opportunities in the market for products that are sustainable. Social benefits are also improved through quality of life, resilience of communities is increased and compliance with the regulations by organizations are increased thereby avoiding penalties and fines (Zutshi & Sohal, 2004). The strategies used in EM are EIA which involves sustainable resource management, prevention of pollution by implementing measures to avert the occurrence of pollution, the use of cleaner production technology, minimization of wastes, and promotion of improvement that is continuous toward optimizing environmental performance (Prajogo et al., 2014). The principles of EM include prevention, precaution, sustainability, and participation. EM ensures that resources are used to meet current needs without affecting their presence for future generations in meeting their needs, it emphasizes the importance of pollution prevention and damage to the environment before it happens, and it advocates for approaches that are precautionary thereby taking action to prevent harm to the environment by ensuring further participation of all stakeholders, industries, communities, individuals and government agencies (Reed et al., 2014). Challenges in the implementation of EM are lack of data, priorities that may be conflicting, and limited resources (Raymond et al., 2010).

2.2 Environmental Management Systems (EMS)

It is an approach that is systematic and well-structured at managing issues occurring in the environment and its impacts to achieve sustainable development (McAleer, 2001). It focuses on the identification and management of the environmental aspects of activities of organizations, services, and products. It may assist organizations in creating a sustainable approach to managing their environmental impacts, obeying environmental regulations, and reducing all costs that may relate to environmental risks (Da Fonseca, 2015). Inspections are done regularly by regulatory organizations to ensure compliance with environmental regulations and EMS standards (Dahlström et al., 2003). The approach and implementation of EMS can reduce

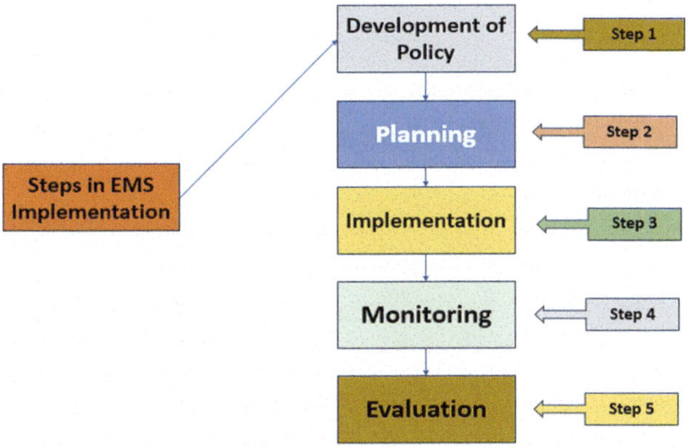

Fig. 1 Steps in EMS implementation

energy consumption by industries, reduce waste, and improve the efficiency of operation and cost saving. It also helps organizations to show commitment to the sustainability of the environment which enhances their reputation and provides an advantage that is competitive with others (Prevention, 1998). Development of policy, planning, implementation, monitoring, and evaluation are the steps in the implementation of EMS (Hilson & Nayee, 2002) (Fig. 1).

2.3 Life Cycle Assessment (LCA)

It is a systematic approach used to assess the environmental consequences a product, process, or service can have on the environment; starting from its initial extraction of the material when it was raw to the end of its life when it will be disposed of. At every stage of the product life cycle, LCA takes into cognizance the implications on the environment, extraction method of raw material, manufacturing, utilization, distribution, and end-of-life management. It offers a perspective that is comprehensive about the impacts of products and processes on the environment, thereby guiding companies and policymakers to make sustainable decisions and minimize the burden on the environment. The steps in conducting LCA are the definition of goal and scope, inventory of life cycle, impact assessment of life cycle, interpretation and improvement, and decision making (Chang et al., 2014; ISO, 2006) (Fig. 2).

2.4 Environmental Risk Assessment (ERA)

It is a process that is structured to assess the potential negative impacts human activities have on the environment. It is achieved by the identification, analysis, and evaluation of risks that relate to specific actions or substances to enhance improved decision-making and reduce environmental damage (Covello et al., 1993). It focuses on the assessment of the probability and severity of adverse effects on various

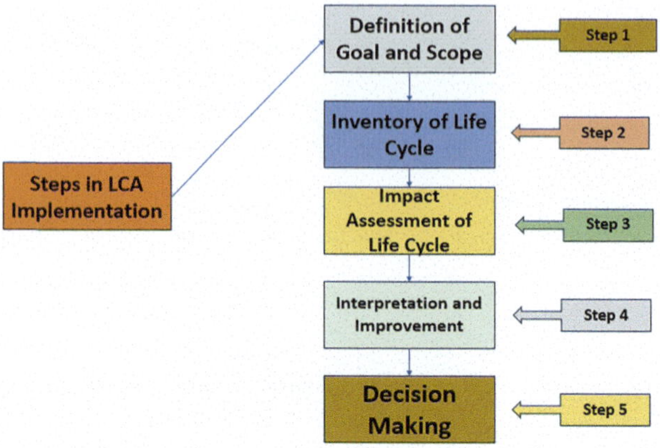

Fig. 2 Steps in LCA implementation

aquatic, terrestrial, and atmospheric systems (ISO, 2006). Its role is vital in supporting sustainable development, guiding management practices, and ensuring the safety of ecosystems and human health. The application of ERA cuts across different domains, such as industrial activities, the development of infrastructures, chemical substances, organisms that are genetically modified, and incidents of pollution. The steps involved in conducting ERA are the identification of hazards, assessment of exposure, assessment of effects, and management of risks (Ecological Risk Assessment, 1998).

2.5 Environmental Performance Indicators (EPI)

They are the measurements employed to evaluate the performance of the organization in the environment, processes, products, or services (Azapagic & Perdan, 2000). They can provide data that is quantitative or qualitative, enabling the assessment, monitoring, and reporting of impacts on the environment, usage of resources, and endeavors toward sustainability (Ogwu et al., 2022, 2023). It serves as a tool valuable for tracking progress, goals establishment, identifying areas that require a boost, and communicating the performance of the environment to stakeholders. The context and objective determine the environmental performance indicator that will be used at a particular time (UN Global Impact, 2022). Some EPS that are commonly used is consumption of energy, emission of greenhouse gases, usage of water, generation of wastes, rate of recycling, impact on biodiversity, compliance by the environment, eco-efficiency, and environmental certifications (Azapagic & Perdan, 2000).

2.6 Green Procurement

It involves the integration of environmental factors into the purchase decisions of an organization. It is also called sustainable or environmentally responsible procurement. This approach aims to minimize the environmental impacts that are associated with the acquisition of goods and services by taking into consideration issues such as efficiency in energy, conservation of resources, reduction in wastes, prevention of pollution, and the use of eco-friendly technologies or materials (Standardization, 2017; Obahiagbon & Ogwu, 2023a, b, 2024a, b). The implementation of this approach requires collaboration among the stakeholders, suppliers, sustainability team, and management of an industry (Seuring & Müller, 2008). After this, there is integration of environmental considerations into the policies of procurement, processes, and selection criteria for suppliers as well as monitoring and evaluation of the performance of the environment. The principles and strategies are environmental criteria integration, supplier evaluation, assessment of life cycle, certification and labels of products, and collaboration with suppliers (UN Global Impact, 2022).

2.7 Environmental Policy (EP)

It refers to the steps or actions that are taken by organizations and the government to tackle issues concerning environmental challenges. This may vary largely across the environment and includes issues such as air pollution, climate change, and water pollution (Divan & Rosencranz, 2022). Policies have been dated far back to the nineteenth century, and the creation of national parks in the United States during this period led to the creation of a significant environmental policy (Jaffe et al., 2005). The creation of the United States Environmental Protection Agency (USEPA) in 1970 led to the widespread of policies in the twentieth century with the adoption of the Clean Air Act and Clean Water Act. The United Nations played a significant role in environmental policy reform and modeling (Vijge et al., 2020; Akudugu & Ogwu, 2024); as the 1972 Stockholm Conference led to a landmark in policy development at the international level in the United Nations Conference on the Human Environment in 1972 (Paglia, 2021). During this conference, the United Nations Environment Program (UNEP) was created and plays a major role in the global environmental policy framework (Biermann, 2021).

3 Strategies for Environmental Management in the Global South

The increasing human population in the world today has intensified pressure on agriculture to meet food requirements, raw materials for industrial processes, and biofuels. Countries located proximally to the equator have peculiar issues in agricultural production and environmental management and are majorly impacted by climate change (Thinda et al., 2021; Onyeneke, 2021; Ahmed, 2022). These changes have resulted in several negative effects on the environment which affects productivity, food security, and environmental management (Iyiola et al. 2022a, b, c).

3.1 Agroforestry

Agroforestry management involves the approach of optimizing the integration of agricultural practices and trees. This integration leads to advantages such as efficiently absorbing and utilizing nutrients, water and light, mitigation of environmental degradation, and improving food security thereby resulting to an environment that is sustainable. Several strategies can be employed in agroforestry management. Species selection involves choosing of the correct tree and crop combination in which they can complement each other in aspects of growth rate, nutrient requirements, and ecological benefits (Zinngrebe et al., 2020). The spacing and arrangement of trees and crops can go a long way in maximizing the use of resources and efficiency, minimize competition between the crops and sunlight penetration (Zanzanaini et al., 2017). It is essential to regularly prune trees as this approach can enhance their growth, productivity and shape to reduce competition between species and increase the circulation of air and light penetration (García-Martín et al., 2016). Soil management is very important, and it involves the use of practices such as cover cropping, mulching, and the addition of organic matter to soil to improve the soil fertility, the structure of the soil, and the retention of soil by water which benefits tree growth (Laroche et al., 2020). Water management is crucial to growth of trees and involves the use of irrigation systems when there is reduce water supply, planting along contours and harvesting of water. These activities ensure the adequate moisture for crops and trees most especially in areas that are dry (Mosquera-Losada et al., 2018). The control of pests and diseases are quite important for optimal growth and production of plants.

Systems such as the Integrated Pest Management (IPM) techniques can be used to control pests and diseases without negatively affecting the beneficial organisms in the soil and maintain the balance of the soil ecologically (Buck et al., 2020). A well-planned and coordinated harvesting and utilization of tree products (e.g., nuts, fruits, timber) and products from agriculture can generate optimal economic returns for the farmer (Guillerme et al., 2020). The integration of agroforestry with livestock systems can generate additional income, improve soil fertility by the manure generated, and the management of vegetation by grazing activities of animals. Value addition and access to markets and marketing opportunities for products can increase the profits generated from agroforestry products. Monitoring is quite important as it ensures up-to-date information on the environmental condition variations, performance of crops and trees, monitoring of pests and disease outbreaks, and management practices that can adapt and maintain a resilience system and sustainability.

3.2 Deforestation

Deforestation has a significant effect on environmental management but this can be mitigated using strategies that can reduce its impacts on the environment and promote land use sustainability. The implementation of policies to protect the forests that are remaining can be a measure of management. It involves the use of legal frameworks, mechanisms that ensure the enforcement of regulations, and the involvement of the community in the management of resources. Reforestation and afforestation can be achieved by establishing programs that focus on tree planting in degraded areas. This helps to restore the forest cover and enhance biodiversity conservation, and carbon emission is mitigated (Iyiola et al., 2023a, b, 2024a). Agroforestry is a principle in sustainable land use and it involves the process of planting trees with livestock or crops, thereby sustaining the environment and generating more economic power for the owners of the land. With this, ecosystem services are maintained and biodiversity in the environment is sustained (Liu et al., 2021). Logging must be sustained by initiating selective logging and forestry practices, reducing the impact of logging and certification helps to minimize the damages to the environment and ensures sustainability and viability of resources in the forest (Houghton, 2012, Bausano et al., 2023).

3.3 Sustainable Agriculture

The approach of sustainable agriculture aims at protecting the environment and expanding the natural resources in the environment. This is an approach that is advocated for sustainable environmental management and is dependent on nonrenewable resources. It can involve the growth of single components as practiced in most areas in Africa or a combination of several components (Iyiola et al., 2024b; Lampteym, 2022; Kimathi et al., 2021; Martey et al., 2020; Adenle et al., 2019). Different factors can influence the adoption of a single agricultural sustainable practice (Table 2). These factors can range from socio-economic factors to demographic characteristics.

Apart from the adoption of single components in sustainable agricultural practice, some farmers can adopt multiple approaches to sustainable agriculture toward environmental management. These multiple approaches may include the use of animal manure, maize-legume crop rotation, the use of inorganic fertilizers, and improved seeds. Some multiple approaches are reported by Ehiakpor et al. (2021) and Teklewold et al. (2013). The profitability extent of the adoption of multiple approaches to sustainable agriculture is determined by various factors. Various outcome variables have been used in estimating the impacts of sustainable agriculture practices such as crop yields, household income, labour demand, food security, and the use of agrochemicals (Manda et al., 2016; Amondo et al., 2019; Marenya et al., 2020; Oduniyi & Chagwiza, 2021). Agriculture is the largest industry in the world and when their operations are sustainably managed, critical habitats are protected and restored, watersheds are protected, and the water quality and health of the soil

Table 2 Factors that influence the level of adoption of sustainable agriculture practices

Adoption	Influences	Reference
The decision by rice farmers to adopt soil and water conservation	It is influenced by capital, education, labour constraints, social networks, and conditions of soil on farms	Abdulai and Huffman (2014)
The adoption of improved maize varieties	Education, livestock holdings, land per capita, and market information	Manda et al. (2018)
Drought-resistant maize varieties	Influenced by access to extension, availability of labour, gender and location of farmer	Martey et al. (2020)
Adoption of climate-resistant potato varieties	Access to information, seeds of good quality, training, and variations in agro-ecological zones	Kimathi et al. (2021)

is improved. When agricultural activities are not sustained, there is great danger on the people and the environment.

3.4 Renewable Energy Sources (Solar, Wind Power, and Biomass)

The energy generated from Renewable Energy Sources (RES) can contribute to energy sustainability in terms of social, economic, and environmental components. They can reduce the emissions from pollutants, improve access to clean energy and create opportunities in terms of socioeconomic development. The increase in human population and development in the economy have serious implications for the environment and its management. Energy is a significant component in economic development and its effects are positive on the social, environmental, and economic components of the environment (Algarni et al., 2023). The use of solar energy is a positive light toward environmental management. This is achieved by the utilization of energy from the sun to generate energy useful for domestic and industrial activities. Some other benefits of solar energy are highlighted:

- They help in the reduction of dependence on fuels from fossil sources thereby resulting to clean energy production and a drastic reduction in the negative impacts of the combustion and use of fossil fuels (Jansen et al., 2020).
- They help to mitigate climate change and pollution to the air as their processes do not generate greenhouse emissions as compared to fuels from fossil sources (United Nations, 2016).
- The natural environmental resources are conserved because there is no form of consumption or removal in the use of solar energy, unlike fossil fuels which require the use of finite materials like coal, natural gas, or oil. The habitat is also protected because no form of extraction is done (Degraer et al., 2020).
- The use of land and its impacts is also minimized as solar energy components can be installed in places with existing structures and facilities. This reduces the

fragmentation and degradation of habitats which are principal in the use of fossil fuels.

- Sustainable development is promoted using solar energy because investments are attracted and jobs are created thereby leading to local economic development, and the sustainability of the environment (Bailey et al., 2014).
- Access to energy is also improved as the energy provided is clean and very reliable most especially in areas where there are no grid systems, thereby reducing reliance on generating plants and improving access to energy that is modern, clean, and reduces environmental destruction (Gasparatos et al., 2017).
- It supports the conservation of biodiversity as they can combine features such as wildlife-friendly landscaping, restoration of habitats, and land use all of which can protect biodiversity (Gasparatos et al., 2017).
- It can enhance resilience through the building of energy solutions and systems that are resistant to the events of weather as observed in other energy sources (Lloret et al., 2022).
- Ultimately, the use of solar energy complies with the Global Environmental Goals by contributing to the agreement of reducing greenhouse gas emissions and improving sustainable development as dictated in the Paris Agreement targets (International Energy Agency, 2019; United Nations, 2016).

Wind energy is also a renewable energy source which ensures a clean environment, and less of pollution of air and land (Peters et al., 2020). Its use to produce energy has fewer effects on the environment when compared with other sources of energy (Galparsoro et al., 2022). They do not release gases that can pollute the environment but require water which is needed for cooling turbines (EIA, 2022). Wind turbines used to generate electricity are of various sizes and may require space for installation, fire hazards may occur to turbines from leaking pipes although this is rare. Some noises may occur because of the spin and decline in biodiversity of birds, bats, and some animals as they may be hit by the blades of the turbine (EIA, 2022). They help to mitigate greenhouse gas emissions as none are generated during power generation as compared with fossil fuel-based power plants (Degraer et al., 2020; Hall et al., 2020; Lloret et al., 2022). They support sustainable development, and energy access, facilitate climate resilience, and contribute to the global environmentally sustainable goals (Galparsoro et al., 2022; Jansen et al., 2020; Bailey et al., 2014; Mangi, 2013).

4 Global Environmental Policies

The efforts to tackle environmental problems on the planet depend on international environment policies. They play roles that are essential in global pollution prevention, promoting sustainable development, and conservation of natural resources. The policies can tackle issues in the environment, loss of biodiversity, climate change, and pollution. Countries can collaborate through their various international agreements and conventions that set goals and develop ways for addressing the threats to the environment and create regulations.

4.1 United Nations Framework Convention on Climate Change (UNFCCC)

The UNFCCC was adopted in 1992 and all countries ratified the framework. It elaborates on the significance of a policy to fight climate change internationally and provide a stable platform for stabilizing greenhouse gases in the atmosphere globally and prevent interference by humans which may be harmful to the fragile climatic systems that can support life. With these principles, countries join forces toward the creation of a future that is sustainable and resilient, thereby minimizing the impacts of climate change, safeguarding of societies, and preserving the environment toward sustainability (Hickmann et al., 2021).

4.2 Kyoto Protocol

It is a continuation of UNFCCC, and it counters the danger of climate change through international efforts. It was adopted in 1997 and full operation in 2005. This protocol ensures a reduction in the emissions from developed countries from 2008 to 2012 (Annex I Parties). The aim of this protocol is to reduce greenhouse gas in the atmosphere thereby resulting to creation of a future that is sustainable. The protocol combines innovative market-based emission trading systems which initiate greener practices as well as reduction of emission efforts. Successes have been achieved by this protocol in awareness of global climate action, and groundwork for international frameworks, it faced with diverse challenges (Depledge, 2022).

4.3 Paris Agreement

It was adopted in 2015 under UNFCCC and it represents a significant milestone in the climate change fight. It aims to strengthen the global response of countries and ensure the global temperature is below 2 °C, above the levels of preindustrial and keeping it to 1.5 °C. All nations are expected to intensify efforts in reducing emissions of greenhouse gases, impacts of climate change adaptation, and providing support to Global South countries. This agreement was accepted by all but is faced with issues of absolute compliance by nations (Meinshausen et al., 2022).

5 Addressing Environmental Pollution

There are various holistic approaches to ensure a sustainable management of the Global South environment. There is a need to encourage research activities and innovative approaches so new solutions to address environmental problems can be discovered (Pielke Jr, 2007). This approach can be strengthened by the involvement of scientific research, advancement in technology, and collaborations which are interdisciplinary toward developing a cleaner technology, alternative sources of energy, and methods of controlling and managing the more effective environment (Weber & Peters, 2009). The green infrastructure and planning of urban areas by developing green spaces, green roofs, urban forests, and sustainable transport systems must be prioritized. This will help in the reduction of pollution and improve the quality of air and water in the environment (Weber & Peters, 2009). The responsibility of corporate and social bodies must be promoted through the adoption of sustainable practices in their business approaches, the reduction of pollution in their manufacturing activities and promote transparency in their activities toward the reduction of environmental pollution (Bansal & Roth, 2000). Circular economy transition which involves a model of minimizing wastes in the environment and maximizing the use of resources can reduce environmental pollution (Ghisellini et al. Ghisellini et al., 2016). This approach involves the reduction in raw materials consumption, promoting the recycle and reuse approach, and creation of products that can last long and recyclable. With this, there will be a reduction in the pollution resulting from extraction, production, and disposal of wastes in the environment (Geissdoerfer et al., 2017). Environmental education and awareness promotion is also an important aspect which involves the fusing of environmental education into the curricula of schools which can result to capacity building and strengthening responsibilities among the community (Tilbury & Stevenson, 2002). Also, rehabilitation of ecosystems that are degraded and creation of protected areas for biodiversity conservation can be done to protect habitats that are crucial and threatened by environmental issues (Fig. 3) (Weber & Peters, 2009).

There is need for an efficient taxation system by imposing taxes or fees on disposal of municipal wastes and create locations for landfills and proper management (Weber & Peters, 2009; Kumar, 2012). Economic incentives for business owners and individuals on waste reduction can be created and increase the reuse and recycle approach, which can conserve natural resources and reduce the concentration of greenhouse gas emissions to the environment (Porter & Linde, 1995). Pollution can also be reduced from industrial processes which involve the use of more advanced technologies in manufacturing processes. These technologies can be designed to substitute techniques for pollution control that are older and less efficient. The principles of the Stockholm declaration must be strictly obeyed by all nations (Conca & Dabelko, 2002). The declaration is focused on safeguarding all natural resources at all costs, removing the use of weapons that can cause mass destruction of the environment and preventing the exhaustion of natural and nonrenewable resources, assisting Global South countries in addressing pollution issues, improving the

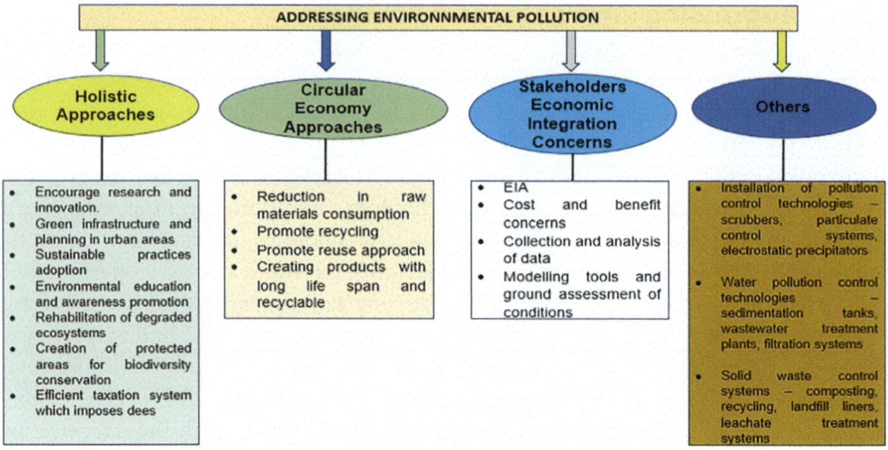

Fig. 3 Highlights on addressing environmental pollution

applications of science and technology in the environment and promoting research focuses on the environment in developing countries (Kumar, 2012; Stockholm, 1973).

There must be collaboration between international countries and the Global South, international organizations, and stakeholders in the environment to address the issue of global environmental pollution (Hisschemöller & Gupta, 1999). This cooperation is essential as issues such as climate change, loss of biodiversity, and pollution of the ocean cannot be solved by individual countries because it is a transboundary issue. Likewise, the sharing of knowledge, resources, and best practices will go a long way in developing policies and programs that are effective for environmental management in the Global South (Ourbak & Magnan, 2018). The information from environmental authorities must be made readily available to the public. This will ensure transparency and accountability in making decisions relating to the environment, ensure the participation of the public in public governance and allow citizens to make contributions toward environmental policy (Sun et al., 2019). The optimization of cost-effective measures by improving the existing control efficiency and regulations which is designed to protect the environment is important. The existing environmental controls can be improved through risk-based approaches, compliance assistance, performance-based approaches, and regulatory reforms (Weber & Peters, 2009; Iyiola et al., 2024c).

Installation of pollution control technologies which is designed for improving the efficiency of industrial processes, waste and emission reduction, and protection of environment and environmental health is essential. Technologies for control of air pollution such as scrubbers, particulate control systems, and electrostatic precipitators can be used to remove pollutants from exhausts of industrial outlets (Leson & Winer, 1991). Technologies for water pollution control are sedimentation tanks, wastewater treatment plants, and filtration systems that can be used to remove pollutants from wastewater from industries before it is released into the environment (Topare et al., 2011). Technologies for solid waste control are recycling, composting systems, landfill liners, and leachate treatment systems which reduce the waste generated and manage waste disposal (Iyiola et al., 2023c; Rana et al., 2015). Incinerators, chemical treatment systems, and solidification processes are technologies used for the removal of hazardous wastes and managing their disposal (Białowiec, 2011). The integration of economic concerns by stakeholders through conducting EIA, cost and benefit analysis, and potential risks for different stakeholders can influence the decision-making processes and ensure maximization of economic benefits from policies occurs. Stakeholders can also dialogue, bringing up possible solutions on the initiation of policies. This can be achieved through focus groups, surveys, or consultations to understand the concerns and perspectives of initiatives (Herremans et al., 2016). Collection and analysis of data from pollutants can go a long way in monitoring levels in the air, water, and soil and tracking emissions from industries and transportation sources (Filippelli et al., 2020). Modeling tools and the ground assessment of social and physical conditions in a location can be used to assess the health and impacts of pollution on the economy of Global South countries (Bai et al., 2021).

6 Conclusion

The increasing challenge of global environmental pollution requires an urgent environmental management approach. The nature of pollutants varies in water, air, soil, and noise as they contribute to the degradation of the environment and human health issues in some cases. Agricultural practices, activities of industries, and transportation are seen as the major anthropogenic activities causing environmental degradation, and measures of adopting cleaner technologies, improved waste management and sustainable farming are some positive aspects. Renewable biological resources are greatly affected by the pollutants in the environment as migration, biodiversity loss, and ecosystem destruction may occur. For combating pollution, PEM and EP are the major strategies to be implemented. Despite the various policies and strategies, there are constraints to this approach such as the economic barriers, insufficient international cooperation and noncompliance by Global South countries to environmental management. Sustainable practices such as agroforestry, sustainable agriculture, clean and renewable energy sources, and avoidance of deforestation are some strategies employed in environmental management. A holistic approach that involves locals, stakeholders, actors, and government is required and must be aligned with the sustainable development goals so nations can enforce a comprehensive EP and PEMs. There is also a need for well-formulated policies and institutional quality that are sensitive to the environment which ultimately leads to management in the Global South.

References

Abdulai, A., & Huffman, W. (2014). The adoption and impact of soil and water conservation technology: An endogenous switching regression application. *Land Economics, 90*, 26–43.

Adenle, A. A., Wedig, K., & Azadi, H. (2019). Sustainable agriculture and food security in Africa: The role of innovative technologies and international organizations. *Technology in Society, 58*, 101143.

Ahmed, M. H. (2022). Impact of improved seed and inorganic fertilizer on maize yield and welfare: Evidence from eastern Ethiopia. *Journal of Agriculture and Food Research, 7*, 100266. https://doi.org/10.1016/j.jafr.2021.100266

Ahmed, A., Mathrani, S., & Jayamaha, N. (2021). An integrated lean and ISO 14001 framework for environmental performance: An assessment of New Zealand meat industry. *International Journal of Lean Six Sigma*. https://doi.org/10.1108/IJLSS-05-2021-0100

Akudugu, M. A., & Ogwu, M. C. (2024). Sustainable development policies and interventions: A bibliometric analysis of the contributions of the academic community. *Journal of Cleaner Production, 434*, Article #139919. https://doi.org/10.1016/j.jclepro.2023.139919

Algarni, S., Tirth, V., Alqahtani, T., Alshehery, S., & Kshirsahar, P. (2023). Contribution of renewable energy sources to the environmental impacts and economic benefits for sustainable development. *Sustainable Energy Technologies and Assessments, 56*, 103098. https://doi.org/10.1016/j.seta.2023.103098

Amondo, E., Simtowe, F., Rahut, D. B., & Erenstein, O. (2019). Productivity and production risk effects of adopting droughttolerant maize varieties in Zambia. *International Journal of Climate Change Strategies and Management, 11*, 570–591. https://doi.org/10.1108/IJCCSM-03-2018-0024

Arneth, A., et al. (2020). Post-2020 biodiversity targets need to embrace climate change. *Proceedings of the National Academy of Sciences, 117*, 30882–30891.

Azapagic, A., & Perdan, S. (2000). Indicators of sustainable development for industry: A general framework. *Process Safety and Environment Protection, 78*(4), 243–261.

Bai, X., et al. (2021). Spatial-temporal variation characteristics of air pollution and apportionment of contributions by different sources in Shanxi province of China. *Atmospheric Environment, 244*, 117926.

Bailey, H., Brookes, K. L., & Thompson, P. M. (2014). Assessing environmental impacts of offshore wind farms: Lessons learned and recommendations for the future. *Aquatic Biosystems, 10*, 8.

Bansal, P., & Bogner, W. C. (2002). Deciding on ISO 14001: Economics, institutions, and context. *Long Range Planning, 35*(3), 269–290.

Bansal, P., & Roth, K. (2000). Why companies go green: A model of ecological responsiveness. *The Academy of Management Journal, 43*(4), 717–736.

Bausano, G., Masiero, M., Migliavacca, M., Pettenella, D., & Rougieux, P. (2023). Food, biofuels or cosmetics? Land-use, deforestation and CO_2 emissions embodied in the palm oil consumption of four European countries: A biophysical accounting approach. *Agricultural and Food Economics, 11*, 35.

Białowiec, A. (2011). Hazardous emissions from municipal solid waste landfills. *Contemporary Problems of Management and Environmental Protection, 9*(9), 7–28.

Bieling, C., & Plieninger, T. (2017). *The science and practice of landscape stewardship*. Cambridge University Press.

Biermann, F. (2021). The future of 'environmental' policy in the Anthropocene: Time for a paradigm shift. *Environmental Politics, 30*(1–2), 61–80.

Bridgewater, P., Régnier, M., & García, R. C. (2015). Implementing SDG 15: Can large-scale public programs help deliver biodiversity conservation, restoration and management, while assisting human development? *Natural Resources Forum, 39*(3–4), 214–223.

Buck, L., Scherr, S., Mecham, J., Trujillo, L., Fleming, M., & Bosco, S. (2020). Using integrated landscape management to scale agroforestry: Examples from Ecuador. *Sustainability Science, 15*, 1401–1415. https://doi.org/10.1007/s11625-020-00839-1

Caldwell, L. K., & Caldwell, L. K. (2020). *International environmental policy: From the twentieth to the twenty-first century*. Duke University Press.

Carmenta, R., Coomes, D. A., DeClerck, F. A. J., Hart, A. K., Harvey, C. A., Milder, J., Reed, J., Vira, B., & Estrada-Carmona, N. (2020). Characterizing and evaluating integrated landscape initiatives. *One Earth, 2*(2), 174–187.

Chang, D., Lee, C., & Chen, C.-H. (2014). Review of life cycle assessment towards sustainable product development. *Journal of Cleaner Production, 83*, 48–60.

Conca, K., & Dabelko, G. D. (2002). *Environmental peacemaking*. Woodrow Wilson Center Press.

Covello, V. T., et al. (1993). *Release assessment. Risk assessment methods: Approaches for assessing health and environmental risks* (pp. 35–89). Plenum Press.

Da Fonseca, L. M. C. M. (2015). ISO 14001: 2015: An improved tool for sustainability. *Journal of Industrial Engineering and Management, 8*(1), 37–50.

Dahlström, K., et al. (2003). Environmental management systems and company performance: Assessing the case for extending risk-based regulation. *European Environment, 13*(4), 187–203.

Das, T. K. (2020). *Industrial environmental management: Engineering, science, and policy*. Wiley.

Degraer, S., et al. (2020). Offshore wind farm artificial reefs affect ecosystem structure and functioning: A synthesis. *Oceanography, 33*, 48–57.

Depledge, J. (2022). The "top-down" Kyoto protocol? Exploring caricature and misrepresentation in literature on global climate change governance. *International Environmental Agreements: Politics, Law and Economics, 22*(4), 673–692.

Divan, S., & Rosencranz, A. (2022). *Environmental law and policy in India: Cases and materials*. Oxford University Press.

Ecological Risk Assessment. (1998). *Guidelines for ecological risk assessment*. Environmental Protection Agency.

Ehiakpor, D. S., Danso-Abbeam, G., & Mubashiru, Y. (2021). Adoption of interrelated sustainable agricultural practices among smallholder farmers in Ghana. *Land Use Policy, 101*, 105142.

EIA. (2022). *Wind explained: Wind Energy and the Environment*. www.eia.gov/energyexplained/wind-energy-and-the-environment.php

Filippelli, G., et al. (2020). New approaches to identifying and reducing the global burden of disease from pollution. *GeoHealth, 4*(4), e2018GH000167.

Galparsoro, I., Menchaca, I., Garmendia, J. M., Borja, A., Maldonado, A. D., Iglesias, G., & Bald, J. (2022). Reviewing the ecological impacts of offshore wind farms. *npj Ocean Sustainability, 1*, 1. https://doi.org/10.1038/s44183-022-00003-5

García-Martín, M., Bieling, C., Hart, A., & Plieninger, T. (2016). Integrated landscape initiatives in Europe: Multi-sector collaboration in multi-functional landscapes. *Land Use Policy, 58*, 43–53.

Gasparatos, A., Doll, C. N. H., Esteban, M., Ahmed, A., & Olang, T. A. (2017). Renewable energy and biodiversity: Implications for transitioning to a green economy. *Renewable and Sustainable Energy Reviews, 70*, 161–184.

Geissdoerfer, M., et al. (2017). The circular economy—A new sustainability paradigm? *Journal of Cleaner Production, 143*, 757–768.

Ghisellini, P., Cialani, C., & Ulgiati, S. (2016). A review on circular economy: the expected transition to a balanced interplay of environmental and economic systems. *Journal of Cleaner Production, 114*, 11–32.

Gibbs, D. (2009). *Sustainability entrepreneurs, ecopreneurs and the development of a sustainable economy*. Greener Management International. Taylor & Francis.

Guillerme, S., Barcet, H., de Munnik, N., Maire, E., & Marais-Sicre, C. (2020). Evolution of traditional agroforestry landscapes and development of invasive species: Lessons from the Pyrenees (France). *Sustainability Science, 15*, 1285. https://doi.org/10.1007/s11625-020-00847-1

Hall, R., João, E., & Knapp, C. W. (2020). Environmental impacts of decommissioning: Onshore versus offshore wind farms. *Environmental Impact Assessment Review, 83*, 106404.

Hauschild, M. Z., & Huijbregts, M. A. (2015). *Introducing life cycle impact assessment*. Springer.

Haywood, L. K., et al. (2021). Waste disposal practices in low-income settlements of South Africa. *International Journal of Environmental Research and Public Health, 18*(15), 8176.

Herremans, I. M., Nazari, J. A., & Mahmoudian, F. (2016). Stakeholder relationships, engagement, and sustainability reporting. *Journal of Business Ethics, 138*, 417–435.

Hickmann, T., et al. (2021). The United Nations framework convention on climate change secretariat as an orchestrator in global climate policymaking. *International Review of Administrative Sciences, 87*(1), 21–38.

Hill, M. K. (2020). *Understanding environmental pollution*. Cambridge University Press.

Hilson, G., & Nayee, V. (2002). Environmental management system implementation in the mining industry: A key to achieving cleaner production. *International Journal of Mineral Processing, 64*(1), 19–41.

Hisschemöller, M., & Gupta, J. (1999). Problem-solving through international environmental agreements: The issue of regime effectiveness. *International Political Science Review, 20*(2), 151–174.

Houghton, R. A. (2012). Carbon emissions and the drivers of deforestation and forest degradation in the tropics. *Current Opinion in Environment Sustainability, 4*, 597–603. https://doi.org/10.1016/j.cosust.2012.06.006

International Energy Agency. (2019). *Offshore Wind Outlook*. https://iea.blob.core.windows.net/assets/495ab264-4ddf-4b68-b9c0-514295ff40a7/Offshore_Wind_Outlook_2019.pdf

ISO I. (2006). 14040. Environmental management—Life cycle assessment—Principles and framework, pp. 235–248.

Iyiola, A. O., & Valbo-Jørgensen, J. (2023). Intrinsic values of the African environment: A sustainable perspective. In S. C. Izah & M. C. Ogwu (Eds.), *Sustainable utilization and conservation of Africa's biological resources and environment* (Sustainable development and biodiversity, vol 888). Springer. https://doi.org/10.1007/978-981-19-6974-4_23

Iyiola, A. O., Ipinmoroti, M. O., Kolawole, A. S., Akinsorotan, A. M., Oyewole, E. O., & Akinrinade, A. J. (2022a). Weather factors and their implications for fish species in Owalla reservoir, Osun State, Nigeria. *Journal of Global Affairs and Sustainable Development (JoGASD), 1*, 93–106.

Iyiola, A. O., Akinrinade, A. J., & Ajayi, F. O. (2022b). Effects of water pollution on biodiversity across coastal regions. In S. C. Izah (Ed.), *Biodiversity in Africa: Potentials, threats and conservation* (Sustainable development and biodiversity 29) (pp. 345–367). Springer. https://doi.org/10.1007/978-981-19-3326-4_13

Iyiola, A. O., Berchie, A., Oyewole, O. O., & Akinrinade, A. J. (2022c). Impacts of climate change on biodiversity in Africa. In S. C. Izah (Ed.), *Biodiversity in Africa: Potentials, threats and conservation* (Sustainable development and biodiversity 29) (pp. 369–394). Springer. https://doi.org/10.1007/978-981-19-3326-4_14

Iyiola, A. O., Kolawole, A. S., & Oyewole, E. O. (2023a). Sustainable alternatives to agrochemicals and their socio-economic and ecological values. In M. C. Ogwu & S. C. Izah (Eds.), *One health implications of agrochemicals and their sustainable alternatives, sustainable development and biodiversity*. Springer.

Iyiola, A. O., Akinsorotan, A. M., & Ewutanure, J. S. (2023b). Impacts of agrochemicals on fish composition in natural waters: A sustainable management approach. In M. C. Ogwu & S. C. Izah (Eds.), *One health implications of agrochemicals and their sustainable alternatives, sustainable development and biodiversity*. Springer.

Iyiola, A. O., Akinsorotan, O. A., Ojeleye, A. E., & Fajimolu, A. O. (2023c). An overview of environmental resources in Africa: Emerging issues and sustainable exploitation. In S. C. Izah & M. C. Ogwu (Eds.), *Sustainable utilization and conservation of Africa's biological resources and environment* (Sustainable development and biodiversity, vol 888). Springer. https://doi.org/10.1007/978-981-19-6974-4_20

Iyiola, A. O., Kolawole, A. S., Ajayi, F. O., Ogidi, O. I., & Ogwu, M. C. (2024a). Sustainable water use and management for agricultural transformation in Africa. In S. Madhav, A. L. Srivastav, S. C. Izah, & E. van Hullebusch (Eds.), *Water resources management for rural development: Challenges and mitigation* (pp. 287–300). https://doi.org/10.1016/B978-0-443-18778-0.00001-5

Iyiola, A. O., Setufe, S. B., Ofori, E., Bilikoni, J., & Ogwu, M. C. (2024b). Biomarkers for the detection of pollutants from the water environment. In S. C. Izah, M. C. Ogwu, & H. Hamidifar (Eds.), *Biomonitoring of pollutants in the Global South*. Springer. https://doi.org/10.1007/978-981-97-1658-6_16

Iyiola, A. O., Kolawole, A. S., Setufe, S. B., Bilikoni, J., Ofori, E., & Ogwu, M. C. (2024c). Fish as a sustainable biomonitoring tool in aquatic environments. In S. C. Izah, M. C. Ogwu, & H. Hamidifar (Eds.), *Biomonitoring of pollutants in the Global South*. Springer. https://doi.org/10.1007/978-981-97-1658-6_12

Izah, S. C., Iyiola, A. O., & Richard, G. (2022). Impacts of pollution on the hydrogeochemical and microbial community of aquatic ecosystems in Bayelsa State, Southern Nigeria. In S. Madhav, V. B. Singh, M. Kumar, & S. Singh (Eds.), *Hydrogeochemistry of aquatic ecosystems* (pp. 283–305). Wiley. https://doi.org/10.1002/9781119870562.ch13. Print ISBN:9781119870531 |Online ISBN:9781119870562.

Izah SC, Iyiola AO, Yarkwan B, Richard G (2023) Impact of air quality as a component of climate change on biodiversity-based ecosystem services. In: Srivastav AL, Dubey AK, Kumar A, Narang SK, Khan MA (eds.) Visualization techniques for climate change with machine learning and artificial intelligence, Elsevier 123–147. https://doi.org/10.1016/B978-0-323-99714-0.00005-4.

Jaffe, A. B., Newell, R. G., & Stavins, R. N. (2005). A tale of two market failures: Technology and environmental policy. *Ecological Economics, 54*(2–3), 164–174.

Jansen, M., et al. (2020). Offshore wind competitiveness in mature markets without subsidy. *Nature Energy, 5*, 614–622.

Ji, X., et al. (2022). Reconsider policy allocation strategies: A review of environmental policy instruments and application of the CGE model. *Journal of Environmental Management, 323*, 116176.

Kanter, D. R., Schwoob, M.-H., Baethgen, W. E., Bervejillo, J. E., Carriquiry, M., Dobermann, A., Ferraro, B., Lanfranco, B., Mondelli, M., Penengo, C., Saldias, R., Silva, M. E., & de Lima, J. M. S. (2016). Translating the sustainable development goals into action: A participatory backcasting approach for developing national agricultural transformation pathways. *Global Food Security, 10*, 71–79.

Kazancoglu, Y. A. (2017). The effectiveness of environmental management systems: A case study in a manufacturing company. *Journal of Cleaner Production, 148*, 734–741.

Kim, D., et al. (2018). Air pollutants and early origins of respiratory diseases. *Chronic Diseases and Translational Medicine, 4*(2), 75–94.

Kimathi, S. M., Ayuya, O. I., & Mutai, B. (2021). Adoption of climate-resilient potato varieties under partial population exposure and its determinants: Case of smallholder farmers in Meru County, Kenya. *Cogent Food & Agriculture, 7*, 66. https://doi.org/10.1080/23311932.2020.1860185

Kolawole, A. S., & Iyiola, A. O. (2023). Environmental pollution: Threats, impact on biodiversity, and protection strategies. In S. C. Izah & M. C. Ogwu (Eds.), *Sustainable utilization and conservation of Africa's biological resources and environment* (Sustainable development and biodiversity, vol 888). Springer. https://doi.org/10.1007/978-981-19-6974-4_14. Print ISBN: 978-981-91-6973-7.

Kumar, P. (2012). *The economics of ecosystems and biodiversity: Ecological and economic foundations* (p. 2012). Routledge.

Lampteym, S. (2022). Agronomic practices in soil water management for sustainable crop production under rain fed agriculture of drylands in sub-Sahara Africa. *African Journal of Agricultural Research, 18*, 18–26. https://doi.org/10.5897/AJAR2021.15822

Laroche, G., Domon, G., & Olivier, A. (2020). Exploring the social coherence of envisioned landscapes featuring agroforestry intercropping systems (AIS) using rural resident visual assessments and perceptions. *Sustainability Science, 15*, 1337. https://doi.org/10.1007/s11625-020-00837-3

Leson, G., & Winer, A. M. (1991). Biofiltration: An innovative air pollution control technology for VOC emissions. *Journal of the Air & Waste Management Association (1995), 41*(8), 1045–1054.

Liu, X., Yu, L., Cai, W., Ding, Q., Hu, W., Peng, D., Li, W., Zhou, Z., Huang, X., Yu, C., & Gong, P. (2021). The land footprint of the global food trade: Perspectives from a case study of soybeans. *Land Use Policy, 111*, 105764.

Lloret, J., et al. (2022). Unravelling the ecological impacts of large-scale offshore wind farms in the Mediterranean Sea. *Science of the Total Environment, 824*, 153803.

López-Pacheco, I. Y., et al. (2019). Anthropogenic contaminants of high concern: Existence in water resources and their adverse effects. *Science of the Total Environment, 690*, 1068–1088.

Manda, J., Alene, A. D., Gardebroek, C., et al. (2016). Adoption and impacts of sustainable agricultural practices on maize yields and incomes: Evidence from rural Zambia. *Journal of Agricultural Economics, 67*, 130–153. https://doi.org/10.1111/1477-9552.12127

Manda, J., Gardebroek, C., Kuntashula, E., Alene, A. (2018) Impact of improved maize varieties on food security in Eastern Zambia: A doubly robust analysis. *Review of Development Economics, 22*(1). https://doi.org/10.1111/rode.12516

Mangi, S. C. (2013). The impact of offshore wind farms on marine ecosystems: A review taking an ecosystem services perspective. *Proceedings of the IEEE, 101*, 999–1009.

Mann, C., Garcia-Martin, M., Raymond, C. M., Shaw, B. J., & Plieninger, T. (2018). The potential for integrated landscape management to fulfil Europe's commitments to the sustainable development goals. *Landscape Urban Plan, 177*, 75–82.

Marenya, P. P., Gebremariam, G., Jaleta, M., & Rahut, D. B. (2020). Sustainable intensification among smallholder maize farmers in Ethiopia: Adoption and impacts under rainfall and unobserved heterogeneity. *Food Policy, 95*, 101941. https://doi.org/10.1016/j.foodpol.2020.101941

Martey, E., Etwire, P. M., & Kuwornu, J. K. M. (2020). Economic impacts of smallholder farmers' adoption of drought-tolerant maize varieties. *Land Use Policy, 94*, 104524. https://doi.org/10.1016/j.landusepol.2020.104524

Matuszak-Flejszman, A. (2009). Benefits of environmental management system in polish companies compliant with ISO 14001. *Polish Journal of Environmental Studies, 18*(3), 411.

McAleer, S. L. (2001). Recommendations for the integration of an occupational health and safety management system and an environmental management system. *Environmental Design*.

Meinshausen, M., et al. (2022). Realization of Paris agreement pledges may limit warming just below 2°C. *Nature, 604*(7905), 304–309.

Mengist, W., Soromessa, T., & Legese, G. (2020). Method for conducting systematic literature review and meta-analysis for environmental science research. *MethodsX, 7*, 100777.

Milder, J. C., Hart, A. K., Dobie, P., Minai, J., & Zaleski, C. (2014). Integrated landscape initiatives for African agriculture, development, and conservation: A region-wide assessment. *World Development, 54*, 68–80.

Mosquera-Losada, M. R., Santiago-Freijanes, J. J., Rois-Díaz, M., Moreno, G., den Herder, M., Aldrey-Vázquez, J. A., Ferreiro-Domínguez, N., Pantera, A., Pisanelli, A., & Rigueiro-Rodríguez, A. (2018). Agroforestry in Europe: A land management policy tool to combat climate change. *Land Use Policy, 78*, 603–613.

Nnaemeka, A. N. (2020). Environmental pollution and associated health hazards to host communities (case study: Niger delta region of Nigeria). *Central Asian Journal of Environmental Science and Technology Innovation, 1*(1), 30–42.

O'Farrell, P. J., & Anderson, P. M. L. (2010). Sustainable multifunctional landscapes: A review to implementation. *Current Opinion in Environmental Sustainability, 2*(1–2), 59–65.

Obahiagbon, E. G., & Ogwu, M. C. (2023a). Sustainable supply chain Management in the Herbal Medicine Industry. In S. C. Izah, M. C. Ogwu, & M. Akram (Eds.), *Herbal medicine phytochemistry* (Reference series in phytochemistry). Springer. https://doi.org/10.1007/978-3-031-21973-3_64-1

Obahiagbon, E. G., & Ogwu, M. C. (2023b). Consumer perception and demand for sustainable herbal medicine products and market. In S. C. Izah, M. C. Ogwu, & M. Akram (Eds.), *Herbal medicine phytochemistry* (Reference series in phytochemistry). Springer. https://doi.org/10.1007/978-3-031-21973-3_65-1

Obahiagbon, E. G., & Ogwu, M. C. (2024a). Organic food preservatives: The shift towards natural alternatives and sustainability in the global South's markets. In M. C. Ogwu, S. C. Izah, & N. R. Ntuli (Eds.), *Food safety and quality in the global south* (pp. 299–329). Springer. https://doi.org/10.1007/978-981-97-2428-4_10

Obahiagbon, E. G., & Ogwu, M. C. (2024b). The nexus of business, sustainability, and herbal medicine. In S. C. Izah, M. C. Ogwu, & M. Akram (Eds.), *Herbal medicine phytochemistry* (Reference series in phytochemistry). Springer. https://doi.org/10.1007/978-3-031-21973-3_67-1

Oduniyi, O. S., & Chagwiza, C. (2021). Impact of adoption of sustainable land management practices on food security of smallholder farmers in Mpumalanga province of South Africa. *GeoJournal, 87*, 4203. https://doi.org/10.1007/s10708-021-10497-0

Ogwu, M. C., Izah, S. C., & Iyiola, A. O. (2022). An overview of the potentials, threats, and conservation of biodiversity in Africa. In S. C. Izah (Ed.), *Biodiversity in Africa: Potentials, threats and conservation, sustainable development and biodiversity 29* (pp. 1–20). Springer. https://doi.org/10.1007/978-981-19-3326-4_1

Ogwu, M. C., Iyiola, A. O., & Izah, S. C. (2023). Overview of African biological resources and environment. In S. C. Izah & M. C. Ogwu (Eds.), *Sustainable utilization and conservation of Africa's biological resources and environment* (Sustainable development and biodiversity, vol 888). Springer. https://doi.org/10.1007/978-981-19-6974-4_1. Print ISBN: 978-981-91-6973-7.

Onyeneke, R. U. (2021). Does climate change adaptation lead to increased productivity of rice production? Lessons from Ebonyi State, Nigeria. *Renewable Agriculture and Food Systems, 36*, 54–68. https://doi.org/10.1017/S1742170519000486

Ourbak, T., & Magnan, A. K. (2018). The Paris agreement and climate change negotiations: Small islands, big players. *Regional Environmental Change, 18*, 2201–2207.

Paglia, E. (2021). The Swedish initiative and the 1972 Stockholm conference: The decisive role of science diplomacy in the emergence of global environmental governance. *Humanities and Social Sciences Communications, 8*(1), 1–10.

Peters, J. L., Remmers, T., Wheeler, A. J., Murphy, J., & Cummins, V. A. (2020). Systematic review and meta-analysis of GIS use to reveal trends in offshore wind energy research and offer insights on best practices. *Renewable and Sustainable Energy Reviews, 128*, 109916.

Pielke, R. A., Jr. (2007). *The honest broker: Making sense of science in policy and politics* (p. 2007). Cambridge University Press.

Porter, M. E., & Linde, C. V. (1995). Toward a new conception of the environment-competitiveness relationship. *The Journal of Economic Perspectives, 9*(4), 97–118.

Prajogo, D., Tang, A. K. Y., & Lai, K.-H. (2014). The diffusion of environmental management system and its effect on environmental management practices. *International Journal of Operations & Production Management, 34*(5), 565–585.

Prevention, P. (1998). *Handbook A*. World Bank Group.

Prince, U., Ewuzie, U., & Onwuka, C. (2020). *Environmental pollution: Causes, effects, and the remedies* (pp. 419–429). Elsevier.

Pullin, A., & Stewart, G. (2007). Guidelines for systematic review in environmental management. *Conservation Biology, 20*, 1647–1656.

Rana, R., Ganguly, R., & Gupta, A. K. (2015). An assessment of solid waste management system in Chandigarh City, India. *Electronic Journal of Geotechnical Engineering, 20*(6), 1547–1572.

Raymond, C. M., et al. (2010). Integrating local and scientific knowledge for environmental management. *Journal of Environmental Management, 91*(8), 1766–1777.

Reed, M. S., et al. (2014). Five principles for the practice of knowledge exchange in environmental management. *Journal of Environmental Management, 146*, 337–345.

Seuring, S., & Müller, M. (2008). From a literature review to a conceptual framework for sustainable supply chain management. *Journal of Cleaner Production, 16*(15), 1699–1710.

Singh, J. (2015). International conference on harmonization of technical requirements for registration of pharmaceuticals for human use. *Journal of Pharmacology and Pharmacotherapeutics, 6*(3), 185–187.

Smith, Z. A., & Jacques, P. (2022). *The environmental policy paradox*. Taylor and Francis.

Song, L., & Zhou, X. (2021). Does the green industry policy reduce industrial pollution emissions?—Evidence from China's national eco-industrial park. *Sustainability., 13*(11), 6343.

Standardization, I. O. (2017). ISO 20400: 2017 Sustainable procurement—guidance.

Stockholm, C. (1973). *United Nations conference on the human environment: Stockholm, Sweden, June 5–16, 1972: A guide to the microfiche edition of the conference bibliography*. Xerox University Microfilms.

Sun, D., et al. (2019). Monitoring effect of transparency: How does government environmental disclosure facilitate corporate environmentalism? *Business Strategy and the Environment, 28*(8), 1594–1607.

Teklewold, H., Kassie, M., & Shiferaw, B. (2013). Adoption of multiple sustainable agricultural practices in rural Ethiopia. *Journal of Agricultural Economics, 64*, 597–623. https://doi.org/10.1111/1477-9552.12011

Thinda, K. T., Ogundeji, A. A., Belle, J. A., & Ojo TO. (2021). Determinants of relevant constraints inhibiting farmers' adoption of climate change adaptation strategies in South Africa. *Journal of Asian and African Studies, 56*, 610–627. https://doi.org/10.1177/0021909620934836

Tilbury, D., & Stevenson, R. B. (2002). *Education and sustainability: Responding to the global challenge*. IUCN.

Topare, N. S., Attar, S., & Manfe, M. M. (2011). Sewage/wastewater treatment technologies: A review. *Scientific Reviews & Chemical Communications, 1*(1), 18–24.

Ukaogo, P. O., Ewuzie, U., & Onwuka, C. V. (2020). Environmental pollution: Causes, effects, and the remedies. In P. Chowdhary (Ed.), *Microorganisms for sustainable environment and health* (pp. 419–429). Elsevier.

UN Global Compact. (2022). *Guide to corporate sustainability: Shaping a sustainable future*. United Nations Global Compact.

United Nations. (2016). Report of the Inter-Agency and Expert Group on Sustainable Development Goal Indicators. (E/CN.3/2016/2/Rev.1). 49. United Nations Economic and Social Council.

Van der Heijden, A., Driessen, P. P., & Cramer, J. M. (2010). Making sense of corporate social responsibility: Exploring organizational processes and strategies. *Journal of Cleaner Production, 18*(18), 1787–1796.

Van Noordwijk, M. (2019). Sustainable development through trees on farms: Agroforestry in its fifth decade. In *World Agroforestry Centre (ICRAF)*.

Vijge, M. J., et al. (2020). Governance through global goals. In *Architectures of earth system governance: Institutional complexity and structural transformation* (pp. 254–274).

Wahab, M. K. A., Iyiola, A. O., & Abdulwahab, U. F. (2023). Mitigating the one health impacts of agrochemicals through sustainable policies and regulations. In M. C. Ogwu & S. C. Izah (Eds.), *One health implications of agrochemicals and their sustainable alternatives* (Sustainable development and biodiversity). Springer. https://doi.org/10.1007/978-981-99-3439-3_8

Waldron, A., Garrity, D., Malhi, Y., Girardin, C., Miller, D. C., & Seddon, N. (2017). Agroforestry can enhance food security while meeting other sustainable development goals. *Tropical Conservation Science, 10*, 1940082917720667.

Weber, C. L., & Peters, G. P. (2009). Climate change policy and international trade: Policy considerations in the US. *Energy Policy, 37*(2), 432–440.

Xiao, Y., & Watson, M. (2017). Guidance on conducting a systematic literature review. *Journal of Planning Education and Research, 39*, 93–112.

Zanzanaini, C., Trần, B. T., Singh, C., Hart, A., Milder, J., & DeClerck, F. (2017). Integrated landscape initiatives for agriculture, livelihoods and ecosystem conservation: An assessment of experiences from South and Southeast Asia. *Landscape Urban Plan, 165*, 11–21.

Zinngrebe, Y., Borasino, E., Chiputwa, B., Dobie, P., Garcia, E., Gassner, A., Kihumuro, P., Komarudin, H., Liswanti, N., Makui, P., Plieninger, T., Winter, E., & Hauck, J. (2020). Agroforestry governance for operationalising the landscape approach: Connecting con-

servation and farming actors. *Sustainability Science, 15*, 1417. https://doi.org/10.1007/s11625-020-00840-8
Zomer, R. J., Neufeldt, H., Xu, J., Ahrends, A., Bossio, D., Trabucco, A., van Noordwijk, M., & Wang, M. (2016). Global tree cover and biomass carbon on agricultural land: The contribution of agroforestry to global and national carbon budgets. *Scientific Reports, 6*(1), 29987.
Zutshi, A., & Sohal, A. (2004). Environmental management system adoption by Australasian organisations: Part 1: Reasons, benefits and impediments. *Technovation, 24*(4), 335–357.

Sustainable Food and Feed Production Techniques in the Global South

R. T. Babalola, C. A. Ogunlade, A. R. Olaniyan,
A. A. Famuwagun, O. O. Akindele, and W. O. Adejuwon

Abstract

The pursuit of providing nutritious food and feed for all segments of society, as outlined in the second Sustainable Development Goal, poses a significant challenge in the Global South, where resources are limited. Understanding sustainable production techniques is paramount for countries in this region, spanning from farm to fork. This chapter delves into the unique climatic conditions of Global South countries and explores the adoption of techniques from the Global North. Addressing disasters and implementing solutions to mitigate their impact is crucial for ensuring food and feed sustainability. Tracing the evolution of food and feed production from ancient times to contemporary practices sheds light on knowledge gaps and informs necessary action plans. Recognizing the importance of adopting good production techniques underscores the urgency for technological advancements in sustainable food and feed production. The involvement of various sectors and the level of education and training of personnel play pivotal roles in establishing sustainable practices. A comprehensive understanding of the production processes is essential for navigating the farm-to-fork journey and identifying areas where additional skills or technology are needed. This

R. T. Babalola (✉) · A. A. Famuwagun
Department of Food Science and Technology, Osun State University, Osogbo, Nigeria
e-mail: racheal.babalola@uniosun.edu.ng

C. A. Ogunlade · A. R. Olaniyan
Department of Agricultural Engineering, Osun State University, Osogbo, Nigeria

O. O. Akindele
Department of Food Engineering, Ladoke Akintola University of Technology, Ogbomosho, Nigeria

W. O. Adejuwon
Department of Food Science and Technology, Obafemi Awolowo University, Ile-Ife, Osun State, Nigeria

M. C. Ogwu et al. (eds.), *Sustainable Bioeconomy Development in the Global South*, https://doi.org/10.1007/978-981-96-0305-3_12

chapter serves as a guide for promoting sustainable food and feed production practices in the Global South.

Keywords

Agriculture · Food · Feed · Sustainable · Production · Technology · Innovation

Abbreviations

SDG Sustainable development goals
IE Intergenerational Equity
CO_2 Carbon dioxide
CA Conservative agriculture
OF Organic Farming
GT Green Technology
PA Precision Agriculture
GPS Global Positioning System
GIS Geographical Information System
VRT Variable-Rate Technology
DNA Deoxyribonucleic acid
GAP Good Agricultural Practices
GDP Gross Domestic Product
AM Agricultural Mechanization
GMP Good Manufacturing Practices
HACCP Hazard Analysis and Critical Control Points

1 Introduction

Food and feed remain the basic life necessities after air and water, neither human beings nor animals can survive for long without intake of balanced/healthy food and feed respectively. The production of quality food and feed in sustainable quantities has accordingly become an atypical task for humans across the Globe (Coker et al., 2015). The process of ensuring sustainable food and feed production is a fundamental part of achieving food security, the second goal of the sustainable development goals (SDGs), and it is instrumental to actualizing other sustainable goals. Sustainable food and feed production systems involve provision for immediate consumption without jeopardizing future production, simply put as meeting the immediate needs alongside the foretelling of satisfying ultimate needs. According to Codex Alimentarius (2024), "Food, means any substance, whether processed, semi-processed or raw, which is intended for human consumption," they are mostly the resultant products of agricultural practices, either crop production or animal husbandry. Alternatively, feed production can be developed from human foods, wastes, and by-products, or they can be independently produced from different raw materials.

Agriculture has since been a nation's backbone to sustainability, and according to Lipper et al. (2014), agriculture is the license to food security and alongside sources of employment and economic growth for cities and countries. Sustainable food and feed production are often underlined by climate change systems for farmland activities (i.e., the planting periods) and poor postharvest processing methods for harvested crops. Enhancing the planting period coupled with sophisticated postharvest technology will offset seasonality and serve as a guide for future productions. Delving into the causes of food and feed insecurity will shed light on the doable measures to achieving sustainability at all levels of production pointing out the responsibilities of all the actors involved in the production (Ogwu, 2019). However, activities agglomerating to climate change should be understudied and eradicated at all costs. Fascinatingly, both human beings and animals are directly and consequentially involved in activities leading to climatic changes thereby causing natural disasters which adversely affect food and feed production.

1.1 Peculiarity of the Global South Countries Climatic Conditions

Climatic conditions refer to the state of the climate which is predetermined by climate-precursor factors like deforestation, livestock and plant farming, excavations, and fossil fuel burning. The climate, often erroneously interchanged as weather is consistently inconsistent due to the various activities of humans having direct and indirect impacts on the temperature. The climate can be defined as the average weather of a geographical location at a specified period while the weather is defined as the fluctuating state of the atmosphere that has been characterized by temperature. For some decades, climate change among other factors affects the food system causing instability in the quantity of food available for countries (UNFCCC, 2011).

The Global South Countries are the set of undeveloped and developed countries characterized by their agricultural activities, which is the driving force of the economy. In their pursuit of attaining a stable economic state, the depth of unregulated agricultural activities in the Global South has led to a rise in the level of gas emission. Their high level of energy intensity coupled with the use of fossil fuels, has continuously contributed to the quantity of global emissions for 30 years, accounting for 78% of the world's gas emissions (Fuhr, 2021).

The imperativeness of agricultural production for subsequent economic stability called for unavoidable greenhouse gas emissions, therefore contributing immensely to climate change. One momentous effect of gas emission is seasonal variability, and the effect of seasonal variability is evident from the beginning from the stage of land preparation to the final stage of harvesting in unmechanized farming systems (Egboduku et al., 2024; Agoh et al., 2024; Kosoe & Ogwu, 2024). Seasonal variabilities have threatened the security of agricultural production globally; nevertheless, developed countries have successfully masked this challenge because of their high adaptation to technological innovations (Asogwa et al., 2022).

Owning to the ever-increasing population, there is a greater aggressiveness to feeding this population, leading to more gas emissions disrupting the climate (Sadowski et al., 2024). Agricultural production had exponentially intertwined gas emissions and climate change, creating an inseparable fuss amidst the trio. Climatic change is a major obstacle to sustainable agricultural production in Global South countries whereas, the Global North countries are booming in agricultural production regardless of the climatic change (Salman et al., 2022).

The existence of regulatory bodies like the provincial and territorial governments in Canada, the United States Environmental Protection Agency, the European parliament, the council of the European Union, and the European Commission in some Global North Countries had significantly sustained their climatic conditions and improved their food security. Most of the developing and underdeveloped countries unlike the developed countries do not have good networking systems linking and regulating the farmland activities with climate change. The establishment of regulatory bodies in the Global South Countries will curtail if not eliminate activities leading to climate change.

1.2 Engaging the Climatic Seasons for Sustainable Food/Feed Production

Summer, autumn, spring, and fall are the climatic seasons characteristics of the developed countries, and the understanding of these seasons geared up their provision of innovative farming systems. Innovative farming systems could be described as the introduction of technological advancements into agricultural practices to eradicate their complete reliance on the climatic seasons, i.e., regardless of the season, continuous agricultural practices are certain. These innovative farming systems also compensated for restrictions on types of agricultural produce based on the soil type (Nizzy & Ogwu, 2024; Ogwu et al., 2024).

On the contrary, the rain and Harmattan (dry) seasons accustomed to most developing countries in Africa had promoted their low agricultural productivity accompanied by selective agricultural produce. Most farm practices are rainfed, implying season variability coupled with spoilage due to reprehensible storage facilities to accommodate the bulk harvest. However, the farmland is also at risk of flooding and prone to erosion due to excessive rainfalls if precautionary measures are not taken. The soil nature has also been hindering production exploration, limiting crop production to the production of compatible products alone.

One of the major defense mechanisms for climate change is the introduction and adaptation of diverse technological innovations (Thapa et al., 2019). The most suitable technological innovations for sustainable agricultural production during the rainy and dry seasons are the elevated farming system (greenhouse/vertical farming methods) and the intelligent irrigation farming system respectively.

1.2.1 Intelligent Irrigation Farming System

The irrigation farming system is differentiated from rainfed farming because of the alternative sources of water aside from rain, that are used on the farmland. It is the best farming system during the dry season because it enables consistency in agricultural production by preventing seasonal shortages which are mostly caused by lack of water during the dry season. Irrigation farming systems include undefined and sophisticated methods of practice, and their flexible nature makes it an affordable farming method for subsistence and commercial farmers.

The intelligent irrigation system is described as the introduction of intelligence to automatic irrigation systems, which is accomplished either with the use of intelligent control techniques like fuzzy logic or neural networks or with the incorporation of significant amounts of data into the analysis to improve irrigation control. The principal purpose of intelligent irrigation systems is to increase agricultural productivity alongside reducing the impact of crops on the environment (Vallejo-Gomez et al., 2023).

1.2.2 Vertical Farming System

This is the process of cultivating crops in vertically stacked layers indoors or within a controlled environment, and it is characterized by higher water and nutrient efficiency, and low usage of pesticides and herbicides, resulting in reduced agricultural pollution. It offers a potential solution to most of the agricultural challenges like land scarcity, climate control, water conservation, and food security. It has been estimated to cater to the projected 70% increase in food demand by the year 2050 (Zhu & Marcelis, 2023).

Vertical farming system is suitable for most annual crops, its mode of operation does not require large land mass and it maximally utilizes the available space. It could be described as a waste management practice because of the recycled materials being often used for the stacking, it likewise allows for the soil type selection. It is an organized farming system that should be introduced and encouraged for all households for sustainable food production.

1.3 Nature's Effect on Food Production

Most of nature's impacts on the climate are triggered directly or indirectly by the actions of man and animals. Modernized agricultural practices like fertilizer application, seedling(s) improvement variations, irrigation practices, etc. are part of the human causative effects on food and feed production (Shirazu et al., 2022). Additionally, the reckless activities involved in globalization and industrialization have also contributed greatly to most of the natural disasters evolving to climatic changes.

Natural disasters are sudden, uncontrollable events or natural accidents, including droughts, earthquakes, forest fires, land fires, landslides, floods, and volcanic eruptions. These are unpredictable events that cause long-term damage to farmland, invariably disrupting the routine agricultural practices thereby affecting the production of food and feed negatively (FSA, 2019; Malau & Darhyati, 2021). Tirivangasi (2018) reported the direct interconnectivity between natural disasters and food security, emphasizing the negative effects of natural disasters from the prevalence of food insecurity to food malnutrition resulting in noncommunicable diseases, increased food prices, rise in the level of poverty, unemployment, etc.

Case Study

The coldest year witnessed so far in Nigeria is the year 2000, after which subsequent increase in environmental temperature has progressively caused a heat uproar predicted to continue until the year 2100. Subsequently, the food production sector has been faced with lower production as the year went by; the storage of vegetables, roots, and tubers was affected by the high environmental temperature; the rate of decomposition was rapid for products like vegetables, onions, tomatoes, yams, etc.; and poultry product growth and production were adversely affected (Idumah et al., 2016). Farmland tillage becomes difficult under high environmental temperatures leading to reduced hectares of farmland for farming production.

Qualitative research is necessary to understand the causative effects, and extent of damages, and to reduce the level of natural disasters'impact on agricultural production.

The process of achieving sustainable agricultural production includes reducing if not eliminating the causes of natural disasters, which will stabilize the climatic condition and, permit the predictive farming system. These could be achieved following these practices:

1.3.1 Intergenerational Equity (IE)

This ensures a favorable climate from generation to generation, ensuring the certainty of future infrastructure, access to equal opportunities with time, and continuous access to the social welfare of the youth. IE was established for the perseverance of better opportunities and choices with generations, it is an act/law to prevent opportunity degeneration. Implementing IE requires a multi-stakeholder approach of young and old people, public and private sectors, of various disciplines in fostering information exchange and transfer of resources (Puaschunder, 2020).

1.3.2 Climate Conscientiousness

This entails provisional solutions to tackling climate vulnerability, controlling carbon growth and energy improvisation for all using non-fossil fuel sources (Puaschunder, 2020). Climate change mitigation is the process of reducing the flow of heat-trapping greenhouse gases into the atmosphere. Fawzy et al. (2020) listed three climate change mitigation approaches, viz.,

1. Conventional mitigation efforts employ decarbonization technologies and every other technology that will reduce CO_2 emissions. Technologies that are painstak-

ingly established with acceptable levels of risks, like carbon capture storage and utilization, renewable energy, nuclear power, efficiency gains, and fuel switching.
2. Negative emissions technologies or carbon dioxide removal methods, involve deploying technologies to capture and confiscate CO_2 from the atmosphere.
3. Radiative forcing geoengineering technologies like circus cloud thinning, surface-based brightening, marine sky brightening, space-based mirrors, aerosol injection, etc. These employ the principle of altering the earth's radiation balance by managing the solar and terrestrial radiation.

1.3.3 A Paradigm Shift

This should be established whose definition of food security is not limited to the availability and accessibility of nutritious and healthy food for all but will incorporate into its definition the protection of the forest. This will negate the general notion that deforestation is essential to creating more agricultural farmlands required to feed the ever-increasing human population (Alencar et al., 2023). Forest protection serves as a natural habitat for rural populations and sources of food, i.e., wild meat, plants, fisheries, and insects (Baudron et al., 2019). Jagger et al. (2022) stated that forestation is a way of alleviating poverty which enhances food insecurity, allowing different traditional agricultural practices in the forest will serve as sources of income.

2 Food Production Practices

Food production has been revolving with time, most of the practices, advantages, and demerits are tabulated in Table 1. These production methods are adequate for curbing hunger but are not sustainable production practices, most of which are to be discontinued for environmental security.

2.1 Sustainable Food Production Practices

Sustainable food practices are the methods that are sufficient to end the pressing hunger at the moment with adequate preparations and considerations for future food production, and theyare described in the following sections.

2.1.1 Biotechnology or Biofortification

This was introduced as a solution to the negative effects of the green revolution; it took place during the twentieth century, it is defined as the transformation of agriculture with the introduction of genetically modified plants and animals, for nutritional improvement (Grauerholz & Owens, 2015). It enhanced the production of basic crops which serves as an improvement on environmental sustainability by promoting environmentally resistant products, it has a high initial investment cost with subsequent cheap maintenance costs and, it provides a network of accessibility to the rural and restricted areas. However, there have been reports of loss of

Table 1 Methods of food production practices

S/N	Production practice	Benefits	Demerit
1	Traditional farming	Biodiversity maintenance Resource utilization	Deforestation Soil degradation Pollution
2	Green revolution	Exponential productivity of farm produce and animal husbandry Hunger alleviation	Biodiversity loss Adverse pollution (air, soil and water)
3	Organic agriculture	Low cost manure process Soil preservation Quick crop development	Greenhouse effect Global warming
4	Biofortification revolution	Enhanced production of basic crops Improvement Environmental resistant products	Loss of biodiversity Destruction of the ecosystem
5	Agricultural intensification	Higher productivity Land use optimization	Risk of antibiotics resistance
6	Conservative agriculture	Low soil disturbance Plant diversification	High equipment cost Exposure to fungal disease

biodiversity coupled with destruction of the ecosystem while maximizing farmland (Soria-Lopez et al., 2023).

2.1.2 Sustainable Agricultural Intensification

Sustainable agricultural intensification is defined as the agriculture system whereby production yields are increased without extending the farming land or causing any harm to the environmental impact (Prosekov & Ivanova, 2018). According to Cassman and Grassini (2020), the aims of sustainable intensification include the transformation of the holistic food chain into a more sustainable production system through the development of good management practices in harnessing natural resources while reducing harmful agricultural practices.

2.1.3 The Traditional Agricultural System

The traditional agricultural system was announced by the United Nations in the year 2014 to promote small-scale or family farming with the sole aim of achieving the second goal of sustainable development goal (SDG), i.e., “zero hunger” while maximizing limited resources. It is characterized by high productivity, mitigation of climate change, low energy consumption, and, conservation of biodiversity consumption. It is regarded as a recycling process that reduces wastage via the incorporation of animal husbandry into crop farming thereby using the animals’ excreta as manure (Thakur et al., 2023).

While the government and all agricultural stakeholders are advocating for biotechnology and sustainable agricultural intensification because of the initial and running capital involved, families should be encouraged to practice traditional agricultural systems on their rooftops or in their backyards.

2.1.4 Conservative Agriculture

Conservative agriculture is one of the food production techniques aimed at ensuring food sustainability. According to Adhikary et al. (2020), the non-sustainability of agricultural practices is evident with constraints like soil erosion, soil salinization, and soil organic matter reduction, which are repercussions of bad agricultural practices (monocropping and intensive tillage). Conservative Agriculture (CA) is a system applicable to both land tillage practices and animal husbandry practices, and it was targeted at eliminating bad agricultural practices and their effects by enhancing natural biological processes.

CA is defined as an agricultural practice that conserves natural resources to achieve maximum products with higher profit margins. It however requires extensive knowledge of the climatic conditions, planting periods, and the soil-crop adaptation mechanisms for proper planning of the agricultural production. González-Sánchez et al. (2016) noted that the European countries despite their technological advancements are yet to implement CA while other countries outside Europe with similar agro-climatic conditions have implemented the system.

CA has been proposed by the Southern African Countries to combat their rainfall variability; it is to be actualized by combining both empirical data and results generated from previous agricultural practices. This is expected to balance the agro-ecology and the socio-economy; the agro-ecology will be maintained by early planting at near optimum planting periods and the socio-economy by lesser-labor adoption.

2.1.5 The Food System

The food system is one of the existing food production techniques encompassing all the food production stakeholders; beginning from seedling selection to the point of consuming the product; the societal and natural environment; the economic sector; and the forestry or fisheries. The interconnectivity of the food system describes the peculiarity of each sector indicating that an overhaul in a sub-sector will affect the entire production system (Nguyen, 2018). The activities of a food system spread across all levels/stages of policymakers from grassroots to global scales, i.e., local, state/district, national, regional, and global (von Braun et al., 2023).

2.1.6 Natural Food Production Methods/Systems

Natural food production methods/systems are the regular food production methods from ancient years, known as conventional agriculture, involving the planting of crops and rearing of animals without modifications or aids. Due to technological knowledge, natural food production has been partly or wholly substituted with alternative methods like alternative crops, permaculture, organic farming, and green technology.

1. *Alternative crops* are the newly introduced crops into an agro-ecosystem to replace the traditional or native crops that are prone to both biotic and abiotic stresses (Hirich et al., 2020). They are neither lesser-known nor underutilized crops, but they are crops that have been reintroduced into a particular agroeco-

system for beneficial farming purposes. These alternative crops vary with geographical locations, and they should be identified, cultivated, and introduced into the production stream.

2. *Permaculture* is a derivative word from either "permanent agriculture" or "permanent culture"; it marks the era of a radical shift from conventional farming systems and alternative food movements (Grauerholz & Owens, 2015). Permaculture as an unconventional reflex is the positive response to environmental crisis, described as intentionally building landscapes to imitate natural designs with the aim of food security. It is a holistic design established based on conventional knowledge, observation of nature, and contemporary scientific findings, in reconstructing the society of centralized power to ordinary people in their communities. It is expected to thrive on Earthcare (care for the planet's general health: soil, water bodies, animals, and man), people care (satisfying man's physical, social, emotional, and intellectual needs), and fair share (equity) ethics (Nath, 2022).
3. *Organic Farming (OF)* is the farming method that exempts the use of chemicals, fertilizers, pesticides, or genetically modified seeds but utilizes composts, biofertilizers, bio-pesticides, and environmentally friendly approaches like crop rotations, mulching, intercropping, etc. in agricultural practices with the sole objective of food production sustainability. OF was built on four major principles; the principle of health (sustainability and enhancement of human, animal, and nature's health); the principle of ecology (sustainability of the ecosystem and cycles); the principle of fairness (equity relationships regarding environments and life opportunities); and principle of care (precautionary care should be ensured for future generations/environment) (Soni et al., 2022; Attia et al., 2023). Mukherjee et al. (2018) and Das et al. (2020) stated that products from OF are antioxidative and shelf-stable; however, Singh et al. (2021) estimated that the products are 20% more expensive than conventional products, thus categorizing them as "elite products."

   Yousuf et al. (2018) argued that OF maintained a fixed resource base which is liable to depletion caused by overexploitation. He proposed the abandonment of OF for a more advanced system, the Green Technology (GT). *Green technology* is a technological advancement that influences product production and processing. Islam et al. (2022) defined GT as a safer system retaining nutrients while reducing and destroying harmful effects of production on the environment by engaging in safer production and utilization of systems, strategies, processes, and materials.

2.2 Prospects of Good Production Techniques

The prospects of good agricultural production techniques could be viewed in light of the sophisticated (modern) techniques recently employed in production processes. According to Sonia et al. (2023), these sophisticated techniques have exponentially revolutionized the agricultural industry via diverse innovative approaches engaged in ameliorating difficulties encountered in the production processes. These techniques include precision agriculture, genetic engineering, and smart farming.

2.2.1 Precision Agriculture (PA)

This is defined as the modern agricultural production technique that utilizes digital devices in monitoring and optimizing the production process. It collects, processes, and analyzes data (chronological, spatial, and individual), generates data output, and correlates data with vital information required for deciding on the estimated variability necessary for the improvement of resources (Husti, 2018; ISPA, 2019) (Fig. 1).

Khadka (2023) emphasized the importance of computer education for farmers as the key tool essential in practicing PA, in addition, identified some sine qua non of PA:

1. Global Positioning System (GPS): A satellite-based navigation system that is used for mapping, it provides information on both the exact location and the real-time data of the climatic conditions of any geographical location.

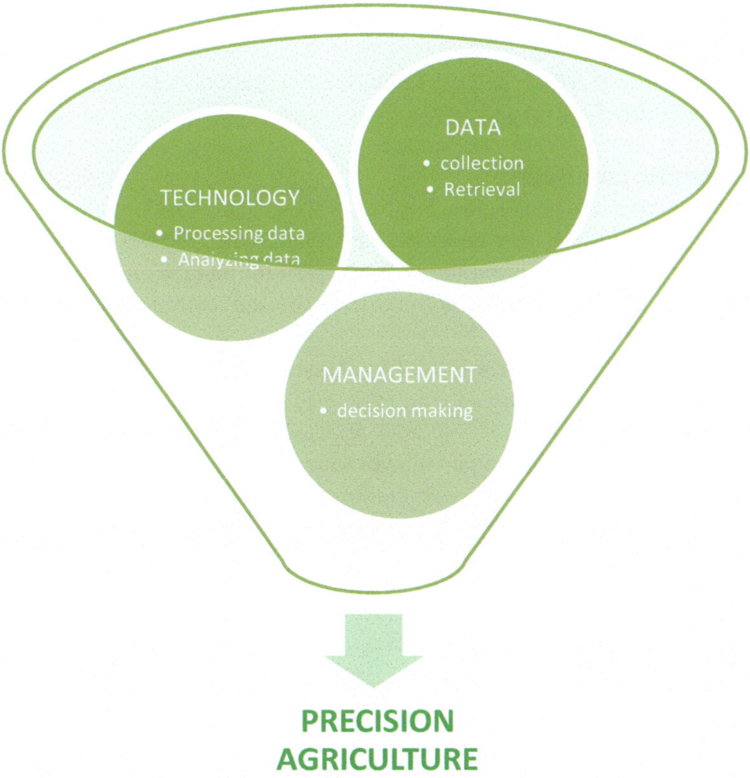

Fig. 1 Features of precision agriculture

2. Geographical Information System (GIS): A database for supplying the geographical information of the farmland, providing data on field topography, surface and subsurface drainages, soil types and testing methods, rates of chemical applications, irrigation systems applicable, and crop yield data.
3. Grid Sampling for Targeted Farming: A sampling method for understanding the spatial variability of the characteristics of soil like pH and nutrient values and the organic matter.
4. Remote Sensor Technology: This includes the implementation of diverse technologies in measuring the environment's climatic conditions such as temperature, humidity, nutrient level, and soil texture and nutrient level, and in managing agricultural processes like drought control, crop identification, stress location, etc.
5. Variable-Rate Technology (VRT): This is an embodiment of farming field equipment with the capability of controlling the input rate of crops precisely.

The effects of PA are the advantages and the disadvantages of employing this optimized technology. These include:

1. Data dependency: The precision of this technique depends on the validity of the data processed. Erroneous data collection from databases will negatively affect the agricultural process.
2. Computer literacy: The ability of the farmers to generate, process, and analyze data is the bedrock of PA success.

2.2.2 Genetic Engineering

Genetic engineering can be described as one of the pillars of modern biotechnology, which allows gene editing, revolutionizes and manipulates the DNA. In agricultural production, it offers innovative solutions necessary for addressing the continuously increasing food demand and offers eco-friendly and extremely improved crop seedlings.

The successful crop modification ability of genetic engineering is a breakthrough for sustainable food production.

- It provides improved nutritional value products for eradicating hidden hunger.
- It increased crop yields thereby forestalling food insecurity.
- It avails the widespread of organic products, thus eliminating the use of chemical pesticides.
- Improvement of the economic state of the country.

2.2.3 Smart Agriculture

Smart agriculture has widely been recognized as smart farming, digital farming, or agriculture 4.0, it is a revolutionized agriculture system that combines budding disrupting digital technologies in agricultural productions. Technologies like artificial intelligence (AI), cyber-physical systems (CPS), Internet of Things (IoT), Wireless Sensor Networks (WSN), Digital Twin (DT), Big Data Analysis (BDA), Cloud

Computing (CC), Autonomous Robotic Systems (ARS), System Integration (SI), Simulation, Augmented Reality (AR), and Additive Manufacturing (AM) are being employed (Aceto et al., 2019; Singh et al., 2022).

Due to the technological complexity of smart agriculture, it is expected to tackle the causal factors of food insecurity menace like climatic/environmental impact, crop variability and losses, seasonality, and human resources shortage caused by rural-urban migration.

2.3 Effect/Importance of Good Production Techniques

Good production techniques for food and feed production could be described as production systems that guarantee zero hunger, increase the gross domestic product, maximize natural resources, and do not jeopardize future generation's provisions. The effect of these good production techniques, in the same vein, as the challenges will be apparent across all sectors, affecting both humans and animals with the creditable climatic conditions. Kharel et al. (2022) noted that Good Agricultural Practices (GAP) are essential in actualizing the "triple wins" (social, economic, and environmental) sustainability.

Krasachat (2023) stated some importance of GAP in ensuring sustainable food production:

1. Certainty of conserving and maintaining essential natural resources via regulations on eco-friendly and societal products.
2. An avenue for enhancing the farmers' production efficiency.
3. High GDP due to increase in export products due to GAP-certified practices of quality standards with no chemical additives either intentional or non-intentional.

2.4 Short- and Long-Term Effects on Humans

Humans, who are the actors of the value chain, are predominantly the beneficiaries of the effects of good agricultural practices at short-term or at long-term offs.

1. Expatriation of chemicals in the form of fertilizers or pesticides guarantees the safety of farmers, processors, and consumers.
2. New sources of income are generated for the producers through the production and utilization of organic manure systems.
3. GAP-certified safety predisposes the farmers to the global world and connects the farmers with foreign traders through product exportation, thus creating new market niche for them (Kharel et al., 2022).

2.5 Technological Effects on Food/Feed Production Techniques

Technological advancement in agricultural production can be best described as agricultural mechanization of production. It has been established that one of the problems of food insecurity in the Global South Countries is the failure to adopt technology in practicing mechanized agriculture. Agricultural mechanization has been identified as a basic element for increasing agricultural productivity, an indispensable tool for enhancing food security, and a weapon for the sustainability of agricultural systems (Mrema et al., 2014; Sims & Kienzle, 2017).

What Is Agricultural Mechanization?
Agricultural mechanization can be described as the application of innovative tools, powered machinery, and technological implementations in the agricultural sector, targeted at increasing the profit margin, reducing human-labor dependency, improving productivity, and maintaining the farmland (Cossar, 2019; Mdoda et al., 2022). Simply defined as replacing ancient agricultural practices with modern technologies that utilize machinery for farming operations in place of human labor, it majorly overcomes the problem of time-consuming associated with human labor caused by fatigue, thus reducing the time frame for production alongside increasing the agricultural products.

FAO (2016) and Mottaleb et al. (2016) emphasized the importance of employing appropriate agricultural mechanization, indicating that there are different stages/types of agricultural mechanization, which can be grouped into:

1. Selective Agricultural Mechanization

Selective agricultural mechanization can be described as the application of mechanized equipment for selective agricultural processes while other processes are still been handled primitively. Most underdeveloped and developing countries like Nigeria and Ghana practice this type of agricultural mechanization. Taiwo and Kumi's (2015) survey of the Ashanti region of Ghana revealed that human labor is still been used extensively for preliminary farm operations like cultivation, planting, spraying, harvesting, and drying while mechanization is only been employed for land clearing and maize shelling. In the same vein, Olatunji et al. (2022) studied the agricultural production in the Lagos part of Nigeria, it was reported that just minute farm operations have been mechanized while the major agricultural processes are handled by human labor with the use of crude tools and antique methodologies.

2. Intensive agricultural mechanization.

Intensive agricultural mechanization could be described as the implementation of mechanical devices coupled with full-blown automation of the entire farming operations from start to finish. Developed countries in the Global South mostly employ intensive agricultural mechanization, and this has been made possible because of their government policies supporting agricultural production.

What Then Is the Best Approach to Agricultural Mechanization?
As depicted in Fig. 2, both groups of agricultural mechanizations are efficient and adaptable; however, decisions on the best approach depend on these factors:

- Land mass acquisitions/land tenure systems
- Topography/Nature of the soil
- Capital/Investment
- Crops/products
- Illiteracy of the farmer

Advantages of agricultural mechanization

It encourages agricultural production of large scales of land

It increases productivity

Disadvantages

Requires high initial investment, high maintenance cost, and technological experts

It increases the rate of unemployment

If not properly managed, it could lead to soil depletion, pollution, and deforestation

It discourages small-scale farming

Future improvements in feed production are expected to intertwine information and technology allowing flow detail to run throughout the production chain, whereby all the actors (farm, feed mill, production plant, retailer, and consumer) are duly connected. This will result in the use of automated devices for weighing and updating the animal's body weight, temperature, and consumption efficiency. Employing this technique will reduce waste, control environmental pollution, provide good

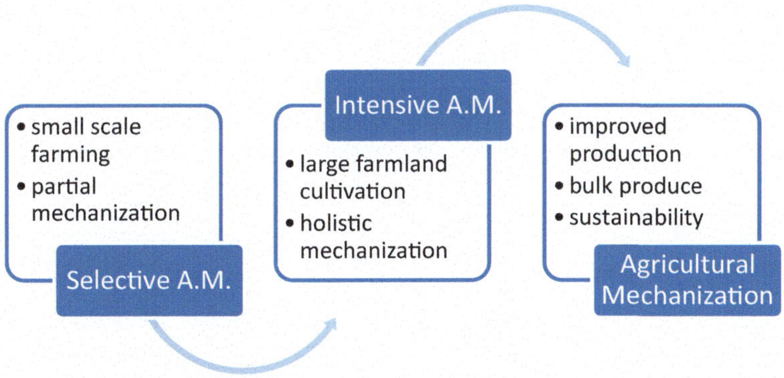

Fig. 2 Features of agricultural mechanization

documentation of the feed and animals, and increase on-farm biosecurity (Coffey et al., 2016).

3 Understanding the Science of Food Production

The connection between science and agricultural production is a direct relationship implying the reliance on one another, although agriculture is a branch of science. The discovery of farming spanning from the Stone Age era till the technological advancement stage has been scientific evolvement. Mourad et al. (2020) recognized and underscored the pertinence of "citizen science" to agriculture; citizen science can be described as a project that requires the active involvement of citizens in the scientific endeavor for generating new knowledge or understanding of the work at hand. This scientific endeavor includes findings and expertise in the fields of agricultural extension programs to collaborative work in defining the aim, stating the scope, and highlighting the methods for scientific research.

The innovative idea of citizen science was to make science accessible to citizens and to equally get contributions from citizens to science. It is a science-society feedback communication expected to discover the lapses in the production sector and involve science in the continuity of agricultural production. Table 2 showcases some of the challenges facing sustainable food production, the effects, and the science-oriented solutions been proffered.

Figure 3 represents the different field programs that constitute citizen science, and the reasonable areas of interacting with the society for sustainable agriculture practices.

Table 2 Impact of science on agricultural challenges

S/N	Challenge	Effects	Science-oriented solution
1	Global warming	Climatic change, droughts, heat waves, floods, erratic precipitation	Renewable energy Energy efficient-use Carbon-neutral society
2	Land and water mass pollution	Product contamination Reduces the soil's nutrient	Organic farming Biotechnology
3	Climatic variability Drought Flood	Scarcity High cost	Irrigational farming Green house Afforestation
4	Soil degradation	Poor yield	Rotational and mixed farming

Fig. 3 Field programs that constitute the citizen science

3.1 Advancements in Food Production in the Global North Countries

United States of America (USA) has been listed as one of the top agricultural producer countries in the world. Dimitri and Effland (2020) pointed out the key favorable factors for this production height attained by the USA, these include:

- Local and regional food systems as important channels for supporting small and medium-sized farms
- Greenhouse production as a production shield during harsh or unfavorable climatic conditions
- Organic food and agriculture production
- Favorable farm and food policies

3.2 Mechanism for the Transfer of Technology Know-How

Technology is the equitable application of knowledge for practical purposes and is divided into two major classifications that are indispensable to the efficient utilization of technology:

1. Material Technology: The purchase, assemblage, and usage of the machines and equipment necessary for mechanization.
2. Knowledge-based Technology: Technical knowledge and basic managerial skills for successful processes (Umar, 2022; Adzenga & Dalap, 2023).

The adaptation of technology for mechanized agriculture is an assured gateway for increased productivity and food sustainability; however, many factors influence the practicability of this mechanization in technology-alien countries. Material technology is supposed to be the least and final stage of mechanization but is the focus and the only concern of most underdeveloped and developed countries. Whereas the greatest concern and the driving force for mechanization must be centered on knowledge-based technology, i.e., "Technology."

Technology know-how could be described as the proper understanding of the tools, devices, and machines for rightful adaptation to the specific field, for subsequent rectification of errors/faults, and manufacturing of machine parts for necessary replacement. This is imperial to mechanization because most of these technological advancements are imported from other countries with absolutely different climatic conditions. A holistic technology transfer will solve the problems attributed to technological advancement like the complexity of the technology and some socio-economic factors (culture, literacy, finance, and age-barrier).

The techniques for knowledge-based transfer of technology include observing the agricultural, knowledge and information system (AKIS) channeled through the subsystems involved in agricultural production vis-a-vis agricultural producers, research and development, education, extension, and support systems (Dirimanova & Radev, 2017).

The agricultural producers should be fully motivated to adopt technological advancement introduced by the research and development division of the particular geographical location. Every actor in the production system must be educated on the importance and effects of technology while the government provides an enabling environment through policies for agricultural land extension and loan schemes to support the farmers.

3.3 Adaptation Schemes for Improved Food/Feed Production in the Global South Countries

The Global South countries attaining to be developed and achieve sustainable food production like the Global North countries are predisposed to adapting the production schemes being practiced in these developed countries. Production scheme adaptation is commensurate to bridging the knowledge gap for improved production. According to Cawley et al. (2023), knowledge is divided into explicit and implicit formats, explicit knowledge is the prevalence format which is typically documented and formed into procedures and rules, whereas implicit or "tacit" knowledge is intransient knowledge accumulated by years of experience but practically been retained by the individual alone.

Farmers' explicit knowledge accrued via research coupled with their perceived optimal approach to production based on their tacit knowledge of their local conditions has been the greatest tool for enhanced food production in the Irish agricultural sector. Owning to the uniqueness of tacit knowledge, it might not be captured with the standard verbal reports but critical in facilitating agricultural innovations, a network-enabling environment for participatory learning is pertinent to transferring this knowledge via apprenticeship format (Cawley et al., 2023).

Nderema et al. (2022) identified 10 modus operandi for tacit knowledge acquisition:

1. Seminars: A peer-oriented meetings organized by researchers to address pertinent topics.
2. Workshops: Interactive opportunities for enhancing knowledge.
3. After Action Reviews: A system designed to review completed projects comprising subject matter professionals, writers, and facilitators.
4. Collaborative Research: A collaboration of researchers, organizations, and institutions on specified objectives.
5. Compilation of Information Materials: Information materials compilation aids in tacit knowledge transfer.
6. Communities of Practice: A network of people with akin agitations interacting regularly and exchanging ideas on best practices in their areas of interest.
7. Job Rotation: Rotational responsibilities create an avenue for team members to share experiences garnered and lessons learned during retrospective meetings after each project.
8. Knowledge Interview: Inquisitive interrogation with an expert in particular agricultural research refreshes their memory and unleashes the mindset to work during their active periods.
9. Mentorship: A mutual benefit method of transferring and increasing knowledge.
10. Storytelling: An educative but relaxed narrative of experiences at work.

Establishing avenues for the continuous acquisition of knowledge by applying the listed modus operandi alongside continuous practices of the production knowledge acquired will help the Global South Countries in sustainable productions.

3.4 Sectoral Involvements in Sustainable Food/Feed Production

In achieving sustainable food production in any country, it is paramount to understand the importance of every actor and the activities of every sector involved. According to Oxford Business Group (OBG, 2022), changes targeted in the agricultural sector are culpable of transforming the entire system for greater productivity while enhancing sustainability. Lessons learned from the pandemic and post-COVID-19 period also emphasized the significance of individual sectors in food crop production and beckoned for massive changes in people's participation (Webb et al., 2020; Witteveen & Velthorst, 2020).

Bolang et al. (2023) studied the participation of individuals, groups, gender, and age involved in agricultural production. The categories involved are:

- The small-holder and large-scale farmers: The small-holder farmers' low productivities are attributed to the inaccessibility of loans for expansion/upgraded services whereas the large-scale farmers have unlimited access to loan facilities.
- Age involvement: Facts pointed to the higher productivity and upturn of the agricultural system by the youth. The youth have been tagged as the major movers of the agricultural system, and this calls for more participation of the youth.
- Literacy: According to Debelo et al. (2020), an individual's ability to the adoption of new and sophisticated technology depends on the higher educational attainment of such personality. Lack of basic education has been identified as a major impede of most Global South Countries farmers.
- Gender: Another identifiable group for agricultural productivity is women; they are mostly the human labour employed for agricultural activities contributing significantly to food security. However, mostly hindered by cultural practices of arable farming systems and marginalized from family/clan farmland inheritance. The involvement of men is without restrictions to land acquisition and loan systems.

4 Outline of All the Sectors Involved in Food Production

The food production unit is a fundamental commission in any country, it is responsible for food security and safety, entwined the public's health and nutrition, and is accountable for the social development of the community/Country. Saddled with all of these knotted roles, food production can be best described and understood by disintegration into the various units/sectors involved. Sadiku et al. (2019) identified food production as a multi-sector involving seven major components, viz., agriculture, food processing, food distribution, regulation, financial services, research and development, and marketing.

4.1 Agriculture Sector

Food production is fixated on all the activities of the agriculture sector comprising land tillage, aquaculture, and animal husbandry. The agriculture sector is the largest employer of labor across the globe, catering for about 40% of the present global population. With about 500 million subsistence farms and farmers, majority of which are still rainfed, providing almost 80% of the food being consumed in most developing world (UN, 2019).

4.2 Food Processing

Food processing is defined as the agglomerate of the unit operations involved in the conversion of agricultural raw material into either intermediate or final products for the consumer. Food processing is primarily aimed at minimizing wastage and preventing redundancy of food products alongside increasing the availability and marketability of food products. It can be paraphrased as the study of food security, food nutrition, and food safety.

Food Security

NABARD (2022) described food security as the progressive evolvement of "freedom from hunger" status to a broad concept accommodating these four indispensable scopes of food availability, food accessibility, utilization, and stability. Food availability refers to the quantity of physical food supplied to a geographical location either from local agricultural production or by importation. According to Parvathamma (2015), food availability is not sufficient to deal with the menace of food insecurity rather that individual has the resources to purchase a sufficient quantity of food. The means of utilizing the purchased food commodity at each household level determines the nutritional accessibility, absorption, and metabolism. Stability refers to the rate of continuity of these first three scopes, i.e., the physical availability of food backed up with the purchasing power and properly cooked using good culinary methods of retaining the nutrients.

A whole food-secure community/country will be free from the effects of food insecurity, which include hunger, malnutrition, and poverty.

Food Nutrition

This is the ability of the consumed meal to functionally maintain good health (physically and psychologically) by utilizing the macro and micronutrients available in the consumed meal. The macronutrients are those required in large amounts in the body, they include carbohydrates, protein, lipids, fiber, and water, while the micronutrients are those required in small quantities like the classes of vitamins and minerals (Singh, 2023).

Food Safety

Food safety can be described as a scientific discipline that certifies that the food product will not harm or hurt the consumers when the food product has been prepared for its intended purpose (Maryam et al., 2017). A food not certified safe is not a food at all, safe food is pertinent to food processing because at every scope of processing (food availability, food accessibility, utilization, and stability), a safe food is required. FAO et al. (2018a, b, c) stated that to warrant food security and nutrition, it is fundamental to keep food safe. It is unequivocally a prerequisite to achieving the sustainable development goals (SDGs).

4.3 Food Distribution

Dos-Santos (2020) noted that agricultural production is not sufficient to end hunger in any community or country, agriculture production trailed with equitable food distribution will end hunger. Agriculture management should be directly involved in the distribution of food, i.e., create models for justifying the cost of logistics activities regardless of the legal requirements, but taking into consideration the food product network from the producer to the consumer in a bid to minimize the number of intermediaries (Arnautović et al., 2020).

An efficient model will reduce the number of middlemen in line from the point of agricultural production to the consumer, for affordable cost price at the consumer's end and a wider profit margin on the producer's side. Food distribution also further emphasizes the first two scopes of food processing, which are food availability and accessibility.

4.4 Regulation

The complexity of food production in conjunction with the chains of handlers involved in the "farm to fork" journey constantly exposes the food to different mishandling technologies, which might lead to substandard products or product adulterations. The fast-growing population of the world has also increased the usage of chemicals in food production for rapid and robust harvests to feed the fast-growing population. All these concerns of widespread unsafe food in the global market will endanger and jeopardize human health if not tamed.

Shafi et al. (2024) iterated the importance of laws and regulations in every country to guarantee the wholesomeness of food products and prohibit the display and sale of unsafe products that could endanger human health. Food safety regulations cover all aspects of food production; they are the set of laws, standards of production, and good manufacturing practices (GMP) enacted to ensure the good product quality and safety of food at all levels of the food production and supply chain. Food regulations comprise different aspects, including

1. Hygiene and sanitation standards for food handlers: These state the proper handling practices to reduce microbial contamination and to prevent cross-contamination of food during handling.
2. Hazard Analysis and Critical Control Points (HACCP): HACCP is a food safety regulation that identifies potential food hazards during processing, implements the necessary control measures at the critical control points, and also monitors and maintains strict adherence to food standards.

3. Codex Alimentarius Commission (CAC): CAC is a consortium of laws from over 99% of Codex members all over the world. It is a compilation of all Codex standards alongside the guidelines and codes of practice, as agreed by the binding countries. It is set to enforce uniformity of product categories across the globe (Adams, 2023).

4.5 Financial Services

Most Global South countries are continuously faced with food insecurity because of their ever-increasing population without commensurate arable land, technology, and financial support for the commercialization of agricultural production. McKinsey Global Institute (2014) reported that Nigeria's agriculture production falls below the global benchmark in other countries, it was candidly stated that the course was lack of access to finance. Furthermore, Osabohien et al. (2020a, b) corroborated the need by government and stakeholders to augment the need for agricultural finance and credit facilities, if agricultural productivity will be increased through diversity.

Agricultural finance covers for investment and infrastructure needs; it is described as part of the rural intervention by stakeholders and governments to finance all agricultural-related activities. It is an inspiring tool that enables the peasant farmers to expand their production land and it equally encourages long-term planning (Miller, 2013).

4.6 Research and Development

Research is an intrusive discovery of methods or ways of solving existing problems or foretelling better alternatives to normal models. Sadiku et al. (2019) stated that research is a reflection of the factors underlining consumers' behavior, customers' buying choices, and their formulation of attitudes and opinions for a deep understanding of their preferred product. Agricultural research has been described as:

1. The best creative invention for increasing food production and reducing the level of poverty in the world.
2. The vital tool for enhancing the input use efficiency regarding soil fertility, pest-resistant cultivars, and water-use efficiency.
3. The recent ICT (Information and Communications Technology) tools contribute radically to the sustainable use of inputs. Tools like GIS (Geographical Information Systems) and remote sensing (Pingali, 2019).

4.7 Marketing

Marketing depicts the cycle of activities involved in the processes of transporting goods from their production units to points of consumption. Agricultural marketing is an activity connecting all the key actors with the processing actions in the transportation of raw materials, goods, and their derivatives. Agricultural marketing serves as the criterion for deciding the production and sale of products, it preempts produce declination with the pre-evaluation questions of what, where, when, and for whom the proposed product should be sold to. It allows a producer-consumer feedback interaction for assessment and improved satisfaction of consumers.

Agricultural marketing is an enormous deal in achieving sustainable food production, it is involved in product launch, product standardization and certification, packaging, and all logistics involved in the transportation of agricultural products across the globe. Due to the interlink of agricultural marketing, considerations should be made for all factors (i.e., natural, biological, and climatic) that affect production, to prevent spoilage of highly perishable products before reaching the consumers (Timur et al., 2020).

4.8 Governmental Policies Consistency

The challenges to food security within a country include land inequality, climate-related issues, deforestation, economic instability, political instability, water scarcity, conflicts, global health crises like the COVID-19 pandemic, unavailability of labor, disrupted food supply chain, resource depletion, biodiversity loss, greenhouse emissions, and resistance to technological advancements (Obahiagbon & Ogwu, 2023a, b, 2024a, b). All of these challenges could be managed by the government exclusively because they are the foremost actor in shaping policies influencing all the dimensions of the aforementioned challenges (Agarwala et al., 2022; Kosoe & Ogwu, 2024; Akudugu & Ogwu, 2024).

Governments are saddled with the responsibility of enacting policies for smooth, efficient, and continuous operations of sustainable food production. Policies typically encompassing range from agricultural subsidies and trade agreements to regulating the environmental practices to ensuring standard compliance of food products. Through strategic policies, the government:

1. Tackles food security challenges at the global, regional, country, and grass-roots levels.
2. Aim for a stable and sufficient food supply for the masses.
3. Provides financial support for farmers.
4. Establishes strategic food reserves for the country.
5. Develop safety nets for crisis-vulnerable communities.
6. Prioritizes environmental health.
7. Encourages long-term resilience.

The government in most Global South countries bases their policy enactment on personal interests leading to discontinuity of policies after each governmental regime with subsequent enactment of new policies at the inception of new regimes. This has caused major limitations in attaining sustainability when compared with the Global North Countries whose policies are continuous regardless of personalities/regimes.

5 Feed Evolvement

The ever-increasing human population has found solace in animals as part of the food requirement, and more animal production requires healthy feed for healthy and fast-growing animals. Animal production hence experienced a radical transition from a free-ranged system to a confined system of practice, and they were thereafter fed with nutritionally balanced diets because of their end-use. This approach began in the eighteenth century and spanned into improved feed production in the nineteenth century when feed industrialization was birthed. This resulted in the dumping of grain mill wastes for better options of grain milling (Coffey et al., 2016).

Feed Production

Feed production improvement to attain increased production and to achieve effective cost units, cutting-edge technology must be implemented. Cutting-edge technology is an innovation that saves both energy and the environment, it is estimated to increase the animal's body weight by 8.1%, and this will be achieved with special feed additives aimed at improving the feed composition without hurting the animal (Cheremushkin et al., 2020).

6 Conclusion

In a bid to actualize sustainable food and feed production, the global south countries need regulatory bodies for the implementation and regulation of various technological innovations in responding to the climate change effect of gas emissions. Over-dependability on the natural climatic seasons should be discouraged with efficient innovative farming. Qualitative research is necessary to understand the causative effects, the extent of damages, and in reducing the level of natural disasters while natural disasters could be managed by introducing intergenerational equity, climate conscientiousness, and a paradigm shift.

While the government and all agricultural stakeholders are advocating for biotechnology and sustainable agricultural intensification because of the initial and running capital involved, families should be encouraged to practice traditional agricultural systems on their rooftops or in their backyards. An enabling environment should be created for obtaining tacit knowledge from the Global North Countries; participation of all sectors of production must be enforced; and regulations must be enacted for checks and balances at all these sectors of production.

References

Aceto, G., Persico, V., & Pescapé, A. (2019). A survey on information and communication technologies for industry 4.0: State-of-the-art, taxonomies, perspectives, and challenges. *IEEE Communications Surveys & Tutorials, 21*(4), 3467–3501.

Adams, R. (2023). Food safety regulations and consumer confidence. *International Journal of Livestock Policy, 2*(1), 15–25.

Adhikary, S., Biswas, B., & Priya, A. (2020). Conservation agriculture: An efficient tool to overcome the drawbacks of conventional agricultural system towards sustainable crop production. *International Journal of Current Microbiology and Applied Sciences, 9*(7), 1333–1340.

Adzenga, J. I., & Dalap, S. L. (2023). Agricultural technology transfer in Nigeria: A review of the challenges and prospects in the contemporary times. *Journal of Agricultural Extension, 28*(1), 21–29.

Agarwala, C., Jemaneh, J., & Kassie, Y. (2022). Government policies and sustainable food systems: Navigating challenges, seizing opportunities, and advancing environmental and social resilience. *Law and Economics, 16*(2), 88–102.

Agoh, E. C., Ogwu, M. C., Chukwuemeka, O. S., & Ekeledo, P. I. (2024). Environmental and human health effects of cassava processing and processing waste. In M. C. Ogwu, S. C. Izah, A. A. Cunha Alves, & babu, S.C. (Eds.), *Plant biology, sustainability and climate change, sustainable cassava* (pp. 203–219). Academic. https://doi.org/10.1016/B978-0-443-21747-0.00001-1

Akudugu, M. A., & Ogwu, M. C. (2024). Sustainable development policies and interventions: A bibliometric analysis of the contributions of the academic community. *Journal of Cleaner Production*, Article # 139919. https://doi.org/10.1016/j.jclepro.2023.139919

Alencar, L., Baldauf, C., Manhães, A. P., & Melo, F. P. (2023). Tradeoffs and synergies between food security and forest cover in Brazilian drylands. *Land Degradation & Development, 34*(17), 5343–5354.

Arnautović, I., Jokić, M., Laban, B., Popović, D., Popović, S., & Nastić, S. (2020). The importance of introducing a direct food distribution manufactured by agricultural production in a single economy. *Poljoprivredna tehnika, 45*(4), 1–7.

Asogwa, J. O., Manasseh, C. O., Abada, F. C., & Okoh, J. O. (2022). Effect of climate variability on crop production: Evidence from selected communities in Rivers state Nigeria. *Journal of Xi'an Shiyou University*. Natural Science Edition. ISSN :1673-06.

Attia, Y., Asiry, K. A., Edris, S., Al-Qurashi, A. D., & Ibrahim, O. H. (2023). Organic agriculture and foods for sustainable food production and safety, an updated review. *Journal of King Abdulaziz University: Meteorology, Environment & Arid Land Agriculture Sciences, 32*(1).

Baudron, F., Tomscha, S. A., Powell, B., Groot, J. C., Gergel, S. E., & Sunderland, T. (2019). Testing the various pathways linking forest cover to dietary diversity in tropical landscapes. *Frontiers in Sustainable Food Systems, 3*, 474255.

Bolang, P. D., Sanyare, F. N., & Gyader, G. N. (2023). Achieving greater participation in agricultural production: A study of government workers in the Upper west region of Ghana. *Heliyon, 9*(11).

Cassman, K. G., & Grassini, P. (2020). A global perspective on sustainable intensification research. *Nature Sustainability, 3*(4), 262–268.

Cawley, A., Heanue, K., Hilliard, R., O'Donoghue, C., & Sheehan, M. (2023). How knowledge transfer impact happens at the farm level: Insights from advisers and farmers in the Irish agricultural sector. *Sustainability, 15*(4), 3226.

Cheremushkin, I. V., Zaveva, O. V., & Apalichina, O. A. (2020, April). Innovation and technological development of the feed industry. In *IOP conference series: Earth and environmental science* (Vol. 459, No. 6, p. 062052). IOP Publishing.

Coffey, D., Dawson, K., Ferket, P., & Connolly, A. (2016). Review of the feed industry from a historical perspective and implications for its future. *Journal of Applied Animal Nutrition, 4*, e3.

Coker, M. A., Agishi, T. V., & Abumbe, G. T. (2015). Science, technology and sustainable food production in Africa: An assessment. *Mediterranean Journal of Social Sciences, 6*(6), 285.

Cossar, F. (2019). *Impact of mechanization on smallholder agricultural production: Evidence from Ghana*.

Das, S., Chatterjee, A., & Pal, T. K. (2020). Organic farming in India: A vision towards a healthy nation. *Food Quality and Safety, 4*(2), 69–76.

Debelo, W. B., Minale, M., Fisha, H., Tedla, A., Eshetu, R., Jafar, M., & Wakgari, G. (2020). Review on roles and challenges of agricultural extension system on growth of agricultural production in Ethiopia. *Journal of Plant Sciences, 8*(6), 189–200.

Dimitri, C., & Effland, A. (2020). From farming to food systems: The evolution of US agricultural production and policy into the 21st century. *Renewable Agriculture and Food Systems, 35*(4), 391–406.

Dirimanova, V. I. O. L. E. T. A., & Radev, T. E. O. D. O. R. (2017). The knowledge transfer in the agricultural sector in South-Central region of Bulgaria. *Bulgarian Journal of Agricultural Science, 23*(3), 505–511.

Dos-Santos, M. J. P. L. (2020). *Equitable food distribution and sustainable development*.

Egboduku, W. O., Egboduku, T., Golohor, O. M., Imarhiagbe, O., & Ogwu, M. C. (2024). Cassava as raw material for sustainable bioeconomy development. In M. C. Ogwu, S. C. Izah, A. A. Cunha Alves, & babu, S.C. (Eds.), *Plant biology, sustainability and climate change, sustainable cassava* (pp. 57–73). Academic. https://doi.org/10.1016/B978-0-443-21747-0.00022-9

FAO. (2018a). *Evaluation of the emergency prevention system (EMPRES) programme in food chaincrises*. FAO.

FAO. (2018b). *Sustainable food systems. Concept and frameworks*. FAO.

FAO (2018c) *The state of world fisheries and aquaculture 2018—meeting the sustainable development goals*. Licence: CC BY-NC-SA 3.0 IGO, Rome.

Fawzy, S., Osman, A. I., Doran, J., & Rooney, D. W. (2020). Strategies for mitigation of climate change: A review. *Environmental Chemistry Letters, 18*, 2069–2094.

Food & Agricultural Organization of the United Nations (FAO), 2016. Sustainable food and agriculture: The role of mechanization in agricultural sustainability. Available on: https://www.fao.org/sustainability/news/detail/en/c/460221/

FSA (Food Security Agency). (2019). *Food security and vulnerability map* [In Bahasa Indonesia] (Jakarta).

Fuhr, H. (2021). The rise of the Global South and the rise in carbon emissions. *Third World Quarterly, 42*(11), 2724–2746.

González-Sánchez, E., Kassam, A., Basch, G., Streit, B., Holgado-Cabrera, A., & Triviño-Tarradas, P. (2016). Conservation agriculture and its contribution to the achievement of agri-environmental and economic challenges in Europe. *AIMS Agriculture and Food, 1*(4), 387–408.

Grauerholz, L., & Owens, N. (2015). Alternative food movements. In *International encyclopedia of the social & behavioral sciences* (1(2)) (pp. 566–572).

Hirich, A., Choukr-Allah, R., & Ragab, R. (2020). *Emerging research in alternative crops* (1st ed.). Springer International Publishing.

Husti, I. (2018). Thoughts and opinions on precision farming. *Mezőgazdasági Technika, 7*, 2–6.

Idumah, F. O., Mangodo, C., Ighodaro, U. B., & Owombo, P. T. (2016). Climate change and food production in Nigeria: Implication for food security in Nigeria. *Journal of Agricultural Science, 8*(2), 74–83.

Islam, F., Saeed, F., Afzaal, M., Ahmad, A., Hussain, M., Khalid, M. A., et al. (2022). Applications of green technologies-based approaches for food safety enhancement: A comprehensive review. *Food Science & Nutrition, 10*(9), 2855–2867.

ISPA. (2019). *ISPA forms official definition of 'precision agriculture'*. Avialable at: https://www.precisionag.com/market-watch/ispa-forms-official-definition-of-precision-agriculture

Jagger, P., Cheek, J. Z., Miller, D., Ryan, C., Shyamsundar, P., & Sills, E. (2022). The role of forests and trees in poverty dynamics. *Forest Policy and Economics, 140*, 102750.

Khadka, D. (2023). Modern technology and scope of precision agriculture (PA). *International Journal of Social Sciences and Management, 10*(4), 83–87.

Kharel, M., Dahal, B. M., & Raut, N. (2022). Good agriculture practices for safe food and sustainable agriculture in Nepal: A review. *Journal of Agriculture and Food Research, 10*, 100447.

Kosoe, E. A., & Ogwu, M. C. (2024). Sustainable cassava processing: Processes, techniques, wastes, and waste streams. In M. C. Ogwu, S. C. Izah, A. A. Cunha Alves, & babu, S.C. (Eds.), *Plant biology, sustainability and climate change, sustainable cassava* (pp. 257–272). Academic. https://doi.org/10.1016/B978-0-443-21747-0.00011-4

Krasachat, W. (2023). The effect of good agricultural practices on the technical efficiency of chili production in Thailand. *Sustainability, 15*(1), 866.

Lipper, L., Thornton, P., Campbell, B. M., Baedeker, T., Braimoh, A., Bwalya, M., et al. (2014). Climate-smart agriculture for food security. *Nature Climate Change, 4*(12), 1068–1072.

Malau, L. R. E., & Darhyati, A. T. (2021, November). The impact of climate change and natural disasters on food security in Indonesia: Lessons learned on preserving forests sustainability. In *IOP conference series: Earth and environmental science* (Vol. 886, No. 1, p. 012090). IOP Publishing.

Maryam, B. M., Datsugwai, M. S. S., & Shehu, I. (2017). The role of biotechnology in food production and processing. *Industrial Engineering, 1*(1), 24–35.

McKinsey Global Institute. (2014). *Nigeria's renewal: Delivering inclusive growth in Africa's largest economy*. McKinsey & Company.

Mdoda, L., Mdletshe, S. T. C., Dyiki, M. C., & Gidi, L. (2022). The impact of agricultural mechanization on smallholder agricultural productivity: Evidence from Mnquma Local Municipality in the Eastern Cape Province. *South African Journal of Agricultural Extension, 50*(1), 76–101.

Miller, C. (2013). Agricultural finance and vulnerability. *United Nations Framework Convention on Climate Change*, no. February 2011, p. 7, 2011.

Mottaleb, K. A., Krupnik, T. J., & Erenstein, O. (2016). Factors associated with small-scale agricultural machinery adoption in Bangladesh: Census findings. *Journal of Rural Studies, 46*, 155–168.

Mourad, K. A., Hosseini, S. H., & Avery, H. (2020). The role of citizen science in sustainable agriculture. *Sustainability, 12*(24), 10375.

Mrema, G., Soni, P., & Rolle, R. S. (2014). A regional strategy for sustainable agricultural mechanization: Sustainable mechanization across agri-food chains in Asia and the Pacific region.

Mukherjee, A., Kapoor, A., & Dutta, S. (2018). Organic food business in India: A survey of companies. *Research in Economics and Management, 3*(2), 72–90.

NABARD. (2022). *Food and nutritional security in India: Charting the way to a robust agri-food system*. https://www.nabard.org/auth/writereaddata/tender/2501230131nrs-35-food-and-nutritional-security-in-india.pdf

Nath, M. K. (2022). Potentialities of permaculture to emerge as an alternative for intensive agriculture—A review. *Indian Journal of Organic Farming, 1*, 16–24.

Nderema, H. W., Njeru, L. K., & Chimoita, E. L. (2022). Effect of use of tacit knowledge transfer techniques on organizational performance of Kenya agricultural and livestock research Organization researchers. *Asian Journal of Agricultural Extension, Economics & Sociology, 40*(10), 492–502.

Nguyen, H. (2018). *Sustainable food systems: Concept and framework*. Rome, Italy.

Nizzy, A. M., & Ogwu, M. C. (2024). Valorization of cassava processing by-products into biofuel for a sustainable environment. In M. C. Ogwu, S. C. Izah, A. A. Cunha Alves, & babu, S.C. (Eds.), *Plant biology, sustainability and climate change, sustainable cassava* (pp. 291–309). Academic. https://doi.org/10.1016/B978-0-443-21747-0.00010-2

Obahiagbon, E. G., & Ogwu, M. C. (2023a). Sustainable supply chain management in the herbal medicine industry. In S. C. Izah, M. C. Ogwu, & M. Akram (Eds.), *Herbal medicine phytochemistry* (Reference series in phytochemistry). Springer. https://doi.org/10.1007/978-3-031-21973-3_64-1

Obahiagbon, E. G., & Ogwu, M. C. (2023b). Consumer perception and demand for sustainable herbal medicine products and market. In S. C. Izah, M. C. Ogwu, & M. Akram (Eds.), *Herbal medicine phytochemistry* (Reference series in phytochemistry). Springer. https://doi.org/10.1007/978-3-031-21973-3_65-1

Obahiagbon, E. G., & Ogwu, M. C. (2024a). Organic food preservatives: The shift towards natural alternatives and sustainability in the global South's markets. In M. C. Ogwu, S. C. Izah, & N. R. Ntuli (Eds.), *Food safety and quality in the global south* (pp. 299–329). Springer. https://doi.org/10.1007/978-981-97-2428-4_10

Obahiagbon, E. G., & Ogwu, M. C. (2024b). The nexus of business, sustainability, and herbal medicine. In S. C. Izah, M. C. Ogwu, & M. Akram (Eds.), *Herbal medicine phytochemistry* (Reference series in phytochemistry). Springer. https://doi.org/10.1007/978-3-031-21973-3_67-1

Ogwu, M. C. (2019). Understanding the composition of food waste: An "-omics" approach to food waste management. In A. P. Gunjal, M. S. Waghmode, N. N. Patil, & P. Bhatt (Eds.), *Global initiatives for waste reduction and cutting food loss* (pp. 212–236). IGI Global. https://doi.org/10.4018/978-1-5225-7706-5.ch011

Ogwu, M. C., Odozi, I. P., Ahonsi, O. C., Uleanya, K. O., & Odozi, E. B. (2024). Organic acid production from cassava. In M. C. Ogwu, S. C. Izah, A. A. Cunha Alves, & babu, S.C. (Eds.), *Plant biology, sustainability and climate change, sustainable cassava* (pp. 395–418). Academic. https://doi.org/10.1016/B978-0-443-21747-0.00009-6

Olatunji, K. O., Ogunkunle, O., Adisa, O. A., & Taiwo, A. (2022). Effect of agricultural mechanization on production and farmers economy in Nigeria: A case study of Lagos state. *International Journal of Engineering Research in Africa, 60*, 211–231.

Osabohien, R., Osuagwu, E., Osabuohien, E., Ekhator-Mobayode, U. E., Matthew, O., & Gershon, O. (2020a). Household access to agricultural credit and agriculturalproduction in Nigeria: A propensity score matching model. *South African Journal of Economic and Management Sciences, 22*(1), a2688.

Osabohien, R., Ufua, D., Moses, C. L., & Osabuohien, E. (2020b). Accountability inagricultural governance and food security in Nigeria. *Brazilian Journal of Food Technology, 23*, e2019054.

Oxford Business Group. (2022). *The Agriculture in Africa 2022 Focus Report*. www.oxfordbusinessgroup.com

Parvathamma, G. L. (2015). An assessment on food security in developing economies-problems and policy initiatives. *IOSR Journal of Economics and Finance, 6*, 01–10.

Pingali, P. (2019). Policies for sustainable food systems. In *Sustainable food and agriculture* (pp. 509–521). Academic.

Prosekov, A. Y., & Ivanova, S. A. (2018). Food security: The challenge of the present. *Geoforum, 91*, 73–77.

Puaschunder, J. M. (2020). Governance & climate justice: global south & developing nations. https://doi.org/10.1007/978-3-319-63281-0

Sadiku, M. N., Musa, S. M., & Ashaolu, T. J. (2019). Food industry: An introduction. *International Journal of Trend in Scientific Research and Development, 3*(4), 128–130.

Sadowski, A., Wojcieszak-Zbierska, M. M., & Zmyślona, J. (2024). Agricultural production in the least developed countries and its impact on emission of greenhouse gases–An energy approach. *Land Use Policy, 136*, 106968.

Salman, M., Long, X., Wang, G., & Zha, D. (2022). Paris climate agreement and global environmental efficiency: New evidence from fuzzy regression discontinuity design. *Energy Policy, 168*, 113128.

Shafi, A., Ahmad, F., Mohammad, Z. H., & Zaidi, S. (2024). Food Laws and Regulations related to food security. In *Microbial biotechnology in the food industry: Advances, challenges, and potential solutions* (pp. 129–151). Springer International Publishing.

Shirazu, A., Doke, D. A., & Yahaya, A. K. (2022). Perceived effects of rainfall and temperature variability on yields of cereal crops in the Mion District of Northern Ghana. *Ghana Journal of Geography, 14*(3), 28–54.

Sims, B., & Kienzle, J. (2017). Sustainable agricultural mechanization for smallholders: What is it and how can we implement it? *Agriculture, 7*(6), 50.

Singh, R. (2023). Balanced diet. In *Advance in food and nutrition*. ABS Books.

Singh, M., Chaure, N. K., & Parihar, S. S. (2021). Organic farming for sustainable agriculture. *Indian Journal of Organic Farming, 1*(1), 1–8.

Singh, G., Kalra, N., Yadav, N., Sharma, A., & Saini, M. (2022). Smart agriculture: a review. *Siberian Journal of Life Sciences and Agriculture, 14*(6), 423–454.

Soni, R., Gupta, R., Agarwal, P., & Mishra, R. (2022). Organic farming: A sustainable agricultural practice. *Vantage: Journal of Thematic Analysis, 3*(1), 21–44.

Sonia, S., Muhammad, S. S., Shahla, A., Asif, N., Muhammad, S., Muhammad, K. M., Aftab, U., Umer, H., & Qaisar, A. (2023). Application of modern techniques in agriculture production: A comprehensive review section. *European Chemical Bulletin, 12*(Special Issue 13), 1722–1731.

Soria-Lopez, A., Garcia-Perez, P., Carpena, M., Garcia-Oliveira, P., Otero, P., Fraga-Corral, M., et al. (2023). Challenges for future food systems: From the green revolution to food supply chains with a special focus on sustainability. *Food Frontiers, 4*(1), 9–20.

Taiwo, A., & Kumi, F. (2015). Status of agricultural mechanization in Ghana: A case study of maize producing farmers in Ejura/Sekyedumase district, Ashanti region. *International Research Journal of Engineering and Technology, 2*(9), 36–43.

Thakur, P. K., Gautam, V. K., Deoli, V., & Agrawal, A. (2022). Traditional agriculture system: A need of hours for modern world. *Acta Scientific Agriculture, 7.1*(2023), 19–21.

Thapa, G., Kumar, A., & Joshi, P. K. (2019). *Agricultural transformation in Nepal.* Springer.

Timur, N., Jursinbay, B., Sarsengaliy, B., Kuatbay, I., & Berdak, M. (2020). The importance of agricultural marketing services in the development of agriculture of the republic of Karakalpakstan. *Journal of Critical Reviews, 7*(10), 2910–2914.

Tirivangasi, H. M. (2018). Regional disaster risk management strategies for food security: Probing Southern African Development Community channels for influencing national policy. *Jamba Journal of Disaster Risk Studies, 10*, 1–7.

Umar, I. S. (2022). *Adoption of agricultural technologies; how far have farmers gone?* (Inaugural lecture series 99). University Seminar and Colloquium Committee, Federal University of Technology, Minna, 56p.

UNFCCC. (2011). United Nations framework convention on climate change working group II: Climate change impacts

United Nations-UN. (2019). *Sustainable development goals.* Retrieved from https://www.un.org/sustainabledevelopment/sustainable-consumption-production.

Vallejo-Gomez, D., Osorio, M., & Hincapie, C. A. (2023). Smart irrigation systems in agriculture: A systematic review. *Agronomy, 13*(2), 342.

Von Braun, J., Afsana, K., Fresco, L. O., Hassan, M. H. A., & Torero, M. (2023). Food system concepts and definitions for science and political action. In *Science and innovations for food systems transformation* (pp. 11–17).

Webb, A., McQuaid, R., & Rand, S. (2020). Employment in the informal economy: Implications of the COVID-19 pandemic. *International Journal of Sociology and Social Policy, 40*(9/10), 1005–1019. https://doi.org/10.1108/IJSSP-08-2020-0371

Witteveen, D., & Velthorst, E. (2020). Economic hardship and mental health complaints during COVID-19. *Proceedings of the National Academy of Sciences, 117*(44), 27277–27284. https://doi.org/10.1073/pnas.2009609117

Yousuf, O., Titikshya, S., & Singh, A. (2018). Organic food production through green technology: An ideal way of sustainable development. *The Pharma Innovation Journal, 7*(6), 160–163.

Zhu, X. G., & Marcelis, L. (2023). Vertical farming for crop production. *Modern Agriculture, 1*(1), 13–15.

China's Research Progress in Advanced Technologies for Biological Agriculture

Xun Wei, Yilin Zhao, Feifei Chen, Liang Chen, and Xiangyuan Wan

Abstract

Biological agriculture, a new model utilizing modern biotechnology to enhance the production efficiency and sustainability of traditional agriculture, is both a strategic high ground in international agricultural competition and a crucial sector for China's bioeconomy. In recent years, modern biotechnological approaches such as genetic engineering, protein engineering, enzyme engineering fermentation engineering, and synthetic biotechnology have made certain progress in developing new varieties of high-yield, quality-efficient, and environmentally friendly fertilizers, healthy and green feed, as well as safe pesticides with little residue. In the future, there are five major challenges and opportunities facing the

X. Wei (✉)
Research Institute of Biology and Agriculture, Shunde Innovation School, University of Science and Technology, Beijing, Beijing, China

Zhongzhi International Institute of Agricultural Biosciences, Beijing, China

Environmental Economics and Natural Resources Group, Wageningen University & Research, Wageningen, the Netherlands
e-mail: weixun@ustb.edu.cn

Y. Zhao · X. Wan (✉)
Research Institute of Biology and Agriculture, Shunde Innovation School, University of Science and Technology Beijing, Beijing, China

Zhongzhi International Institute of Agricultural Biosciences, Beijing, China
e-mail: wanxiangyuan@ustb.edu.cn

F. Chen
Research Institute of Biology and Agriculture, Shunde Innovation School, University of Science and Technology Beijing, Beijing, China

L. Chen
Institute of Animal Sciences, Chinese Academy of Agricultural Sciences, Beijing, China

M. C. Ogwu et al. (eds.), *Sustainable Bioeconomy Development in the Global South*, https://doi.org/10.1007/978-981-96-0305-3_13

development of biological agriculture, namely, constant expansion of the field of agricultural biology, the increasing importance of common technologies, platforms, and applications across plants and animals, the increasing importance of Big Data and AI, and increasingly intense international competition and biosecurity issues.

Keywords

Agricultural biotechnology · Biotechnological breeding · Biofertilizer · Biological feed · Biopesticide · Biosafety

Abbreviations

AC21	21st Century Agriculture
AI	Artificial intelligence
DNNGP	Deep learning methods using plant omics data for whole-genome prediction
gcHap	Gene function haplotypes
GPT	Generative pre-trained transformers
NLP	Natural language processing
QTN	Quantitative trait nucleotide

1 Introduction

In the face of current challenges such as food security and climate change, all countries are looking for technical and economic models to support sustainable development, and biotechnology and bioeconomy are feasible solutions (Wei et al., 2021). Since the reform and opening up, China has achieved historic progress in the field of agriculture and rural areas, including the advancement of agricultural modernization, improvement of farmers' living standards, and continuous enhancement of urban-rural integration development. Currently, there is a new trend in economic and social development shifting from "solving food and clothing needs" to "promoting dietary diversity." Issues such as low agricultural production efficiency and adverse impacts on resources and the environment urgently need to be addressed. Modern biotechnologies such as genetic engineering, cellular engineering, fermentation engineering, and enzyme engineering are increasingly mature, enabling the cultivation of new agricultural varieties with superior comprehensive traits, the development of green and efficient agricultural inputs, the integration of agricultural production, life, and ecological functions, and the realization of sustainable agricultural development. With the deep integration of modern biotechnology and traditional agriculture, the industrialization level of agricultural biotechnology is continuously improving, gradually forming a new model of empowering traditional agriculture with biotechnology called "biological agriculture"(Li et al., 2006) (Fig. 1).

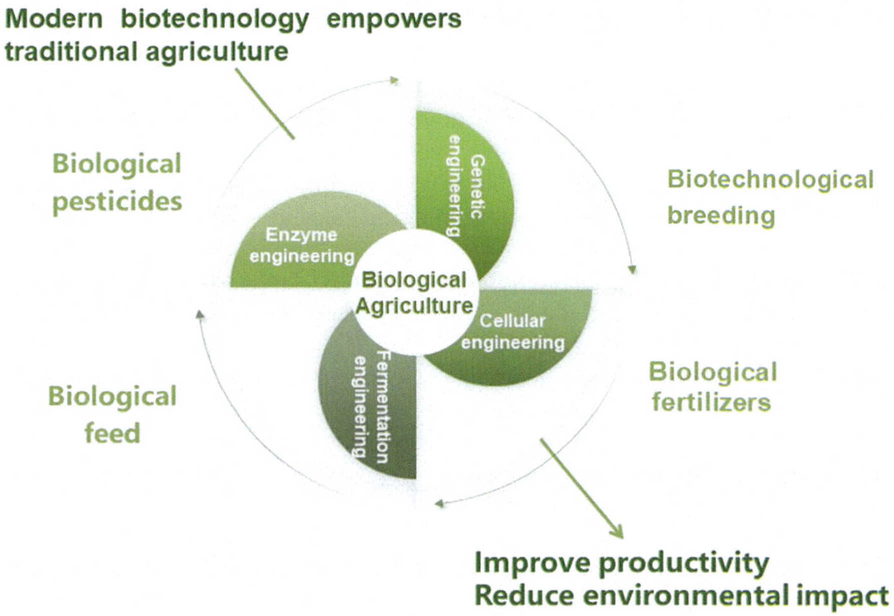

Fig. 1 Connotation and extension of biological agriculture

From a global perspective, governments of various countries have prioritized agricultural biotechnology, issuing policies and regulations while increasing innovation investment, to seize the commanding heights of international agricultural competition in the field of biological agriculture (Nwanze et al., 2004). In recent years, biological agriculture has become a new economic growth point in developed countries and regions such as Europe and the United States, with significant benefits in improving agricultural industry efficiency and benefits, protecting resources and the environment, and mitigating the impact of climate change.

On May 10, 2022, the National Development and Reform Commission issued the "14th Five-Year Plan for the Development of the Bio-Economy" (Chen, n.d.), clearly identifying biological agriculture as one of the four key priority areas for development. Among them, biotechnological breeding, biological fertilizers, biological feed, and biological pesticides are the main focus directions for the current and future periods, aiming to better guarantee the production supply of important agricultural products such as grains and to adapt to increasingly diversified food consumption demands.

This chapter reviews the research progress and application status of cutting-edge technologies in biological agriculture in four aspects: the creation of new agricultural varieties, the development of biological fertilizers, the development of feed, and the research and development of pesticides. It discusses the major challenges facing the development of biological agriculture and proposes corresponding strategies and recommendations.

2 Biotechnological Breeding

Biotechnological breeding primarily refers to the use of genetics, cell biology, and modern biotechnology to cultivate new biological varieties. In a broad sense, biotechnological breeding includes various breeding techniques such as hybridization, molecular markers, whole-genome selection, gene editing, and transgenesis, applicable to breeding work across different biological categories such as plants, microorganisms, and animals. Under the biotechnology regulatory system in China, biotechnological breeding mainly refers to the activity of creating new varieties through biotechnological techniques such as transgenesis, gene editing, RNA interference, and synthetic biology.

Currently, countries worldwide are intensifying efforts to research and develop genetic diversity, aiming to develop new biotechnological breeding technologies. With the continuous development of breeding technologies, China has cultivated over seventy thousand new varieties, undergoing six to seven rounds of large-scale variety turnover. This has achieved a coverage rate of over 96% for excellent varieties, with improved varieties contributing to over 40% of crop yields. This has effectively supported the steady increase in food production and quality, making significant contributions to national food security. Biotechnological breeding is the "core" technology of modern agriculture and the core technology for addressing bottleneck issues in China's seed industry.

2.1 Gene Cloning and Functional Analysis with Significant Breeding Application Value

In recent years, significant breakthroughs have been made in gene cloning and functional analysis with significant breeding application potential. This is crucial for enhancing crop yield, resistance, quality, and provides new strategies and approaches for addressing global food security and sustainable agricultural development (Zhang et al., 2022a).

Regarding rice, significant progress has been made in research on high-temperature resistance. Major genes such as high-temperature sensor TT3.1, chloroplast protein TT3.2, and high-temperature resistant gene *TT2* have been discovered, revealing the role of calcium signaling and wax metabolism in high-temperature resistance. This provides a theoretical basis and breeding resources for rice to cope with climate warming. Genes such as *GRF4*, *DELLA*, and *NGR5* have been discovered, revealing their synergistic regulatory mechanisms in rice growth and nitrogen metabolism (Li et al., 2018; Wu et al., 2020). This is significant for improving nitrogen fertilizer utilization efficiency and maintaining favorable traits. In wheat, researchers have successfully cloned major resistance genes *Fhb7* from *Thinopyrum elongatum* and susceptible gene *TaPsIPK1* leading to stripe rust infection, significantly enhancing wheat resistance to diseases and providing new ideas for disease-resistant breeding (Wang et al., 2020a; Wang et al., 2022). In maize, it has been found that the *KRN2* gene in maize and the *OsKRN2* gene in rice are subject to convergent selection, regulating the yield of maize and rice (Chen et al., 2022). Studies on the maize gene *Thp9-T* have shown that it can increase the protein content and nitrogen utilization efficiency of maize seeds, providing an important foundation for increasing crop yield and improving breeding efficiency (Huang et al., 2022). In soybeans, the overexpression of three key genes *AtVDE*, *AtPsbS*, and *AtZEP* has improved photosynthetic efficiency and yield (De Souza et al., 2022). Two years of field trials have shown an average increase in soybean yield by 24.5%, with a maximum increase of 33%. In cotton, significant breakthroughs have been made in fiber quality improvement. Using advanced sequencing technology has improved the accuracy and completeness of the genomes of *Gossypium herbaceum*, *Gossypium arboreum*, and *Gossypium hirsutum*, identifying key genes such as *KCS6*, which significantly improves the length of upland cotton fibers, providing new targets for cotton breeding, and is expected to improve fiber quality and yield (Huang et al., 2020).

2.2 Elucidation of Mechanisms for Fine Regulation of Complex Traits

Building on the cloning and analysis of important genes, research into the mechanisms for fine regulation of complex traits has progressed steadily (Gao et al., 2021). The fine regulation of the immune response mechanism in rice by the *ROD1* gene has been studied. ROD1 is a calcium ion sensor that initiates immune response suppression by recognizing calcium ion signals. Its activity is regulated by the E3 ubiquitin ligases RIP1 and APIP6. ROD1 is significant for rice growth and development, especially for the development of panicles. The rice blast fungus simulates the function of ROD1 by secreting the effector protein AvrPiz-t, manipulating the immune suppression system of rice. This study not only reveals the regulatory mechanism of key immune suppression signaling pathways in rice but also provides new insights into breeding more stable and disease-resistant rice varieties (Gao et al., 2021). The hormone biosynthesis and signaling mechanisms during the elongation of rice stems have been elucidated (Nagai et al., 2020). It has been found that two key factors, ACE1 and DEC1, interact with gibberellin to antagonistically regulate internode growth in rice. This finding is of great significance for understanding the regulatory mechanisms of stem elongation in rice and even grass plants, providing a new perspective on the genetic regulation of rice height.

2.3 Conservation, Development, and Utilization of Germplasm Resources

The identification of germplasm characteristics is essential for identifying potential sources of important traits, providing crucial information and methods for improving the genetic potential of crops. By analyzing 1844 core germplasm resources of foxtail millet, the first genetic variation map of Setaria was created, and the first high-quality genomes of pseudo-cereals and C4 crops were constructed (He et al., 2023). Population structure analysis of 560 soybean cultivars identified 13 maturity groups (Li et al., 2022). In cotton, through in-depth research on upland cotton germplasm resources, the potential of homologous genes in fiber quality improvement has been revealed, and the concept of high-quality cotton design breeding based on modularization of the genome has been proposed. Sorghum has been an important raw material for brewing since ancient times (You et al., 2023). By resequencing 244 Chinese sorghum cultivars, three genetic populations, the Northern, the Southern, and Chishui, have been identified (Zhang et al., 2022b), and an important selective clearance region containing homologous genes related to grain size, pericarp thickness, and inflorescence structure has been determined.

2.4 Optimization and Application of Advanced Biotechnological Breeding Technologies

In recent years, significant achievements have been made in the application of gene editing technology in crop improvement, mainly due to the continuous optimization of advanced gene editing tools such as CRISPR/Cas. The use of these technologies has greatly improved crop disease resistance and stress tolerance. Creating a new gene, *RBL1Δ12*, by cloning a broad-spectrum disease-resistant gene *RBL1*, significantly enhances rice resistance to multiple diseases (Sha et al., 2023).Yu et al. proposed a new strategy for rapid domestication of alien tetraploid wild rice, aiming to cultivate new polyploid rice crops to increase grain yield and environmental adaptability. Materials with improved traits, such as reduced shattering, shortened awns, and reduced plant height, have been obtained through gene editing based on the assembly of the reference genome of alien tetraploid rice (Yu et al., 2021). Knocking out the *AT1* gene using gene editing technology can improve the survival rate of wheat in saline-alkali soil (Zhang et al., 2023).

Whole-genome selection is a method of breeding selection using whole-genome data to predict crop phenotype characteristics and accelerate breeding processes by evaluating genetic markers across the entire genome. In recent years, related research has comprehensively revealed the natural variation characteristics of gene function haplotypes (gcHap) in Asian cultivated rice for the first time and explored the application of whole-genome selection (Zhang et al., 2021a). Furthermore, deep learning methods using plant omics data for whole-genome prediction (DNNGP) can efficiently integrate crop big data and apply it in breeding (Wang et al., 2023a, b). These methods not only accelerate the progress of crop genetic research but also improve the accuracy of crop phenotype prediction, providing effective tools for intelligent breeding.

Synthetic biology mainly focuses on creating new biosynthetic pathways and functions. By designing and constructing new gene networks, new traits can be introduced or existing traits improved in crops. For example, a new high-efficiency rice variety created using synthetic biology technology introduced four enzymes into rice chloroplasts, creating a CO_2 concentration mechanism similar to that of C4 plants, significantly improving rice photosynthetic efficiency, biomass, and yield (Wang et al., 2020b).

Artificial intelligence (AI) is gradually being applied in crop breeding. AI can analyze large amounts of genetic and environmental data, helping to predict crop traits, optimize breeding strategies, and manage farmland. Based on the newly created comprehensive rice quantitative trait nucleotide (QTN) map and inferred QTN effects, researchers have developed a genome navigation system called RiceNavi for QTN aggregation and optimization of breeding routes. It has been successfully applied to the genetic improvement of high-quality indica rice varieties (Wei et al., 2021). CropGPT is an intelligent breeding method that combines GPT (generative pre-trained transformers) and NLP (natural language processing) technologies. This method relies on interdisciplinary collaboration, including breeders, biologists, mathematicians, and computer scientists, involving the selection of superior germplasm resources, the analysis of genotypes and phenotypes using high-throughput technologies, and the optimization of breeding recommendations through big data models. Its aim is to enhance breeding prediction accuracy through iterative optimization and data integration, thus strengthening intelligent breeding capabilities (Zhu et al., 2023).

In conclusion, the engineering application of biotechnological breeding technology provides more possibilities for crop improvement, which helps meet the growing global food demand, increases agricultural sustainability, reduces environmental impact, and has profound implications for food security and global agriculture.

3 Biological Fertilizers

Biological fertilizers are microbial agents made primarily of active microorganisms (Okur, 2018). They are a type of auxiliary fertilizer that has specific fertilizing effects on crops and is environmentally friendly. Against the backdrop of global climate change, food crises, and depletion of natural resources, biological fertilizers have been widely used in agricultural production due to their advantages such as being green and efficient.

In the past two decades, research on biological fertilizers has mainly focused on microbial strains, microbial–plant interactions, and abroad, there is more attention to the mechanism of microbial fertilizer action and related studies on sustainable agricultural development. In recent years, China has focused more on research related to crop yield and the application of microbial fertilizers. As of August 2023, a total of 10,175 microbial fertilizer products have been registered in China (Table 1).

Table 1 China's microbial fertilizer registered products as of August 2023

Microbial fertilizer	Number
Microbial inoculants	2239
Bio-organic fertilizer	3052
Other fertilizers	4884

3.1 Large-Scale Exploration and Evaluation of Core Microbial Resources

The large-scale exploration and evaluation of core microbial resources are the core content of biological fertilizer production (Rai et al., 2023). This process involves isolating beneficial microorganisms from soil or other biological resources, identifying and evaluating their functions, ecological characteristics, and their impact on plant health. With the continuous development of related research, China has discovered an increasing number of microbial resources, and the variety of biological fertilizers has also increased. Currently, the Ministry of Agriculture and Rural Affairs of China has registered nine types of microbial fertilizer products: silicate fertilizers, organic matter decomposition agents, phosphate-solubilizing agents, mycorrhizal agents, soil remediation agents, photosynthetic bacteria agents, microbial agents, nitrogen-fixing agents, and rhizobia agents. According to statistics from the Ministry of Agriculture and Rural Affairs, the cumulative number of registered microbial fertilizer products in China has increased from 149 in 2007 to over 9990 in 2022. The effective microbial strains in related products have transitioned from single-effect to compound multi-effect types. In addition, China's microbial fertilizer production has reached 30 million tons per year, with an annual total output value of approximately 40 billion yuan. Microbial fertilizer agents are applied in all provinces of China, with the most applications in the central region, followed by the northern and northwestern regions, and fewer applications in the northeastern and southern regions. Microbial fertilizers have been applied in the production processes of more than 30 major crops, with an accumulated application area exceeding 500 million mu (about 33.3 million hectares). Among them, the application volume in cereal crops is the highest, followed by fiber and oil crops.

China's efficient microbial strain resources are relatively scarce, and there is an urgent need to strengthen the exploration of microbial strain resources. Currently, there is relatively little systematic research on microbial strain resources. Although the number of strains continues to increase, there are also issues of uneven quality, especially the lack of certain specific functions of efficient microbial strain resources. Examples include efficient phosphate-solubilizing microbial strains that dissolve difficult-to-dissolve phosphorus in soil and improve phosphorus fertilizer utilization efficiency, nitrogen-fixing microbial strains that are efficient in nitrogen fixation and resistant to nitrogen stress, and potassium-dissolving microbial strains that efficiently activate difficult-to-dissolve potassium in soil and improve potassium fertilizer utilization efficiency.

3.2 Development and Application of Novel Cultivation Technologies

With the gradual maturity of novel cultivation technologies such as in-situ cultivation, cell sorting, single-cell sequencing, and microfluidic cultivation, various combined "non-targeted" and "targeted" isolation technologies and methods for "uncultured/

difficult-to-culture microorganisms"have created technical conditions for exploring new agricultural microbial resources and provided a foundation for obtaining excellent microbial strains. In-situ cultivation technology can simulate the natural growth conditions of microorganisms, directly cultivate microorganisms in their original environment, and help capture microbial strains that are difficult to obtain under traditional cultivation conditions. Microfluidic cultivation technology can control the growth environment of microorganisms using microchannels and microstructures, precisely control and observe the growth of individual microbial cells, and then accurately screen, isolate, and cultivate target microorganisms. Cell sorting technology separates and isolates single cells from microbial populations, conducts personalized research on specific cell types or strains, and then identifies and classifies important microorganisms with potential application value in biological fertilizer production. Single cell sequencing technology can analyze the genomic information and function of individual microbial cells, reveal the genetic diversity and metabolic potential of different cells in microbial populations, and help determine specific strains or metabolic pathways that are useful in biological fertilizer production. In addition, nanotechnology applied in biological fertilizer production reduces nutrient loss and improves nutrient utilization efficiency (Rai et al., 2023). The development and application of these novel cultivation technologies help discover, cultivate, and utilize potential beneficial microbial strains in biological fertilizer production, providing a more sustainable and efficient path for the innovation and optimization of biological fertilizers.

3.3 Analysis of the Impact Mechanism of Microbial Physiological States

Microorganisms are the core of biological fertilizer efficacy, and analyzing the impact mechanism of their physiological states is of great significance for creating new multifunctional biological fertilizers and realizing the efficacy and potential of biological fertilizers. Biological fertilizers can directly provide nutrients for crops or regulate crop growth through the combination of microorganisms and plants, producing enzymes, crop hormones, and other substances (Kumar et al., 2022). For example, the metabolism of beneficial microorganisms in biological fertilizers can continuously provide nutrients needed for crop growth. The reproduction of microorganisms in biological fertilizers can promote the conversion of organic residues and humus in soil, thereby effectively improving soil structure (Zhang et al., 2021b). In addition, microbial metabolism can induce the photosynthetic capacity and intercellular substance transport capacity of plants, thereby promoting crop growth. Analyzing the impact mechanism of microbial physiological states and clarifying the role of microorganisms and their application conditions are the basis for producing multifunctional biological fertilizers. For example, Syngenta Group relies on national-level scientific research platforms, uses microbial agents as carriers, and connects the interactive effects of "biological fertilizers-soil health-efficient nutrient" to develop five series of products: agricultural microbial agents, bio-complex fertilizers, bio-organic-inorganic fertilizers, bio-organic fertilizers, and microbial soil improvement products.

4 Biological Feed

China is a major producer of livestock and poultry and feed in the world, with feed production reaching 321 million tons in 2023 (Tian et al., 2022) The development of new, ecological, and environmentally friendly feeds has always been a hot topic for sustainable development in animal husbandry. Biological feed research began in the 1980s. Currently, biological feed has shown a rapid and efficient trend in both research and practical application. Since 2010, the number of related patents has increased rapidly worldwide, entering a period of research explosion (Fig. 2). Biological feed mainly includes feed enzyme preparations, feed amino acids and vitamins, prebiotics (direct feeding of microorganisms), oligosaccharides for feed, natural plant extracts, bioactive oligopeptides, etc. (Ma et al., 2023). In recent years, modern biotechnologies such as synthetic biology, genetic engineering, fermentation engineering, and enzyme engineering have been widely applied in feed development. Biological feeds not only have the characteristics of being green and safe but also have higher nutritional value, which is conducive to economically efficient and high-yield farming, with broad development prospects. Additionally, biological feeds will play an important role in addressing issues such as food safety, feed resource shortages, and environmental pollution. Microbial preparations, enzyme preparations, feed amino acids, and fermented biological feeds have been widely recognized by academia, industry, and farmers for promoting nutrient absorption and conversion, improving feed utilization, palatability, substituting for some antibiotics, and expanding protein feed sources.

4.1 Mechanisms Related to Biological Feed

Biological feed technology mainly involves studying the effects on reducing antinutritional factors, improving utilization of nutrients and energy, gastrointestinal

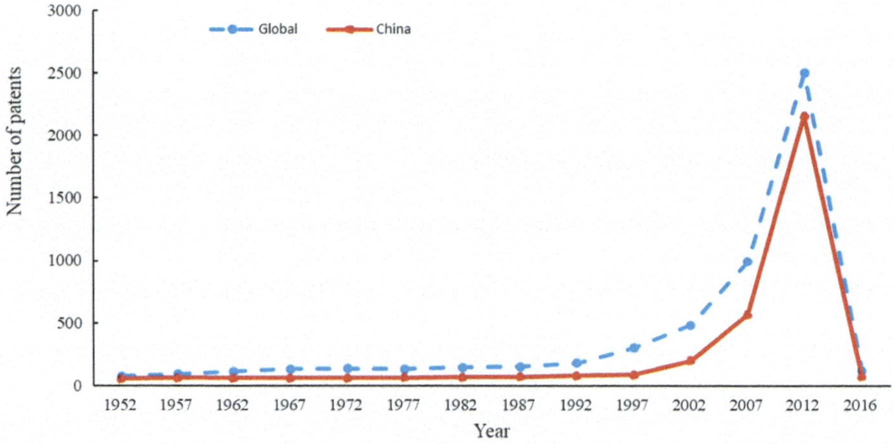

Fig. 2 Global and Chinese trends in biological feed patent applications

microecological environment of livestock and poultry and the production environment to develop biological feed additives that can regulate the structure of the intestinal flora. By introducing beneficial microbial populations into biological feeds, such as probiotics and prebiotics, the intestinal microbial composition can be optimized, enhancing the ability to resist harmful microorganisms and thus improving the health of livestock and poultry. Previous studies mostly focused on exploring the beneficial mechanisms of different strains (Ma et al., 2023). Lactic acid bacteria, for example, are the main probiotics used in silage preservation, stimulating the secretion of digestive enzymes, organic acids, vitamins, and other bioactive substances in the gastrointestinal tract of livestock and poultry (Missotten et al., 2015). Lactic acid bacteria can effectively lower intestinal pH by producing large amounts of lactic acid and acetic acid, promoting the growth of intestinal villi, increasing the absorption area of the small and large intestines, and strengthening the intestinal barrier through competitive inhibition and the production of adhesive substances (Van Zyl et al., 2020).

Biological feed can produce a wide range of unsaturated fatty acids and aromatic compounds that impart unique flavors, enhance palatability, and increase feed intake for animals. Additionally, biological feed generates proteases, amylases, cellulases, phytases, water-soluble vitamins, amino acids, and various growth-promoting factors upon metabolism within the digestive system of animals, to improve growth performance. Moreover, biological feed may establish a dense microbial barrier in the intestines of animals, which effectively prevents the absorption of harmful substances and waste.

The active ingredients in biological feeds can also enhance the immune function of animals. Research has shown that probiotics participate in immune regulation by promoting the differentiation of immature T cells, enhancing the host's immune function (Ren et al., 2019). After feed fermentation, beneficial microorganisms in the feed can effectively increase, reducing the colonization of intestinal bacteria (such as coliforms and *Salmonella*), and improving the microbial ecosystem in the intestines of livestock and poultry (Sugiharto & Ranjitkar, 2019). As a non-specific immunomodulator, probiotics and metabolically active substances in fermented feed can stimulate the development of the immune organs of the intestines of livestock and poultry, activate intestinal immune mechanisms, enhance the cellular and humoral immune capabilities of animal bodies, and reduce the occurrence of diseases (Sugiharto, 2016; Sugiharto et al., 2017).

Through in-depth research on the impact mechanisms of biological feeds on immune cells, the development of biological feed formulations that enhance the immune system of livestock and poultry can increase their resistance to diseases. Enzymes produced by biological feeds, such as proteases, amylases, cellulases, phytases, water-soluble vitamins, amino acids, and various growth-promoting factors, can promote the growth and development of livestock and poultry to optimal levels. Additionally, biological fermentation feeds can establish dense microbial barriers in the intestines of livestock and poultry, effectively preventing the absorption of harmful substances and waste (Jin et al., 2017; Yan et al., 2019). For example, two-stage fermentation of corn-soybean mixed feed using *Bacillus subtilis* and

Bacillus coagulans can improve the digestibility of feed for pigs (Shi et al., 2017), while high-concentration fermented feeds can prevent the proliferation of pathogens (such as Enterobacteriaceae) along the gastrointestinal tract of piglets (Lyberg et al., 2006).

4.2 Innovation in Enzyme Preparations and Active Peptide Technology

Technological innovations in biological feeds mainly include enzyme preparations and active peptides (Morlighem & Radis-Baptista, 2019) By studying the mechanisms of action of different types of enzymes in the digestive tracts of livestock and poultry, efficient enzyme preparations can be developed to promote the degradation of indigestible components in feed. Developed countries have basically achieved genetic engineering technology for major feed amino acids and enzyme preparations. China has also formed a research and development group for biocatalysis and microbial transformation. According to statistics from the China Biotechnology Industry Association, China's enzyme preparation production reached 229,000 tons in 2022, with the output value of 3.5 billion yuan. The application of enzyme preparations is crucial in the field of biological feeds. Proteases, amylases, β-glucanases, cellulases, and pectinases in complex enzyme preparations can supplement the deficiency of endogenous enzymes in animal bodies and have effects such as promoting growth, aiding digestion, and enhancing immunity. Complex enzyme preparations can also increase the digestibility of animal proteins, reduce the excretion of nitrogen elements in the form of ammonia, and reduce the heavy metal content in the feces of beef cattle, thereby reducing related environmental pollution and the carbon footprint of the industry (Yi et al., 2020).

Research on bioactive peptides focuses on exploring the mechanisms by which active peptides affect the growth, immunity, and disease resistance of livestock and poultry. Adding antimicrobial peptides to diets can improve the growth performance, nutrient retention, intestinal morphology, and immunity of piglets, while also reducing harmful microorganisms in animals (Wu et al., 2012). For example, peptides derived from fungal mycelium, such as NZ2114 and MP1102, can effectively kill *Streptococcus suis* (Zhao et al., 2019). This technological innovation is expected to provide new ideas for the formulation of biological feeds, thereby improving the sustainability and economic benefits of livestock and poultry production.

Modern biotechnologies such as synthetic biology and metabolic engineering, with enzyme engineering technology as the core, applied in the field of biological feeds, can increase the digestibility and utilization of feed grains, improve palatability, reduce resistance, and remove toxic and harmful substances. For example, a method was developed for enhancing the catalytic efficiency of glycoside hydrolase based on deep neural networks and molecular evolution (Liu et al., 2023). At the same time, biological enzymes can efficiently convert and utilize industrial and agricultural by-products and waste, turning waste into treasure to improve resource utilization efficiency and reduce environmental pollution and carbon emissions (Kholif et al., 2017). The application of biological feeds is expected to fill more than 50% of the feed gap in China, effectively alleviating issues such as competition for food between humans and animals and the safety of animal products in the long run.

5 Biopesticides

In recent years, with the increasing attention to green pest control in agriculture and the growing concern for food and environmental safety, biopesticides have emerged as a new product for the sustainable development of agriculture and the promotion of ecological harmony, owing to their advantages of high efficiency, low toxicity, safety, and environmental friendliness. Currently, biopesticides mainly include biochemical pesticides, botanical pesticides, microbial pesticides, and the expansion and utilization of natural enemies. The development and expansion of biopesticides are of great significance to China's food security, food safety, environmental protection, the construction of ecological civilization, and the promotion of a healthy China.

5.1 Analysis of Mechanisms and Product Innovation of Biochemical Pesticides

Biochemical pesticides refer to a category of pesticides that interfere with or affect the physiological processes of pests, weeds, or pathogens through biochemical means to achieve control purposes. Currently, the biochemical pesticides applied in China include chemical signaling substances, insect growth regulators, insect antagonists, plant growth regulators, and other biochemical pesticides. These biochemical pesticides can be of natural origin or chemically synthesized but do not have a direct lethal effect on target organisms. Insect pheromone biological control technology utilizes insect pheromone substances to interfere with and affect the communication between individual insects, aiming to trap or disrupt the mating of pests, including sex pheromones, aggregation pheromones, alarm pheromones, and insect source attractants. Insect growth regulators mainly simulate or disrupt the endocrine system of insects, thereby interfering with their growth and development processes, cause physiological dysfunction, and ultimately lead to death or abnormal reproduction behavior. Insect antagonists, such as organophosphorus antagonists and neonicotinoids, block insect neural transmission, causing paralysis and ultimately death. Plant growth regulators are small molecule substances like plant hormones. Upon entering the plant, they exhibit chemical properties like plant hormones, facilitating growth and development regulation, plant type adjustment, root and stem enlargement, induction and control of plant sex differentiation, improvement of stress resistance, yield increase, and quality improvement. Biochemical pesticides include gibberellins, cytokinins, naphthylacetic acid, and amino oligosaccharides. Notably, biochemical pesticides distinguish themselves from traditional chemically synthesized organic pesticides by regulating or interfering with the growth, reproduction, and behavior of plants or pests rather than through toxic mechanisms. Additionally, biochemical pesticides exhibit robust selectivity and pose minimal risks to humans, animals, crops, and the environment.

5.2 Extraction and Utilization of Active Ingredients in Botanical Pesticides

Botanical pesticides are secondary metabolites synthesized by plants through different metabolic pathways. These compounds have diverse chemical structures and unique mechanisms of action, which can directly kill, inhibit the growth of, or remove pests, pathogens, or weeds. Compared to traditional chemical pesticides, botanical pesticides are new, efficient, residue-free, and pollution-free "green pesticides" (Lengai et al., 2020). Key characteristics include their natural composition, easy decomposition after application, minimal environmental impact, diverse main components, low propensity for resistance development, safety for beneficial organisms, suitability for large-scale cultivation, and economic development costs. Over 400,000 secondary metabolites have been identified from plants. Currently, lignans, alkaloids, flavonoids, terpenes, and other secondary metabolites have been found to have physiological activities as botanical pesticides. China has registered a total of 24 active ingredients for botanical pesticides, amounting to 227 registrations, including matrine, sophoridine, pyrethrin, matrine, berberine, brucine, matrine, eugenol, thymol, catechin, etc. Matrine has the largest number of registered products, reaching 115 (as insecticides and fungicides), followed by berberine (27), rotenone (23), and meliacin (17, as insecticides and fungicides). The extraction and utilization of active ingredients in botanical pesticides represent a multifaceted endeavor spanning botany, chemistry, biology, and agricultural science. By effectively extracting and utilizing the active ingredients in plant resources, more environmentally friendly and sustainable agricultural control methods can be developed, reducing dependence on chemical pesticides and promoting sustainable agricultural development.

5.3 Analysis of the Mechanism of Action of Microorganisms and their Secondary Metabolites

In agricultural production, microorganisms and their secondary metabolites are mainly used to control plant pests and diseases or regulate plant growth and development. So far, more than 100 strains of *Bacillus thuringiensis* have been isolated and identified, mainly used to control larvae of Lepidoptera, Coleoptera, and Diptera, occupying 70% of the global biopesticide market. The Bt protein produced by *Bacillus thuringiensis*, mainly through Cry proteins, forms pores on the cell membrane of insects or induces cell death. In recent years, more and more microbial pesticides with potential applications have been discovered. For example, *Talaromyces flavus* exhibits strong inhibitory effects on various anthracnose diseases; *Bacillus thuringiensis* induces systemic responses in plants by simultaneously activating salicylic acid, jasmonic acid, and ethylene signaling pathways, regulating clubroot in Brassica rapa (Wang et al., 2020c). Microbial secondary metabolites such as peptides, siderophores, chitinases, amylases, proteases, polyphenols, flavonoids, etc., have also been applied in the control of plant diseases. In addition, over 700 species of viruses have been isolated, mainly used to control Lepidoptera, Homoptera, Coleoptera, Diptera, and Orthoptera. Notably, nucleopolyhedrovirus and granulovirus are widely used in the control of pests in vegetables and field crops, achieving good results. These microorganisms and their metabolites play a role in biological control through direct antagonism, the production of secondary metabolites, and the induction of host plant resistance. In the future, further exploration is needed in this field, studying the combined use of microorganisms and compounds to achieve better and longer-lasting biological control effects.

5.4 Expansion and Utilization of Natural Enemy Organisms

Improving the deployment and efficacy of natural predators stands a focus in biopesticide research, aiming to control pest populations by utilizing naturally existing predatory organisms to reduce damage to crops and ecosystems. Compared to chemical pesticide control, natural enemy control is not only environmentally friendly but also improves the quality and safety of agricultural products.

The expansion of natural enemies involves studying the ecological habits and reproductive laws of natural enemies, cultivating and increasing the population of natural enemies in agricultural fields and ecosystems to achieve effective and sustainable control of pests and diseases. Strategies to promote the expansion of natural enemies include: protecting and improving habitats to maintain the stability of ecosystems; adopting comprehensive agricultural management methods to provide more suitable environments for natural enemies; introducing and releasing natural enemies; providing abundant food resources for natural enemies; and regularly monitoring the population and distribution of natural enemies and pests to understand the interactions between natural enemies and pests. By comprehensively utilizing these methods to promote the expansion of natural enemies, their effectiveness in ecosystems can be improved, thus achieving natural biological control of pests.

The utilization of natural enemies needs to be based on a deep understanding of agricultural ecosystems. Based on the life history and behavioral habits of target pests, natural enemies most suitable for controlling specific pests are determined. Currently, only two natural enemies have been registered by the Pesticide Institute of the Ministry of Agriculture and Rural Affairs of China, both of which are *Trichogramma dendrolimimatsumura*, mainly used to control Lepidoptera pests. In addition, other natural enemies have been widely applied in the field. For example, *Harmonia axyridis* is effectively used to control aphids and spider mites; *Encarsia formosa Gahan* is effectively used to control greenhouse whiteflies on greenhouse tomatoes; *Phytoseiulus persimilis* is used to continuously and effectively control two-spotted spider mites on greenhouse beans.

5.5 Safety Risk Control of Antibiotics for Agricultural Use

Agricultural antibiotics, derived from microbial fermentation, serve as valuable fungicides enhancing agricultural product quality. However, their use presents notable safety concerns, including antibiotic residues, resistance, and environmental pollution. Therefore, implementing scientifically reasonable safety risk control measures is essential to maintaining food safety and ecological balance (Imarhiagbe et al., 2023; Kosoe et al., 2023).

Primarily, the core of safety risk control lies in the rational use of agricultural antibiotics. Users should follow the antibiotic usage regulations issued by agricultural authorities, select appropriate types, doses, and application methods of antibiotics, avoid abuse and overuse, and thereby reduce the residual levels of antibiotics in agricultural products. Secondly, establishing a sound monitoring, testing, and management system is critical to ensuring the safety of agricultural products. Regularly monitor the residual levels of antibiotics in agricultural products and promptly publicize the monitoring results. For products exceeding the standards, source tracing is needed, and corresponding measures should be taken to ensure that agricultural products on the market comply with hygiene standards. In addition, exploring alternatives is also an important direction, such as using plant extracts and probiotics to reduce dependence on antibiotics and lower the safety risks of antibiotics. The safety risk control of agricultural antibiotics requires cooperation from all parties, including agricultural practitioners, regulatory authorities, research institutions, etc., through comprehensive application, rational use, monitoring and testing, management, technological innovation, etc., to better ensure food safety, mitigate the development of antibiotic resistance, and provide support for the sustainable development of agriculture, bioeconomy product design, supply, and consumption (Obahiagbon & Ogwu, 2023a, b, 2024a, b).

6 Challenges and Prospects of Biological Agriculture

With the deepening of research and practice in agricultural biotechnology, the development and application of agricultural biotechnology will face a more complex innovation environment in the new round of technological revolution and industrial transformation, requiring strengthened top-level design and response to risks and challenges (Fig. 3).

Fig. 3 Outlook on future biological agriculture

6.1 The Constant Expansion of Novel Biotechnologies in the Field of Agricultural Biology

As early as the beginning of the twentieth century, biotechnology was anticipated to replace information technology as the main force driving the new technological revolution and promoting economic and social development (Fukuyama & Wagner, 2000). With the development and widespread application of technologies such as stem cell technology and synthetic biology, the boundaries of agricultural biology in ensuring food security, life safety, and ecological security continue to expand. Firstly, by utilizing the characteristics of plants such as drought resistance, salt tolerance, and efficient nutrient utilization, combined with research results on microbial interactions, productivity in desert areas, saline-alkali land, etc., can be improved while preventing land and grassland degradation (Gorter et al., 2020). Secondly, integrating technologies such as plant tissue culture, stem cell technology, and synthetic biology can achieve the protection of endangered species and enrich biodiversity. Thirdly, various functional microbial preparations are widely used in improving breeding environments and pollution control (Tagg et al., 2024), such as manure treatment agents (bacillus, fungi, cellulose-degrading bacteria, etc.), and aquatic breeding water quality regulation agents (photosynthetic bacteria, nitrifying bacteria, denitrifying bacteria, etc.).

6.2 The Increasing Importance of Common Technologies, Platforms, and Applications Across Plants

Compared to information technology and industrial technology, biotechnology is more versatile. Genome sequencing, gene editing, stem cell technology, etc., can all be applied to plants, animals, and even humans. Cross-species basic research, platform construction, and product development can improve the innovation efficiency of the entire biological agriculture sector. For example, in the field of quantitative genetics, breaking through the related commonality theory research between existing crops and even plants and animals, and using gene editing, whole-genome selection, and liquid-phase chip detection as starting points to develop molecular breeding methods across crops.

6.3 The Increasing Importance of Big Data Science, Artificial Intelligence, etc., in the Development of Agricultural Biology

In 1999, American scholar Jeremy Rifkin proposed that in the twentieth century, information science and life science would end the situation of no intersection in the past forty years and gradually merge into a strong force leading technological changes and economic development, becoming the cornerstone of the biotechnology era (Rifkin, 1998). In recent years, the deep integration of information technology and biotechnology has created important growth points for biological agriculture applications (Wang et al., 2023b). In the future, agricultural biotechnology will further combine with new technologies such as big data science, artificial intelligence, and bionics on the basis of big data science, and apply major breakthroughs in interdisciplinary sciences such as brain–machine integration, human–machine hybridization, and biological fusion to promote sustainable development in agriculture, improve the level of agricultural informatization, electrification, bipolarization, and intelligence, and improve the quality of human life.

6.4 Increasingly Intense International Competition

Major developed countries in the world have systematically and prospectively deployed basic research and core technology fields in biotechnology since the twentieth century. In 1985, the United States took the lead in formulating the "Human Genome Project" and established the Advisory Committee on Biotechnology & twenty-first Century Agriculture (AC21) in 2003, responsible for the impact of biotechnology on the food-agriculture system and providing countermeasure suggestions for the application of biotechnology in agriculture. In recent years, the United States has invested nearly half of its civilian technology funds in research and development of biotechnology and medicine. The German government named 2001 as the "Year of Life Sciences." The European Union and Japan have issued special action plans for biotechnology and listed biotechnology as a priority industry for development, becoming an important lever for developed countries to overcome economic downturns, especially since the outbreak of the COVID-19 pandemic, when numerous social capitals have also turned to the development and application of biotechnology. Compared to industries such as information technology, the biological agriculture industry has a lower level of monopoly, providing a possibility for a breakthrough in the "bottleneck" dilemma for many developing countries, which brings more possibilities for international competition in this field.

6.5 Increasing Importance of Biosecurity Issues

As the role of biotechnology in promoting human development becomes increasingly prominent, biosecurity has gradually become an important part of national security. In 2018, the United States issued the "National Biodefense Strategy," explicitly stating that the greatest threat facing the United States is biological threats (Departments of Defense et al., n.d.). Major infectious disease epidemics and other events have prompted many countries to adjust the focus of national security to ensure biosecurity. Firstly, it is necessary to protect biological resources. The foundation of the development of biological agriculture lies in biological resources, including genes. The development potential of biological agriculture depends to a large extent on the degree of development and utilization of biological resources. Therefore, protecting genetic diversity and biological diversity is an important guarantee for technological innovation and industrial development in biological agriculture. Secondly, it is necessary to strengthen laboratory biosecurity. Strengthening the operation and maintenance management of high-level biosecurity laboratories through hardware construction, personnel training, and other means to prevent biosecurity problems such as laboratory leaks caused by operation. Thirdly, it is necessary to optimize the technical application supervision system. Reasonably regulate the application of biotechnology, while steadily promoting the industrialization process of biotechnological breeding, prevent the use of synthetic biology and other technologies for the synthesis of dangerous organisms, and reduce biosecurity risks and threats from a systemic perspective.

7 Conclusion

As an important technical mean in China's agricultural field, biological breeding has achieved remarkable results in recent years. With the continuous development of breeding technologies, China has cultivated more than 70,000 new varieties, experienced six to seven rounds of large-scale variety changes, achieved over 96% coverage of excellent varieties, and seen the contribution rate of improved varieties to crop yield reach more than 40%. This has effectively supported the steady improvement of grain production and quality, making great contributions to national food security. Meanwhile, advanced biological breeding technologies such as gene editing technology, whole-genome selection, synthetic biology, and artificial intelligence (AI) have injected new vitality into the development of biological breeding through the comprehensive application of these technologies. In particular, the application of gene editing technology in crop improvement has achieved remarkable results in recent years. However, biological breeding also faces many challenges in its development process. Firstly, breeding resources are increasingly scarce, making it more difficult to mine and utilize excellent genes. Secondly, there is a lack of technological innovation in biological breeding, with a certain gap compared with the international advanced level. In addition, the biological breeding industry chain is not perfect, and the conversion rate of results is low, which limits the wide application of breeding technology. Therefore, the future development direction of biological breeding is destined to include deepening research on genomics, exploring and utilizing excellent gene resources; innovating biological breeding technologies to improve breeding efficiency; establishing a perfect biological breeding industry chain to promote the conversion of results; and strengthening policy support for biological breeding and increasing investment. In conclusion, biological breeding plays an important role in China's agricultural development, and the engineering application and development of these biological breeding technologies provide more possibilities for crop improvement. This will help meet the global growing demand for food, increase agricultural sustainability, reduce environmental impact, and thus have a far-reaching impact on food security and global agriculture.

Against the backdrop of global climate change, food crises, and the depletion of natural resources, biological fertilizers have been widely used in agricultural production due to their green and environmentally friendly advantages, as well as high fertilizer efficiency. The scale of China's biological fertilizer industry continues to expand, product categories are becoming increasingly diverse, and significant achievements have been made in the research and development of functional microbial strains. According to the statistics of the Ministry of Agriculture and Rural Affairs, the cumulative registered number of microbial fertilizer products in China has increased from 149 in 2007 to more than 9990 in 2022, and the effective strains of related products have changed from single type to compound multi-effect type. China's microbial fertilizer output has reached 30 million tons/year, with an annual output value of about 40 billion yuan, providing strong support for agricultural production. At the same time, biological fertilizers have played an important role in improving soil structure, increasing soil fertility, reducing the use of chemical fertilizers, and ensuring the quality and safety of agricultural products. However, China's biological fertilizers also face many challenges in their development process. Firstly, there is a relative shortage of efficient microbial strain resources, which urgently requires strengthening the exploration of microbial strain resources. Secondly, the effectiveness of biological fertilizers is greatly affected by environmental factors and has poor stability, which affects farmers' enthusiasm for use. In addition, the promotion of biological fertilizers is insufficient, and market awareness needs to be improved. For the future development of biological fertilizers, it is necessary to strengthen basic research on biological fertilizers, screen and cultivate efficient and stress-resistant functional microbial strains to improve product quality; innovate biological fertilizer production processes to improve production efficiency; promote the application technology of biological fertilizers and increase publicity to improve farmers' recognition and enthusiasm for using biological fertilizers; strengthen the combined use of biological fertilizers with chemical fertilizers, organic fertilizers, and other agricultural inputs to achieve comprehensive benefits; and strengthen policy support to encourage technological innovation and transformation of biological fertilizer industry to promote sustainable industrial development.

The mechanism of biological feed technology is divided into two parts. One part is to develop biological feed additives that can regulate the structure of intestinal flora by studying the role of the micro-ecological environment and production environment of livestock and poultry gastrointestinal tract. The other part is to develop a biological feed formula that can enhance the immune system of livestock and poultry through an in-depth study of the influence mechanism of biological feed on immune cells. Its technological innovation mainly lies in the aspects of enzyme preparations and active peptides. High-efficiency enzyme inhibitors can promote the degradation of ingredients that are difficult to digest in feed. The compound enzyme preparations can not only promote growth, help digestion and increase immunity by supplementing the deficiency of in-body enzymes in animal bodies but also improve the digestibility of animal protein and reduce environmental pollution and industrial carbon footprint. The research on bioactive peptides focuses on the exploration of the mechanism of the influence of active peptides on the growth, immunity and disease resistance of livestock and poultry. Technological innovation is expected to provide new ideas for the formula of biological feed, thus improving the sustainability and economic benefits of livestock and poultry production.

Although there is no international unified standard for the definition and registration scope of biological pesticides, their characteristics of low toxicity and side effects, safety and good environmental compatibility have been recognized. Biological pesticides have become the hottest development focus and field of the global pesticide industry. Biological pesticides mainly include biochemical pesticides, botanical pesticides, microbial pesticides, and the propagation and utilization of natural enemy organisms. Its mechanism of action covers interfering with or influencing the physiological processes of pests, weeds or pathogens through biochemical means, extracting or directly using the active ingredients in plant resources to make pesticides, using the antagonistic effects of microorganisms and their metabolites, and using natural enemies already existing in nature to control pests based on in-depth research on agricultural ecosystems. Although most biological pesticides are considered safe, green and efficient, agricultural antibiotics, as one of the biological pesticides, easily cause safety risks such as drug resistance while improving the quality of farm products. The production research and development costs of biological pesticides are still higher than those of chemical pesticides. For all biological pesticide categories, accelerating the screening of active ingredients, strengthening technological innovation, developing safer, lower-cost, and more convenient products, and establishing and improving monitoring, detection and management systems are the keys to the rapid development of biological pesticides.

References

Applying lysozyme, alkaline protease, and sodium hypochlorite to reduce bioclogging during managed aquifer recharge: A laboratory study-Web of Science Core Collection [WWW Document]. (n.d.). https://elib.ustb.edu.cn/https/77726476706e69737468656265737421e7f2439321236b597b068aa9d6562f34899051d9fc85a85327/wos/woscc/full-record/WOS:000931440900001. Accessed 25 Jul 2024.

Chen, W., Chen, L., Zhang, X., Yang, N., Guo, J., Wang, M., Ji, S., Zhao, X., Yin, P., Cai, L., Xu, J., Zhang, L., Han, Y., Xiao, Y., Xu, G., Wang, Y., Wang, S., Wu, S., Yang, F., Jackson, D., Cheng, J., Chen, S., Sun, C., Qin, F., Tian, F., Fernie, A. R., Li, J., Yan, J., & Yang, X. (2022).

Convergent selection of a WD40 protein that enhances grain yield in maize and rice. *Science, 375*, eabg7985. https://doi.org/10.1126/science.abg7985

De Souza, A. P., Burgess, S. J., Doran, L., Hansen, J., Manukyan, L., Maryn, N., Gotarkar, D., Leonelli, L., Niyogi, K. K., & Long, S. P. (2022). Soybean photosynthesis and crop yield are improved by accelerating recovery from photoprotection. *Science, 377*, 851–854. https://doi.org/10.1126/science.adc9831

Departments of Defense, Health and Human Services, Homeland Security. (n.d.). National biodefense strategy.

Fukuyama, F., & Wagner, C. S. (2000). *Information and biological revolutions: Global governance challenges — Summary of a study group*. RAND Corporation.

Gao, M., He, Y., Yin, X., Zhong, X., Yan, B., Wu, Y., Chen, J., Li, X., Zhai, K., Huang, Y., Gong, X., Chang, H., Xie, S., Liu, J., Yue, J., Xu, J., Zhang, G., Deng, Y., Wang, E., Tharreau, D., Wang, G.-L., Yang, W., & He, Z. (2021). Ca2+ sensor-mediated ROS scavenging suppresses rice immunity and is exploited by a fungal effector. *Cell, 184*, 5391–5404.e17. https://doi.org/10.1016/j.cell.2021.09.009

Gorter, F. A., Manhart, M., & Ackermann, M. (2020). Understanding the evolution of interspecies interactions in microbial communities. *Philosophical Transactions of the Royal Society B, 375*, 20190256. https://doi.org/10.1098/rstb.2019.0256

He, Q., Tang, S., Zhi, H., Chen, J., Zhang, J., Liang, H., Alam, O., Li, H., Zhang, H., Xing, L., Li, X., Zhang, W., Wang, H., Shi, J., Du, H., Wu, H., Wang, L., Yang, P., Xing, L., Yan, H., Song, Z., Liu, J., Wang, H., Tian, X., Qiao, Z., Feng, G., Guo, R., Zhu, W., Ren, Y., Hao, H., Li, M., Zhang, A., Guo, E., Yan, F., Li, Q., Liu, Y., Tian, B., Zhao, X., Jia, R., Feng, B., Zhang, J., Wei, J., Lai, J., Jia, G., Purugganan, M., & Diao, X. (2023). A graph-based genome and pan-genome variation of the model plant Setaria. *Nature Genetics, 55*, 1232–1242. https://doi.org/10.1038/s41588-023-01423-w

Huang, G., Wu, Z., Percy, R. G., Bai, M., Li, Y., Frelichowski, J. E., Hu, J., Wang, K., Yu, J. Z., & Zhu, Y. (2020). Genome sequence of Gossypium herbaceum and genome updates of Gossypium arboreum and Gossypium hirsutum provide insights into cotton A-genome evolution. *Nature Genetics, 52*, 516–524. https://doi.org/10.1038/s41588-020-0607-4

Huang, Y., Wang, H., Zhu, Y., Huang, X., Li, S., Wu, X., Zhao, Y., Bao, Z., Qin, L., Jin, Y., Cui, Y., Ma, G., Xiao, Q., Wang, Q., Wang, J., Yang, X., Liu, H., Lu, X., Larkins, B. A., Wang, W., & Wu, Y. (2022). THP9 enhances seed protein content and nitrogen-use efficiency in maize. *Nature, 612*, 292–300. https://doi.org/10.1038/s41586-022-05441-2

Imarhiagbe, O., Okafor, A. C., Ikponmwosa, B. O., & Ogwu, M. C. (2023). Sustainable agricultural Pest control strategies to boost food and socioecological security: The allelopathic strategy. In M. C. Ogwu & S. Chibueze Izah (Eds.), *One health implications of agrochemicals and their sustainable alternatives* (Sustainable development and biodiversity) (pp. 637–657). Springer. https://doi.org/10.1007/978-981-99-3439-3_23

Jin, W., Ye, H. A. N., Zhao, J., Zhou, Z., & Huan, F. A. N. (2017). Consuming fermented distillers' dried grains with solubles (DDGS) feed reveals a shift in the faecal microbiota of growing and fattening pigs using 454 pyrosequencing. *Journal of Integrative Agriculture, 16*, 900–910. https://doi.org/10.1016/S2095-3119(16)61523-X

Kosoe, E. A., Achana, G. T. W., & Ogwu, M. C. (2023). Meta-evaluation of the one health implication on food Systems of agrochemical use. In M. C. Ogwu & S. Chibueze Izah (Eds.), *One health implications of agrochemicals and their sustainable alternatives* (Sustainable development and biodiversity) (pp. 387–409). Springer. https://doi.org/10.1007/978-981-99-3439-3_14

Kumar, S., Sindhu, S. S., & Kumar, R. (2022). Biofertilizers: An ecofriendly technology for nutrient recycling and environmental sustainability. *Current Research in Microbial Sciences, 3*, 100094. https://doi.org/10.1016/j.crmicr.2021.100094

Lengai, G. M. W., Muthomi, J. W., & Mbega, E. R. (2020). Phytochemical activity and role of botanical pesticides in pest management for sustainable agricultural crop production. *Scientific African, 7*, e00239. https://doi.org/10.1016/j.sciaf.2019.e00239

Li, S., Tian, Y., Wu, K., Ye, Y., Yu, J., Zhang, J., Liu, Q., Hu, M., Li, H., Tong, Y., Harberd, N. P., & Fu, X. (2018). Modulating plant growth–Metabolism coordination for sustainable agriculture. *Nature, 560*, 595–600. https://doi.org/10.1038/s41586-018-0415-5

Li, C., Wang, W., Pan, Y., Liu, F., He, J., Liu, C., Cao, J., Zhang, X., Zhao, J., & Gai, J. (2022). Germplasm sources, genetic richness, and population differentiation of modern Chinese soybean cultivars based on pedigree integrated with genomic-marker analysis. *Frontiers in Plant Science, 13*, 945839. https://doi.org/10.3389/fpls.2022.945839

Liu, H., Guan, F., Liu, T., Yang, L., Fan, L., Liu, X., Luo, H., Wu, N., Yao, B., Tian, J., & Huang, H. (2023). MECE: A method for enhancing the catalytic efficiency of glycoside hydrolase based on deep neural networks and molecular evolution. *Scientific Bulletin, 68*, 2793–2805. https://doi.org/10.1016/j.scib.2023.09.039

Lyberg, K., Lundh, T., Pedersen, C., & Lindberg, J. E. (2006). Influence of soaking, fermentation and phytase supplementation on nutrient digestibility in pigs offered a grower diet based on wheat and barley. *Animal Science, 82*, 853–858. https://doi.org/10.1017/ASC2006109

Ma, L., Wang, L., Zhang, Z., & Xiao, D. (2023). Research progress of biological feed in beef cattle. *Animals, 13*, 2662. https://doi.org/10.3390/ani13162662

Missotten, J. A., Michiels, J., Degroote, J., & De Smet, S. (2015). Fermented liquid feed for pigs: An ancient technique for the future. *Journal of Animal Science and Biotechnology, 6*, 1–9. https://doi.org/10.1186/2049-1891-6-4

Morlighem, J.-É. R. L., & Radis-Baptista, G. (2019). The place for enzymes and biologically active peptides from marine organisms for application in industrial and pharmaceutical biotechnology. *Current Protein & Peptide Science, 20*, 334–355. https://doi.org/10.2174/1389203719666180926121722

Nagai, K., Mori, Y., Ishikawa, S., Furuta, T., Gamuyao, R., Niimi, Y., Hobo, T., Fukuda, M., Kojima, M., Takebayashi, Y., Fukushima, A., Himuro, Y., Kobayashi, M., Ackley, W., Hisano, H., Sato, K., Yoshida, A., Wu, J., Sakakibara, H., Sato, Y., Tsuji, H., Akagi, T., & Ashikari, M. (2020). Antagonistic regulation of the gibberellic acid response during stem growth in rice. *Nature, 584*, 109–114. https://doi.org/10.1038/s41586-020-2501-8

Obahiagbon, E. G., & Ogwu, M. C. (2023a). Sustainable supply chain management in the herbal medicine industry. In S. C. Izah, M. C. Ogwu, & M. Akram (Eds.), *Herbal medicine phytochemistry* (Reference Series in phytochemistry). Springer. https://doi.org/10.1007/978-3-031-21973-3_64-1

Obahiagbon, E. G., & Ogwu, M. C. (2023b). Consumer perception and demand for sustainable herbal medicine products and market. In S. C. Izah, M. C. Ogwu, & M. Akram (Eds.), *Herbal medicine phytochemistry* (Reference Series in Phytochemistry). Springer. https://doi.org/10.1007/978-3-031-21973-3_65-1

Obahiagbon, E. G., & Ogwu, M. C. (2024a). Organic food preservatives: The shift towards natural alternatives and sustainability in the global south's markets. In M. C. Ogwu, S. C. Izah, & N. R. Ntuli (Eds.), *Food safety and quality in the global south* (pp. 299–329). Springer. https://doi.org/10.1007/978-981-97-2428-4_10

Obahiagbon, E. G., & Ogwu, M. C. (2024b). The nexus of business, sustainability, and herbal medicine. In S. C. Izah, M. C. Ogwu, & M. Akram (Eds.), *Herbal medicine phytochemistry* (Reference series in phytochemistry). Springer. https://doi.org/10.1007/978-3-031-21973-3_67-1

Okur, N. (2018). A review: Bio-fertilizers- power of beneficial microorganisms in soils. *Biomedical Journal of Scientific & Technical Research, 4*. https://doi.org/10.26717/BJSTR.2018.04.0001076

Rai, P. K., Rai, A., Sharma, N. K., Singh, T., & Kumar, Y. (2023). Limitations of biofertilizers and their revitalization through nanotechnology. *Journal of Cleaner Production, 418*, 138194. https://doi.org/10.1016/j.jclepro.2023.138194

Ren, D., Wang, D., Liu, H., Shen, M., & Yu, H. (2019). Two strains of probiotic lactobacillus enhance immune response and promote naive T cell polarization to Th1. *Food and Agricultural Immunology, 30*, 281–295. https://doi.org/10.1080/09540105.2019.1579785

Rifkin, J., 1998. The biotech century.

Sha, G., Sun, P., Kong, X., Han, X., Sun, Q., Fouillen, L., Zhao, J., Li, Y., Yang, L., Wang, Y., Gong, Q., Zhou, Y., Zhou, W., Jain, R., Gao, J., Huang, R., Chen, X., Zheng, L., Zhang, W., Qin, Z., Zhou, Q., Zeng, Q., Xie, K., Xu, J., Chiu, T.-Y., Guo, L., Mortimer, J. C., Boutté, Y., Li, Q., Kang, Z., Ronald, P. C., & Li, G. (2023). Genome editing of a rice CDP-DAG synthase confers multipathogen resistance. *Nature, 618*, 1017–1023. https://doi.org/10.1038/s41586-023-06205-2

Shi, C., Zhang, Y., Lu, Z., & Wang, Y. (2017). Solid-state fermentation of corn-soybean meal mixed feed with Bacillus subtilis and enterococcus faecium for degrading antinutritional factors and enhancing nutritional value. *Journal of Animal Science and Biotechnology, 8*, 50. https://doi.org/10.1186/s40104-017-0184-2

Sugiharto, S. (2016). Role of nutraceuticals in gut health and growth performance of poultry. *Journal of the Saudi Society of Agricultural Sciences, 15*, 99–111. https://doi.org/10.1016/j.jssas.2014.06.001

Sugiharto, S., & Ranjitkar, S. (2019). Recent advances in fermented feeds towards improved broiler chicken performance, gastrointestinal tract microecology and immune responses: A review. *Animal Nutrition, 5*, 1–10. https://doi.org/10.1016/j.aninu.2018.11.001

Sugiharto, S., Yudiarti, T., Isroli, I., Widiastuti, E., & Kusumanti, E. (2017). Dietary supplementation of probiotics in poultry exposed to heat stress–A review. *Annals of Animal Science, 17*, 591–604. https://doi.org/10.1515/aoas-2016-0062

Tagg, A. S., Sperlea, T., Hassenrück, C., Kreikemeyer, B., Fischer, D., & Labrenz, M. (2024). Microplastic-antifouling paint particle contamination alters microbial communities in surrounding marine sediment. *Science of the Total Environment, 926*, 171863. https://doi.org/10.1016/j.scitotenv.2024.171863

Tian, M., Feng, Y., He, X., Zhang, D., Wang, W., & Liu, D. (2022). Mycotoxins in livestock feed in China - Current status and future challenges. *Toxicon, 214*, 112–120. https://doi.org/10.1016/j.toxicon.2022.05.041

Van Zyl, W. F., Deane, S. M., & Dicks, L. M. (2020). Molecular insights into probiotic mechanisms of action employed against intestinal pathogenic bacteria. *Gut Microbes, 12*, 1831339. https://doi.org/10.1080/19490976.2020.1831339

Wang, H., Sun, S., Ge, W., Zhao, L., Hou, B., Wang, K., Lyu, Z., Chen, L., Xu, S., Guo, J., Li, M., Su, P., Li, X., Wang, G., Bo, C., Fang, X., Zhuang, W., Cheng, X., Wu, J., Dong, L., Chen, W., Li, W., Xiao, G., Zhao, J., Hao, Y., Xu, Y., Gao, Y., Liu, W., Liu, Y., Yin, H., Li, J., Li, X., Zhao, Y., Wang, X., Ni, F., Ma, X., Li, A., Xu, S. S., Bai, G., Nevo, E., Gao, C., Ohm, H., & Kong, L. (2020a). Horizontal gene transfer of *Fhb7* from fungus underlies *Fusarium* head blight resistance in wheat. *Science, 368*, eaba5435. https://doi.org/10.1126/science.aba5435

Wang, L.-M., Shen, B.-R., Li, B.-D., Zhang, C.-L., Lin, M., Tong, P.-P., Cui, L.-L., Zhang, Z.-S., & Peng, X.-X. (2020b). A synthetic photorespiratory shortcut enhances photosynthesis to boost biomass and grain yield in Rice. *Molecular Plant, 13*, 1802–1815. https://doi.org/10.1016/j.molp.2020.10.007

Wang, M., Geng, L., Sun, X., Shu, C., Song, F., & Zhang, J. (2020c). Screening of bacillus thuringiensis strains to identify new potential biocontrol agents against Sclerotinia sclerotiorum and Plutella xylostella in Brassica campestris L. *Biological Control, 145*, 104262. https://doi.org/10.1016/j.biocontrol.2020.104262

Wang, N., Tang, C., Fan, X., He, M., Gan, P., Zhang, S., Hu, Z., Wang, X., Yan, T., Shu, W., Yu, L., Zhao, J., He, J., Li, L., Wang, J., Huang, X., Huang, L., Zhou, J.-M., Kang, Z., & Wang, X. (2022). Inactivation of a wheat protein kinase gene confers broad-spectrum resistance to rust fungi. *Cell, 185*, 2961–2974.e19. https://doi.org/10.1016/j.cell.2022.06.027

Wang, K., Abid, M. A., Rasheed, A., Crossa, J., Hearne, S., & Li, H. (2023a). DNNGP, a deep neural network-based method for genomic prediction using multi-omics data in plants. *Molecular Plant, 16*, 279–293. https://doi.org/10.1016/j.molp.2022.11.004

Wang, Y., Fan, J., Yu, S., Cai, S., Guo, X., Zhao, C., & 1. Beijing Key Laboratory of Digital Plant, Beijing Research Center for Information Technology in Agriculture, China National Engineering Research Center for Information Technology in Agriculture (NERCITA), Beijing 100097, China, 2. College of Information Engineering, Northwest A&F University, Yangling

712100, Shaanxi, China, 3. Beijing PAIDE Science and Technology Development Co., Ltd., Beijing 100097, China. (2023b). Research advance in phenotype detection robots for agriculture and forestry. *International Journal of Agricultural and Biological Engineering, 16*, 14–25. https://doi.org/10.25165/j.ijabe.20231601.7945

Wei, X., Qiu, J., Yong, K., Fan, J., Zhang, Q., Hua, H., Liu, J., Wang, Q., Olsen, K. M., Han, B., & Huang, X. (2021). A quantitative genomics map of rice provides genetic insights and guides breeding. *Nature Genetics, 53*, 243–253. https://doi.org/10.1038/s41588-020-00769-9

Wu, S., Zhang, F., Huang, Z., Liu, H., Xie, C., Zhang, J., Thacker, P. A., & Qiao, S. (2012). Effects of the antimicrobial peptide cecropin AD on performance and intestinal health in weaned piglets challenged with Escherichia coli. *Peptides, 35*, 225–230. https://doi.org/10.1016/j.peptides.2012.03.030

Wu, K., Wang, S., Song, W., Zhang, J., Wang, Y., Liu, Q., Yu, J., Ye, Y., Li, S., Chen, J., Zhao, Y., Wang, J., Wu, X., Wang, M., Zhang, Y., Liu, B., Wu, Y., Harberd, N. P., & Fu, X. (2020). Enhanced sustainable green revolution yield via nitrogen-responsive chromatin modulation in rice. *Science, 367*, eaaz2046. https://doi.org/10.1126/science.aaz2046

Yan, J., Zhou, B., Xi, Y., Huan, H., Li, M., Yu, J., Zhu, H., Dai, Z., Ying, S., & Zhou, W. (2019). Fermented feed regulates growth performance and the cecal microbiota community in geese. *Poultry Science, 98*, 4673–4684. https://doi.org/10.3382/ps/pez169

Yi, R., Pan, Y., Long, X., Tan, F., & Zhao, X. (2020). Enzyme producing activity of probiotics and preparation of compound enzyme. *Journal of Chemistry, 2020*, 1–8. https://doi.org/10.1155/2020/9140281

You, J., Liu, Z., Qi, Z., Ma, Y., Sun, M., Su, L., Niu, H., Peng, Y., Luo, X., Zhu, M., Huang, Y., Chang, X., Hu, X., Zhang, Y., Pi, R., Liu, Y., Meng, Q., Li, J., Zhang, Q., Zhu, L., Lin, Z., Min, L., Yuan, D., Grover, C. E., Fang, D. D., Lindsey, K., Wendel, J. F., Tu, L., Zhang, X., & Wang, M. (2023). Regulatory controls of duplicated gene expression during fiber development in allotetraploid cotton. *Nature Genetics, 55*, 1987–1997. https://doi.org/10.1038/s41588-023-01530-8

Yu, H., Lin, T., Meng, X., Du, H., Zhang, J., Liu, G., Chen, M., Jing, Y., Kou, L., Li, X., Gao, Q., Liang, Y., Liu, X., Fan, Z., Liang, Y., Cheng, Z., Chen, M., Tian, Z., Wang, Y., Chu, C., Zuo, J., Wan, J., Qian, Q., Han, B., Zuccolo, A., Wing, R. A., Gao, C., Liang, C., & Li, J. (2021). A route to de novo domestication of wild allotetraploid rice. *Cell, 184*, 1156–1170.e14. https://doi.org/10.1016/j.cell.2021.01.013

Zhang, F., Wang, C., Li, M., Cui, Y., Shi, Y., Wu, Z., Hu, Z., Wang, W., Xu, J., & Li, Z. (2021a). The landscape of gene–CDS–haplotype diversity in rice: Properties, population organization, footprints of domestication and breeding, and implications for genetic improvement. *Molecular Plant, 14*, 787–804. https://doi.org/10.1016/j.molp.2021.02.003

Zhang, J., Cook, J., Nearing, J. T., Zhang, J., Raudonis, R., Glick, B. R., Langille, M. G. I., & Cheng, Z. (2021b). Harnessing the plant microbiome to promote the growth of agricultural crops. *Microbiological Research, 245*, 126690. https://doi.org/10.1016/j.micres.2020.126690

Zhang, H., Zhou, J.-F., Kan, Y., Shan, J.-X., Ye, W.-W., Dong, N.-Q., Guo, T., Xiang, Y.-H., Yang, Y.-B., Li, Y.-C., Zhao, H.-Y., Yu, H.-X., Lu, Z.-Q., Guo, S.-Q., Lei, J.-J., Liao, B., Mu, X.-R., Cao, Y.-J., Yu, J.-J., Lin, Y., & Lin, H.-X. (2022a). A genetic module at one locus in rice protects chloroplasts to enhance thermotolerance. *Science, 376*, 1293–1300. https://doi.org/10.1126/science.abo5721

Zhang, L., Ding, Y., Xu, J., Gao, X., Cao, N., Li, K., Feng, Z., Cheng, B., Zhou, L., Ren, M., Lu, X., Bao, Z., Tao, Y., Xin, Z., & Zou, G. (2022b). Selection signatures in Chinese sorghum reveals its unique liquor-making properties. *Frontiers in Plant Science, 13*, 923734. https://doi.org/10.3389/fpls.2022.923734

Zhang, H., Yu, F., Xie, P., Sun, S., Qiao, X., Tang, S., Chen, C., Yang, S., Mei, C., Yang, D., Wu, Y., Xia, R., Li, X., Lu, J., Liu, Y., Xie, X., Ma, D., Xu, X., Liang, Z., Feng, Z., Huang, X., Yu, H., Liu, G., Wang, Y., Li, J., Zhang, Q., Chen, C., Ouyang, Y., & Xie, Q. (2023). A Gγ protein regulates alkaline sensitivity in crops. *Science, 379*, eade8416. https://doi.org/10.1126/science.ade8416

Zhao, F., Yang, N., Wang, X., Mao, R., Hao, Y., Li, Z., Wang, X., Teng, D., Fan, H., & Wang, J. (2019). In vitro/vivo mechanism of action of MP1102 with low/nonresistance against Streptococcus suis type 2 strain CVCC 3928. *Frontiers in Cellular and Infection Microbiology, 9*, 48. https://doi.org/10.3389/fcimb.2019.00048

Zhu, W., Han, R., Shang, X., Zhou, T., Liang, C., Qin, X., Chen, H., Feng, Z., Zhang, H., Fan, X., Li, W., & Li, L. (2023). The CropGPT project: Call for a global, coordinated effort in precision design breeding driven by AI using biological big data. *Molecular Plant*, S1674205223004094. https://doi.org/10.1016/j.molp.2023.12.015

Bioeconomy and Health Outcomes in Global South: A Case Study of Nigeria

Omotola Fadekemi Ajayi
and Temidayo Oladiran Akinbobola

Abstract

It is stated that having good health is a primary right of human and a key component of sustainable development. The UN identified health and well-being as one of the sustainable development goals due to the significance of health outcomes for national development. However, to improve the health outcomes of citizens, governments adopted a new approach named bioeconomy. Bioeconomy is identified as an effective way to enhance human health. The term "bioeconomy" refers to all industries and systems that rely on the production of food, feed, medicines, vaccines, and energy using biobased resources. Many nations, especially those in the Global South, have embraced the bioeconomy to provide medications, vaccines, and diagnostic test kits that will improve human health. The Nigerian government also embraced the bioeconomy to improve health outcomes Nigeria is now able to produce test kits, such as TRI-MALARIA, TRI-SYPHILIS, and TRI-HBsAg test kits, thanks to the application of biotechnology. Secondary data from 1980 to 2020 were used in the study. The Generalized Method of Moment approach was used in this investigation. It was discovered that Nigerian health outcomes are significantly and negatively impacted by the bioeconomy. The study concludes that the bioeconomy does not improve health

O. F. Ajayi (✉)
Department of Economics, College of Art, Management and Social Sciences, Chrisland University, Abeokuta, Nigeria
e-mail: oajayi@chrislanduniversity.edu.ng

T. O. Akinbobola
Department of Economics, Faculty of Social Sciences, Obafemi Awolowo University, Ife, Nigeria
e-mail: tbobola@oauife.edu.ng

M. C. Ogwu et al. (eds.), *Sustainable Bioeconomy Development in the Global South*, https://doi.org/10.1007/978-981-96-0305-3_14

outcomes in Nigeria. The study recommends that the government should grant incentives to the private sector to encourage more investors into industry, resulting in increase in the production of biomedicines.

Keywords

Health outcomes · Bioeconomy · Sustainable development goals · Sustainable development · Biotechnology

Abbreviations

ALCUE	A Latin American and Caribbean Knowledge-Based Bio-Economy
ARDL	Autoregressive Distributed Lag
CART	Chimeric Antigen Receptor
CIAT	Center for Tropical Agriculture
CIRAD	French Agricultural Research Centre for International Development
EU	European Union
GDP	Gross Domestic Products
GMM	Generalized Method of Moments
GMO	Genetically Modified Organism
NABDA	National Biotechnology Development Agency
NBMA	National Biosafety Management Agency
NBRDA	National Biotechnology Research and Development Agency
OLS	Ordinary Least Square
POC	Point-of-Care
SAP	Structural Adjustment Program
VECM	Vector Error Correction Model
WHO	World Health Organization

1 Introduction

It is impossible to overstate the importance of health to people, country production, and growth. Human development is recognized as having a strong link to health (Mishra et al., 2015). A society cannot function without a healthy population, and good health is recognized as a fundamental human right. Living well is built on being healthy, and every economy's growth depends on it. The development and growth of an economy depend on a healthy populace (Alhassan & Abdu, 2017; Ajayi & Akinbobola, 2020; UN, 2020). This is because a healthy population can lead to higher labor productivity level, higher savings due to longer life expectancies, higher rates of school attendance, and the development of children's cognitive abilities, all of which will increase GDP.

The World Health Organization (2018) defines health as a condition of whole physical, mental, and social well-being rather than only the absence of illness or disability. According to the Mexican Commission's (2004) report on macroeconomics and health, a person's ability to reach their full potential during their lifetime is referred to as their state of health. This further explains that good health increases human capital, fosters individual production potential, and speeds up economic progress (Nkpoyen et al., 2014). This makes sound health desirable and is vital for the achievement of *sustainable development* (Onisanwa et al., 2018). Health outcome refers to how an individual's health has changed as a result of various interventions used to enhance the individual's health (Lee & Leung, 2014). Securing a healthy life and promoting the well-being of individuals is crucial for building a booming society (UN, 2020). This significance of a healthy population to the country's growth made the United Nations identify good health and well-being as one of the sustainable development goals (SDGs). However, many factors have been identified to influence the health outcomes of individuals of any economy (Ali & Ahmad, 2014; Whitman et al., 2022). These factors include income, education, healthcare, climate conditions, inflation, government expenditure on health, income, and population, among others.

Moreover, it has been argued that there is a need for the health of individuals to be improved to achieve good health which is the third on the United Nations' list of goals to be achieved by 2030. It has further been argued by scholars and agencies (Serageldin, 1999; Panel, 2017; World Health Organization, 2018) that there is a need for governments of countries, including Nigeria, to adopt new techniques or methods that will help solve the problem of bad health, which further enables the achievement of sustainable development.

Bioeconomy has been identified by scholars to provide an effective way of improving health outcomes for humans as it helps to solve the global challenges which include bad health, food insecurity among others facing the world as a whole (Zilberman, 2014; IFPRI reports 2017; Maciejczak, 2017; Lyytimäki et al., 2018). This approach named bioeconomy is gaining attraction among governments and policy makers worldwide.

According to the IndBioeconomy Report, countries view the bioeconomy as a platform for tackling issues related to food insecurity and nutrition, climate change, fossil fuel, and fostering innovation to build a healthy society. The shift to a sustainable bioeconomy is viewed as a critical strategy by an increasing number of national and International Organizations and actors as a means of resolving contemporary issues and conflicts (Philp, 2018; IACGB, 2020). To address the Grand Societal Challenges of the European Union, accomplish the SDGs, and aid in the post-COVID-19 recovery, a sustainable bioeconomy is therefore regarded as a crucial component of international and regional political agendas.

Bioeconomy is also known as the transition of an economy from a fossil fuel-based economy to a biobased economy (Wang et al., 2022; Egboduku et al., 2024). The term "bioeconomy" refers to the process of producing food, energy, medicines and industrial goods by converting renewable bioresources in a sustainable and environmentally friendly manner (DG Research, 2005). An economy that relies on biotechnology to boost economic output is known as a bioeconomy (OECD, 2009). Biotechnology is explained to be the basis for bioeconomy. Furthermore, it has been discussed that there are many ways through which bioeconomy influences human health. This includes pharmaceutical production from natural products that are crucial to human healthcare, particularly in anti-infective agents and cancer therapeutics (Haines, 2021). Biotechnology has significantly improved human disease control; from antibiotics to treat bacterial infections to anti-HIV/AIDS drugs like azidothymidine (AZT), anti-malarial substances, and innovative vaccinations that have saved millions of lives, biotechnology has made significant strides in this area (Gartland et al., 2013). However, it has been argued by scholars that biotechnology will be a driving force in the field of health (Awais et al., 2010).

Also, Nigeria faces numerous challenges that affect population health, like other developing countries (Izah et al., 2022). To improve the health outcomes of individuals to achieve sustainable development, the Nigerian government also adopted bioeconomy through a biotechnological process. Bioeconomy can therefore be traced in Nigeria via biotechnological processes such as genetic engineering, vaccine production, and test kit development which are directly related to human health. Advancement in Biotechnology offers the choice to deliver marvelous health advantages (Gulzar et al., 2019).

However, despite the opinion of scholars (such as Graff et al., 2006; Carter et al., 2018) that biotechnology is a vital tool for meeting the expanding population's medical demands, recent literature has not yet offered empirical proof of how the bioeconomy affects health outcomes in industrialized and global south countries, including Nigeria. Therefore, owing to the importance of bioeconomy for the improvement of human health outcomes, it is essential to examine the impact of bioeconomy on the health outcomes of Nigeria's growing population. This is because a healthy population is important for sustainable development to be achieved and it is connected to some SDGs which include zero poverty (SDG 1), food security (SDG2), education (SDG4), and economic growth (SDG 8). Hence, this study examines the effect of bioeconomy on health outcomes in Nigeria for the period of 1980 to 2020.

2 Bioeconomy in the Selected Global South Countries

2.1 Bioeconomy in China

China's bioeconomic practices emerged somewhat later, with an early emphasis on biotechnological invention and the growth of the bioindustry. Following the 18th National Congress of the Chinese Communist Party, significant advancements were achieved, the bioindustry's scope increased, and its composition and purposes grew increasingly rational. The bioeconomy was confirmed as a significant national development plan with the release of the "14th Five-Year Plan for Bioeconomy

Development" recently (Wei et al., 2022). Every economic form has a primary and dominant nation. China's agricultural technology and organizational techniques were very sophisticated in the era of the agricultural economy, and the country's overall GDP once contributed to more than 20% of the global economy for an extended period (Musonda et al., 2021).

The Chinese government has announced 14th-year plan for Bioeconomy Development, which describes the bioeconomy as being driven by the advancement of biotechnology and life sciences to provide a sustainable scheme for the expansion of human society. Protecting, developing, and using bioresources is its foundation (Zhang et al., 2022). China's bioeconomy was said to have grown at a rate of 20% per year between 2013 and 2015. The objectives of China's bioeconomy development are outl-ined in the "14th Five-Year Plan for Bioeconomy Development" (2021–2025). The "14th Five-Year Plan for Bioeconomy Development" (2021–2025), a strategic priority document, outlines the objectives for China's bioeconomy development in four primary fields, namely, biomedicine, bioagriculture, biomanufacturing, and biosecurity. The strategy states that China would move quickly to advance the fields of bioenergy, bioinformatics, biomedicine, agricultural biotechnology, and environmental protection (Zhang et al., 2022). The "14th Five-Year Plan for Bioeconomy Development" set high standards for biomedicine in response to the changing demands from "treatment-centered medicine" to "health-centered medicine" especially in terms of applying biology, biochemistry, and genetic engineering, principles in the development of novel pharmaceuticals and medical devices. According to projections, the Chinese biomedical market is predicted to develop at a rate of more than 20% annually and surpass RMB 800 billion by 2025, making it one of the most promising pharmaceutical markets globally. About 18.6% of the medical market's total sales in 2020 came from the biological drug industry. The National Medical Product Administration's "2021 Drug Evaluation Report," which was made public in June 2022, states that 47 novel medications were examined and approved in 2021, a record-breaking number. In July 2022, about 57,398 biomedical industries were established. The COVID-19 vaccine was produced by these biomedical businesses more than 5.5 billion doses per year (2020 Report on Drug Administration and Supervision Statistics, in Chinese). The National Medical Products Administration authorized its CAR-T medications which are both the first and second ever produced made for clinical use in 2021 which were produced Fosun Kite and JW Therapeutics. The Healthy China approach depends heavily on new biomedicines, which are also crucial for tackling the issues posed by an aging population and safeguarding people's lives and health.

2.2 Bioeconomy in Columbia

According to Consejo et al., Colombia describes the bioeconomy as a world where biomass and biodiversity are efficiently and sustainably managed to create new, high-value products, services, and processes through innovation and knowledge. To encourage the sustainable use of the country's biodiversity, Colombia launched its first National Biotechnology Program in 1991 (Canales & Gómez González, 2020). The program's objectives included growing the biotechnology industry and creating and producing biofuel. In 1994, Colombia became a signatory to the Convention on Biological Diversity. The environment in Colombia is ideal for the growth of a bioeconomy.

The nation's megadiverse status and abundance of biomass, coupled with a rapidly consolidating science and technical know-how, especially in the fields of life sciences, engineering, and ICTs, as well as its firsthand experience in several significant industries and sectors, including food, pharmaceuticals, cosmetics, GMO use, bioinputs, and biofuels, are some of these conditional factors. Moreover, some business sectors, particularly the agricultural sector, have innovative initiatives or experience.

A study conducted in 2015 discovered that 34 Colombian businesses might be seen as emblematic of the growth of bioeconomic activities in three primary pathways: bioenergy, biotechnology, and biodiversity, mostly used in the context of health (Henry et al., 2018). The businesses are categorized into nine sub-sectors namely genetically modified organisms (GMOs), biodiesel, bioremediation, pharmaceutical industry, and applications in health among others. Colciencias, in collaboration with CIRAD and CIAT, arranged the National Bioeconomy Forum in April 2017. At the event, experts from several countries discussed the state of the bioeconomy in Latin America as well as some noteworthy cases from each country.

Thus, three working sessions were held at the forum which are the pathways to Colombia's bioeconomy. These include bioenergy and biorefineries; biotechnology and biodiversity for sustainable production; and bioprocesses and bioproducts. The discussion was to examine governmental policies that assist the bioeconomy in the nation's socioeconomic and environmental growth. The purpose of these gatherings was to complete certain recommendations for creating plans of action for the growth of Colombia's bioeconomy.

From 2011 to 2017, Colombia took part in the ALCUE KBBE project, a regional endeavor in collaboration with the European Commission. Using selected company case studies, the study evaluated Colombia's bioeconomy's advantages and disadvantages in bioenergy, biotechnology, and biodiversity (Hodson de Jaramillo et al., 2019) state that social and environmental sustainability measures are necessary for Colombia's bioeconomy, which is primarily export-oriented.

In 2018, the national policy explicitly used the phrase "bioeconomy." The bioeconomy plays a crucial role in creating new economic prospects through green growth policies (Consejo Nacional de Política Económica y Social, 2009). Three bioeconomy targets are outlined in the policy for 2030: an increase of 86% in bioinnovative enterprises, a target of 180% for spin-offs and start-ups, and a 19% increase in the production of biocompanies. Important industries for the growth of the bioeconomy were identified by Biointropic (2018), including chemicals, pharmaceuticals, food and drink, healthcare, agriculture and animal husbandry, and cosmetics and cleaning products (Canales & Gómez González, 2020). Colombia unveiled its plan for a new economy at the 2016 World Economic Forum, with agro-industry, services, and tourism serving as the main engines of growth (Henry et al., 2018). This complements the DNP's [Colombia's National Planning Department] Green Growth Mission, which acknowledges that the nation's economic growth is mostly dependent on natural resources and suggests production sustainability methods that prioritize resource efficiency and social inclusion (Agua et al., 2017; Henry et al., 2018).

2.3 Bioeconomy in India

The development of biotech products, processes, and technologies about agriculture, food and nutritional security, affordable healthcare and wellness, environmental safety, clean energy, biofuel, and biomanufacturing is strongly prioritized in the National Biotechnology Development Strategy of India, which was published in 2015. The government has strengthened institutional frameworks that ensure sustainability and innovation scaling up through its initiatives. Focus has been placed on technology-driven research to improve millions of people's lives and means of subsistence by offering accessible solutions to public health problems that impact the community as a whole.

All facets of the biotechnology industry, including novel protein therapeutics, stem cells, systems biology, synthetic biology, enhanced hybrids in agriculture, clean technologies, sustainability, and renewable energy from biological sources, were important considerations for this bioeconomy. 2019 was an excellent year for India's bioeconomy. The BioPharma and Bioservices sector performed well, particularly in contract research and manufacturing services, which helped the economy. The biopharma segment continues to drive the bioeconomy forward. It alone has generated 58% of the whole bioeconomy's worth. Diagnostics and medical equipment account for half of the biopharma business. Vaccines make for 30% of the bioeconomy's value, with biotherapeutics accounting for the remainder. The Indian bioeconomy is estimated to be worth $80.12 billion in 2021 (January–December). The bioeconomy experienced 14% growth in 2020. In 2021, India's bioeconomy is expected to contribute approximately 2.6% of its GDP.

Despite the epidemic, India's bioeconomy thrived as the BioPharma sector met the country's vaccine and testing demands. The COVID-19 immunization and testing drive has accounted for over a fifth of the bioeconomy. In India, around 1.45 billion doses were delivered until 2021. India certified COVID-19 vaccines for emergency use in January, demonstrating exceptional performance by all norms. India administered 1 billion COVID-19 vaccination doses in around 9 months. In 2021, the COVID-19 bioeconomy was valued at $14.56 billion. The value of COVID-19 vaccines was $8.7 billion, while testing cost $5.9 billion. A COVID-19 vaccination dose typically costs $6 on average, and a sample test costs $11. BioPharma still holds the biggest share of the bioeconomy as it accounted for 49% of bioeconomy.

An estimated $39.4 billion was the segment's total economic contribution. Approximately 52.8% of the biopharmaceutical market was made up of the diagnostics segment, while 26% was made up of the therapeutics segment. What remained was the vaccinations component, which did not include the COVID-19 vaccines. One of the top markets for diagnostics in India is the in-vitro diagnostics market. The performance of the industry is being driven by elements such as the high incidence of chronic illnesses, the growing usage of point-of-care (POC) diagnostics, and the growing acceptance and understanding of companion diagnostics and personalized medicine.

The value of the Biotherapeutics segment is projected to reach $1.05 billion. The main contributors include medications for diabetes, cancer, infections, and cardiology (*India bioeconomy report (IBER),* 2022). The key products in the Biotherapeutics category are drugs for diabetes, cancer, infections, and cardiology. The segment's projected value is $1.05 billion. In 2022, the diagnostic market in India showed substantial growth, driven by favorable macroeconomic conditions.

The overall bioeconomic value of Diagnostics, excluding medical equipment, increased to $10.8 billion. India completed approximately 600–700 million diagnostics tests annually, with a significant daily average of over two million tests. The monthly value of diagnostics within the bioeconomy hit $0.9 billion, signifying a 1.5-fold increase over 2018 (*India bioeconomy report (IBER),* 2023).

2.4 Bioeconomy in Nigeria

Bioeconomy is embraced in Nigeria via biotechnological processes such as fermentation, breeding, attenuation process for vaccine production, genetic engineering, etc. in solving these global challenges. Biotechnology application in Nigeria can be dated back to the preparation of the local alcoholic drinks such as otinka/burukutu, palm wine, and non-alcoholic drinks like kunu "tiger nut milk," and the breeding of animals and plants which is described as the age of domestication (Kolawole et al., 2007; Oluwajoba et al., 2013; Atter et al., 2014). Modern biotechnology can be dated back to the era of the green revolution between the 1950s and 1960s (Nzimiro, 1985; Chahal & Gosal, 2002; Ivase et al., 2019).

The application of biotechnology led to the introduction of the Structural Adjustment Program (SAP) in 1986, the establishment of the National Centre for Genetic Resources and Biotechnology in 1987, and the National Committee on Biotechnology in 1993 to provide a master plan. This development led to the establishment of the National Biotechnology Development Agency (NABDA) in 2001 which is directed toward the development, designing, and implementation of the nation's biotechnology policy (Ohimain, 2013; Ivase et al., 2019), and National Biosafety Management Agency (NBMA) in 2011. NABDA and NBMA were established to aid the achievement of biotechnology policy set goals, some achievements were made through the National Biotechnology Development Agency and National Biosafety Management Agency.

Table 1 Outlook on the ongoing project in medical biotechnology of NBRDA

Project title	Purpose	Status	Remarks
Short tandem repeat analysis of DNA from ethnic nationalities in Nigeria	To determine the allele distributions for the core 16 genetic loci, DNA profiles of independent donors from the chosen ethnic population at the instance (i.e., 34 donors from each population) were generated.	Early stage	Ongoing
Factors influencing the effective quantal immune reactions to primary HIV-1 and HCV isolate from resistant Nigerian individuals.	Develop both preventive vaccines for HCV/HIV	Early stage	Ongoing
To accurately detect and distinguish salmonella, a diagnostic test based on PCR is being developed.	To create a rapid assay that combines bacterial pre-enrichment of a stool sample with PCR to distinguish between different salmonella species utilizing non-coding RNAs as the target.	Early stage	Ongoing
Design and creation of HCV and COVID-19 viral antigen diagnostic test strips	To create a reliable detection technique that can accurately distinguish HCV from other illnesses	Early stage	Ongoing

Source: NABDA, 2024
Authors' Computation, 2024

Test kits were developed which include the TRI-HCG Pregnancy Rapid test kit, TRI-MALARIA test kit, TRI-HBsAg test kit for Hepatitis B, and TRI-SYPHILIS test kit for syphilis detection. These projects are in the commercialization phase. Stem cell/bone marrow transplant technology is also developed to cure patients with sickle cell at the University of Benin Teaching Hospital. Efforts have been made to ensure that this technology is available in every part of the country (Table 1).

3 Related Literature on Determinants of Health Outcomes

Authors	Title	Methodology	Findings
Kamiya (2010)	Effect of the health system and social-economic factors on health outcomes in developing countries	System generalized method of moments (system GMM)	It found that per capita income and sanitation have a significantly improves health outcomes.
Awais et al. (2016)	Current status of biotechnology in health	Descriptive technique	The study showed that health improvement can only be achieved through the application of biotechnology.

(continued)

Authors	Title	Methodology	Findings
Mtui (2011)	Status of biotechnology in selected eastern and central African countries		It is found that more vigorous practicals are needed for the countries to benefit from biotechnology in the aspect of health improvement. No actual information on the impact of biotechnology on health.
Shahbaz et al. (2016)	Determinants of life expectancy in Pakistan (1972–2012).	ARDL VECM granger-causality	It found that urbanization, food supplies, and health expenditure have a positive and significant effect on life expectancy and were adversely affected by economic misery and illiteracy.
Sede and Ohemeng (2015)	Socio-economic determinants of life expectancy in Nigeria (1980–2011)	VECM analysis and granger causality test	According to the study, life expectancy is unaffected by per capita income or level of education. It also discovered that life expectancy and per capita income have a bidirectional causal relationship, but government health spending and life expectancy have a unidirectional relationship.
Gupta and Chaphalkar (2016)	Role of biotechnology in human health care	Content analysis	It found that biotechnology has improved human health and reduced environmental pollution.
Afzal et al. (2016)	Role played by biotechnology human health	Descriptive analysis	The study found that biotechnology has improved human health, Well-being, and life expectancy have increased globally via biotechnology.
Hassan et al. (2016)	Determinants of life expectancy in 108 emerging countries.	Fixed effect model	It was discovered that life expectancy is positively impacted by education.
Edeme et al. (2017)	Public health expenditure and outcomes in Nigeria.	Ordinary Least Square	It is discovered that long-term association exists between health outcomes and government spending.
Ketenci and Murthy (2018)	Life expectancy and its determinants in the United States	Dynamic ordinary Least Square (DOLS)	It found that per capita income and education improves life expectancy.
Etikan and Babtope (2019)	Life expectancy and socio-economic factors in Nigeria	Ordinary Least Square (OLS)	It found that access to clean water, sanitation, and government expenditure positively affect life expectancy while private expenditure has a negative effect on health outcomes (life expectancy).

(continued)

Authors	Title	Methodology	Findings
Lawal et al. (2021)	Inflation and life expectancy in Nigeria: A causal analysis	VAR autoregressive lag order and granger causality test.	The granger causality conclusion showed that there is no unidirectional causal relationship between life expectancy and inflation rate in Nigeria across the years of investigation, but there is a causal relationship between life expectancy and exchange rate.
Akintunde et al. (2013)	Social-economic factors and life expectancy in sub-Sahara Africa.	Generalized method of moments (GMM)	It discovered that life expectancy is improved by GDP per capita, health spending per capita, and educational attainment. The incidence of HIV/AIDS and CO2 emissions have a detrimental effect on life expectancy.
Arikpo et al. (2019)	Macroeconomic determinants of life expectancy in Nigeria.	Ordinary least squares technique	The study found that the strongest impact on life expectancy is caused by income disparity. It further found that income per capita and government expenditure on health also have significant effects on life expectancy.
Aigheyisi (2020)	Does agricultural productivity affect Nigerian life expectancy determinants?	ARDL	The study discovered that while higher agricultural productivity has a short-term positive influence on life expectancy, it has a long-term detrimental effect. It also finds that life expectancy is negatively impacted by inflation and unemployment in the short and long terms. It is discovered that real per capita income has a negative long-run effect that is statistically negligible, but a negative short-run effect that indicates an uneven distribution of income.
Chireshe and Ocran (2020)	The relationship between health outcomes and healthcare spending in sub-Saharan Africa (SSA)	Fixed effect GMM	The findings show that life expectancy rises in direct proportion to total healthcare spending per capita.
Muhammad (2021)	Foreign health aids and Nigerian health outcomes.	ARDL	The study's findings demonstrated that both the short- and long-term health outcomes for Nigeria are improved by international medical aid.

(continued)

Authors	Title	Methodology	Findings
Şentürk and Ali (2021)	Life expectancy and socio-economic factors of the Turkish population	ARDL bound test	The results showed that education, purchasing power, and economic development improves life expectancy in Turkey.
Mehedi et al.	Factor affecting life expectancy in five SAARC countries.	Panel data estimation method	The analysis showed schooling and sanitation improve life expectancy. However, the overall fertility rate, the number of people living in cities, and CO2 emissions all have a detrimental effect on life expectancy.
Rubi et al. (2021)	The correlation between Bangladesh's GDP, population, and life expectancy.	Multiple linear regression model Artificial neural network (ANN)	The study revealed that GDP enhances life expectancy and population also affects life expectancy.
Green et al. (2021)	The COVID-19 pandemic's effects on adult health outcomes and social determinants of health.	Systematic review	The study found that low-income and racial minority vulnerable groups are particularly vulnerable to COVID-19 and have been disproportionately affected by it. It also discovered that COVID-19 has made gender inequality and familial violence worse, which has negatively impacted women's Well-being.
Galvani-Townsend et al. (2022)	Social determinant of longevity	Descriptive analysis	It found that public health spending can lead to high longevity.
Angell et al. (2022)	Population health outcomes in Nigeria and other west African countries	Systematic analysis	The study found that health outcomes remain poor. The study did not account for bioeconomy as a determinant of health outcomes.
Nwani (2022)	Air pollution trajectories and life expectancy in Nigeria	VECM	The VECM result showed that air pollution had a statistically significant negative impact on life expectancy. The granger causality result showed a one-way causal relationship between air pollution and life expectancy. The study did not account for bioeconomy as a determinant of health outcomes.

(continued)

Authors	Title	Methodology	Findings
Rahman et al. (2022)	The determinants of life expectancy	Panel corrected standard errors (PCSE) Feasible general least square (FGLS Pairwise granger causality	The study found that environmental degradation has a negative effect on life expectancy (LE) while LE is improved by health expenditure, clean water, and sanitation. It further discovered that there is unidirectional causality from carbon emissions to life expectancy and bidirectional causalities between drinking water and life expectancy and sanitation and life expectancy.
Liu and Zhong (2022)	Health expenditures, renewable energy life, and expectancy in China	VECM approach	The study discovered a long-term correlation between life expectancy, healthcare costs, and renewable energy. It also found that health spending and renewable energy improve life expectancy in China.
Batti et al.	Application of biotechnology in the modern healthcare system	Content analysis	Global health disparities could be lessened and better health outcomes could be achieved with the help of healthcare biotech.
Mohapatra (2022)	The impact of health expenditures, health infrastructure on health status in SAARC countries	Panel generalized Least Square	The primary influencing factor was discovered to be out-of-pocket expenses.
Setyadi et al. (2023)	*Determinants of modeling the life expectancy of males and females in Indonesia*	*Generalized method of moment (GMM)*	*The study found that GRDP* per capita, *literacy, unemployment, and poverty improve the life expectancy of males and females in Indonesia.*
Nandi et al. (2023)	The effect of socioeconomic variables on life expectancy in Bangladesh.	*Path analysis* Bivariate analysis	The results of the study showed that life expectancy is highly correlated with the employment rate, GNI, and age dependence ratio. Bivariate analysis also showed a significant correlation between all factors and life expectancy.
Roffia et al. (2023)	Factors influencing life expectancy at birth in 36 OECD nations	Fixed-effects multiple regression analysis	This study demonstrated that healthcare expenditure and healthcare system determine how long people live in a given nation.
Onisanwa et al. (2024)	Life expectancy determinants in Nigeria	Auto-regressive distributed lag (ARDL) bound test approach.	The study revealed that GDP per capita, adult literacy rate, total fertility and urbanization are positively and significantly connected with life expectancy. The effect of CO_2 emission on life expectancy at birth is negative and significant.

(continued)

Authors	Title	Methodology	Findings
Mimi et al. (2024)	Economic, environmental, and demographic factors and life expectancy in 36 high-income and 65 middle-income nations.	Panels corrected standard errors (PCSEs) regression	The study discovered that GDP per capita, population growth, financial development, food production, and fertility rates strengthen life expectancy.
Azimi and Rahman (2024)	Food insecurity, environmental degradation and health outcome.	GMM	The study indicated life expectancy is negatively affected by the prevalence of undernourishment, inequality in per capita calorie intake, and CO_2 emissions significantly. It also found that per capita kilocalorie supply, per capita GDP, per capita health expenditures, and urbanization significantly improve life expectancy.
Ţarcă et al. (2024)	The impact of CO_2 emissions, the number of doctors, and GDP per capita on healthcare spending on life expectancy.	Fixed effects	The study discovered that rising CO_2 emissions shorten life expectancy at birth, but rising health spending, the number of doctors per 10,000 people, and per capita income all enhance life expectancy.

Source: Authors' Computation, 2024

4 Model Specification, Data, and Methodology

4.1 Model Specification

Grossman (2000) identified stock of knowledge, i.e., education as a primary determinant of health. The efficiencies of production are determined by the consumer's stock of knowledge. This is because the more educated people may be more efficient at investing in health (Grossman, 2000). It is further explained that health can be inherited and it is a durable capital good that can depreciate over time, and the gradual natural deterioration of health can be partially offset through investment in health as medical care service is combined with other inputs to produce new health. Health is identified to depend on several factors, some of which can be influenced by the individuals. Hence, health can be produced. A production function consists of inputs and outputs from which a nation's health can be derived (Ali & Ahmad, 2014). According to Grossman (1972), a health production function can be theoretically written as:

$$\text{HE} = f(\text{X}) \quad (1)$$

where X is a vector of distinct inputs to the health production function, and HE is a measure of individual health outcomes. These inputs include income, education, community environment, nutrient intake, etc. However, this theoretical model is designed for health analysis at the micro-level. To analyze health production at the macro level, the elements of the vector by per capita variables are re-arranged into subsectoral vectors of economic, social, and environmental factors. Equation (1) can therefore be written as:

$$\mathrm{HE} = f(\mathrm{Y,S,V}) \tag{2}$$

Where per capita economic variables are represented by Y vector, per capita social variables are represented by a vector of S and per capita environment variables are represented by a vector of V. Equation (2) can further be written as:

$$HE = f\left(y_1, y_2, \ldots\ldots y_n, s_{1,} s_2, \ldots\ldots s_n, v_1, v_2, \ldots\ldots v_n\right) \tag{3}$$

However, Case and Deaton (2005) argue that technology that can stimulate health repair is superlative. Gupta and Chaphalkar (2016) and Awais et al. among others identified that bioeconomy can improve health outcomes. Therefore, the health production function can be written as:

$$\mathrm{HE} = f\left(\mathrm{y}, geh, \mathrm{pop}, edu, \mathrm{bioe}, cli, rex\right) \tag{4}$$

$$\mathrm{HE} = y^{\alpha_2}\ geh^{\alpha_3}\mathrm{pop}^{\alpha_4} edu^{\alpha_5}\mathrm{bioe}^{\alpha_6}\ cli^{\alpha_7}\ rex^{\alpha_8} \tag{5}$$

Writing eq. (5) in its linear, we have:

$$\mathrm{HE}_t = \alpha_1 \mathrm{he}_{t-1} + \alpha_2 y_t + \alpha_3 \mathrm{geh}_t + \alpha_4 pop_t + \alpha_5 edu_t + \alpha_6 \mathrm{bioe}_t + \alpha_7 cli_t + +\alpha_8 rex_t + \mu_t \tag{6}$$

Where *t* is time, HE is the log of health outcomes, *y* is the log per capita income, pop is log of population growth rate, rex is the log of the real exchange rate, edu is the log of education, geh is log of government expenditure on health, bioe is the log of bioeconomy, cli is climate change, and α_1, α_2, $\alpha_{3,}\alpha_4$, α_5, α_6, α_7, $\alpha_{8,}$ are coefficients of independent variables. A priori expectation is that per capita income, government expenditure on health, education, and bioeconomy will have a positive relationship with sustainable development while population growth, climate change, and the real exchange rate will have an inverse relationship with sustainable development. Mathematically, from Eq. (6), the a priori expectation is given as:

$$\alpha_1 > 0; \alpha_2 > 0; \alpha_3 > 0; \alpha_4 > 0; \alpha_5 > 0; \alpha_6 > 0; \alpha_4 < 0; \alpha_7 < 0 \,\mathrm{and}\, \alpha_8 < 0$$

4.2 Data

The data used were sourced from both World Bank Development Indicators (WDI) 2021 and the Central Bank of Nigeria Statistical Bulletin (2021). Bioeconomy is sourced from CBN while real exchange rate, per capita income, education, and climate change were sourced from WDI. Health outcomes is proxied using life expectancy which was sourced from WDI, and bioeconomy is measured by an index of value-added, export, and investment into bioeconomy sector of the economy following SAT-BBE Consortium (2014) and Bracco and Flammini (2018). This measurement of bioeconomy is also based on the availability of data.

4.3 Measurement of Variables

Variables	Description	Measurement	Sources
Bioeconomy (BE)	Bioeconomy is an economy in which biotechnology stimulates economic outputs to a large extent.	It is measured by an index of value-added, export, and investment (SAT-BBE Consortium, 2014; Bracco & Flammini, 2018)	CBN (2021)
Education (EDU)	Education is a measure of literacy in an economy.	Primary school enrolment.	WDI (2021)
Climate change (CLI)	Climate change is any significant change in the expected average weather of a region.	Carbon dioxide (CO_2)	WDI (2021)
Population growth rate (POP)	This is the measure of the increase in the number of individuals in a population.	The annual growth rate of the population.	WDI (2021)
Government expenditure on health (GEH)	This refers to the proportion of GDP that the government spends on health.	It is measured as the proportion of GDP spent on health.	CBN (2021)
Health outcomes (HE)	This refers to the state of complete physical, mental, and social Well-being.	Life expectancy	WDI (2021)
Per capita income (Y)	Real per capita income is the income earned per person in a particular country.	It is measured by dividing the total national income by the population.	WDI (2021)
Real exchange rate (REX)	A real effective exchange rate is the value of a currency relative to a weighted average of multiple foreign currencies.	It is measured as a nominal effective exchange rate divided by a price deflator.	WDI (2021)

Source: Authors' Computation, 2024

4.4 Methodology

The study used Hansen's (1982) Generalized Method of Moments (GMM). This is justified given that using OLS to estimate the model results in skewed and inconsistent estimates when the lagged dependent variable is included among the independent variables due to simultaneity and endogeneity issues among the model's variables. The suitability of the employed instruments has a significant impact on the dependability of the GMM estimates. Theoretically, a variable must be connected to explanatory variables but unconnected to the residual term to be utilized as an instrument. According to the literature, an instrument can be created using the lag of explanatory variables. The assumption that past values of variables are uncorrelated with error tern underlies the use of lag values of explanatory variables as instruments.

For the models, the Sargan test value should be closer to one, indicating that the instrument is valid and that the null hypothesis can be accepted. To validate and defend the usage of the GMM model, it is crucial to assess whether there is a serial correlation in the residual. Therefore, the study uses the Arellano and Bond serial correlation AR (1) and AR (2) tests, devised by Arellano and Bond (1991), following the literature on GMM. Specifically, first-order and second-order serial correlation are tested for in the model using the AR (1) and AR (2) tests.

5 Results, Interpretation and Discussion

5.1 Unit Root

There are two types of time series data: stationary and non-stationary. Nevertheless, a misleading regression will be produced when a stationary time series variable is regressed on a non-stationary time series variable, and a meaningless regression will be produced when a non-stationary variable is regressed on a non-stationary variable (Gujarati & Porter, 2009). Likewise, a meaningless regression cannot help in developing the right policy for an economy to solve either micro or macroeconomic problems. A time series variable must be stationary so that its behavior can be studied over several periods, if not stationary its period can only be studied for the period under consideration. It is, therefore, important to test if a time series variable is stationary or not. The unit root test is usually conducted to know if a time series variable is stationary at a level or not. As a result, the unit root tests were conducted using the Philips–Perron (PP) test and the Augmented Dickey–Fuller (ADF) test in order to prevent erroneous regression estimates. The results are shown in Table 2.

From Table 3 climate change (CLI), real exchange rate (REX), health outcomes (HS), education (EDU), government expenditure on health (GEH), per capita income (Y) are integrated of order one I(1), which indicates that the variables are I(1) using Augmented Dickey–Fuller test. It further shows that bioeconomy (BIOE) and population growth (POP) the unit root test reveals that these variables are integrated of order zero. Philips–Perron test also reveals education (EDU), climate change (CLI), health outcomes (HE), unemployment (UNEM), per capita income (Y), government expenditure on health (GEH), and real exchange rate (REX) are integrated of order one which indicates that the variables are I(1) and capital stock (CAP), population growth (pop) and bioeconomy (BIOE) are integrated of order zero that is I(0).

Table 2 Unit root results

Augmented Dickey–Fuller (ADF) test				Philips–Perron (PP) test		
Variable	Level	First Difference	Status	Level	First Difference	Status
EDU	−1.852163 (0.3504)	−5.188067* (0.0003)	I(1)	−1.921037 (0.3194)	−5.173075* (0.0001)	I(1)
CLI	−2.171002 (0.2197)	−6.602008* (0.0000)	I(1)	−2.200143 (0.2095)	−6.5980878* (0.0000)	I(1)
HE	−0.539060 (0.8719)	−3.576650* (0.0113)	I(1)	2.570772 (1.0000)	−3.702337* (0.0081)	I(1)
POP	−3.141642** (0.0320)	–	I(0)	−3.592684** (0.0106)		I(0)
Y	−0.250022 (0.9228)	−4.923529*** (0.0003)	I(1)	−0.589897 (0.08609)	−4.920665*** (0.0003)	I(1)
REX	−2.033669 (0.2718)	−4.664120* (0.0006)	I(1)	−2.192334 (0.2122)	−4.528309* (0.0009)	I(1)
BIOE	−2.712024*** (0.0813)		I(0)	−2.820273*** (0.0649)		I(0)
GEH	−1.386465 (0.5764)	−9.832980* (0.0000)	I(1)	−0.859057 (0.7896)	−17.98388* (0.0001)	I(1)

Note that the meaningful level is * = 1%, ** = 5%, and *** = 10%. The automatic maximum lag length based on Schwarz information criterion is applied to the augmented Dickey–Fuller (ADF) test, while the automatic maximum lag length based on Newey–West Bandwidth is applied to the Philips–Perron (PP) test

Table 3 The effect of bioeconomy on health outcomes in Nigeria

Dependent variable: health outcomes		
Independent variables	Coefficient	*p* value
HE (−1)	0.991 (38.966)	[0.000]*
CLI	−0.017 (−7.079)	[0.000] *
EDU	−0.005 (−1.316)	[0.203]
BIOE	−0.001 (−4.317)	[0.000] *
Y	0.006 (−0.885)	[0.386]
POP	−0.048 (−1.660)	[0.113]
REX	−0.001 (−1.975)	[0.062]
Instrument rank = 12 Sargan test = 0.458 AR (1) = 0.440 AR (2) = 0.380		

Note: 1. * = 1%. 2. *() are t-statistics*. 3. [] are p values 4. *Source: Author's computation, 2024*

5.2 Results and Interpretation

Before the model estimation is carried out, it is essential to test for the validity of the instruments adopted in the model. To test the validity of instruments in the models, the study examined the instrument rank. The instrument rank is greater than the number of GMM estimated parameters (explanatory variable), this indicates that it could be adopted to calculate the Sargan test of over-identifying restrictions (Gujarati & Porter, 2009).

The effect of Nigeria's bioeconomy on health outcomes is seen in Table 3. The result from Table 3 shows that bioeconomy (bioe) has a negative but significant effect on health outcomes ($t = -4.317$, $p < 0.05$). This implies that the government has not invested in bioeconomy to the level that it will promote health outcomes. Likewise, the result shows that climate change affect health outcomes negatively and significantly ($t = -4.226$, $p < 0.05$). This means an increase in climate change impedes health outcomes.

Education (EDU) negatively but insignificantly affects health outcomes ($t = -1.316$, $p > 0.05$). This implies that a decrease in educational level impedes health outcomes. The real exchange rate (REX) has a negative but insignificant effect on health outcomes ($t = 1.862$, $p > 0.05$). This explains that an increase in real exchange impedes health outcomes. Likewise, the increase in population (POP) negatively impacts health outcomes ($t = -1.660$, $p > 0.05$). This shows that an increase in population growth rate (pop) hinders health outcomes.

Similarly, Table 3 shows that per capita income (y) has a positive and insignificant effect on health outcomes ($t = 0.885$, $p > 0.05$). The lagged value of health outcomes has a positive and significant effect on health outcomes ($t = 38.966$, $p < 0.05$). This implies that achieving good health outcomes in the previous year does guarantee good health in the current year. The statistical verification of the result is revealed by the Sargan test and autocorrelation test. The Sargan test of over-identifying restriction has a value of 0.458, which is closer to one and it shows that the instruments are valid. Arellano-Bond test is also conducted and both AR (1) and AR (2) reveal that there is no autocorrelation in the model with the probability values of 0.440 and 0.380 respectively.

5.3 Discussion of Findings

The study finds that bioeconomy (bioe) has a negative) effect (−0.001) on health outcomes in Nigeria. This is contrary to the a priori expectation that bioeconomy promotes outcomes. This could be attributed to new greenhouse emissions that could be triggered by bioeconomy and stimulate harmful effects on the environment through quality and availability of water, quality of soil, and deforestation (Haines, 2021). This negative effect could also be attributed to food vs. fuel competition, as the same natural resources are used in the production of food and fuel (Bourguignon, 2017; Ogwu, 2019). For instance, in Nigeria, cassava is used in the production of biofuel and at the same time used in the production of food such as bread, fufu, garri which can lead to food vs. fuel competition. This is also contrary to Awais et al. (2010), Sharad et al. (2010), Afzal et al. (2016), and Gupta and Chaphalkar (2016) who found that bioeconomy promotes health outcomes.

Similarly, education (edu) is also found to have a negative effect (−0.005) on health outcomes. This is contrary to the a priori expectation that education spurs health outcomes. Health outcome is expected to be influenced positively by education. The findings from the study are contrary to studies by Kamiya (2010), Paul and Mehera (2016), and Hassan who found that education promotes health outcomes. This could be attributed to the lack of educational facilities required for the enhancement of learning, lack of skilled teachers, and lack of basic knowledge of health. This explains that low or lack of education may hinder some individuals from avoiding sedative jobs and taking necessary precautions. The study is supported by Sede and Ohemeng (2015), Ketenci and Murthy (2018), and Akintunde et al. (2013) who found that education negatively affects life expectancy which is a measure of health outcomes.

Climate change (cli) exerts negative effect (−0.017) on health outcomes. This supports the a priori expectation that climate change negatively affects health outcomes. This implies that human activities which include deforestation, agriculture, and burning of fossil fuels, among others, could have a negative effect on the environment which further impedes health outcomes. Climate change negatively affects both human mental and physical health. This finding is consistent with the view of Koubi (2019), Akintunde et al. (2013), and Onisanwa et al. (2024) that climate change has a negative effect on life expectancy as a measure of health outcomes. This indicates that there is a need for the Nigerian government to increase its allocation for programs that are geared toward the mitigation of the negative effects of climate change.

Population growth (POP) has a negative effect (−0.048) on health outcomes. This supports the a priori expectation that an increase in population will negatively affect health outcomes. This implies that healthcare resources in Nigeria do not correspond to the growth rate of the population which negatively affects health outcomes. This corroborates with Onwuka (2006) and Güney (2017) who found that an increase in population affects health outcomes in Nigeria and developing countries negatively as the demand for health services increases with low production of health services. However, this study contradicts the study by Mimi et al. (2024).

The study further found that health outcomes is positively affected by per capita income (y) but insignificant. This supports the a priori expectation that per capita income promotes health outcomes. This implies that per capita income in Nigeria is significantly low to extensively promote health outcomes of individuals in Nigeria. Therefore, there is a need for the government to design policies that will enhance the increase in per capita income.

The findings further reveal that the real exchange rate (rex) has a negative effect on health outcomes. This is consistent with the a priori expectation that the increase in the real exchange rate has a negative effect on health outcomes in Nigeria. This implies that an increase in the real exchange rate impedes health outcomes through an increase in importation prices. This is because the devaluation of currency increases the costs of importation which further leads to an increase in healthcare services. After all, the equipments and drugs are not produced within the country. This supports studies by Shahbazi and Asadi (2014) and Sede and Ohemeng (2015) who found that the real exchange rate has a negative effect on health which is a component of sustainable development.

5.4 Challenges and Opportunities of Medical Biotechnology

According to the investigation carried out by the study, bioeconomy through biotechnology application to health does not improve health outcomes in Nigeria despite the establishment of the National Biotechnology Development Agency (NABDA) and biotechnology development centers across the country. Some factors have been identified as the challenges to the application of biotechnology to health to derive the expected impact of bioeconomy on health outcomes.

I. Biotechnology application involves a huge capital to set up, whereas in Nigeria almost all the standard and well-equipped medical research and diagnostic centers are privately owned, which has made it difficult for the poor to have access to these services for them to be in good health. This has also been an obstacle for small-scale entrepreneurs to undertake such business as it requires huge capital to enter the industry. The DNA labs are an example of a private medical biotechnology center in Nigeria that is faced with the problem of financial constraints (Bello et al., 2023). To overcome the challenge of financial constraints and at the same time encourage private investors in medical biotechnology, the government should provide tax incentives in the form of tax holidays, investment allowances, tax credits, and reductions in tax rates, which could lead to an increased and improved healthcare technologies, creation of job opportunities, increased per capita income, thereby improving standard of living, and henceforth increased economic growth.

II. Lack of effective and efficient collaboration: There is no partnership for the private sector that engaged in the medical biotechnology sector. This has deterred new investors from entering the market. Ineffective collaboration can also impede the growth of the current company in the sector, as demonstrated by the indigenous center's (DNA Labs) experience (Bello et al., 2023). To effectively capitalize on the economic opportunities presented by medical biotechnology, emerging nations should allocate resources toward establishing a research and development-friendly environment. One tactic that is frequently used is the establishment of public–private partnerships (PPPs).

III. Lack of adequate experts in the field of biotechnology: There are yet to be sufficient experts in the field of biotechnology and adequate employees in healthcare centers to enable the maximum utilization of biotechnology in the healthcare sector of the economy (Bello et al., 2023). For instance, the problem of inadequate employees was experienced during the COVID-19 break out.

IV. Nigeria has not yet offered solid foundations for medical biotechnology advancements that are consistent with the nation's social, environmental, and economic sustainability.

However, there are still opportunities for medical biotechnology in Nigeria. First, the increase in Nigeria's population is a great requirement for achieving the benefits that bioeconomy has to offer. This is because the range of the population that is employable is increasing; however, there is a need for the government to invest in research and development (R&D) which is the bedrock of innovation. This will help enable the government to overcome the challenge of an inadequate supply of skilled manpower.

Second, in terms of biopharmaceutical drugs, Nigeria is rich in bioresources that can be used in the production of drugs and vaccines. The use of this abundance of bioresources will help in the reduction of capital needed for the production of drugs. Lastly, the government should provide a conducive environment for both foreign and domestic investors to invest in medical biotechnology.

6 Conclusion

Despite the efforts that the government geared toward the transformation of the economy to bioeconomy through the application of biotechnology and establishment of the National Biotechnology Development Agency (NABDA), the health outcomes in Nigeria have not improved. The study found that bioeconomy has a negative effect on health outcomes in Nigeria which can be attributed to the challenges of medical biotechnology. The study therefore concludes that bioeconomy does not promote health outcomes in Nigeria. The study therefore recommends that for the negative effect of bioeconomy on health outcomes to be off-set, the government should encourage private–public health partnerships by granting incentives to the private entrepreneurs in the health sector to promote the use of biodrugs and biotest kits.

References

Afzal, H., Zahid, K., Ali, Q., Sarwar, K., Shakoor, S., Nasir, U., & Nasir, I. A. (2016). Role of biotechnology in improving human health. *Journal of Molecular Biomarkers and Diagnosis, 7*(06).

Agua, L., De Aguas, E. E. E. T., & Colombia, A. E. (2017). *Misión De Crecimiento Verde*.

Aigheyisi, O. (2020). *Determinants of life expectancy in Nigeria: Does agricultural productivity matter?* (pp. 194–210).

Ajayi, O. F., & Akinbobola, T. O. (2020). Health status and economic growth. In *Decent work and economic growth* (pp. 521–530). Springer.

Akintunde, T. S., Olomola, P. A., & Oladeji, S. I. (2013). Population dynamics and economic growth in Sub-Saharan Africa. *Journal of Economics and Sustainable Development, 04*(13), 148–157.

Alhassan, A., & Abdu, M. (2017). Socio-economic determinants of health status in Nigeria: A case study of Ilorin Metropolis. *Lafia Journal of Economics and Management Sciences, 2*(1), 277–292.

Ali, A., & Ahmad, K. (2014). *The impact of socio-economic factors on life expectancy for Sultanate of Oman: An empirical analysis.*

Angell, B., Sanuade, O., Adetifa, I. M., Okeke, I. N., Adamu, A. L., Aliyu, M. H., Ameh, E. A., Kyari, F., Gadanya, M. A., Mabayoje, D. A., & Yinka-Ogunleye, A. (2022). Population health outcomes in Nigeria compared with other west African countries, 1998–2019: A systematic analysis for the Global Burden of Disease Study. *The Lancet, 399*(10330), 1117–1129.

Arellano, M., & Bond, S. (1991). Some tests of specification for panel data: Monte Carlo evidence and an application to employment equations. *The Review of Economic Studies, 58*(2), 277–297.

Arikpo, D. I., Eke, F. A., & Obafemi, F. N. (2019). Determinants of life expectancy in Nigeria: A macroeconomic analysis. *International Journal of Social Science and Economic Research, 4*(4), 1–15.

Atter, A., Obiri-Danso, K., & Amoa-Awua, W.K. (2014). *Microbiological and chemical processes associated with the production of burukutu a traditional beer in Ghana.*

Awais, M., Pervez, A., Yaqub, A., Sarwar, R., Alam, F., & Siraj, S. (2010). Current status of biotechnology in health. *American-Eurasian Journal of Agricultural & Environmental Sciences, 7*(2), 210–220.

Azimi, M. N., & Rahman, M. M. (2024). Food insecurity, environment, institutional quality, and health outcomes: Evidence from South Asia. *Globalization and Health, 20*(1), 21.zqa.

Bello, M. S., Aminu, Z. A., Kwaru, M. A., & Mustapha, S. H. (2023). Challenges in establishing innovation in the Biotechnologyector: A Case of Dna Lab, Nigeria. *The International Journal of Business & Management, 11*(4), 16–20.

Biointropic. (2018). *Bioeconomía como fuente de nuevas industrias basadas en el capital natural de Colombia*. Medellín.

Bourguignon, D. (2017). *Bioeconomy. Challenges and opportunities*

Bracco, S., & Flammini, A. (2018). Assessing the contribution of bioeconomy to countries' economy. In *A brief review of national frameworks.*

Canales, N., & Gómez González, J. (2020). *Policy dialogue on a bioeconomy for sustainable development in Colombia*. Stockholm Environment Institute.

Carter, D. J., Glaziou, P., Lönnroth, K., Siroka, A., Floyd, K., Weil, D., Raviglione, M., Houben, R. M., & Boccia, D. (2018). The impact of social protection and poverty elimination on global tuberculosis incidence: A statistical modelling analysis of Sustainable Development Goal 1. *The Lancet Global Health, 6*(5), e514–e522.

Case, A., & Deaton, A. (2005). Health and wealth among the poor: India and South Africa compared. *American Economic Review, 95*(2), 229–233.

Chahal, G. S., & Gosal, S. S. (2002). *Principles and procedures of plant breeding: Biotechnological and conventional approaches*. Alpha Science Int'l Ltd..

Chireshe, J., & Ocran, M. K. (2020). Health care expenditure and health outcomes in sub-Saharan African countries. *African Development Review, 32*(3), 349–361.

Consejo Nacional de Política Económica y Social. (2009). *Política Nacional de Ciencia, Tecnología e Innovación*. CONPES 3582. https://minciencias.gov.co/sites/default/files/upload/reglamentacion/conpes-3582-2009.pdf

DG Research. (2005). *New perspective based bioeconomy*. DG Research, Conference report. http://edz.bib.unimannhemin.de/daten/edzbra/05/kbbe_conferencereportpdf. Accessed 18 2020

Edeme, R. K., Emecheta, C., & Omeje, M. O. (2017). Public health expenditure and health outcomes in Nigeria. *American Journal of Biomedical and Life Sciences, 5*(5), 96–102.

Egboduku, W. O., Egboduku, T., Golohor, O. M., Imarhiagbe, O., & Ogwu, M. C. (2024). Cassava as raw material for sustainable bioeconomy development. In M. C. Ogwu, S. C. Izah, A. A. Cunha Alves, & babu, S.C. (Eds.), *Plant biology, sustainability and climate change, sustainable cassava* (pp. 57–73). Academic Press. https://doi.org/10.1016/B978-0-443-21747-0.00022-9

Etikan, I., & Babtope, O. (2019). A basic approach in sampling methodology and sample size calculation. *Med Life Clinic, 1*(2), 1006.

Galvani-Townsend, S., Martinez, I., & Pandey, A. (2022). Is life expectancy higher in countries and territories with publicly funded health care? Global analysis of health care access and the social determinants of health. *Journal of Globalization and Health, 12*.

Gartland, K. M., Bruschi, F., Dundar, M., Gahan, P. B., Magni, M. V., & Akbarova, Y. (2013). Progress towards the 'Golden Age'of biotechnology. *Current Opinion in Biotechnology, 24*, S6–S13.

Graff, G., Roland-Holst, D., & Zilberman, D. (2006). Agricultural biotechnology and poverty reduction in low-income countries. *World Development, 34*(8), 1430–1445.

Green, H., Fernandez, R., & MacPhail, C. (2021). The social determinants of health and health outcomes among adults during the COVID-19 pandemic: A systematic review. *Public Health Nursing, 38*(6), 942–952.

Grossman, M. (1972). On the concept of health capital and the demand for health. *Journal of Political Economy, 80*(2), 223–255.

Grossman, M. (2000). The human capital model. In *Handbook of health economics* (Vol. 1, pp. 347–408). Elsevier.

Gujarati, D. N., & Porter, D. C. (2009). *Basic econometrics* (Vol. 5, pp. 1–15).

Gulzar, N., Saleem, I. M., Rafiq, S., & Nadeem, M. (2019). Therapeutic potential of probiotics and prebiotics. In *Oral health by using probiotic products*. IntechOpen.

Güney, T. (2017). Population growth and sustainable development in developed-developing countries: AN IV (2SLS) APPROACH GELİŞMİŞ-GELİŞMEKTE OLAN ÜLKELERDE NÜFUS ARTIŞI VE SÜRDÜRÜLEBİLİR KALKINMA: BİR IV (2SLS) YAKLAŞIMI. *Population, 22*(4), 1255–1277.

Gupta, A., & Chaphalkar, S. R. (2016). Role of biotechnology in human health care. *Emergent Life Sciences Research, 2*, 67–72.

Haines, A. (2021). Health in the bioeconomy. *The Lancet Planetary Health, 5*(1), e4–e5.

Hassan, F. A., Minato, N., Ishida, S., & Mohamed Nor, N. (2016). Social environment determinants of life expectancy in developing countries: A panel data analysis. *Global Journal of Health Science, 9*(5), 105.

Henry, G., Hodson, E., Aramendis, R., Trigo, E., & Rankin, S. (2018). *Bioeconomy: An engine for integral development of Colombia*.

Hodson de Jaramillo, E., Trigo, E., Henry, G., Aramendis Ramírez, R. H., Castaño, A., Coremberg, A., et al. (2019). La bioeconomía. Nuevo marco para el crecimiento sostenible en América Latina: Bioeconomy. In *New framework sustainable growth in Latin America*. Editorial Pontificia Universidad Javeriana.

IACGB. (2020). *Global bioeconomy policy report (IV): A decade of bioeconomy policy development around the world*. Secretariat of the Global Bioeconomy Summit 2020.

India Bioeconomy Report (IBER). (2022).

India Bioeconomy Report (IBER). (2023).

Ivase, T. J. P., Ali, J. B., Moveh, S., Dodo, Y. A., Otitolaiye, V. O., & Ogenyi, B. (2019). Current status and challenges of agricultural biotechnology in Nigeria: A concise review. *Materials and Organisms, 6*(9).

Izah, S. C., Ovuru, K. F., & Ogwu, M. C. (2022). Lassa fever in Nigeria: Social and ecological risk factors exacerbating transmission and dustainable management strategies. *International Journal of Tropical Diseases, 5*(2), 065. https://doi.org/10.23937/2643-461x/1710065

Kamiya, Y. (2010). *Determinants of health in developing countries: Cross-country evidence*. Osaka School of International Public Policy (OSIPP) Discussion Paper.

Ketenci, N., & Murthy, V. N. (2018). Some determinants of life expectancy in the United Statesresults from cointegration tests under structural breaks. *Journal of Economics and Finance, 42*, 508–525.

Kolawole, O. M., Kayode, R. M. O., & Akinduyo, B. (2007). Proximate and microbial analyses of burukutu and pito produced in Ilorin, Nigeria. *African Journal of Biotechnology, 6*(5), 587.

Koubi, V. (2019). Climate change and conflict. *Annual Review of Political Science, 22*, 343–360.

Lawal, N. A., Osinusi, K. B., & Bisiriyu, S. O. (2021). Inflation and life expectancy in Nigeria: A causal analysis. *Acta Universitatis Danubius. Œconomica, 17*(5).

Lee, A. M., & Leung, S. S. K. (2014). *Health outcomes*. Encyclopedia of Quality of Life and Well-Being Research.

Liu, H., & Zhong, K. (2022). Relationship between health spending, life expectancy and renewable energy in China: A new evidence from the VECM approach. *Frontiers in Public Health, 10*, 993546.

Lyytimäki, J., Antikainen, R., Hokkanen, J., Koskela, S., Kurppa, S., Känkänen, R., & Seppälä, J. (2018). Developing key indicators of green growth. *Sustainable Development, 26*(1), 51–64.

Maciejczak, M. (2017). Bioeconomy as a complex adaptive system of sustainable development. *Marketing, 2*(2), 7–10.

Mimi, M. B., Kibria, M. G., & Selim, M. M. I. (2024). How do economic, health, environmental and demographic factors affect life expectancy?–A novel attempt for developed and developing economies. *International Journal of Sustainable Development & World Ecology*, 1–16.

Mishra, P., et al. (2015). Impact of health on economic productivity. *International Journal of Management and Commerce Innovation, 3*(1), 477–481.

Mohapatra, S. (2022). Health expenditures, health infrastructure and health status in SAARC countries: A panel data analysis. *Vikalpa, 47*(3), 205–216.

Mtui, G. Y. (2011). Status of biotechnology in Eastern and Central Africa. *Biotechnology and Molecular Biology Review, 6*(9), 183–198.

Muhammad and Maryam Ahmad. (2021). Health aid and health outcomes in Nigeria: The role of governance. *Health Economics and Outcomes Research, 7*(4), 171.

Musonda, F., Millinger, M., & Thrän, D. (2021). Optimal biomass allocation to the Germanbioeconomy based on conflicting economic and environmental objectives. *Journal of Cleaner Production, 309*, 127465.

Nandi, D. C., Hossain, M. F., Roy, P., & Ullah, M. S. (2023). An investigation of the relation between life expectancy & socioeconomic variables using path analysis for Sustainable Development Goals (SDG) in Bangladesh. *PLoS One, 18*(2), e0275431.

Nkpoyen, F., Bassey, G. E., & Uyang, F. A. (2014). Health capital and poverty reduction in rural Cross River State, Nigeria. *Journal of Education and Research, 2*(5), 357–372.

Nwani, S. E. (2022). Air pollution trajectories and life expectancy in Nigeria. *International Journal of Social Economics, 49*(7), 1049–1070.

Nzimiro, I. (1985). *The green revolution in Nigeria or modernization of hunger*. Zim Pan.

OECD. (2009). *The bioeconomy to 2030: Designing a policy agenda*. Main findings and policy conclusions. www.oecd.org/futures/longtermtechnologicalsocietalchallenges/42837897

Ogwu, M. C. (2019). Understanding the composition of food waste: An "-omics" approach to food waste management. In A. P. Gunjal, M. S. Waghmode, N. N. Patil, & P. Bhatt (Eds.), *Global initiatives for waste reduction and cutting food loss* (pp. 212–236). IGI Global. https://doi.org/10.4018/978-1-5225-7706-5.ch011

Ohimain, E. I. (2013). A review of the Nigerian biofuel policy and incentives (2007). *Renewable and Sustainable Energy Reviews, 22*, 246–256.

Oluwajoba, S. O., Akinyosoye, F. A., & Oyetayo, V. O. (2013). In vitro screening and selection of probiotic lactic acid bacteria isolated from spontaneously fermenting Kunu-Zaki. *Advances in Microbiology, 3*(04), 309.

Onisanwa, I., Oloruntimilehin, N. I., Ikeh, F. E., & Ndangra, D. P. (2024). Social, economic and environmental determinants of life expectancy in Nigeria. *Tanzanian Economic Review, 14*(1), 69–96.

Onisanwa, I. D., Sunday, B. S. I., & Adaji, M. O. (2018). Healthcare financing and health status analysis in Nigeria. *Amity Journal of Healthcare Management, 3*(2), 31–42.

Onwuka, E. C. (2006). Another look at the impact of Nigeria's growing population on the country's development. *African Population Studies, 21*(1).

Panel, G. (2017). Urban diets and nutrition: Trends, challenges and opportunities for policy action. *Policy Brief, 9*.

Paul, P. K., & Mehera, C. (2016). Impacts of education on sustainable development: A micro study in Burdwan District of West Bengal, India. *American Journal of Educational Research, 4*(7), 551–555.

Philp, J. (2018). The bioeconomy, the challenge of the century for policy makers. *New Biotechnology, 40*, 11–19.

Rahman, M. M., Rana, R., & Khanam, R. (2022). Determinants of life expectancy in most polluted countries: Exploring the effect of environmental degradation. *PLoS One, 17*(1), e0262802.

Roffia, P., Bucciol, A., & Hashlamoun, S. (2023). Determinants of life expectancy at birth: A longitudinal study on OECD countries. *International Journal of Health Economics and Management, 23*(2), 189–212.

Rubi, M. A., Bijoy, H. I., & Bitto, A. K. (2021, July). Life expectancy prediction based on GDP and population size of Bangladesh using multiple linear regression and ANN model. In *2021 12th international conference on computing communication and networking technologies (ICCCNT)* (pp. 1–6). IEEE.

SAT-BBE Consortium. (2014). Annotated bibliography on qualitative and quantitative models for nalysing the bioeconomy. In *Systems analysis tools framework for the EU bio-based economy strategy*.

Sede, P. I., & Ohemeng, W. (2015). Socio-economic determinants of life expectancy in Nigeria (1980–2011). *Health Economics Review, 5*(1), 1–11.

Şentürk, İ., & Ali, A. (2021). Socioeconomic determinants of gender-specific life expectancy in Turkey: A time series analysis. *Sosyoekonomi, 29*(49), 85–111.

Serageldin, I. (1999). Biotechnology and food security in the 21st century. *Science, 285*(5426), 387–389.

Setyadi, S., Didu, S., Indriyani, L., Fitri, A. K., & Wiidiastuti, A. (2023). Modeling life expectancy in Indonesia using system GMM model. *Review of Applied Socio-Economic Research, 25*(1), 83–98.

Shahbaz, M., Loganathan, N., Mujahid, N., Ali, A., & Nawaz, A. (2016). Determinants of life expectancy and its prospects under the role of economic misery: A case of Pakistan. *Social IndicatorsResearch, 126*, 1299–1316.

Shahbazi, K., & Asadi, F. (2014). Impact of exchange rate on imports of medicines and medical equipment. *The Journal of Economic Policy, 6*(11), 35–54.

Sharad, S., Jaya, D., & Jha, A. K. (2010). Role of biotechnology in sustainable development. *International Journal of Research in Ayurveda and Pharmacy (IJRAP), 1*(1), 43–46.

Țarcă, V., Țarcă, E., & Moscalu, M. (2024, June). Social and economic determinants of life expectancy at birth in Eastern Europe. In *Healthcare* (Vol. 12(11), p. 1148). MDPI.

UN. (2020). *Why-it-matters-2020.pdf* (un.org). Accessed August 13 2023.

Wang, T., Yu, Z., Ahmad, R., Riaz, S., Khan, K. U., Siyal, S., Chaudhry, M. A., & Zhang, T. (2022). Transition of bioeconomy as a key concept for the agriculture and agribusiness development: An extensive review on ASEAN countries. *Frontiers in Sustainable Food Systems, 6*, 998594.

Wei, X., Luo, J., Pu, A., Liu, Q., Zhang, L., Wu, S., et al. (2022). From biotechnology to bioeconomy: A review of development dynamics and pathways. *Sustainability, 14*(16), 10413.

Whitman, A., De Lew, N., Chappel, A., Aysola, V., Zuckerman, R., & Sommers, B. D. (2022). Addressing social determinants of health: Examples of successful evidence-based strategies and current federal efforts. *Office of Health Policy, 1*, 1–30.

World Health Organization. (2018). *The state of food security and nutrition in the world 2018: Building climate resilience for food security and nutrition*. Food & Agriculture Org.
World Health Organization. (2020). Determinants of health. *Health Impact Assessment (HIA)*.
Zhang, X., Zhao, C., Shao, M. W., Chen, Y. L., Liu, P., & Chen, G. Q. (2022). The roadmap of bioeconomy in China. *Engineering Biology, 6*(4), 71–81.
Zilberman, D. (2014). Fellows address: The economics of sustainable development. *American Journal of Agricultural Economics, 96*(2), 385–396.

Rice Processing Techniques and Efficiency Differentials: A Case Study from Nigeria

Omobolaji O. Obisesan, Olorunfemi O. Ogundele, and Oluremi J. Akintayo

Abstract

The link between processing techniques and efficiency differentials among rice processors significantly influences the quality of processed rice. However, limited information exists on this connection, necessitating this research review. Research articles on rice processing, techniques of rice processing, and efficiency of processing were studied to understand the differentials among techniques. This research review was able to deduce that efforts to enhance rice processing efficiency in Nigeria have been underway, driven by government programs, infrastructure investment, and mechanization promotion. This has been fueled by initiatives such as the Rice Transformation Agenda, the Anchor Borrowers' Program, etc. seeking to boost domestic production and reduce reliance on imports. Rice processors' socioeconomic characteristics (age of processors, educational status, experience in processing, membership of association) and processing characteristics (distance to processing centers, access to paddy, access to credit, etc.) were major determinants of the efficiency of rice processing in the country. Additionally, we found out that private sector investments in modern milling facilities and technology adoption have improved efficiency and quality standards in the industry. Despite progress, challenges persist. Inadequate infrastructure, finance limitations, high energy, spare parts and transportation costs, and losses during processing have been found to hamper optimal efficiency. Therefore, addressing these hurdles necessitates collaborative efforts involving rice processors' groups, government, the private sector, other stakeholders, and development partners fostering sustainable growth in the rice value chain.

O. O. Obisesan (✉) · O. J. Akintayo
Institute of Agricultural Research and Training (O.A.U), Moor Plantation, Apata, Ibadan, Oyo State, Nigeria

O. O. Ogundele
Nigerian Institute of Social and Economic Research (NISER), Ibadan, Oyo State, Nigeria

M. C. Ogwu et al. (eds.), *Sustainable Bioeconomy Development in the Global South*, https://doi.org/10.1007/978-981-96-0305-3_15

Keywords

Efficiency differentials · Processing efficiency · Processing techniques · Rice processors · Nigeria

Abbreviation

AE	Allocative Efficiency
DEA	Data Envelopment Analysis
DFA	Distribution Free Approach
DRS	Decreasing Returns to Scale
EE	Economic Efficiency
FAO	The Food and Agriculture Organization
FDH	Free Distribution Hull
GMA	Gross Margin Analysis
MFC	Marginal Fixed Cost
MVP	Marginal Value Product
PFE	Partial Frontier Efficiency
RGAN	Rice Growers Association of Nigeria
ROI	Return On Investment
RTFM	Recursive Thick Frontier Model
RTS	Returns to scale
SDG	Sustainable Development Goals
SPFF	Stochastic Production Frontier Function
TE	Technical Efficiency
TFA	Thick Frontier Approach

1 Introduction

Rice is one of the staple crops that has been identified as crucial for achieving Sustainable Development Goal 2 (SDG 2) capable of putting an end to hunger. Its processing however faces hurdles as a result of many constraining factors, among which are inadequate processing and management, thereby resulting in production losses (Akolgoa and Asumboya, 2016). Despite a notable increase in rice production, with the area cultivated growing from approximately 3.1 million hectares (7.7 million acres) in 2015 to 5.9 million hectares in 2018, rice production witnessed a decrease of 5.3 million hectares by the year 2020 (Mwai, 2022). Nigeria still relies on imports due to the failure to meet domestic demand (Obisesan et al., 2019). Processing inefficiencies, not limited to equipment limitations, technology adopted, and resource allocation, contribute to substantial post-harvest losses (Danbaba et al., 2019). The Food and Agriculture Organization's 2020 prediction highlighted the end of the era of cheap grain supply, particularly rice, due to factors like low oil prices, increased consumption, and climate change (FAO, 2020). Nigeria, especially, is facing challenges with rising grain costs, and this in part is caused by communal disputes in some farming communities in the country and farmer–herder clashes in some others. Despite governmental initiatives such as the Rice Transformation Agenda and the Agricultural Transformation Agenda (Adesina, 2012), the Anchor Borrowers' Program, etc. seeking to boost domestic rice production and reduce reliance on imports, there remained a significant gap between harvested paddy and milled rice equivalent, leading to increased rice imports. Various stakeholders, including associations and state governments, have prioritized rice processing (Obinna & Nzekor, 2018). However, partnerships and initiatives have not adequately addressed the gap between local production and demand (Wang et al., 2023).

In Nigeria, rice processing plays a significant role in the country's agricultural landscape and food security (Chukwujekwu et al., 2023). Rice is a staple food in Nigeria, with domestic production striving to meet the growing demand fueled by population growth and changing dietary preferences (Obinna and Nzekor, 2018). Understanding the various rice processing techniques and efficiency differentials of techniques of rice processing is therefore crucial for enhancing productivity, improving food security, and promoting economic development in the country (Okoh et al., 2021; Olasehinde et al., 2022). The aims and objectives of this case study on rice processing techniques and efficiency differentials in Nigeria is therefore to: (i) assess the socioeconomic implications of efficiency differentials in rice processing, (ii) profile the existing methods employed in rice processing across various regions of Nigeria; and (iii) identify disparities in efficiency levels among different rice processing techniques and facilities.

2 Overview of Rice Processing in Nigeria

Rice processing in Nigeria typically involves several stages, including harvesting, threshing, milling, parboiling (for some varieties), polishing, and packaging (Lawal et al., 2023). Each stage presents opportunities for technological advancements and efficiency improvements (Tondo, 2017). Historically, rice processing in Nigeria has been dominated by traditional methods, which often involve manual labor and rudimentary tools (Ogundele and Okoruwa, 2006). These methods are characterized by low efficiency, high labor intensity, and inconsistent quality (Lançon et al., 2003). Traditional methods are still prevalent, especially in rural areas where access to modern processing technologies is limited (The World Bank Group, 2019). In recent years, there has been a growing adoption of mechanized rice processing techniques in Nigeria (Obinna and Nzekor, 2018). Mechanized processing involves the use of modern machinery and equipment for tasks such as threshing, milling, and polishing. Mechanization offers the potential for significant efficiency gains, increased output, and improved quality control compared to traditional methods (Christiaensen,

2017). In Nigeria, rice is often harvested manually, especially in smallholder farms. However, larger commercial farms may use mechanized harvesters for efficiency. Manual harvesting can be labor-intensive and time-consuming while some argue that it allows for greater control during the harvesting process (Dissanayake et al., 2012; Okunola et al., 2019; Udemezue and Agwu, 2018). After harvesting, rice grains are separated from the stalks through threshing. Traditional methods of threshing in Nigeria involves beating the harvested rice stalks against a hard surface or using animals to trample over them. However, mechanized threshers are increasingly being adopted to improve efficiency and reduce labor requirements (Lawal et al., 2023).

Proper drying is crucial to prevent mold growth and preserve the quality of rice grains. In Nigeria, sun drying is the most common method, where harvested rice is spread out on mats or concrete platforms to dry under the sun. However, variations in weather conditions and inadequate drying facilities can lead to uneven drying and lower-quality rice. Parboiling is another common practice in rice processing in Nigeria, particularly for certain varieties like parboiled rice (Akinwale et al., 2016). Parboiling involves partial cooking of still rice in its husk before milling, which results in improved nutritional quality, reduced breakage during milling, and enhanced flavor (The World Bank, 2020). Efficient parboiling techniques can contribute to higher-quality rice products and increased market competitiveness (The World Bank, 2020). Rice milling involves removing the husk, bran, and germ layers from the rice grains to produce polished white rice. In Nigeria, both traditional hand-operated mills and mechanized mills are used for rice milling (Obisesan et al., 2019). Mechanized mills offer higher efficiency and throughput capacity compared to traditional methods but require significant capital investment (Adisa et al., 2020). Once milled, rice is typically packaged into bags or containers for distribution and sale. Proper packaging is essential to maintain the quality and freshness of rice during transportation and storage (Olayemi et al., 2010).

3 Rice Processing Techniques

In Nigeria, various rice processing techniques are employed, ranging from traditional methods to semi-modern also tagged tradmodern and mechanized approaches. Some of the commonly used rice processing techniques in Nigeria include traditional hand-pounding methods, semi-mechanized processing, and fully mechanized processing. In the traditional hand-pounding method, rice grains are separated from the straw manually using wooden sticks or by beating the harvested rice stalks against a hard surface in a process called threshing. Winnowing involves tossing the mixture of rice grains and chaff in the air using shallow baskets or trays and allowing the wind to blow away the lighter chaff while the heavier grains fall back into the basket. De-husking and polishing of rice grains are done manually using stone or wooden mortar and pestle in the process called milling.

Mechanical threshers are used to separate rice grains from the stalk traw, reducing the labor required compared to traditional methods in the semi-mechanized processing method. Fans or blowers are employed to separate rice grains from chaff, improving the efficiency of winnowing. Semi-mechanized rice mills equipped with hullers, polishers, and graders are used to de-husk, polish, and sort rice grains. While machines perform most of the tasks, some manual intervention may still be required. The fully mechanized processing method makes use of automated threshers with conveyor belts, and separating mechanisms are used to efficiently separate rice grains from the straw. Pre-cleaning machines remove impurities such as stones, debris, and immature grains from the harvested paddy rice. Parboiling is a pre-processing technique where paddy rice is soaked, steamed, and dried before milling. This process involves soaking the rice in hot water, steaming it under pressure to gelatinize the starch, and then drying it to reduce breakage during milling. Parboiling improves the nutritional quality, texture, and cooking properties of rice and is commonly practiced in Nigeria, especially in regions where parboiled rice is preferred. Modern rice mills feature high-capacity hullers and polishers that remove the husk and bran layers from the rice grains, resulting in polished white rice. Optical sorters and mechanical graders are employed to sort rice grains based on size, shape, and color, ensuring uniformity and quality. While automated packaging machines fill and seal rice bags, readying them for distribution to markets.

Consequently, as regards value-added processing, some rice processing facilities in Nigeria engage in activities, such as fortification, where essential nutrients are added to the rice to enhance its nutritional content. Other value-added processes include packaging rice in branded bags, producing rice-based snacks or beverages, and diversifying product lines to cater to specific consumer preferences. These processing techniques vary in their level of sophistication, efficiency, and cost-effectiveness, and their adoption often depends on factors such as scale of operation, available resources, market demand, and technological advancements. While traditional methods persist, there is a growing trend toward mechanization and modernization in the Nigerian rice processing industry. This is driven by the need to improve efficiency, quality, and competitiveness as research institutes and other rice stakeholders continue to work on ways to improve and add value to rice processed in the country.

3.1 Efficiency Differentials in Rice Processing

Efficiency differentials in rice processing can arise due to various factors, including technological capabilities, infrastructure, access to resources, and scale of operation (Akintayo and Rahji, 2017). Farms and processing facilities equipped with modern, efficient machinery and equipment tend to have higher productivity and lower processing costs compared to those relying on manual labor or outdated technologies (Michels and Musshoff, 2022). Access to reliable power supply, transportation networks, and storage facilities can significantly influence processing efficiency (Agbigbe, 2016; Obahiagbon and Ogwu, 2023a, b, 2024a, b). In Nigeria, inadequate infrastructure, particularly in rural areas, can pose challenges to efficient rice processing and distribution (Okoh et al., 2021). The availability or unavailability of inputs such as seeds, fertilizers, and water, as well as access to credit and extension services, can impact rice production and processing efficiency (Temnyalov, 2023). Smallholder farmers and processors may face constraints in accessing these resources, limiting their efficiency. Larger commercial rice farms and processing facilities often benefit from economies of scale, allowing them to spread fixed costs over larger production volumes and achieve higher efficiency levels compared to small-scale operators (Obinna and Nzekor, 2018).

4 Bioeconomic Perspectives of Rice Processing

Rice processing techniques and efficiency differentials in Nigeria when properly harnessed will contribute to a sustainable bioeconomy by optimized resource use, waste reduction, mitigating environmental impact, empowering communities, enhancing resilience to climate change, expanding market opportunities, and fostering supportive policy environments (Kosoe et al., 2023; Akudugu and Ogwu, 2024). These efforts are essential for achieving long-term socioeconomic prosperity while safeguarding environmental integrity and ensuring the well-being of present and future generations. These efforts toward a sustainable bioeconomy can be achieved through several interconnected mechanisms among which are: Resource Optimization, Waste Reduction and Valorization, Environmental Impact Mitigation, Economic Empowerment, Climate Resilience, Market Expansion and Diversification, Policy support, and institutional strengthening.

Sustainable processing techniques, such as water-saving technologies and climate-smart agricultural practices, will help build resilience to climate change. These techniques mitigate the adverse effects of climate variability on rice production while ensuring food security and livelihood stability for rural communities. Likewise, efficient processing enables the production of value-added rice products, such as parboiled rice and fortified rice which cater for diverse consumer preferences and market demands. By diversification and expanding market reach, the rice sector can capture new market segments, increase export earnings, and reduce dependence on imported rice, thereby enhancing food sovereignty and economic resilience. Furthermore, government policies and institutional frameworks that promote sustainable rice processing practices are essential for driving the transition toward a bio-based economy (Egboduku et al., 2024; Ogwu, 2019). Policy measures, such as subsidies for renewable energy adoption, incentives for eco-certification, and investments in research and development, facilitate the adoption of sustainable technological practices across the rice value chain, fostering innovation, collaboration, and continuous improvement.

4.1 Case Study of Rice Processing Techniques and Efficiency Differentials from Nigeria

Case Study 1: Traditional Processing Methods in Rural Nigeria

Location: Kano State, Northern Nigeria.
Facility Type: Small-scale rice processing cooperative.
Background: The cooperative comprises local farmers who pool resources to process their harvested paddy rice. The processing techniques involved include threshing, winnowing, parboiling and milling. Manual threshing using traditional tools such as wooden sticks and beating drums, hand winnowing is used to separate rice grains from chaff. Parboiling is done in pots, drums and local tanks and the time taken to parboil varies. Stone or wooden mortar and pestle is used for de-husking and milling. Rice processed is bagged based on available sacks, bags or containers or sold directly at the processing sites to middle men or retailers. The manual techniques of rice processing results in significant grain loss during threshing and milling, leading to lower grain recovery rates. These labor-intensive methods prolong the processing time, particularly during peak harvesting seasons. Hence, the cooperative's processing capacity is constrained by the availability of manual labor and reliance on rudimentary equipment. The cooperative faces challenges in recruiting and retaining skilled labor due to seasonal fluctuations in demand and low wages. There are inconsistencies in grain quality and high levels of impurities posing the challenges of meeting market standards. Limited access to electricity, water, and transportation infrastructure further hinders operational efficiency.

Case Study 2: Mechanized Processing Plant in Urban Nigeria

Location: Lagos State, Southern Nigeria.
Facility Type: Large-scale commercial rice mill Ikorodu, Lagos State.
Background: The mill is a modern facility established by a private investor to capitalize on the growing demand for processed rice in urban markets. The processing techniques involve cleaning, de-husking, parboiling, drying, milling, polishing, and packaging. Automated cleaning machines are used to remove impurities and foreign matter from the paddy rice, while high-capacity de-husking machines efficiently remove husks from the grains. Parboiling of the paddy is carried out inside parboiling tanks and this varies in number of hours. Polishing machines, also automated enhance the appearance and marketability of the polished rice. Automated packaging lines pack the processed rice into standardized bags for distribution. The modern techniques of rice processing ensures minimal grain loss, resulting in higher grain recovery rates (Mohammed et al., 2019), the mechanized processing lines operate at high speeds, significantly reducing processing time and increasing throughput capacity. Strict quality control measures are adhered to, to ensure consistent grain quality and purity

in order to meet consistent market standards. Challenges faced are the initial investment in machinery and infrastructure is substantial, requiring access to significant financial resources. Furthermore, recruiting and retaining skilled technicians to operate and maintain complex machinery is a challenge, particularly in a competitive labor market. Consequently, reliance on electricity for operation makes the mill vulnerable to power outages and fluctuations in energy supply; likewise, high cost of petrol and diesel used in powering some engines used in processing rice is another major factor hindering smooth processing processes.

5 Methodological Reviews

In Nigeria, efforts have been made to improve rice processing efficiency through various initiatives, including government support programs, investment in infrastructure, and promotion of mechanization (Ndubueze-Ogaraku and Ogbonna, 2016). The Nigerian government's Rice Transformation Agenda and the Anchor Borrowers' Program, for example, aim to increase domestic rice production and processing capacity to reduce reliance on rice imports (Abiola and Adefabi, 2022). Additionally, private sector investments in modern rice milling facilities and the adoption of improved processing technologies have contributed to enhancing efficiency and quality standards in the Nigerian rice industry. However, challenges such as inadequate infrastructure, limited access to finance, high energy costs, farmer–herder clashes, and post-harvest losses continue to hinder optimal efficiency in rice processing in Nigeria. Addressing these challenges requires concerted efforts from government, private sector stakeholders, and development partners to promote sustainable development of the rice value chain. Studying rice processing techniques and efficiency differentials in Nigeria provides valuable insights into the dynamics of the rice industry and the factors influencing productivity and competitiveness. By addressing key challenges and leveraging opportunities for improvement, Nigeria can enhance its rice processing efficiency and contribute to food security and economic development. To this intent, this study is therefore aimed at exploring rice processing, rice processing techniques, and efficiency differentials.

5.1 Concept of Efficiency

The concept of efficiency focuses on the comparative performance of processes used to transform given inputs into outputs. Production efficiency is defined as the ability to achieve maximum output per unit of input used (Che Omar et al., 2019). Farrell (1957) identified three types of efficiency: Technical, Allocative, and Economic. Efficiency in production concerns the effectiveness of processes in transforming inputs into outputs. It can be defined as the ability to achieve maximum output per unit of input utilized. Farrell (1957) identified three types of efficiency: Technical, Allocative, and Economic (Table 1).

Table 1 Concept of efficiency

Production economics		
Concept of efficiency		
Technical	Allocative	Economic
Ability of an economic unit to produce the maximum attainable output from given inputs, irrespective of demand and market prices	Involves choosing the optimal sets of inputs	The ability to produce a preset quantity of output at minimal cost using a given level of technology. It is achieved when both technical and allocative efficiencies are attained simultaneously, resulting in maximum profits

Source: Okoh et al. (2021); Wang et al. (2023); Kolawole and Michael (2021)

Technical efficiency (TE) is measured by the elasticity value, with a value greater than one indicating inefficiency, and values between zero and one indicating increasing efficiency. Mathematically, it can be expressed as the observed output divided by the frontier output. Technical efficiency ranges from 0 to 1 (Okoh et al., 2021; Wang et al., 2023). Allocative efficiency involves choosing the optimal sets of inputs. Mathematically, it can be expressed as the ratio of marginal value product to price, with a value of 1 indicating efficiency. The allocative efficiency rule of thumb states that when the marginal value product equals the marginal factor cost, it signifies allocative efficiency. If the marginal value product is greater than the marginal factor cost, it indicates underutilization of inputs, while if it is less than the marginal factor cost, it signifies overutilization of inputs (Kolawole & Michael, 2021). Economic efficiency is the ability to produce a preset quantity of output at minimal cost using a given level of technology. It is achieved when both technical and allocative efficiencies are attained simultaneously, resulting in maximum profits. Economic efficiency ranges from 0 to 1 and occurs when both technical and allocative efficiencies are optimized (Obianefo et al., 2023).

5.2 Description of the Three Efficiency Types

5.2.1 Technical Efficiency

Khan et al. (2016) refer to an economic unit's ability to produce the maximum attainable output from given inputs, based on available technologies, regardless of the demand and market prices of inputs and outputs. Nwalieji (2016) views technical efficiency through its elasticity value, defined as the percentage change in output resulting from a percentage change in input. If producers are operating with an elasticity value greater than one, technical efficiency has not been achieved (Perez et al., 2007). Technical efficiency is attained when producers achieve elasticity values between zero and one (Akanbi et al., 2020). This indicates that as the TE ratio approaches one, the degree of technical efficiency increases. According to the stochastic frontier production function, the technical efficiency of processors is defined as the ratio of the observable output to its corresponding frontier output, given the levels of inputs (Ayinde et al., 2012).

Therefore, the technical efficiency of a processor *i* can be represented as:

$$TEi = Yi / Yi* \tag{1}$$

$$TEi = h(Xi,\beta)\exp(vi - ui) / h(Xi,\beta)\exp(vi)\exp(-ui) \tag{2}$$

This is expressed further:

$$TEi = (E(Yi / UiXi)) / (E(Yi / Ui,Xi)) \tag{3}$$

$$TEi = E((-Ui) / Ei) \tag{4}$$

$$TEi = \exp(-ui), \tag{5}$$

That is: TEi = Technical efficiency of processor i = TE; Observable output = Yi;

Frontier output = Yi^*; Xi = inputs; ε = error-terms, $vi-ui$ = decomposed error terms; while the Technical efficiency of a producer/processor/farmer ranges from 0 to 1 (Ogundele and Okoruwa, 2006).

5.2.2 Allocative Efficiency

This suggests the selection of optimal sets of inputs, occurring at the point where the marginal value product equals the marginal factor cost, i.e., MVP = MFC (Earfan Ali and Samad, 2013). It can also be referred to as price efficiency since it allows decision makers to choose inputs in a cost-minimizing manner.

Mathematically, allocative efficiency could be expressed as:

$$MVPxi = Pxi \tag{6}$$

$$MVPxi / Pxi = 1 = ki \tag{7}$$

Where

$$MVPxi = MPxi.Py \tag{8}$$

$$MPx = b.y / x \tag{9}$$

If $ki = 1$, Eq. (7) implies efficient utilization of inputs, with $ki > 1$ input is not yet efficient and more needs to be added; while $ki < 1$ portrays overutilization (Hidayah and Susanto, 2013). The allocative efficiency rule of thumb could also be put as when MVP = MFC is allocative efficiency. MVP > MFC = underutilization of inputs; hence, there is room for increased utilization of input; while MVP < MFC = overutilization of inputs, inputs should be reduced (Adesimi, 1982; Izekor and Alufohai, 2014).

5.2.3 Economic Efficiency

Economic efficiency bears on the capacity to produce a preset quantity of output at minimal cost using the given level of technology. It is elicited when technical and allocative components of efficiency are multiplied. Economic efficiency occurs when there is both technical and allocative efficiency. The simultaneous achievement of both efficiencies (allocative and technical) occurs when price relationships are engaged to indicate maximum profits. All three measures (technical, allocative, and economic efficiencies) are delimited between zero and one (Murillo-Zamorano, 2004).

5.2.4 Efficiency Measurement

Efficiency assessment methods (Table 2) can be categorized into parametric and non-parametric approaches (Murillo-Zamorano, 2004). Non-parametric methods, such as Data Envelopment Analysis (DEA), Free Disposal Hull (FDH), and Partial Frontier Efficiency (PFE), do not assume a specific functional form for the production frontier. They do not incorporate random disturbances and are less restrictive in terms of frontier shape. However, they also provide less structure for the frontier (Hidayah and Susanto, 2013). Parametric methods, such as the Stochastic Frontier Function Approach (SFFA), Distribution Free Approach (DFA), and Thick Frontier Approach (TFA), allow for deviations from the frontier. They decompose the error term to distinguish between technical efficiencies and random shocks, such as labor or capital performance variations (Noonari et al., 2015). These methods provide more flexibility but also introduce more assumptions compared to non-parametric methods. For instance, the Stochastic Production Frontier Function (SPFF) permits deviations from the frontier and allows for the decomposition of the error term to differentiate between technical efficiency and random shocks. It represents the output level as a function of inputs, parameters to be estimated, and error terms (Obianefo et al., 2021). While the Recursive Thick Frontier Model (RTFM) is used to estimate technology functions using standard regression techniques. It considers total costs relative to total outputs and is useful for estimating production or cost functions. However, it relies heavily on pooled data and does not adjust for outliers (Tipi et al., 2009).

The Distribution Free Approach (DFA) specifies the functional form for the frontier while separating inefficiencies and random errors. It does not make strong assumptions about the format of inefficiencies and random errors and assumes stability inefficiencies over time. On the other hand, the Data Envelopment Analysis (DEA) determines the efficiency of decision-making units by comparing their input-output combinations to the frontier. It does not involve error terms and requires assumptions about the regularity properties of the production frontier. While the Free Disposal Hull (FDH) is a variation of DEA, it generates large estimates of

Table 2 Efficiency measurement methods

Efficiency measurement methods	
Parametric methods	Non-parametric methods
Stochastic production frontier	Data envelopment analysis
Recursive Thick Frontier Model	Free Distribution Hull
Distribution Free Approach	
Thick Frontier Approach	

Sources: Awerije and Rahman (2017); Miassi et al. (2023)

average efficiencies compared to DEA and is particularly useful with graphical representations of data frontier (Awerije and Rahman, 2017; Miassi et al., 2023).

5.2.5 Empirical Reviews

This empirical review showcases various studies on rice efficiency and efficiency differentials in Nigeria. In a study conducted by Lawal et al., 2023 in North Central Nigeria, the comparative profitability and technical efficiency of small-scale rice farmers was investigated. The focus was on those with and without access to enhanced production technology. The multistage sampling technique was used to survey 1500 small-scale rice farmers with technology access and an equal number without, totaling 3000 farmers. Their analysis utilized various statistical and econometric tools, including descriptive statistics, budgetary analysis, stochastic production frontier, and F-Chow test statistics. The results revealed that farmers with technology access were, on average, younger (36 years) compared to those without access (46 years). Additionally, a higher proportion of technologically equipped farmers (65.4%) had formal education, compared to their counterparts (53.2%). Both groups demonstrated profitability in rice production, albeit with differing levels of technical efficiency: 81.1% for technologically equipped farmers versus 52.7% for those without. Significant factors influencing rice production among technologically equipped farmers included land size, labor, fertilizer, and agrochemical usage, while for those without access, key factors were land size, labor, and agrochemicals. Technical inefficiency in both groups was influenced by factors such as education, age, land size, experience, household size, and extension services. Farmers with access to technology faced major challenges including poor credit facilities, input shortages, inadequate rainfall, and high labor costs. Conversely, farmers without access struggled with issues such as soil fertility, reluctance to adopt innovation, credit constraints, labor costs, and planting calendar instability. The F-Chow test highlighted the significant impact of technology access on technical efficiency, productivity, and profitability, with a calculated F* Value of 26.44, far exceeding the critical value of 2.495. This underscores the importance of providing enhanced seed varieties, fertilizers, and chemicals to technologically equipped farmers at affordable rates, along with timely extension services. Encouraging cooperative memberships for access to credit facilities can further facilitate technology adoption, enhancing production capacity, income, and livelihoods in the region.

Kusumaningsih (2023) researched the technical efficiency of rice farming and the role of mobile phone usage using a stochastic frontier analysis approach. The study aimed to explore whether mobile phone usage contributes to improving technical efficiency in rice farming. A production function incorporating labor, harvest area, and seed inputs was employed, utilizing stochastic frontier analysis to assess the level of technical efficiency in rice farming and the potential impact of mobile phone usage on efficiency. The study drew on data from 2120 rice farm households obtained from the Indonesia Family Life Survey (IFLS)-5 conducted in 2014. The results from the inefficiency model revealed a significant correlation between mobile phone usage and decreased inefficiency in rice farming. Other factors contributing to this decreased inefficiency included farmers' education levels and the implementation of irrigation systems. Interestingly, farming rice on rented land was associated with increased inefficiency compared to farming on owned land, suggesting greater efficiency in the latter scenario. The production function model indicated a technical efficiency level of 0.59 for rice farming in Indonesia. Additionally, the size of the harvest area and the use of superior seeds were positively and significantly correlated with rice output, while the number of laborers did not exhibit a significant correlation with production levels. These findings highlight the potential of mobile phone technology to enhance efficiency in rice farming practices, underscoring its importance in agricultural development efforts in Indonesia.

Rice plays a crucial role in combating food insecurity, especially in developing nations like Benin (Miassi et al., 2023). Their study's aim was to assess the factors influencing technical efficiency in rice production in Benin and to identify associated constraints. Technical efficiency levels were measured using the Data Envelopment Analysis (DEA) method, and subsequent analysis was conducted via Tobit regression models to determine the factors affecting efficiency among rice producers. Production impediments were identified through Kendall's W concordance test and descriptive statistics. The results revealed a technical efficiency rate of 51%. Inferential analysis highlighted several key factors impacting technical efficiency, including the age of rice producers, household size, access to agricultural credit, and input utilization. Additionally, the study underscored prevalent challenges faced by rice producers, notably limited access to agricultural credit and adequate equipment, as well as susceptibility to climatic disruptions. Given these findings, agricultural development policies in Benin must address the determinants of technical efficiency in rice production. Measures aimed at enhancing access to credit, improving the availability of agricultural equipment, and bolstering resilience against climatic risks can significantly improve the efficiency and productivity of rice producers in the country.

In their research, Okoh et al. (2021) examined the disparity between the supply and demand of rice in Nigeria, which was attributed to discernible differences in the effectiveness of resource allocation among rice farmers. Their study investigated the factors influencing the efficiency of resource utilization and profitability among lowland rice farmers in Enugu State, Nigeria. A multistage sampling approach was used to gather data from 300 smallholder rice farmers across the six agricultural zones of the state. Profitability was assessed using gross margin (GM) analysis, while efficiency in rice farming was evaluated through the marginal value product-marginal factor cost (MVP-MFC) method. Additionally, the Stochastic Frontier Cost Function was employed to determine the factors affecting resource use efficiency among lowland rice farmers in Enugu State. The GM analysis revealed that rice production is indeed profitable, with an average rate of return on investment (ROI) of 2.80. However, the MVP-MFC analysis indicated that all hypothesized input factors were being over-utilized, signaling significant resource use inefficiency among lowland rice farmers in the state. Notably, education and age emerged as the sole socioeconomic variables impacting the allocative efficiency of rice farmers. Consequently, the study advocates for farm-level policies aimed at incentivizing younger adults, who are more inclined to adopt innovations and enhance efficiency. It also stresses the importance of investing in extension education for advisory services to facilitate improved resource utilization efficiencies.

Obianefo et al. (2023) investigated the technology gap in efficiency among small-scale rice processors in Anambra State, Nigeria, using a comprehensive research approach. The study surveyed 100 small-scale rice processors with structured questionnaires for data collection. The analysis incorporated descriptive statistics, stochastic frontier analysis, stochastic meta-frontier model, and inferential statistical techniques. Key variables in rice processing included paddy, firewood, water, and huller. Economic-specific factors such as grading costs and processing assets significantly contributed to inefficiency. The results showed that the technology gap efficiency of small-scale rice processors in Anambra State closely aligns with the frontier output, suggesting the use of optimal technology available in the industry. However, the average efficiency index for processors was 0.506, indicating that their output falls short of its potential by 49.4%. Furthermore, the technology gap efficiency for participating and non-participating processors was 0.924 and 0.983, respectively, indicating gaps of 8.0% and 2.0% that need to be closed. Factors contributing to these gaps include high processing equipment costs, input expenses, and inadequate infrastructure. In conclusion, the study highlights the low average meta-technical efficiency (0.498) of small-scale rice processors in Anambra State, emphasizing the need to enhance efficiency through technological advancements, training, and infrastructure development to optimize industrial production capacity.

The study of Kolawole and Michael (2021) explored the economics of rice production among small-holder female rice farmers in Adamawa State, Nigeria. The research aimed to understand their socioeconomic characteristics, analyze their technical and allocative efficiencies, and evaluate their overall economic efficiencies. Primary data were gathered from 180 randomly selected female rice farmers in Adamawa State using a semi-structured questionnaire. The data were analyzed using descriptive statistics and stochastic frontier function analysis. The results show that factors such as farming experience, education level, and access to credit facilities influence the farmers' technical inefficiency. Similarly, allocative inefficiency is affected by education level, household size, and access to credit. Education and access to credit facilities are identified as common factors impacting both technical and allocative efficiency among female rice farmers in the area. The study concludes that the mean economic efficiency of the female farmers is 0.6, indicating potential for improvement. Therefore, it recommends that the government and other stakeholders in the agricultural sector should focus on enhancing women's access to credit facilities and providing relevant agricultural information to improve rice production efficiency.

In another study by Tondo (2017) investigate the comparative economic aspects of rice processing methods in Benue State, Nigeria, both modern and traditional approaches were analyzed. A random sampling technique was used to select 63 respondents. The study aimed to identify the primary inputs used in rice processing, assess the costs and returns associated with each method, and identify the major challenges hindering the adoption of modern rice processing techniques. Primary data were collected using structured questionnaires, and the analysis utilized descriptive statistics, gross margin analysis, and the Cobb-Douglas production model. The findings revealed that modern rice processors were predominantly male (74.6%), while traditional processors were mostly female (75.4%). In terms of economic performance, the gross margin for modern rice processing methods was significantly higher at ₦16,770.00 per 100 kg of rice compared to ₦4143.00 per 100 kg of rice for traditional methods (with 1 USD equivalent to ₦1200). Additionally, modern processing methods demonstrated superior milling capacity, processing 200 kg/min compared to 50 kg/min for traditional methods. The study identified several barriers to the adoption of modern processing methods, including lack of awareness, limited capital, inadequate infrastructure, and a shortage of skilled technical workers. To address these challenges, the study recommended government intervention to subsidize the costs of modern technology equipment for processors. It also suggested the formation of cooperative groups among processors to facilitate skills training, access to inputs, and other essential resources necessary for business enhancement.

The investigation by Okunola et al. (2019) aimed to explore the factors influencing rice processing and quality in Ekiti State, Nigeria. The study focused on nine major towns across six Local Government Areas known for rice production. A structured questionnaire was administered to gather data from 300 rice processors in the region. Key findings from the survey include the demographic distribution of processors, with 47% being male and 53% female. The majority of processors (66.7%) reported having family sizes ranging from 4 to 7 persons. In terms of education, 63.78% had only primary education, while 22.67% had no formal education. Additionally, the survey revealed that 70% of rice was sourced from middlemen, and 80% of processors relied on personal savings as capital. Regarding processing equipment, 70% of milling machines were obsolete single-pass types, leading to a high incidence of broken grains. Furthermore, 90% of processors used diesel engines to power mills, while only 7% had access to adequate storage facilities. Surprisingly, none of the processors had access to a rice de-stoner. These findings highlight the processing deficiencies in Ekiti State and provide valuable insights for improving rice processing practices in the region. Addressing issues such as outdated equipment, inadequate storage facilities, and the lack of essential machinery like rice de-stoners could significantly enhance the quality and efficiency of rice processing in the area.

In their book chapter, Osabuohien et al. (2018) explored the significance of indigenous institutional arrangements in rice production and processing activities in Ogun State, Nigeria, using key informant interviews (KIIs) as the primary research method. The analysis highlights agricultural financing as the most significant challenge affecting rice production and processing in the region. Additionally, the study reveals that certain rice-producing areas benefit from the presence of rice farmers' clusters, which provide access to modern rice processing machinery such as winnowers, threshers, and destoners. The Rice Growers Association of Nigeria (RGAN) plays a crucial role in coordinating these activities in Ogun State through its executive committee, representing an indigenous institutional arrangement. The chapter recommends that the government prioritize the implementation of an agricultural transformation agenda to support the welfare of rice farmers and promote the development of the rice value chain. Active involvement of rice farmers through organizations like RGAN is essential for achieving these goals. Various opportunities were identified, including the establishment of a "pool of land" to foster clustering and increase production, facilitating access to inputs and technical support, and enhancing market access to improve returns on farming endeavors.

5.2.6 Result and Overview of Rice Processing Techniques and Efficiency Differentials in Nigeria

An overview of rice processing techniques and efficiency differentials, as shown in Table 3 below, provides a snapshot of the research review results. It highlights current practices and identifies potential areas for improvement in the near future.

6 Conclusion

Rice processing techniques and efficiency differentials in Nigeria plays a significant role in promoting a sustainable bioeconomy through several interconnected factors. Efficient processing techniques, such as modern machinery and streamlined workflows, have been found to optimize resource use, including water, energy, and raw materials. By minimizing resource consumption per unit of output, these techniques reduce environmental impact and contribute to the sustainable management of natural resources. Furthermore, enhanced processing efficiency leads to reduced waste generation during rice processing. By-products such as rice husks, bran, and straw can be valorized through various means, including biomass energy production, animal feed production, and composting. This reduces the environmental burden of waste disposal while creating additional revenue streams and contributing to the circular economy. In contrast, traditional rice processing methods, such as open-air drying and manual husking, can have negative environmental consequences, including air pollution and habitat degradation. Transitioning to more efficient and eco-friendly processing techniques, such as mechanical drying and automated milling, will help mitigate these impacts, leading to improved air quality, soil health, and biodiversity conservation. Examining efficiency differentials among rice processors in Nigeria provides critical insights into the dynamics of rice processing within this study's context. Through this case study, we have observed variations in efficiency levels among processors, which can significantly impact the overall productivity and sustainability of the country's rice industry. Understanding these differentials is essential for stakeholders, including policymakers, farmers, and processors, as it can inform targeted interventions and strategies to improve efficiency and optimize resource allocation and utilization throughout the rice processing value chain. Addressing inefficiencies and promoting best practices will enhance the competitiveness of rice processing operations, contribute to food security, and ultimately support the achievement of broader agricultural development goals in the country. This understanding of rice processing techniques and efficiency differentials in Nigeria will contribute to promoting a sustainable bioeconomy by optimizing

Table 3 Result and overview of rice processing and efficiency differentials in Nigeria

Rice processing techniques	Activities	Limitations	Improvement/upscaling
Traditional hand-pounding methods	Manual use of hands and sticks for harvesting, threshing. Use of hands for winnowing Mortar and pestle for milling Old pots and pans for parboiling Drying of grains on floors, mats, roadsides etc.	Broken grains, low quality grains. Laborers usually come down with nasal infection and eye problems	Improving on manual harvesting methods Provision of small-scale or cottage equipment for threshing, winnowing and parboiling at affordable prices. Sensitization on the need to use locally fabricated materials as base for sun drying rice paddy. Making locally fabricated parboiling tanks affordable and available. Provision of more co-operatives led initiatives for milling, de-stoning and packaging
Semi-mechanized processing	Use of both hands and semi-improved machines throughout the rice processing value chain	While machines perform most of the tasks, some manual intervention may still be required.	Provision of more co-operatives led machine initiatives for all the rice processing activities at a reduced cost.
Mechanized processing	Full use of automated machines in the rice processing value chain	This process is expensive and difficult for small holder processors to buy into	Provision of more co-operatives led machine initiatives for all the rice processing activities at a reduced cost. Training and re-training of small-scale processors on use and maintenance of rice processing machines
Value-added processing	This is a collaborative rice processing that is still upcoming in the country	Highly sophisticated and technical	There is need for research institutes and other stakeholder organizations on rice development to wedge in on rice improvements and value addition.

Source: Adisa et al. (2020); Akanbi et al. (2020); Obianefo et al. (2023); Kolawole and Michael (2021)

resource use, reducing waste, mitigating environmental impact, empowering communities, enhancing resilience to climate change, expanding market opportunities, and fostering supportive policy environments. These efforts are essential for achieving long-term socioeconomic prosperity while safeguarding environmental integrity and ensuring the well-being of present and future generations. Based on the review of rice processing techniques and efficiency differentials in Nigeria, we suggest the following: technology adoption, training and capacity building, infrastructure development, financial support, workforce development, and energy security. This involves exploration of alternative energy sources of energy such as solar power and biomass to reduce dependence on the national grid and ensure uninterrupted operations.

References

Abiola, A., & Adefabi, R. A. (2022). Rural structural transformation and agricultural productivity in Nigeria. *Athens Journal of Business & Economics, 8*(2), 119–138. https://doi.org/10.30958/ajbe.8-2-2

Adesimi, A. A. (1982). Resource use productivity and efficiency in maize production in the farm settlements of Western Nigeria. *Ife Journal of Agriculture, 4*, 19–25.

Adesina, A. (2012). *Agricultural transformation agenda repositioning agriculture to drive Nigeria'seconomy*.November,28.http://www.emrc.be/documents/document/20121205120841-agri2012-special_session-tony_bello-min_agric_nigeria.pdf

Adisa, B. O., Famakinwa, M., & Adeloye, K. A. (2020). Adoption of rice post-harvest technologies among smallholder farmers in Osun State, Nigeria. *Contemporary Agriculture, 69*(1–2), 20–26. https://doi.org/10.2478/contagri-2020-0004

Agbigbe, W. A. (2016). *The impact of transportation infrastructure on Nigeria's economic development*. ProQuest Dissertations and Theses, p. 192. http://ezphost.dur.ac.uk/login?url=https://search.proquest.com/docview/1830760935?accountid=14533%0Ahttp://openurl.ac.uk/ukfed:dur.ac.uk?genre=dissertations+%26+theses&issn=&title=The+Impact+of+Transportation+Infrastructure+on+Nigeria%27s+Economic+Develop

Akanbi, S. O., Alarape, W. I., & Olatunji, O. S. (2020). Economic implication of contract farming on small-scale rice farmers in Kwara State, Nigeria. *Agrosearch, 19*(2), 26–40. https://doi.org/10.4314/agrosh.v19i2.3

Akintayo, O. I., & Rahji, M. A. Y. (2017). Technical efficiency in rice ecologies of north central Nigeria: Implications for national self-sufficiency. *Uganda Journal of Agricultural Sciences, 17*(1), 99. https://doi.org/10.4314/ujas.v17i1.9

Akinwale, J. A., Ladele, A. A., & Olajide, B. R. (2016). Yield differentials among participating and non-participating rice farmers in public/private partnership in Benue and Ogun States, Nigeria. *Agrosearch, 16*(1), 25–39. https://doi.org/10.4314/agrosh.v16i1.3

Akolgoa, J. A., & Asumboya, G. (2016). Economic analysis of rice processing by women in Navrongo municipality of upper east region of Ghana. *International Journal of Basic Applied Research, 26*(2), 221–230. https://gssrr.org/index.php/JournalOfBasicAndApplied/article/view/5433

Akudugu, M. A., & Ogwu, M. C. (2024). Sustainable development policies and interventions: A bibliometric analysis of the contributions of the academic community. *Journal of Cleaner Production*, 139919. https://doi.org/10.1016/j.jclepro.2023.139919

Awerije, B. O., & Rahman, S. (2017). Technical, cost and allocative efficiency of processing cassava into gari in delta state, Nigeria. In *Handbook on Cassava: Production, potential uses and recent advances* (pp. 201–216).

Ayinde, O. E., Akanbi, S. O., & Omotesho, O. A. (2012). *Efficiency differential of government and non government assisted rice farms: A case study of Kwara State, Nigeria.*

Che Omar, S., Shaharudin, A., & Tumin, S. A. (2019). The status of the paddy and rice industry in Malaysia. In *Khazanah Research Institute*. http://www.krinstitute.org/assets/contentMS/img/template/editor/20190409_RiceReport_Full Report_Final.pdf

Christiaensen, L. (2017). Agriculture in Africa—Telling myths from facts: A synthesis. *Food Policy, 67*, 1–11. https://doi.org/10.1016/j.foodpol.2017.02.002

Chukwujekwu, A., Obianefo, I. C., Ezeano, C. A. I., & Chinwendu E. A. (2023). Technology Gap Efficiency of Small-Scale Rice Processors in Anambra State, Nigeria. *Sustainability in Enterprise Productivity and Innovation, 15*(6), 4840. https://doi.org/10.3390/su15064840

Danbaba, N., Idakwo, P. Y., Kassum, A. L., Bristone, C., Bakare, S. O., Aliyu, U., Kolo, I. N., Abo, M. E., Mohammed, A., Abdulkadir, A. N., Nkama, I., Badau, M. H., Kabaraini, M. A., Shehu, H., Abosede, A. O., & Danbaba, M. K. (2019). Rice postharvest technology in Nigeria: An overview of current status, constraints and potentials for sustainable development. *Open Access Library Journal, 06*(08), 1–23. https://doi.org/10.4236/oalib.1105509

Dissanayake, C. A. K., De Silva, S., Wasala, W. M. C., & Thilakarathne, B. M. K. S. (2012). An investigation of the rice processing village programme. *Tropical Agricultural Research, 24*(1), 91–98. https://doi.org/10.3390/ijerph8061957

Earfan Ali, K. M., & Samad, Q. A. (2013). A metafrontier production function for estimation of technical efficiencies of wheat farmers under different technologies. *Journal of Asian Scientific Research, 3*(9), 933–944.

Egboduku, W. O., Egboduku, T., Golohor, O. M., Imarhiagbe, O., & Ogwu, M. C. (2024). Cassava as raw material for sustainable bioeconomy development. In M. C. Ogwu, S. C. Izah, A. A. Cunha Alves, & babu, S.C. (Eds.), *Plant biology, sustainability and climate change, sustainable cassava* (pp. 57–73). Academic Press. https://doi.org/10.1016/B978-0-443-21747-0.00022-9

Farrell, M. J. (1957). The measurement of productive efficiency. *Journal of the Royal Statistical Society: Series A (General), 120*(3), 253–281. https://doi.org/10.2307/2343100

Food and Agriculture Organization of the United Nations. (2020). Crop prospects and food situation #1, March 2020. In *Crop prospects and food situation quarterly global report, no.1, March 2020.* (Issue March. https://doi.org/10.4060/ca8032en

Hidayah, I., & Susanto, A. N. (2013). Economies of scale and allocative efficiency of rice farming at west Seram regency, Maluku Province, Indonesia. *Asian Economic and Financial Review, 3*(5), 624–634. https://econpapers.repec.org/article/asiaeafrj/v_3a3_3ay_3a2013_3ai_3a5_3ap_3a624-634_3aid:3a1036.htm

Izekor, O., & Alufohai, G. O. (2014). Production elasticities, return to scale and allocative efficiency in yam production in Edo State, Nigeria. *Agrosearch, 14*(2), 179–190. https://doi.org/10.4314/agrosh.v14i2.8

Khan, S., Mohammed, N., Baten, M., & Ramli, R. (2016). Technical, allocative, cost, profit and scale efficiencies in Kedah, Malaysia rice production: A data envelopment analysis. *Journal of Agricultural and Biological Science, 11*(8).

Kolawole, A. A., & Michael, A. (2021). Economic analysis of rice production by small-holder women farmers in Adamawa State, Nigeria. *Croatian Review of Economic, Business and Social Statistics, 7*(1), 1–12. https://doi.org/10.2478/crebss-2021-0001

Kosoe, E. A., Achana, G. T. W., & Ogwu, M. C. (2023). Regulations and policies for herbal medicine and practitioners. In S. C. Izah, M. C. Ogwu, & M. Akram (Eds.), *Herbal medicine phytochemistry* (Reference series in phytochemistry). Springer. https://doi.org/10.1007/978-3-031-21973-3_33-1

Kusumaningsih, N. (2023). The technical efficiency of rice farming and mobile phone usage: A stochastic frontier analysis. *Food Research, 7*(1), 93–103. https://doi.org/10.26656/fr.2017.7(1).595

Lançon, F., Erenstein, O., Akande, S. O., Titilola, S. O., & et al. (2003). Rice processing in Nigeria: A survey. *Abidjan* .... https://www.hubrural.org/IMG/pdf/nigeria-rice-processing-survey.pdf

Lawal, A. F., Gudugi, A. S. I., Paiko, M. A., Bwala, A. M., Sule, B. M., Bello, A. M., Ubandoma, A., & Garba, Y. (2023). Comparative profitability and technical efficiency of small-scale rice farmers with and without access to improved production technologies in north Central Nigeria. *Journal of Sustainable Development in Africa, 25*(2), 1–21.

Miassi, Y. E., Akdemir, Ş., Dossa, K. F., & Omotayo, A. O. (2023). Technical efficiency and constraints related to rice production in West Africa: The case of Benin Republic. *Cogent Food & Agriculture*. Taylor & Francis Online, *9*(1). https://doi.org/10.1080/23311932.2023.2191881

Michels, M., & Musshoff, O. (2022). A tobit regression model for the timing of smartphone adoption in agriculture. *Heliyon, 8*(11), e11272. https://doi.org/10.1016/j.heliyon.2022.e11272

Mohammed, U. A., Ibrahim, S., Hayatu, M., & Mohammed, F. A. (2019). Rice (Oryza Sativa L.) production in Nigeria: Challenges and prospects. *Dutse Journal of Pure and Applied Sciences (DUJOPAS), 5*(2), 67–75.

Murillo-Zamorano, L. (2004). Economic efficiency and frontier techniques. *Journal of Economic Surveys, 18*(1), 33–77. https://doi.org/10.1111/j.1467-6419.2004.00215.x

Mwai, P. (2022, February 22). Nigeria rice: Is the government exaggerating production figures? *BBC News*. https://www.bbc.com/news/60324939

Ndubueze-Ogaraku, M. E., & Ogbonna, M. C. (2016). Analysis of technical efficiency and its determinants in rice production: Evidence from Abia State, Nigeria M. *Nigerian Agricultural Policy Research Journal, 1*(1).

Noonari, S., Shah ST, A. M., & Peerzado, M. B. (2015). Economic analysis of poultry egg production in Quetta District Balochistan. *Journal of Fisheries & Livestock Production, 03*(03), 1–8. https://doi.org/10.4172/2332-2608.1000137

Nwalieji, H. U. (2016). Comparative profit analysis of rice production enterprise among farmers in Anambra and Ebonyi States, Nigeria. *Asian Journal of Agricultural Extension, Economics and Sociology, 8*(3), 1–11.

Obahiagbon, E. G., & Ogwu, M. C. (2023a). Sustainable supply chain management in the herbal medicine industry. In S. C. Izah, M. C. Ogwu, & M. Akram (Eds.), *Herbal medicine phytochemistry* (Reference series in phytochemistry). Springer. https://doi.org/10.1007/978-3-031-21973-3_64-1

Obahiagbon, E. G., & Ogwu, M. C. (2023b). Consumer perception and demand for sustainable herbal medicine products and market. In S. C. Izah, M. C. Ogwu, & M. Akram (Eds.), *Herbal medicine phytochemistry* (Reference series in phytochemistry). Springer. https://doi.org/10.1007/978-3-031-21973-3_65-1

Obahiagbon, E. G., & Ogwu, M. C. (2024a). Organic food preservatives: The shift towards natural alternatives and sustainability in the global South's markets. In M. C. Ogwu, S. C. Izah, & N. R. Ntuli (Eds.), *Food safety and quality in the global south* (pp. 299–329). Springer. https://doi.org/10.1007/978-981-97-2428-4_10

Obahiagbon, E. G., & Ogwu, M. C. (2024b). The nexus of business, sustainability, and herbal medicine. In S. C. Izah, M. C. Ogwu, & M. Akram (Eds.), *Herbal medicine phytochemistry* (Reference series in phytochemistry). Springer. https://doi.org/10.1007/978-3-031-21973-3_67-1

Obianefo, C. A., Ezeano, I. C., Isibor, C. A., & Ahaneku, C. E. (2023). Technology gap efficiency of small-scale rice processors in Anambra State, Nigeria. *Sustainability in Enterprise Productivity and Innovation, 15*(6), 4840. https://doi.org/10.3390/su15064840

Obianefo, C. A., Ng'ombe, J. N., Mzyece, A., Blessing Masasi, N. J. O., & Anumudu, O. O. (2021). Technical efficiency and technological gaps of rice production in Anambra State, Nigeria. *Agriculture, 11*(12), 1240. https://doi.org/10.3390/agriculture11121240

Obinna, L. O., & Nzekor, E. C. (2018). Adoption and utilization of modern rice processing technology packages among small-sclae rice processors in Ebonyi State, Nigeria. *Journal of Agricultural Economics, Extension and Science, 4*(1), 97–110.

Obisesan, O. O., Salman, K. K., Adenegan, K. O., & Obi-Egbedi, G. O. (2019). Choice of processing techniques among rice processors in Nigeria. *World Journal of Entrepreneurship, Management and Sustainable Development, 15*(1), 70–83. https://doi.org/10.1108/wjemsd-07-2018-0060

Ogundele, O. O., & Okoruwa, V. O. (2006). Technical efficiency differentials in rice production technologies in Nigeria. In *African economic research consortium research papers* (issue April). ftp://41.215.20.26/RePEc/aer/wpaper/RP154.pdf

Ogwu, M. C. (2019). Understanding the composition of food waste: An "-omics" approach to food waste management. In A. P. Gunjal, M. S. Waghmode, N. N. Patil, & P. Bhatt (Eds.), *Global initiatives for waste reduction and cutting food loss* (pp. 212–236). IGI Global. https://doi.org/10.4018/978-1-5225-7706-5.ch011

Okoh, T. C., Opata, P. I., & Umaru, I. I. (2021). Determinants of resource use efficiencies among lowland rice farmers of Enugu State, Nigeria. *Agro-Science, 21*(1), 61–67. https://doi.org/10.4314/as.v21i1.10

Okunola, A. A., Adekanye, T. A., Adewumi, A. D., & Ashamu, F. F. (2019). Assessment of rice processing operations in Ekiti State, Nigeria. *International Journal of Civil Engineering and Technology, 10*(2), 2197–2204.

Olasehinde, T. S., Qiao, F., & Mao, S. (2022). Performance of Nigerian rice farms from 2010 to 2019: A stochastic metafrontier approach. *Agriculture (Switzerland), 12*(7), 1–13. https://doi.org/10.3390/agriculture12071000

Olayemi, F. F., Adegbola, J. A., Bamishaiye, E. I., & Daura, A. M. (2010). Assessment of post-harvest challenged of small scale farm holders of tomatoes, bell and hot peppers in some local government areas of Kano State, Nigeria. *Bayero Journal of Pure and Applied Sciences, 3*(2), 39–42.

Osabuohien, E. S., Okorie, U. E., & Osabohien, R. A. (2018). *Rice production and processing in Ogun State, Nigeria: Qualitative insights from farmers' association* (pp. 188–216). Food Systems Sustainability and Environmental Policies in Modern Economies. https://doi.org/10.4018/978-1-5225-3631-4.ch009

Perez, J. P., Gil, J. M., & Sierra, I. (2007). Technical efficiency of meat sheep production system in Spain. *Journal of Small Ruminant Research, 69*, 237–241.

Temnyalov, E. (2023). An information theory of efficient differential treatment. *American Economic Journal: Microeconomics, 15*(1), 328–358. https://doi.org/10.1257/mic.20200400

The World Bank. (2020). *In Central Benin, parboiled rice production is breathing new life into rural communes*. World Bank Feature Story. https://www.worldbank.org/en/news/feature/2020/01/28/in-central-benin-parboiled-rice-production-is-breathing-new-life-into-rural-communes

The World Bank Group. (2019). *Future of food: Harnessing digital technologies to improve food system outcomes* (No. 08). http://hdl.handle.net/10986/31565

Tipi, T., Yildiz, N., Nargeleçekenler, M., & B. Ç. (2009). Measuring the technical efficiency and determinants of efficiency of rice (Oryza sativa) farms in Marmara region, Turkey. *New Zealand Journal of Crop and Horticultral Science, 37*(2), 121–129.

Tondo, D. T. (2017). Comparative economic analysis of rice processing methods in Benue State, Nigeria. *International Journal of Environment, Agriculture and Biotechnology, 2*(6), 2776–2782. https://doi.org/10.22161/ijeab/2.6.2

Udemezue, J. C., & Agwu, A. E. (2018). Improved rice varieties, production technologies and processing constraints in Nigeria. *Direct Research Journal of Agriculture and Food Science, 6*(12), 360–373. https://doi.org/10.26765/DRJAFS.2018.2688

Wang, X., Li, J., Li, J., Chen, Y., Shi, J., Liu, J., & Sriboonchitta, S. (2023). Temporal and spatial evolution of rice productivity and its influencing factors in China. *Agronomy, 13*(4). https://doi.org/10.3390/agronomy13041075

Impact Assessment of Biobased Products in the Global South

Adams Ovie Iyiola, Olamigoke Olawale Akingba, Sulaiman Kajogbola Alimi, Ademola Michael Akinsorotan, Samuel Ayeh Osei, Christy Funmilayo Idowu, Sylvester Chibueze Izah, and Matthew Chidozie Ogwu

A. O. Iyiola (✉)
Department of Fisheries and Aquatic Resources Management, Faculty of Renewable Natural Resources Management, College of Agriculture and Renewable Natural Resources, Ejigbo Campus, Osun State University, Ejigbo, Osun State, Nigeria
e-mail: adams.iyiola@uniosun.edu.ng

O. O. Akingba
Nigerian Centre for Energy and Environment (NCEE) Administrative Block Complex, University of Benin, Ugbowo Campus, Benin City, Edo State, Nigeria
e-mail: akingba.o@ncee.org.ng

S. K. Alimi
Department of Environmental Management, School of Computing, Engineering and Built Environment, Glasgow Caledonian University, Glasgow, Scotland
e-mail: salimi300@caledonian.ac.uk

A. M. Akinsorotan
Department of Fisheries and Aquaculture, Federal University, Oye-Ekiti, Ekiti State, Nigeria
e-mail: ademola.akinsorotan@fuoye.edu.ng

S. A. Osei
Department of Fisheries and Aquatic Resources Management, Faculty of Bioscience, University of Development Studies, Tamale, Ghana

C. F. Idowu
National Biotechnology Research and Development Agency, Bioresources Development Centre, Ogbomoso, Oyo State, Nigeria
e-mail: Idowu.funmilayo@nabda.gov.ng

S. C. Izah
Department of Microbiology, Faculty of Science, Bayelsa Medical University, Yenagoa, Bayelsa State, Nigeria
e-mail: sylvester.izah@bmu.edu.ng

M. C. Ogwu
Goodnight Family Department of Sustainable Development, Appalachian State University, Boone, NC, USA
e-mail: ogwumc@appstate.edu

M. C. Ogwu et al. (eds.), *Sustainable Bioeconomy Development in the Global South*, https://doi.org/10.1007/978-981-96-0305-3_16

Abstract

Biobased products, derived from renewable biological resources, play a crucial role in reducing greenhouse gas emissions, enhancing resource efficiency, and improving the livelihoods of local communities. Understanding their impacts is essential for evaluating their contribution to economic development and sustainability. However, the production and adoption of biobased products in the Global South come with challenges, including social, environmental, and economic implications that require careful monitoring and assessment. The Environmental Impact Assessment (EIA) of biobased products focuses on critical issues such as land-use change, soil health, water usage, biodiversity conservation, and carbon sequestration. By assessing these factors, sustainable production practices can be identified, ecosystem management enhanced, and land-use planning improved, thereby maximizing the benefits of the Global South's ecosystems. The Social Impact Assessment (SIA) of biobased products examines how rural communities interact with these products, addressing issues such as gender equality, food security, labor rights, and land ownership. Large-scale bioenergy projects or agro-industrial developments can displace smallholder farms, leading to social conflicts over land rights. The Economic Impact Assessment evaluates the contributions of biobased products to job creation, income generation, value chain development, and overall economic growth in the Global South. Infrastructure investments, technology transfer, and human capital development are key to the successful implementation of EIAs, which can foster economic diversification, reduce dependence on fossil fuels, and promote sustainable livelihoods. These assessments provide valuable insights for policymakers, investors, and stakeholders, helping guide decisions that align with the Sustainable Development Goals (SDGs). Integrating social, economic, and environmental considerations into decision-making is essential for harnessing the full potential of biobased products, driving sustainable development, and benefiting both people and the environment in the Global South.

Keywords

Impact Assessment · Global South · Sustainability · Greenhouse gases · Biobased products

Abbreviations

BIC	Biobased Industries Consortium
EIA	Environmental Impact Assessment
GDP	Gross Domestic Product
GHG	Greenhouse gases
LUC	Land-use change
PET	Polyethylene terephthalate
RFS	Renewable Fuel Standard

rPET Recycled Polyethylene terephthalate
SIA Social Impact Assessment

1 Introduction

Products classified as biobased have their primary feedstock derived entirely or partially from biological sources, such as wood and crops. Three product tiers can be used to summarize them: food and feed, biobased materials, and bioenergy (De Meester, 2013). The world's society was heavily dependent on biobased products before the industrial revolution, specifically up until the early 1800s. For instance, ivory and cow horn were utilized for more robust applications (like musical instruments), and natural fibers (like cotton, wool, and silk) were employed in textile materials for thousands of years. The increased use of fossil fuels and materials as feedstock resulted in several environmental issues, such as the effects of climate change, the exhaustion of natural fossil resources, and improper waste management (Iyiola et al., 2024b). Governments became concerned about these environmental problems and promoted the usage of biobased goods. One example of this is the European Union's Renewable Energy Directive (European Commission, 2012). A more biobased economy in which biomass is the primary source of commodities and energy (Vandermeulen et al., 2011, 2012) has been able to flourish because of these activities (Shen et al., 2009).

The advantages of biobased products for the environment are widely acknowledged, including reduced greenhouse gas emissions, lower carbon footprint, and decreased dependence on finite resources (European Commission, 2012). Hoekman and Broch (2010) conducted a comprehensive review of the environmental implications of biofuels and highlighted their potential to mitigate climate change and improve air quality. However, concerns have been raised regarding the indirect land-use changes and environmental impacts associated with increased biofuel production. Consumer behavior plays a crucial role in shaping the market for biobased products. Studies have shown that consumers are increasingly inclined toward environmentally friendly products and are willing to pay a premium for sustainability. Government policies and incentives play a significant role in promoting the development and adoption of biobased products. These include renewable energy mandates, tax incentives, research funding, and procurement preferences for biobased products (U.S. Environmental Protection Agency, 2020). For example, the Renewable Fuel Standard (RFS) in the United States mandates the blending of renewable fuels into transportation fuels, creating a market for biobased fuels (U.S. Environmental Protection Agency, 2020). Similarly, the European Union's Renewable Energy Directive promotes the use of renewable energy sources, including biofuels and biogas. Biobased products, derived from renewable biomass resources, have gained increasing attention as alternatives to conventional products made from fossil fuels. The Global South, comprising regions with diverse ecosystems and significant agricultural resources, holds great potential for the production and utilization of biobased products. However, as with any industrial activity, the production and utilization of biobased products can have environmental implications that need to be carefully assessed. This chapter therefore explores the environmental impact assessment (EIA) of biobased products in the Global South, considering factors such as land-use change, biodiversity loss, water consumption, greenhouse gas emissions, and socioeconomic impacts.

1.1 Economic Importance of Biobased Products

Biobased products can provide several benefits to society, the economy, and the environment. They can drive innovation, and promote sustainability and circular practices in society. Some of these benefits are climate mitigation and adaptation, food supply and security, green industry, and human health. The development of these products can reduce dependence on fossil-based resources such as natural gas, crude oil, coal, and other raw materials which are advocated in the European Union Countries. Some of these products and examples are presented in Table 1. In a nutshell, the economic importance of these products dwells in their ability to drive sustainable development, and employment opportunities, reduce poverty, and promote economic growth while preserving the natural environment.

Table 1 Economic importance of some biobased products

Products	Uses	Example	References
Biobased chemicals	For the manufacture of paints, cosmetics, adhesives, and lubricants	Lignin—Found in plants and serves as resins for wood products, surface-active agents for coatings and paints	Yue (2022)
Packaging or construction materials	As insulation materials in construction and made from fungi, agricultural waste, microbial cells and enzymes	Eucalyptus cellulose is used to produce biobased and biodegradable glitter	Biddy et al. (2016)
Textiles	Production of sustainable textile fibers	Polymers from wood or algae or fermentation of agricultural and food residues	Aydin et al. (2019)
Batteries	For use in automobiles	Refined lignin derived from wood is used with a scalable model for batteries	Mishra et al. (2023)
Biocontrol agents	To increase the natural plant defenses to protect crops	Biocontrol agents based on fungi to replace the use of chemical pesticides	Mandree et al. (2023)
Detergents	To allow washing at lower temperatures, save energy, and cost.	Enzymes from biobased products	Mandree et al. (2023), Cavka et al. (2013)

Most of the global south countries possess natural resources and biodiversity that can be harvested for production of biobased products. The potentials cover the manufacturing, forestry, agriculture, energy, and healthcare sectors.

- *Forestry sector:* Biobased products are important in sustainable forest management and conservative efforts as revenue can be generated from the production of paper, pulp, and forest-based products (Maximo et al., 2022). With this, the ecosystem services and biodiversity are preserved. Employment can also be created with the establishment of wood-processing industries and biorefineries which can result to economic growth in local areas with abundant forests (Balea et al., 2020).
- *Manufacturing industries:* The adoption of biobased materials can benefit Global South countries as bioplastics derived from sugarcane, bagasse, cassava, or corn starch can be used in place of conventional plastics manufactured from fossil fuels (Delheimer, 2023). These biobased plastics are useful in packaging, construction, and reduction of the environmental footprint of production and disposal of plastics. Manufacturing companies may also diversify products using materials from agricultural residues, algae, or lignocellulosic biomass (Folks, 2019).
- *Energy sector:* Opportunities such as rural electrification, easy energy assesses, and decentralized energy production are benefitted by Global South countries through the development of biobased energy solutions such as biogas, biofuels, and biomass-based generation of power (Loomis, 2023). These projects can mitigate climate change by reducing GHG and promoting the sequestration of carbon through sustainable land practices.
- *Healthcare and pharmaceutical industries:* Biobased products can be used in the manufacture of herbal medicines, nutraceuticals, and bioactive compounds (Deore et al., 2019). Most species of plants in the Global South have several medicinal uses and can be used in the treatment of different ailments and health issues (Das & Pal, 2020). Furthermore, the cultural heritage of local communities is preserved as the natural oils, fibers, and extracts used in traditional medicine are obtained from the locality (Kasoju et al., 2023).

2 Impact Assessment of Biobased Products

2.1 Environmental Trend of Biobased Products

The process of reducing the substantial greenhouse gas emissions from the chemical industry is one of the major environmental benefits in the Global South (Fig. 1). This has been practiced in the US biobased goods business which is an important industry that has increased the use of biobased feedstocks, and supported research on the accompanying environmental advantages. A significant trend is the movement toward 100% biobased polyethylene terephthalate (PET) bottles, as opposed to 30% biobased PET bottles, that is used by several large producers of plastic

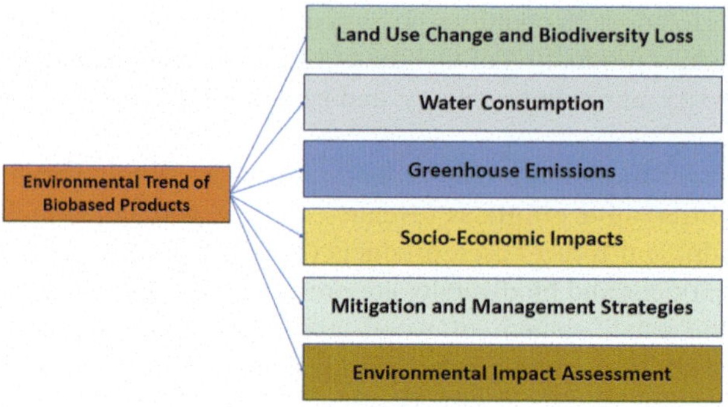

Fig. 1 Environmental trend of biobased products

bottles and consumers worldwide. A case is the debuted prototype of 100% plant-based recycled PET (rPET) containers by Coca-Cola container in October 2021, which was produced from commercially scalable technology and sugarcane-based feedstocks except for the labels and bottle cap (Packaging Gateway, 2023). Furthermore, a three-way collaboration between Nestle, Danone, and US-based Origin Materials creates and markets a PET bottle manufactured from biobased materials on a commercial scale. The advantages of switching to 100% biobased PET for the environment have been of recent focus by the scientific community, with the resulting reductions in greenhouse gases GHG emissions most especially in Global South countries that are highly vulnerable.

2.1.1 Land-Use Change and Biodiversity Loss

A major concern with the development of biobased products in the Global South is the alteration in land use and the effects on biodiversity. Natural ecosystems like grasslands and forests that are converted to agricultural land to produce biomass feedstocks may lose their biodiversity because of habitat fragmentation and destruction (Kluts et al., 2020). Furthermore, by decreasing habitat variability and encouraging the spread of invasive species, the monoculture cultivation of some crops for biobased products can worsen the loss of biodiversity. Land use refers to the human use of a specific area for a specific purpose and a specific amount of time and is commonly represented through units of area and time (e.g., ha.yr) (FAO, 1997). Land-use change (LUC), also known as land transformation, describes how humans have changed a piece of land's features, such as its flora, fauna, soil, and soil surface, from its original state to a modified state. These shifts can be drastic—like when a naturally occurring tropical rainforest is cleared for farming—or more gradual—like when agricultural output shifts from intensive to extensive. Additionally, land use and land-use change have the potential to have environmental effects that are referred to as land-use impacts. Depending on the product being studied and where it is located, a variety of land-use consequences may occur. For example, the loss of biodiversity resulting from the deforestation of a tropical rainforest for soybean farming. Additional impacts include those that influence ecosystem services.

Consequently, it can be shown that the environmental consequences (and benefits) of biobased products should be examined using holistic environmental assessment approaches, which are based on a life cycle viewpoint, to give useful information (Iyiola et al., 2023a, b, c).

2.1.2 Water Consumption

Water needs for irrigation, processing, and other uses are frequently high during the manufacturing of biobased goods. The growing demand for water from the biobased product sectors can intensify water rivalry and scarcity in parts of the Global South where water resources are already limited or under stress. Gupta et al. (2018) reported that inappropriate water management techniques can also result in water pollution and ecosystem damage. Examples of these practices include excessive irrigation and agrochemicals contaminating water bodies.

2.1.3 Greenhouse Gas Emissions

While biobased products are often promoted as more environmentally friendly alternatives to fossil fuel-based products due to their renewable nature, the production processes involved can still result in greenhouse gases (GHG) emissions. For instance, the land-use change, deforestation, and biomass burning associated with biobased product cultivation can release GHG and carbon dioxide into the atmosphere thereby contributing to climate change. Additionally, emissions may arise from the processing, transportation, and end-of-life disposal of biobased products, depending on factors such as energy sources and waste management practices (Kok & Kolk, 2019).

2.1.4 Socioeconomic Impacts

The socioeconomic effects of biobased product development on local populations can be both favorable and unfavorable in the Global South. On the one hand, the emergence of biobased industries has the potential to reduce poverty, boost economic expansion, and generate job opportunities (Lal et al., 2020). Additionally, smallholder farmers and rural households involved in the cultivation and processing of biomass may be able to generate additional revenue streams through the usage of biobased product value chains. However, there are worries that large-scale biobased product ventures may result in the displacement of traditional livelihoods, an uneven distribution of benefits, and conflicts over land tenure (Borras Jr & Franco, 2018).

2.1.5 Mitigation and Management Strategies

Several measures can be used to mitigate and control the environmental implications of biobased goods in the Global South. Lal et al. (2020) reported regulation and policy by creating and implementing laws, rules, and guidelines that support the sustainable production and use of biobased products while taking social justice, land tenure rights, and environmental protections into account. Additionally, to reduce land degradation and biodiversity loss related to the production of biobased products, promote sustainable land management techniques such as agroforestry, crop rotation, and integrated pest control (Reed et al., 2021). Furthermore, to minimize the effects of producing biobased products on water resources, water consumption should be decreased by using water-saving technology, increasing irrigation efficiency, and establishing water management rules (Iyiola & Valbo-Jørgensen, 2023). To ensure that local communities, indigenous peoples, and other stakeholders'interests and concerns are met, there is a need to consider the development and implementation of biobased products and promote inclusive decision-making processes and stakeholder involvement (Le et al., 2017; Kosoe et al., 2023; Akudugu & Ogwu, 2024).

2.1.6 Environmental Impact Assessment

Conducting impact assessments of biobased products in the Global South is important to identify and address potential environmental risks, and opportunities related to their production and utilization. Through the consideration of various factors, including land-use change, biodiversity loss, water consumption, greenhouse gas emissions, and socioeconomic impacts, stakeholders can devise and execute mitigation and management strategies aimed at advancing sustainable development and safeguarding the sustainability of the biobased product industries in the Global South (Iyiola et al. 2022a, b, c).

Biobased materials can cause environmental consequences during their whole life cycle. Although, it is important to reduce land usage and its effects on GHG emissions, eutrophication, and stratospheric ozone depletion. Some approaches that can be taken are:

- The use of organic wastes and agricultural and forest residues to increase the feedstock base.
- Constructing integrated biorefineries to facilitate a more thorough utilization of biomass for the generation of fuels, heat, energy, and biobased materials.
- Carbon cascading, which involves using biomass initially for materials, and then for energy at the end of the life cycle of product. Increasing yields and maximizing agricultural production are critical in emerging nations.

To date, biobased materials are produced largely by small-scale pilot plants. The progress achieved in biotechnology, technological learning, up-scaling of production facilities, and integration of biobased processes will reduce the environmental impacts and price of biobased materials. Despite the increased growth rates in the use of biobased materials, it will require a while for it to meet up with the market energy demand as fossil fuel prices (Shen et al., 2009).

2.2 Social Impacts of Biobased Products

Biobased products have significant social impacts in the environment. The global movement toward sustainable development in recent times has resulted in a heightened emphasis on biobased products as substitutes for traditional materials sourced from fossil fuels (Izah et al., 2022; Iyiola et al. 2024a, b, c). A potential answer to environmental problems like resource depletion and climate change can be provided by biobased goods, which are made from renewable resources including plants, algae, and waste materials. The effects of social impacts on the environment are largely unexplored as compared with the effects the biobased products have on the environment and the economy. An in-depth analysis of the societal effects of biobased products shows they relate to creation, utilization, and disposal of biobased goods. The products can also affect the societal issues such as employment, health and safety, equity, and community development. The production and utilization of biobased products contribute to job creation across various sectors, including agriculture, manufacturing, and research. Small-scale farmers and rural communities benefit from new markets for biomass feedstocks and value-added biobased products, leading to increased income and livelihood opportunities.

Biobased product value chains often involve local communities, fostering social cohesion and development. Community-based initiatives for biomass cultivation, processing, and distribution empower local stakeholders and promote sustainable resource management practices (Birch et al., 2010). Furthermore, biofuel production as cleaner alternatives to fossil fuels improves air quality and reduces respiratory illnesses associated with air pollution. Moreover, stakeholder engagement and participatory decision-making processes ensure that diverse voices are heard in shaping the bioeconomy agenda. Biobased products can support cultural preservation efforts by promoting traditional knowledge and practices related to sustainable resource management and indigenous biotechnologies (Bhullar et al., 2020). The process of incorporating indigenous perspectives into the development of biobased product and governance respects cultural diversity and strengthens the resilience of communities (Boillat & Berkes, 2013).

2.2.1 Employment Opportunities

In the rural and agricultural areas most especially, the ability of biobased goods to generate new jobs is one of their main social benefits. From farming and harvesting to processing and manufacturing, the production of biobased feedstocks, such as crops and biomass, can provide jobs at every stage of the agricultural value chain. Hess et al. (2016) reported that areas with an abundance of biomass resources, the growth of the bioenergy and bioproduct businesses have increased employment opportunities. Furthermore, the establishment of biorefineries and biobased manufacturing facilities can boost regional economies by bringing steady, well-paying employment to places that have historically relied on extractive industries.

2.2.2 Health and Safety Considerations

Potential effects on human health and safety are a significant component of the social impact of biobased goods. Even though biobased goods are frequently marketed as more environmentally friendly than fossil fuels, it is important to consider any potential health hazards, especially during the manufacturing and use of these products (Izah et al., 2023). Similarly, contaminants including particulate matter and volatile organic compounds can be released during the combustion of biomass to produce energy, which can worsen respiratory conditions and other health concerns (Adamson et al., 2020). Therefore, it is crucial to assess and mitigate the potential health and safety risks associated with the production and use of biobased products.

2.2.3 Equity and Social Justice

The adoption of biobased products can also have implications for equity and social justice, particularly in the areas of easy access to resources and benefits. For instance, bioenergy projects that are large scaled may compete with food production for land and water resources, potentially exacerbating food insecurity and displacement of marginalized communities (Preston et al., 2017). Similarly, the distribution of economic benefits from biobased industries may not always be equitable, with profits often accruing to the large corporations rather than the local communities or producers that are small-scaled and cannot compete with large corporations. It is therefore essential that policymakers must consider the potential equity implications of biobased product development and implement strategies to ensure that the benefits are distributed fairly among all stakeholders.

2.2.4 Community Development and Resilience

Despite the potential challenges of biobased products, they can offer opportunities for community development and resilience, particularly in rural and underserved areas. By diversifying local economies and creating new markets for agricultural and forestry products, biobased industries can help to awaken communities that are low-incomed and reduce dependence on volatile commodity markets. Furthermore, the production of biobased products often requires collaboration among diverse stakeholders, including farmers, entrepreneurs, researchers, and policymakers, fostering social capital and collective action (Klerkx et al., 2012). Therefore, investments in biobased product development have the potential to strengthen the social fabric of communities and enhance their capacity to adapt to environmental and economic challenges.

2.2.5 Industrial Objectives for Biobased Products

There are ways in which plants can be modified for use in the industries to create biobased goods. Despite these variations in production methods, they have the following benefits:

- The renewables can be used as chemical feedstocks.
- They can replace hazardous chemicals.
- They use less energy and chemicals.

The first and process related step is the positioning products in the market in line with these green chemistry principles.

The second step involves the use of technology, and more specifically the growing role of biotechnology. It can be defined as any technological application that uses biological systems, living organisms, or derivatives thereof, to make or modify products or processes for specific use.

The third step involves the industrial focus on biobased commodities which includes reducing reliance on petrochemicals, which make up 90% of current organic chemical manufacture.

2.3 Economic Impact Assessment of Biobased Products

Researchers and policymakers have paid serious attention to the economic impact assessment of biobased goods because it has a great potential to promote sustainable economic growth and address issues relating to the environment (Fig. 2). The aim of a biobased economy is to minimize the petrochemicals and mined minerals usage by substituting with materials and production processes that are derived from biological sources. Learning from The Netherlands' Ministry of Agriculture, Nature, and Food Quality, they expressed keen interest in supporting biobased product development as an added measure of improving agricultural output and, consequently, the income of farmers. A request was made to Wageningen (AFSG and LEI) to investigate the current market for biobased products and, consequently, the potential for the biobased economy (Government of the Netherlands, 2024). This was aimed as ascertaining the extent to which such support should be maintained, or even reinforced as deemed fit. The project team made the decision to base its analysis on comprehensive data, starting with the specific product. The products are then grouped into categories based on market sectors and within product groups based on the method of manufacture (Wageningen University and Research [WUR], 2020).

Biobased products which are derived from renewable biological resources, have gained increasing attention due to their ability to reduce environmental concerns, and reduce the dependence on the use of finite fossil fuels. Biobased products cover a range of items, including biofuels, bioplastics, biochemicals, and biomaterials (Kamm et al., 2006). Production processes covers renewable biomass feedstocks into usable products through various technologies such as fermentation, enzymatic processes, and thermochemical conversion. The regulation of the government

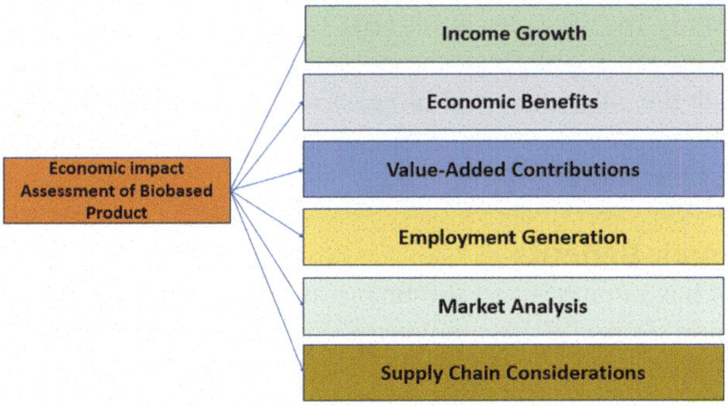

Fig. 2 Economic impact assessment of biobased products

regulations, technological developments, preference of customers, and market rivalry all have an impact on the biobased product industry. Several studies have examined the economic contributions of biobased products, highlighting their potential to create jobs, stimulate economic growth, and enhance rural development (Biotechnology Innovation Organization, 2017).

Biobased products have an impact on market development and growth as it increases the demand for biobased products and opens new market opportunities and revenue streams (Dammer et al., 2013). Moreover, biobased products contribute to Gross Domestic Product (GDP) and economic growth. Direct and indirect employment opportunities are gotten because of the impact of biobased products (Beames et al., 2021). Additionally, it plays a role in rural development and reducing poverty increasing skills development and training. Biobased products attract investments in biobased industries with goes a long way to foster innovation and encourage investment (Johnson et al., 2021). Additionally, biobased products have fostered research and development of new products and technologies which improves productivity. Biobased products have increased exports of biobased products (Daystar et al., 2021). Biobased products also play a big role in improving balance of trade and foreign exchange earnings that enhanced global competitiveness (Bennich & Belyazid, 2017). Biobased products have reduced the dependence on fossil fuels where there are savings on fuel imports and foreign exchange also there is reduced vulnerability to price volatility and improved energy security (Antar et al., 2021; Yamakawa et al., 2018). Additionally, biobased products have enhanced supply chain management. Biobased product has added to government revenue and taxation by the increased tax revenues from biobased industries and the improved government revenue and fiscal stability (Egboduku et al., 2024; Obahiagbon & Ogwu, 2023a, b, 2024a, b). Also biobased products have enhanced public services and infrastructure development. Biobased products have improved livelihoods and income opportunities which in a long run enhance community development and social cohesion and reduce poverty and inequality (Bastos Lima & Palme, 2021).

2.3.1 Income Growth

The adoption of biobased products can result to increased income levels for farmers, processors, stakeholders, and actors that are involved in biomass production and utilization. A study by Wesseler et al. (2019) analyzed the income effects of biobased innovations and highlighted the potentials for income growth in rural communities through the valorization of biomass resources. However, the distributional impacts of biobased transitions must be carefully assessed to ensure equitable outcomes for all stakeholders.

2.3.2 Economic Benefits

Several studies have emphasized the financial advantages of producing and using biobased products. Popp (2016) analyzed the economic impact of European Union biobased industries and revealed a contribution that is significant to the gross domestic product (GDP), creation of job, and regional development. Similarly, research conducted by Scarlat et al. (2018) emphasized the potential of biobased chemicals to enhance economic competitiveness and reduce reliance on imported fossil-based products. The process of assessing the economic impact of biobased products entails determining how these products affect several economic factors, such as trade balance, investment, employment, income, and GDP.

2.3.3 Value-Added Contributions

Biobased products can contribute to value creation within the economy through the development of new industries, technologies, and value chains. Jongeneel et al. (2020) in a study, examined the value-added effects of biobased innovations in the Netherlands, and emphasized the importance of integrated policy frameworks and supportive institutional arrangements to unlock the full economic potential of biobased sectors.

2.3.4 Employment Generation

Biobased products have the potential to create new employment opportunities across the value chain, from agricultural production and biomass processing to manufacturing and distribution. A study by Popp et al. (2019) found that the bioeconomy, including biobased products, contributes to job creation in rural areas, where agricultural biomass serves as a primary feedstock. The expansion of biobased industries can stimulate rural economies, reduce unemployment rates, and enhance social welfare.

2.3.5 Market Analysis

Market analysis plays a crucial role in investigating the economic dynamics of biobased products. Bio-based Industries Consortium (BIC) (2020) reported that the global market for biobased products is projected to experience substantial growth in the coming years, and this growth is driven by increasing consumer demand for sustainable alternatives and supportive government policies. Additionally, Sagar and Kartha (2007) reiterated the significance of market incentives and regulatory frameworks in boosting development of biobased industries.

2.3.6 Supply Chain Considerations

The understanding of supply chains is important in assessing the economic impact of biobased products. Gnansounou (2019) examined the cost competitiveness of biobased products throughout their supply chain, and emphasized the importance of feedstock availability, processing technologies, and logistics in determining economic viability of a biobased product. Furthermore, Cherubini et al. (2009) highlighted the significance of life cycle assessments in evaluating the overall economic and environmental sustainability of biobased products.

2.4 Social Impact Assessment of Biobased Products

Biobased products have played a role in the creation of jobs and income opportunities improving rural development and poverty reduction and increasing skills development and training (Marting Vidaurre et al., 2020) The role of biobased products has been massive as it has contributed to community development by improved infrastructure and public services and enhancing community engagement and social cohesion (Galanakis et al., 2022). Moreover, biobased products have reduced exposure to harmful chemicals and pollutants enhancing the quality of life and life expectancy (Van Schoubroeck et al., 2018) Biobased products has increased education and awareness as increased knowledge and understanding of biobased products is needed (Hairon Azhar et al., 2022). Moreover, biobased products have enhanced environmental awareness and sustainability (Falcone et al., 2019; Morone et al., 2021). The use of biobased products contributes to cultural heritage and traditional knowledge as it preserves and promotes traditional knowledge and empowers indigenous communities.

2.5 Environmental Impacts Assessment of Biobased Products

Biobased products play a role in climate change mitigation by the reduction of greenhouse gas emissions and carbon sequestration and storage (Carus, 2017). Biobased products conserve fossil fuels and minerals and improve the efficient use of renewable resources. The use of biobased products reduces water pollution as chemical runoff and water contamination is reduced improving the quality of water and aquatic ecosystems (Cakmak et al., 2022). Biobased products ensure sustainable land use and management practices and the preservation of natural habitats and biodiversity. Additionally, biobased products reduce emissions of harmful gases and particulate matter improving air quality and human health. Biobased products reduced waste generation and disposal and improved recycling and upcycling practices (Balu et al., 2022). Moreover, the use of biobased products ensures a reduction of chemical use and soil contamination improving soil health and fertility (Kurniawati et al., 2023).

3 Biobased Products and Their Functions in the Global South

Biobased products are goods and materials derived from renewable biological resources such as plants, animals, and microorganisms (Lee et al., 2022). These products are designed to replace or complement traditional fossil-based products, offering a more sustainable and environmentally friendly alternative (Fiorentino et al., 2019). Biobased products include biofuels (e.g., ethanol, biodiesel), bioplastics (e.g., biodegradable plastics, biocomposites), biochemical (e.g., biodegradable detergents, personal care products), biopharmaceuticals, etc. Moreover, biobased products can help reduce dependence on finite resources and promote sustainable resource management and can help minimize pollution from fossil-based products and promote eco-friendly waste management (Fiorentino et al., 2019). The biobased products industry has the potential to create new job opportunities and stimulate local economies. The growing importance of biobased products is driven by the need for sustainable and environmentally friendly solutions to address global challenges such as climate change, resource depletion, and environmental pollution (Leong et al., 2021).

The Global South refers to countries in Africa, Asia, Latin America, and the Pacific, characterized by developing economies, limited resources and infrastructure, high population growth rates, etc. (Kowalski, 2021). Unique challenges in the Global South faces include poverty and inequality, limited access to education, healthcare, and technology, dependence on fossil fuels and non-renewable resources, etc. (Cantarero, 2020). The Global South is vulnerable to climate change and environmental degradation (Ngcamu, 2023). On the contrary, the global south offers diverse opportunities such as rapid economic growth and industrialization and the increasing investment in renewable energy and sustainable technologies. Moreover, there is a growing demand for sustainable and eco-friendly products and the abundance of natural resources, biodiversity, etc. (Mondal & Palit, 2022). Biobased products can address some of the unique challenges in the Global South, by providing sustainable energy and reducing dependence on fossil fuels, creating employment and income opportunities in rural areas, and improving access to healthcare and sanitation (Bonoli, 2023). However, biobased products have impacts and can also pose risks, such as land grabbing and displacement of local communities and the depletion and pollution of water, additionally it creates unfair labor practices and exploitation (Hasenheit et al., 2016).

3.1 Biofuels

Biofuels play a significant role in decarbonizing transportation by offering a low carbon solution for existing technologies such as light duty vehicles. The demand of biofuels is increasing (Goldemberg et al., 2014). Capturing CO_2 from biofuels is relatively cheap compared to other bioenergy and carbon capture options. Biofuels reduces the dependence on fossil fuels and mitigates climate change (Rial, 2024). Examples of such are ethanol, and biodiesel.

3.2 Bioplastics

Bioplastics are made from renewable agricultural products. Bioplastics can be biodegradable or biobased some are resins or both (Garrison et al., 2016). Some bioplastics have better performance properties than traditional plastics. Bioplastics contribute to sustainability and a circular economy. Bioplastics are used in packaging, consumer goods, medicine, etc. (Ibrahim et al., 2021). Examples are, e.g., biodegradable plastics and biocomposites.

3.3 Biofertilizers

Biofertilizers enhance the absorption of phosphorus, zinc, and other minerals by plants, create a physical barrier against pest and simultaneously boost healthy plant growth (Khan et al., 2023). Biofertilizers promotes sustainable agriculture and reduces the use of chemicals. Examples are organic fertilizers and biopesticides.

3.4 Biopharmaceuticals

Biopharmaceuticals are medical drugs that are produced using living cells (Kesik-Brodacka, 2018). Both plants and microorganisms are sources of important secondary metabolites that have the potential to act as biopharmaceuticals. The USA is the leading country in biopharmaceutical research, but other countries are also making progress. There is a need for creation of a circular economy to advance biopharmaceutical ethics and policy (Veleva et al., 2017).

3.5 Bioremediation Agents

Bioremediation agents are biobased products used to clean up environmental pollutants like microorganisms that degrade pollutants and plants that absorb heavy metals (Ahmad et al., 2023). Bioremediation agents include oil-degrading bacteria, heavy metal accumulating plants, and PCB-degrading fungi. The use of bioremediation agents is a growing market, with increasing demand in the Global South particularly in countries with rapid industrialization and environmental concerns (Hussain et al., 2021). Bioremediation agents play a role in restoring ecosystems and promoting a sustainable environment. Examples are microorganisms and enzymes.

3.6 Food and Feed Ingredients

Food and feed ingredients are biobased products derived from renewable biomass sources like plant-based ingredients, microbial ingredients, and algal ingredients. These biobased ingredients are used in various applications including human nutrition, animal feed, food processing, and beverages (Ogwu, 2019; Kumar et al., 2022). The demand for biobased food and feed ingredients is growing in the global south is driven by factors like population growth, urbanization, and increasing demand for sustainable and healthy products. Examples are, e.g., protein-rich ingredients and functional foods.

3.7 Cosmetics and Personal Care Products

The demand for biobased cosmetics and personal care products is growing in the global south as there is the need to increase awareness of sustainability and environmental concerns. Moreover, there is a growing demand for natural and organic products and governments are putting initiatives to promote the use renewable resources (He et al., 2023). Countries like Brazil, India, and South Africa are expected to lead the growth in biobased cosmetics and personal care products market in the global south (Mandree et al., 2023). Examples are natural ingredients, and sustainable packaging.

3.8 Bioconstruction Materials

The use of biobased products in building and construction derived from renewable biomass sources such as bamboo, straw, hemp, etc. is a demand in the global south (Le et al., 2023). There is an increase on sustainable practices and green building practices. Moreover, there is an expansion in the construction industry and infrastructure development. The demand for biobased bioconstruction is also driven by government incentives and policies in promoting renewable natural resources. These biobased materials offer numerous benefits (Heinrich, 2019). Their use ensures sustainability and reduced carbon footprint (Kunič, 2016). The use of bioconstruction materials improves thermal insulation and energy efficiency (Yadav & Agarwal, 2021). Additionally, they are resistant to pest and decay. These bioconstruction materials are locally sourced and support the local economy (Muhit et al., 2024). Examples are bioplastics and biocomposites.

4 Conclusion

The economic impact assessment of biobased products is a multifaceted endeavor that requires consideration of various factors, including market dynamics, supply chain considerations, and regulatory frameworks. While challenges exist, the growing recognition of the economic benefits associated with biobased products underscores their potential to contribute to sustainable economic development. However, the social impacts of biobased products are multifaceted and complex, with both potential benefits and challenges for society. While biobased products offer opportunities for job creation, community development, and environmental sustainability, they raise questions about equity, health and safety, and resource allocation. Therefore, it is essential for policymakers, industry stakeholders, and researchers to carefully consider the social dimensions of biobased product development and implementation. By addressing these concerns and maximizing the social benefits of biobased products, we can move toward a more sustainable and equitable future for all.

References

Adamson, S., Ma, L., & McDow, W. (2020). Health effects of biomass burning. In *Biomass burning* (pp. 287–305). Springer.

Ahmad, A., Mustafa, G., Rana, A., & Zia, A. R. (2023). Improvements in bioremediation agents and their modified strains in mediating environmental pollution. *Current Microbiology, 80*(6), 208.

Akudugu, M. A., & Ogwu, M. C. (2024). Sustainable development policies and interventions: A bibliometric analysis of the contributions of the academic community. *Journal of Cleaner Production*, 139919. https://doi.org/10.1016/j.jclepro.2023.139919

Antar, M., Lyu, D., Nazari, M., Shah, A., Zhou, X., & Smith, D. L. (2021). Biomass for a sustainable bioeconomy: An overview of world biomass production and utilization. *Renewable and Sustainable Energy Reviews, 139*, 110691.

Aydin, J., Wennmann, M., & Zhang, Y. (2019). *Bio-based PEC compositions as binders for fiber-based materials, textiles, woven and nonwoven materials*. US 2019/0226159A1.

Balea, A., Fuente, E., Concepcion Monte, M., Merayo, N., Campano, C., Negro, C., & Blanco, A. (2020). Industrial application of nanocelluloses in paper making: A review of challenges, technical solutions and market perspectives. *Molecules, 25*, 526.

Balu, R., Dutta, N. K., & Roy Choudhury, N. (2022). Plastic waste upcycling: A sustainable solution for waste management, product development, and circular economy. *Polymers, 14*(22), 4788.

Bastos Lima, M. G., & Palme, U. (2021). The bioeconomy–biodiversity nexus: Enhancing or undermining nature's contributions to people? *Conservation, 2*(1), 7–25.

Beames, A., Goedhart, J., & Kanellopoulos, A. (2021). *Biobased economy: Critical foundation for achieving sustainable development goals* (pp. 49–59). Decent Work and Economic Growth.

Bennich, T., & Belyazid, S. (2017). The route to sustainability—Prospects and challenges of the bio-based economy. *Sustainability, 9*(6), 887.

Bhullar, L., et al. (2020). Revitalizing indigenous languages through the integration of traditional knowledge in the bioeconomy. *Sustainability Science, 15*(2), 425–437.

Biddy, N. J., Scarlara, C., & Kinchin, C. (2016). *Chemicals from biomass: A market assessment of bioproducts with near-term potential*. National Renewable Energy Laboratory.

Bio-based Industries Consortium (BIC). (2020). *Market analysis for bio-based products in Europe*.

Biotechnology Innovation Organization. (2017). *Economic impact of the biobased products industry: A report to the Congress of the United States of America*.

Birch, E. L., et al. (2010). Economic and social implications of biofuels development: A case study of Georgia's first-generation biofuels industry. *Journal of Agriculture, Food Systems, and Community Development, 1*(4), 127–143.

Boillat, S., & Berkes, F. (2013). Perception and interpretation of climate change among Quechua farmers of Bolivia: Indigenous knowledge as a resource for adaptive capacity. *Ecology and Society, 18*(4), 21.

Bonoli, A. (2023). *Circular Economy and Circular Bioeconomy as pillars of the ecological transition for a worldwide enhanced sustainability*.

Borras, S. M., Jr., & Franco, J. C. (2018). Global land grabbing and trajectories of agrarian change: A preliminary analysis. *The Journal of Peasant Studies, 45*(1), 1–34.

Cakmak, E. K., Hartl, M., Kisser, J., & Cetecioglu, Z. (2022). Phosphorus mining from eutrophic marine environment towards a blue economy: The role of bio-based applications. *Water Research, 219*, 118505.

Cantarero, M. M. V. (2020). Of renewable energy, energy democracy, and sustainable development: A roadmap to accelerate the energy transition in developing countries. *Energy Research & Social Science, 70*, 101716.

Carus, M. (2017). Biobased economy and climate change—Important links, pitfalls, and opportunities. *Industrial Biotechnology, 13*(2), 41–51.

Cavka, A., Guo, X., Tang-S-J, et al. (2013). Production of bacterial cellulose and enzyme from waste fiber sludge. *Biotechnology for Biofuels, 6*, 25. https://doi.org/10.1186/1754-6834-6-25

Cherubini, F., et al. (2009). Energy-and greenhouse gas-based LCA of biofuel and bioenergy systems: Key issues, ranges and recommendations. *Resource Conservation and Recycling, 53*(8), 434–447.

Dammer, L., Carus, M., Raschka, A., & Scholz, L. (2013). *Market developments of and opportunities for biobased products and chemicals.*

Das, S., & Pal, M. (2020). Review – Non-invasive monitoring of human health by exhaled breath analysis: A comprehensive review. *Journal of the Electrochemical Society, 167*, 037562.

Daystar, J., Handfield, R., Golden, J. S., McConnell, E., & Pascual-Gonzalez, J. (2021). An economic impact analysis of the US biobased products industry. *Industrial Biotechnology, 17*(5), 259–270.

De Meester, S. (2013). *Life cycle assessment in biorefineries: Case studies and methodological development* (p. 292). Faculty of Bioscience Engineering, Ghent University.

Delheimer, S. (2023). *Science and engineering for a biobased industry and economy*. National Institute of Food and Agriculture, United States Department of Agriculture. www.nifa.usda.gov/about-nifa/impacts/science-engineering-biobased-industry-economy

Deore, A. B., Dhumane, J. R., Wagh, R., & Sonawane, R. (2019). The stages of drug discovery and development process. *Asian Journal of Pharmaceutical Research and Development, 7*, 62–67. https://doi.org/10.22270/ajprd.v7i6.616

Egboduku, W. O., Egboduku, T., Golohor, O. M., Imarhiagbe, O., & Ogwu, M. C. (2024). Cassava as raw material for sustainable bioeconomy development. In M. C. Ogwu, S. C. Izah, A. A. Cunha Alves, & babu, S.C. (Eds.), *Plant biology, sustainability and climate change, sustainable cassava* (pp. 57–73). Academic Press. https://doi.org/10.1016/B978-0-443-21747-0.00022-9

European Commission. (2012). *Innovating for sustainable growth: A bioeconomy for Europe*. Publications Office of the European Union.

Falcone, P. M., González García, S., Imbert, E., Lijó, L., Moreira, M. T., Tani, A., & Morone, P. (2019). Transitioning towards the bio-economy: Assessing the social dimension through a stakeholder lens. *Corporate Social Responsibility and Environmental Management, 26*(5), 1135–1153.

FAO. (1997). *State of the World's Forest* (p. 200). Food and Agriculture Organization.

Fiorentino, G., Zucaro, A., & Ulgiati, S. (2019). Towards an energy efficient chemistry. Switching from fossil to bio-based products in a life cycle perspective. *Energy, 170*, 720–729.

Folks, E. (2019). How will bio-based products affect the manufacturing industry? Manufacturing Tomorrow. *Online Trade Magazine 4.0*. www.manufacturingtomorrow.com/article/2019/08/how-will-bio-based-products-affect-the-manufacturing-industry/13827

Galanakis, C. M., Brunori, G., Chiaramonti, D., Matthews, R., Panoutsou, C., & Fritsche, U. R. (2022). Bioeconomy and green recovery in a post-COVID-19 era. *Science of the Total Environment, 808*, 152180.

Garrison, T. F., Murawski, A., & Quirino, R. L. (2016). Bio-based polymers with potential for biodegradability. *Polymers, 8*(7), 262.

Gnansounou, E. (2019). Cost competitiveness of biobased products for the EU chemicals market. *Biofuels, Bioproducts and Biorefining, 13*(5), 1185–1197.

Goldemberg, J., Mello, F. F., Cerri, C. E., Davies, C. A., & Cerri, C. C. (2014). Meeting the global demand for biofuels in 2021 through sustainable land use change policy. *Energy Policy, 69*, 14–18.

Government of the Netherlands. (2024). *Ministry of Agriculture, Nature, and Food Quality*. www.government.nl/ministries/ministry-of-agriculture-and-food-quality

Gupta, J., van Asselt, H., & Yu, X. (2018). Negotiating global environmental and social issues: The domestic institutional foundations of Brazil's biofuel diplomacy. *International Environmental Agreements: Politics, Law and Economics, 18*(1), 97–115.

Hairon Azhar, N. N., Ang, D. T. C., Abdullah, R., Harikrishna, J. A., & Cheng, A. (2022). Bio-based materials riding the wave of sustainability: Common misconceptions, opportunities, challenges and the way forward. *Sustainability, 14*(9), 5032.

Hasenheit, M., Gerdes, H., Kiresiewa, Z., & Beekman, V. (2016). *D2. 2: Summary report on the social, economic and environmental impacts of the bioeconomy*. European Union.

He, J., Iqbal, W., & Su, F. (2023). Nexus between renewable energy investment, green finance, and sustainable development: Role of industrial structure and technical innovations. *Renewable Energy, 210*, 715–724.

Heinrich, L. A. (2019). Future opportunities for bio-based adhesives–advantages beyond renewability. *Green Chemistry, 21*(8), 1866–1888.

Hess, J. R., Wright, C. T., Kenney, K. L., Stevens, D. M., & Phillips, S. B. (2016). Renewable energy and economic development: A review of biogas and electricity projects. *Bioenergy Research, 9*(1), 48–62.

Hoekman, S. K., & Broch, A. (2010). Environmental implications of biofuels. *Annual Review of Environment and Resources, 35*, 163–189.

Hussain, K., Haris, M., Qamar, H., Hussain, T., Ahmad, G., Ansari, M. S., & Khan, A. A. (2021). Bioremediation of waste gases and polluted soils. *Microbial Rejuvenation of Polluted Environment, 2*, 111–137.

Ibrahim, N. I., Shahar, F. S., Sultan, M. T. H., Shah, A. U. M., Safri, S. N. A., & Mat Yazik, M. H. (2021). Overview of bioplastic introduction and its applications in product packaging. *Coatings, 11*(11), 1423.

Iyiola, A. O., Akinrinade, A. J., & Ajayi, F. O. (2022b). Effects of water pollution on biodiversity across coastal regions. In S. C. Izah (Ed.), *Biodiversity in Africa: Potentials, threats and conservation* (Sustainable development and biodiversity) (Vol. 29, pp. 345–367). Springer. https://doi.org/10.1007/978-981-19-3326-4_13

Iyiola, A. O., Akinsorotan, A. M., & Ewutanure, J. S. (2023b). Impacts of agrochemicals on fish composition in natural waters: A sustainable management approach. In M. C. Ogwu & S. C. Izah (Eds.), *One health implications of agrochemicals and their sustainable alternatives, sustainable development and biodiversity*. Springer.

Iyiola, A. O., Akinsorotan, O. A., Ojeleye, A. E., & Fajimolu, A. O. (2023c). An overview of environmental resources in Africa: Emerging issues and sustainable exploitation. In S. C. Izah & M. C. Ogwu (Eds.), *Sustainable utilization and conservation of Africa's biological resources and environment* (Sustainable development and biodiversity) (Vol. 888). Springer. https://doi.org/10.1007/978-981-19-6974-4_20

Iyiola, A. O., Berchie, A., Oyewole, O. O., & Akinrinade, A. J. (2022c). Impacts of climate change on biodiversity in Africa. In S. C. Izah (Ed.), *Biodiversity in Africa: Potentials, threats and conservation* (Sustainable development and biodiversity) (Vol. 29, pp. 369–394). Springer. https://doi.org/10.1007/978-981-19-3326-4_14

Iyiola, A. O., Ipinmoroti, M. O., Kolawole, A. S., Akinsorotan, A. M., Oyewole, E. O., & Akinrinade, A. J. (2022a). Weather factors and their implications for fish species in Owalla reservoir, Osun State, Nigeria. *Journal of Global Affairs and Sustainable Development (JoGASD), 1*, 93–106.

Iyiola, A. O., Kolawole, A. S., Ajayi, F. O., Ogidi, O. I., & Ogwu, M. C. (2024a). Sustainable water use and management for agricultural transformation in Africa. In S. Madhav, A. L. Srivastav, S. C. Izah, & E. van Hullebusch (Eds.), *Water resources management for rural development: Challenges and mitigation* (pp. 287–300). Elsevier. https://doi.org/10.1016/B978-0-443-18778-0.00001-5

Iyiola, A. O., Kolawole, A. S., & Oyewole, E. O. (2023a). Sustainable alternatives to agrochemicals and their socio-economic and ecological values. In M. C. Ogwu & S. C. Izah (Eds.), *One health implications of agrochemicals and their sustainable alternatives, sustainable development and biodiversity*. Springer.

Iyiola, A. O., Kolawole, A. S., Setufe, S. B., Bilikoni, J., Ofori, E., & Ogwu, M. C. (2024c). Fish as a sustainable biomonitoring tool in aquatic environments. In S. C. Izah, M. C. Ogwu, & H. Hamidifar (Eds.), *Biomonitoring of pollutants in the global south*. Springer. https://doi.org/10.1007/978-981-97-1658-6_12

Iyiola, A. O., Setufe, S. B., Ofori, E., Bilikoni, J., & Ogwu, M. C. (2024b). Biomarkers for the detection of pollutants from the water environment. In S. C. Izah, M. C. Ogwu, & H. Hamidifar (Eds.), *Biomonitoring of pollutants in the global south*. Springer. https://doi.org/10.1007/978-981-97-1658-6_16

Iyiola, A. O., & Valbo-Jørgensen, J. (2023). Intrinsic values of the African environment: A sustainable perspective. In S. C. Izah & M. C. Ogwu (Eds.), *Sustainable utilization and conservation of Africa's biological resources and environment* (Sustainable development and biodiversity) (Vol. 888). Springer. https://doi.org/10.1007/978-981-19-6974-4_23

Izah, S. C., Iyiola, A. O., & Richard, G. (2022). Impacts of pollution on the hydrogeochemical and microbial community of aquatic ecosystems in Bayelsa State, Southern Nigeria. In S. Madhav, V. B. Singh, M. Kumar, & S. Singh (Eds.), *Hydrogeochemistry of aquatic ecosystems* (pp. 283–305). Wiley. https://doi.org/10.1002/9781119870562.ch13. Print ISBN: 9781119870531 |Online ISBN: 9781119870562.

Izah, S. C., Iyiola, A. O., Yarkwan, B., & Richard, G. (2023). Impact of air quality as a component of climate change on biodiversity-based ecosystem services. In A. L. Srivastav, A. K. Dubey, A. Kumar, S. K. Narang, & M. A. Khan (Eds.), *Visualization techniques for climate change with machine learning and artificial intelligence* (pp. 123–147). Elsevier. https://doi.org/10.1016/B978-0-323-99714-0.00005-4

Johnson, C., Sierra, A. R., Dettmer, J., Sidiropoulou, K., Zicmane, E., Canalis, A., et al. (2021). The Bio-Based Industries Joint Undertaking as a catalyst for a green transition in Europe under the European Green Deal. *EFB Bioeconomy Journal, 1*, 100014.

Jongeneel, R., Bartels, J., Polman, N., & van Horne, P. (2020). Economic analysis of biobased innovations: The case of The Netherlands. *Sustainability, 12*(12), 4855.

Kamm, B., Gruber, P. R., & Kamm, M. (2006). Biorefineries—Multi-product processes. *Advances in Biochemical Engineering/Biotechnology, 100*, 167–202.

Kasoju, N., Remya, N. S., Sasi, R., Sujesh, S., Soman, B., Kesavadas, C., Muraleedharan, C. V., Harikrishna Varma, P. R., & Behari, S. (2023). Digital health: Trends, opportunities and challenges in medical devices, pharma, and biotechnology. *CSI Transactions on ICT, 11*, 11–30. https://doi.org/10.1007/s40012-023000380-3

Kesik-Brodacka, M. (2018). Progress in biopharmaceutical development. *Biotechnology and Applied Biochemistry, 65*(3), 306–322.

Khan, A., Panthari, D., Sharma, R. S., Punetha, A., Singh, A. V., & Upadhayay, V. K. (2023). Biofertilizers: A microbial-assisted strategy to improve plant growth and soil health. In *Advanced microbial techniques in agriculture, environment, and health management* (pp. 97–118). Academic Press.

Klerkx, L., Van Mierlo, B., Leeuwis, C., & Smits, R. (2012). Evolution of systems approaches to agricultural innovation: Concepts, analysis and interventions. In *Agricultural innovation systems* (pp. 457–483). Springer.

Kluts, I., Van Loy, M., & Muys, B. (2020). Greening the use of lignocellulosic biomass for bioenergy or biomaterials: A review of potential feedstock materials and strategies for sustainable management. *Biomass and Bioenergy, 137*, 105543.

Kok, M., & Kolk, A. (2019). Tensions and paradoxes in sustainable global value chain governance: The case of sustainable cocoa in Ghana. *Journal of Cleaner Production, 226*, 959–969.

Kosoe, E. A., Achana, G. T. W., & Ogwu, M. C. (2023). Regulations and policies for herbal medicine and practitioners. In S. C. Izah, M. C. Ogwu, & M. Akram (Eds.), *Herbal medicine phytochemistry* (Reference series in phytochemistry). Springer. https://doi.org/10.1007/978-3-031-21973-3_33-1

Kowalski, A. M. (2021). Global south-global north differences. In *No poverty* (pp. 389–400). Springer.

Kumar, M., Tomar, M., Punia, S., Dhakane-Lad, J., Dhumal, S., Changan, S., et al. (2022). Plant-based proteins and their multifaceted industrial applications. *LWT, 154*, 112620.

Kunič, R. (2016). Forest-based bioproducts used for construction and its impact on the environmental performance of a building in the whole life cycle. In *Environmental impacts of traditional and innovative forest-based bioproducts* (pp. 173–204).

Kurniawati, A., Toth, G., Ylivainio, K., & Toth, Z. (2023). Opportunities and challenges of bio-based fertilizers utilization for improving soil health. *Organic Agriculture, 13*(3), 335–350.

Lal, P., Mahtolia, A., & Kumar, A. (2020). Bioplastics: A greener alternative for a sustainable future. *Journal of Polymers and the Environment, 28*(3), 695–708.

Le, D. L., Salomone, R., & Nguyen, Q. T. (2023). Circular bio-based building materials: A literature review of case studies and sustainability assessment methods. *Building and Environment, 110774.*

Le, H. T., Vu, T. H., & Trinh, T. N. (2017). Integrating stakeholders into environmental impact assessment in Vietnam: Existing barriers and future directions. *Environmental Impact Assessment Review, 64*, 9–17.

Lee, J. Y., Lee, S. E., & Lee, D. W. (2022). Current status and future prospects of biological routes to bio-based products using raw materials, wastes, and residues as renewable resources. *Critical Reviews in Environmental Science and Technology, 52*(14), 2453–2509.

Leong, H. Y., Chang, C. K., Khoo, K. S., Chew, K. W., Chia, S. R., Lim, J. W., et al. (2021). Waste biorefinery towards a sustainable circular bioeconomy: A solution to global issues. *Biotechnology for Biofuels, 14*, 1–15.

Loomis, J. J. (2023). *Bioeconomy strategies in the Global South should not copy the Global North.* Illuminen. www.illuminem.com/illuminemvoices/bioeconomy-strategies-in-the-global-south-should-not-copy-those-of-the-global-north

Mandree, P., Thopil, G. A., & Ramchuran, S. (2023). Potential opportunities to convert waste to bio-based chemicals at an industrial scale in South Africa. *Fermentation, 9*(10), 908.

Marting Vidaurre, N. A., Vargas-Carpintero, R., Wagner, M., Lask, J., & Lewandowski, I. (2020). Social aspects in the assessment of biobased value chains. *Sustainability, 12*(23), 9843.

Maximo, Y. I., Hassegawa, M., Verkerk, P. J., & Missio, A. L. (2022). Forest bioeconomy in Brazil: Potential innovative products from the Forest sector. *Land, 11*(8), 1297. 10:3390/land11081297

Mishra, K., Siwal, S. S., Nayaka, S. C., Guan, Z., & Thakur, V. K. (2023). Waste-to-chemicals: Green solutions for bioeconomy markets. *Science of the Total Environment, 997*, 164006.

Mondal, S., & Palit, D. (2022). Challenges in natural resource management for ecological sustainability. In *Natural resources conservation and advances for sustainability* (pp. 29–59). Elsevier.

Morone, P., Caferra, R., D'Adamo, I., Falcone, P. M., Imbert, E., & Morone, A. (2021). Consumer willingness to pay for bio-based products: Do certifications matter? *International Journal of Production Economics, 240*, 108248.

Muhit, I. B., Omairey, E. L., & Pashakolaie, V. G. (2024). A holistic sustainability overview of hemp as building and highway construction materials. *Building and Environment, 111470.*

Ngcamu, B. S. (2023). Climate change effects on vulnerable populations in the Global South: A systematic review. *Natural Hazards, 118*(2), 977–991.

Obahiagbon, E. G., & Ogwu, M. C. (2023a). Sustainable supply chain management in the herbal medicine industry. In S. C. Izah, M. C. Ogwu, & M. Akram (Eds.), *Herbal medicine phytochemistry* (Reference series in phytochemistry). Springer. https://doi.org/10.1007/978-3-031-21973-3_64-1

Obahiagbon, E. G., & Ogwu, M. C. (2023b). Consumer perception and demand for sustainable herbal medicine products and market. In S. C. Izah, M. C. Ogwu, & M. Akram (Eds.), *Herbal medicine phytochemistry* (Reference series in phytochemistry). Springer. https://doi.org/10.1007/978-3-031-21973-3_65-1

Obahiagbon, E. G., & Ogwu, M. C. (2024a). Organic food preservatives: The shift towards natural alternatives and sustainability in the global South's markets. In M. C. Ogwu, S. C. Izah, & N. R. Ntuli (Eds.), *Food safety and quality in the global south* (pp. 299–329). Springer. https://doi.org/10.1007/978-981-97-2428-4_10

Obahiagbon, E. G., & Ogwu, M. C. (2024b). The nexus of business, sustainability, and herbal medicine. In S. C. Izah, M. C. Ogwu, & M. Akram (Eds.), *Herbal medicine phyto-*

chemistry (Reference series in phytochemistry). Springer. https://doi.org/10.1007/978-3-031-21973-3_67-1
Ogwu, M. C. (2019). Understanding the composition of food waste: An "-omics" approach to food waste management. In A. P. Gunjal, M. S. Waghmode, N. N. Patil, & P. Bhatt (Eds.), *Global initiatives for waste reduction and cutting food loss* (pp. 212–236). IGI Global. https://doi.org/10.4018/978-1-5225-7706-5.ch011
Packaging Gateway. (2023). *Coca-Cola's 100% plant-based bottle*. Packaging Gateway. https://www.packaging-gateway.com/projects/coca-colaplant-based-bottle
Popp, J. (2016). Employment, economic output and investments of the bioeconomy in the EU. *Journal of Agricultural and Food Industrial Organization, 14*(1), 1–11.
Popp, J., Harangi-Rákos, M., & Gabnai, Z. (2019). Employment impacts of bioeconomy development: A review. *Sustainability, 11*(2), 407.
Preston, F., Silveira, S., & Hino, M. (2017). Biofuels, land access and rural livelihoods in Mozambique. *Development and Change, 48*(1), 38–59.
Reed, J., Barlow, J., Carmenta, R., D'Angelo, E., & Dawoe, E. (2021). Political economy of sustainable palm oil in Ghana: Food sovereignty, political settlements, and land-use planning. *World Development, 137*.
Rial, R. C. (2024). Biofuels versus climate change: Exploring potentials and challenges in the energy transition. *Renewable and Sustainable Energy Reviews, 196*, 114369.
Sagar, A. D., & Kartha, S. (2007). Bioenergy: Markets and policies in India, China, and Southeast Asia. *Annual Review of Environment and Resources, 32*, 101–137.
Scarlat, N., et al. (2018). The role of biomass and bioenergy in a future bioeconomy: Policies and facts. *Environmental Development, 27*, 26–38.
Shen, L., Haufe, J., & Patel, M. K. (2009). *Product overview and market projection of emerging bio-based plastics*. PRO-BIP 2009. Utrecht University.
U.S. Environmental Protection Agency. (2020). *Renewable Fuel Standard Program (RFS)*.
Van Schoubroeck, S., Van Dael, M., Van Passel, S., & Malina, R. (2018). A review of sustainability indicators for biobased chemicals. *Renewable and Sustainable Energy Reviews, 94*, 115–126.
Vandermeulen, V., Prins, W., Nolte, S., & Van Huylenbroeck, G. (2011). How to measure the size of a bio-based economy: Evidence from Flanders. *Biomass and Bioenergy, 35*, 4368–4375.
Vandermeulen, V., Van der Steen, M., Stevens, C. V., & Van Huylenbroeck, G. (2012). Industry expectations regarding the transition toward a biobased economy. *Biofuels, Bioproducts and Biorefining, 6*, 453–464.
Veleva, V., Bodkin, G., & Todorova, S. (2017). The need for better measurement and employee engagement to advance a circular economy: Lessons from Biogen's "zero waste" journey. *Journal of Cleaner Production, 154*, 517–529.
Wageningen University and Research [WUR]. (2020). *Dutch supermarkets provides insight into food wastes*. www.wur.nl/en/research-results-/research-institutes/food-biobased-research/show-fbr/dutch-supermarkets-provide-insight-into-food-waste-f00dwa5.htm
Wesseler, J., Scatasta, S., & Nillesen, E. (2019). The income effect of bio-based innovations in the European Union. *Sustainability, 11*(8), 2314.
Yadav, M., & Agarwal, M. (2021). Biobased building materials for sustainable future: An overview. *Materials Today: Proceedings, 43*, 2895–2902.
Yamakawa, C. K., Qin, F., & Mussatto, S. I. (2018). Advances and opportunities in biomass conversion technologies and biorefineries for the development of a bio-based economy. *Biomass and Bioenergy, 119*, 54–60.
Yue, J. (2022). Green process intensification using microreactor technology for the synthesis of biobased chemicals and fuels. *Chemical Engineering and Processing-Process Intensification, 177*, 109002.

Biobased Resource Recovery Techniques in the Global South

Matthew Chidozie Ogwu, Adams Ovie Iyiola, Waheed Olujide Fawole, and Sylvester Chibueze Izah

Abstract

This chapter discusses the critical roles of biobased resource recovery techniques in fostering sustainable development across the Global South. As nations in Africa, Southeast Asia, and Latin America face escalating challenges related to waste management, environmental degradation, and economic growth, biobased resource recovery offers a viable pathway to address these issues. The chapter provides an overview of key resource recovery techniques such as anaerobic digestion, composting, bioplastics production, bioenergy conversion, microbial fuel cells, and biochar production, which convert organic waste into valuable resources like energy, biofuels, bioproducts, and soil amendments. The chapter further analyzes the current landscape of resource recovery in the Global South, highlighting the socioeconomic, technological, and policy challenges that limit the widespread adoption of these techniques.

M. C. Ogwu (✉)
Goodnight Family Department of Sustainable Development, Appalachian State University, Boone, NC, USA
e-mail: ogwumc@appstate.edu

A. O. Iyiola · W. O. Fawole
Department of Fisheries and Aquatic Resources Management, Faculty of Renewable Natural Resources Management, College of Agriculture and Renewable Natural Resources, Ejigbo Campus, Osun State University, Ejigbo, Osun State, Nigeria
e-mail: adams.iyiola@uniosun.edu.ng; adams.iyiola@uniosun.edu.ng

S. C. Izah
Department of Community Medicine, Faculty of Clinical Sciences, Bayelsa Medical University, Yenagoa, Bayelsa State, Nigeria

Department of Microbiology, Faculty of Science, Bayelsa Medical University, Yenagoa, Bayelsa State, Nigeria
e-mail: Sylvester.izah@bmu.edu.ng

M. C. Ogwu et al. (eds.), *Sustainable Bioeconomy Development in the Global South*, https://doi.org/10.1007/978-981-96-0305-3_17

Infrastructure deficits, economic constraints, regulatory barriers, and social acceptance remain significant hurdles. Through case studies of successful projects from countries like Kenya and India, the chapter showcases the potential of biobased recovery processes in transforming waste into a resource while creating green jobs, supporting rural development, and enhancing environmental sustainability. A key focus is placed on the role of technological innovations and partnerships in advancing biobased recovery solutions. Innovations such as biotechnology, waste-to-energy conversion, and bioengineering are explored as transformative tools for scaling resource recovery efforts. Furthermore, the chapter underscores the importance of integrating circular economy principles within the bioeconomy to achieve sustainable waste management and resource efficiency. To overcome existing barriers, the chapter calls for enhanced policy frameworks, regional cooperation, and investment in capacity building. It emphasizes the importance of North–South and South–South collaboration in scaling biobased resource recovery technologies and practices. Finally, the chapter highlights the potential for biobased resource recovery to contribute to achieving the United Nations Sustainable Development Goals, particularly in areas related to environmental conservation, climate resilience, and sustainable economic growth.

Keywords

Circular economy · Bioeconomy · Anaerobic digestion · Bioenergy · Biochar · Waste-to-energy · Resource recovery · Sustainability · Global South · Green jobs

Abbreviations

CE	Circular economy
CBE	Circular bioeconomy
WtE	Waste-to-energy
AD	Anaerobic digestion
MFC	Microbial fuel cells
SDGs	Sustainable Development Goals
SMEs	Small and medium enterprises

1 Introduction

Biobased resource recovery refers to the process of extracting valuable materials from organic wastes, agricultural residues, and industrial by-products, using biological and biotechnological methods. This practice is central to the circular bioeconomy, a concept that promotes sustainable production and consumption by minimizing waste and regenerating natural systems. Biobased resource recovery techniques convert biological waste into high-value products such as biofuels, bioplastics, biochemicals, biofertilizers, and renewable energy (Kalmykova et al., 2018). As the global demand for resources increases and the environmental impacts of waste intensify, biobased resource recovery offers an innovative solution to meet both sustainability and economic goals. It is particularly important in regions with resource scarcity and waste management challenges, such as the Global South.

The Global South, comprising regions such as Africa, Latin America, and Southeast Asia, faces substantial socioeconomic and environmental challenges due to rapid industrialization, urbanization, and population growth (Amigun et al., 2017). In these regions, waste management systems are often inefficient, leading to significant environmental degradation, public health risks, and economic losses. The circular bioeconomy, through biobased resource recovery, can help mitigate these challenges by converting organic waste into useful resources and reducing the reliance on non-renewable inputs (McCormick & Kautto, 2013). For example, converting agricultural residues into bioenergy or bioproducts can enhance energy security, reduce greenhouse gas emissions, and create economic opportunities in rural areas (Bourguignon, 2016). The Global South's natural resources have a significant impact on the global economy and environment, making sustainable management crucial for global sustainability. By prioritizing sustainable resource management, the Global South can ensure a resilient and prosperous future for its communities and contribute to global sustainable development. The global south faces significant waste management challenges, and biobased resource recovery techniques can help address these issues by converting waste into valuable resources. Many communities in the Global South lack access to reliable energy sources, and biobased resource recovery techniques can provide sustainable energy solutions. Biobased resource recovery techniques can enhance agricultural productivity by converting agricultural waste into fertilizers, biopesticides, and other valuable products. Biobased resource recovery techniques can create new economic opportunities and jobs in the Global South, contributing to local economic growth and development. Biobased resource recovery techniques can help mitigate environmental pollution by reducing waste and emissions, protecting biodiversity, and promoting sustainable resource use. Biobased resource recovery techniques can help reduce greenhouse gas emissions and promote carbon sequestration, contributing to global efforts to address climate change (Yang et al., 2023). Also, dependence on fossil fuels is reduced thereby promoting energy security and reducing the impact of price volatility. The role of Biobased resource recovery techniques aligns with several Sustainable Development Goals and is a necessary approach. This review therefore highlights the biobased resources and techniques for resource recovery. Information was harnessed from peer-reviewed literature search to derive updated information on the resources and the various techniques available with case studies of where these technologies have been adopted in the Global South.

Biobased resource recovery techniques play a critical role in achieving the United Nations Sustainable Development Goals (SDGs), particularly in areas such as clean energy (SDG 7), sustainable cities (SDG 11), responsible consumption and production (SDG 12), and climate action (SDG 13). These techniques contribute to environmental sustainability by reducing waste, lowering carbon emissions, and conserving biodiversity (Lomillos & De la Fuente, 2020). Economically, they create new industries and markets, promote green jobs, and improve resource efficiency by closing the loop between production and consumption cycles (Zilberman et al., 2018). Additionally, the social benefits include improved public health, poverty reduction, and enhanced food security by utilizing organic waste for agricultural applications (Hochman & Zilberman, 2016). For instance, biogas production from organic waste provides an affordable and renewable energy source, particularly in rural communities lacking access to conventional energy infrastructure (Sovacool et al., 2021). Likewise, converting food waste into biofertilizers can enhance soil fertility and agricultural productivity, contributing to rural development and reducing reliance on chemical fertilizers, which can have adverse environmental effects (Gómez et al., 2021).

The primary aim of this chapter is to examine how biobased resource recovery techniques can be utilized to foster sustainable development in the Global South. It will discuss biobased resource recovery methods that apply to the Global South, such as waste-to-energy systems, bioproduct innovations, and biofertilizer production and analyze their socioeconomic and environmental benefits. The chapter seeks to highlight the transformative potential of biobased resource recovery in promoting a circular bioeconomy and achieving sustainable development across the Global South.

2 Biobased Resource Recovery

Biobased resource recovery refers to the process of converting organic waste materials, including agricultural by-products, food waste, and industrial residues, into valuable resources such as bioenergy, bioplastics, biofertilizers, and biochemicals (Gómez et al., 2021). This process leverages biological and biotechnological techniques to not only reduce waste but also close the loop in resource utilization, promoting sustainability within a circular bioeconomy framework (Kalmykova et al., 2018). In essence, biobased resource recovery transforms waste into economic value, thereby minimizing environmental impacts and maximizing resource efficiency.

A circular bioeconomy incorporates bio-based resource recovery into its core by focusing on regenerating natural systems, reducing dependency on non-renewable resources, and enhancing resilience against environmental challenges. Key concepts in biobased resource recovery include the utilization of biological processes such as anaerobic digestion, fermentation, and composting, as well as the conversion of organic material into renewable energy and high-value bio-products (Zilberman et al., 2018).

2.1 Types of Biobased Resources for Recovery

Biobased resource recovery encompasses a variety of waste streams, including but not limited to:

Agricultural Waste: Crop residues, animal manure, and forestry by-products are rich sources of biobased resources. These materials can be converted into bioenergy (biogas or bioethanol), biofertilizers, and other bio-products (Jain et al., 2020; Egboduku et al., 2024).

Food Waste: Leftover food from households, restaurants, and food processing industries can be transformed into biogas or compost, contributing to the production of renewable energy and soil health improvement (Girotto et al., 2015; Kosoe & Ogwu, 2024).

Biomass: Lignocellulosic biomass, which includes plant-based materials such as wood, grasses, and crop residues, serves as a crucial feedstock for biofuel production, reducing the reliance on fossil fuels and contributing to climate change mitigation (Ogwu et al., 2018; Chime et al., 2018; Börjesson & Gustavsson, 2020).

Industrial By-products: Organic wastes from industries, such as pulp and paper or bio-refineries, can be recovered to produce bioenergy, biobased chemicals, or bio-plastics, further supporting sustainability (Ogwu et al., 2024).

Biobased resource recovery offers profound environmental and economic benefits. Environmentally, it reduces the pressure on landfills, mitigates greenhouse gas emissions, and conserves natural resources by transforming organic waste into reusable materials (Smith et al., 2017). For instance, the conversion of agricultural waste into bioenergy reduces reliance on fossil fuels and lowers carbon emissions, while composting food waste reduces methane emissions associated with landfill disposal (Girotto et al., 2015).

Economically, biobased resource recovery creates new revenue streams by converting waste into marketable products, thus promoting green jobs and stimulating the bioeconomy (Scarlat et al., 2015). The production of bioenergy, biochemicals, and biofertilizers from organic waste supports the development of local industries, particularly in rural areas, and reduces the need for imports of chemical fertilizers and non-renewable energy sources (Hochman & Zilberman, 2016). This not only enhances energy security but also promotes sustainable economic growth, especially in developing regions where waste management infrastructure is limited. Biobased resource recovery plays a pivotal role in reducing environmental degradation while simultaneously providing economic opportunities through the creation of valuable bio-products. As countries increasingly seek sustainable alternatives, the integration of biobased resource recovery into the circular economy offers a pathway to achieving environmental, social, and economic sustainability (Ogwu et al., 2023).

3 Current Landscape of Resource Recovery in the Global South

The Global South faces significant challenges in managing waste and resource recovery. Rapid urbanization, population growth, and industrial expansion have led to a substantial increase in waste generation, often overwhelming existing waste management systems (Hoornweg & Bhada-Tata, 2012). Africa, Southeast Asia, and Latin America all have varying levels of infrastructure for waste management, but many countries cannot implement comprehensive recycling and resource recovery programs. In Africa, for example, only 4% of the waste is recycled, while the majority is either openly dumped or incinerated (Kaza et al., 2018). Southeast Asia faces similar issues, with inadequate waste disposal methods and limited recycling infrastructure, contributing to pollution and environmental degradation (Verma et al., 2019). Latin America, despite having more advanced waste management systems, still faces inefficiencies and lacks widespread implementation of resource recovery techniques (González & Chauhan, 2020).

The adoption of biobased resource recovery techniques in the Global South is influenced by several drivers and barriers. On one hand, the increasing demand for renewable energy, the economic potential of biobased products, and the environmental benefits of reducing waste are key drivers for implementing resource recovery initiatives (Godfrey et al., 2019). The global focus on sustainability and climate change mitigation has also accelerated the push for resource recovery, with international organizations and development agencies providing financial and technical support. However, the region faces substantial barriers, including limited financial resources, inadequate policy frameworks, and a lack of technological expertise. Many countries in the Global South do not have the infrastructure needed to support large-scale resource recovery efforts, and public awareness of the benefits of waste management remains low (Oteng-Ababio et al., 2020). Additionally, political instability and corruption can hinder the development and implementation of effective waste management policies (Wilson et al., 2015).

Kenya has made significant strides in biobased resource recovery through biomass conversion. The country's agricultural sector generates large amounts of organic waste, which can be converted into biogas and biofertilizers. Initiatives like the installation of small-scale biogas plants have improved access to clean energy for rural communities, reducing reliance on firewood and cutting greenhouse gas emissions (Mutungwe et al., 2021). The production of biofertilizers from agricultural waste also helps improve soil fertility, increasing crop yields and promoting sustainable farming practices. In India, agricultural waste reuse has become a key component of resource recovery. The country generates vast quantities of crop residues, including rice husks and wheat straw, which are often burned, contributing to air pollution. In recent years, however, India has initiated efforts to convert these residues into bioenergy and biobased products. Technologies such as anaerobic digestion and pyrolysis are used to produce biogas, biochar, and biofuels from crop residues, providing energy solutions and reducing environmental pollution (Kumar et al., 2019a, b). These efforts are supported by government policies and financial incentives aimed at promoting sustainable agriculture and resource recovery.

4 Biobased Recovery Techniques and Processes in the Global South

Biobased recovery techniques play a critical role in enhancing sustainability by converting organic waste into valuable resources, such as energy, fertilizers, and bio-based materials. Biobased resource recovery processes present a viable solution to the waste management and resource challenges faced by the Global South. By converting organic waste into valuable products such as energy, biofuels, and fertilizers, these processes not only address environmental concerns but also provide economic and social benefits. To scale these initiatives, investments in technology, infrastructure, and supportive policies are necessary. These techniques are particularly relevant in the Global South, where resource recovery can provide essential energy and agricultural inputs while mitigating environmental challenges such as waste accumulation (Table 1).

Anaerobic Digestion

Anaerobic digestion is a biochemical process that converts organic waste into biogas (a mixture of methane and carbon dioxide) through microbial activity in the absence of oxygen (Fig. 1; Tauseef et al., 2013). This technique is widely used for treating agricultural, municipal, and industrial waste. Biogas produced through anaerobic digestion can be used for electricity generation, heating, and even as a transportation fuel. The remaining digestate can be utilized as a nutrient-rich fertilizer, promoting circular bioeconomy practices (Holm-Nielsen et al., 2009). Anaerobic digestion (AD) is a key biobased resource recovery process that breaks down organic waste in the absence of oxygen, resulting in the production of biogas (a mixture of methane and carbon dioxide) and digestate, a nutrient-rich fertilizer. AD is widely implemented in the Global South to manage agricultural and livestock waste, providing an alternative energy source while reducing greenhouse gas emissions. The biogas generated from AD can be used for cooking, heating, and electricity generation, helping to address energy poverty in regions such as Sub-Saharan Africa and Southeast Asia (Lehtomäki et al., 2021). In India, for instance, decentralized AD plants have been established in rural areas, converting cattle manure into biogas and helping households reduce their reliance on firewood (Kumar et al., 2019a, b). Similarly, in Kenya, small-scale biogas plants have become instrumental in promoting energy access and reducing deforestation (Njenga et al., 2019a, b).

Composting

Composting is the controlled aerobic decomposition of organic waste, including food scraps, agricultural residues, and green waste, into nutrient-rich compost. This process enhances soil fertility and improves soil structure, making it a valuable tool for sustainable agriculture (Fig. 2; Bernal et al., 2017). Composting also reduces the

Table 1 Biobased resource recovery processes and techniques with their recent advances and benefits

Biobased recovery process	Description	Technological advances	Environmental benefits	Socioeconomic impact	Key references
Anaerobic digestion (AD)	Breakdown of organic waste in the absence of oxygen, producing biogas and digestate	Enhanced biogas yield with microbial consortia and process optimization	Reduction in methane emissions, renewable energy generation	Provides energy access, reduces deforestation in rural communities	Lehtomäki et al. (2021) and Njenga et al. (2019a, b)
Composting	Aerobic decomposition of organic waste into nutrient-rich compost	Rapid composting using enzyme treatments and microbial inoculants	Reduces landfill waste, enriches soil fertility	Improves agricultural productivity, reduces reliance on synthetic fertilizers	Mekonnen and Hoekstra (2020) and Islam and Hossain (2021)
Bioenergy production (biogas, bioethanol, biodiesel)	Conversion of biomass to energy, such as bioethanol, biodiesel, and biogas	Advances in lignocellulosic biomass conversion and algae-based biofuel production	Reduces reliance on fossil fuels, lowers carbon emissions	Promotes energy security and rural job creation	Bailis et al. (2016) and Goldemberg et al. (2008)
Microbial fuel cells (MFCs)	Electricity generation from organic waste via microbial metabolic processes	Development of multi-chamber MFCs for increased efficiency	Waste treatment with simultaneous energy production	Electricity access in rural and off-grid areas	Pant et al. (2020) and Saravanan et al. (2021)
Biochar production	Pyrolysis of organic waste into biochar, a stable form of carbon	Use of advanced pyrolysis reactors for increased biochar yield	Carbon sequestration, improves soil fertility and water retention	Boosts agricultural yields, climate mitigation	Lehmann et al. (2011) and Woolf et al. (2021)
Algae-based resource recovery	Use of algae to treat wastewater and produce biofuels and bioproducts	Advances in algae cultivation systems and photobioreactors	Reduces nutrient pollution in water bodies, biofuel production	Job creation in algae farming, renewable energy source	Muñoz and Guieysse (2006) and Razzak et al. (2013)
Waste-to-bioenergy conversion	Conversion of organic waste to energy through gasification and pyrolysis	Integration of waste-to-energy plants with urban waste management systems	Reduces landfill waste and carbon emissions	Provides electricity and heating for local communities	Kumar et al. (2019a, b)

Advanced composting techniques	Use of bioreactors and accelerated composting systems	Introduction of mechanical and biological treatments to accelerate decomposition	Enhances soil fertility, reduces methane emissions	Strengthens local food production systems	Mekonnen and Hoekstra (2020)
Bioplastics production from organic waste	Conversion of agricultural and food waste into biodegradable plastics	Innovations in bio-polymer synthesis and scaling bioplastic production	Reduces plastic pollution, supports circular economy	New industries in bioplastic production, reduces fossil-based plastic dependency	Razzak et al. (2013) and Lehmann et al. (2011)
Bioremediation	Use of microorganisms or plants to remove contaminants from soil and water	Genetic engineering of bacteria and fungi for enhanced pollutant degradation	Restores soil and water quality in contaminated areas	Improves public health and agricultural viability	Muñoz and Guieysse (2006) and Pant et al. (2020)

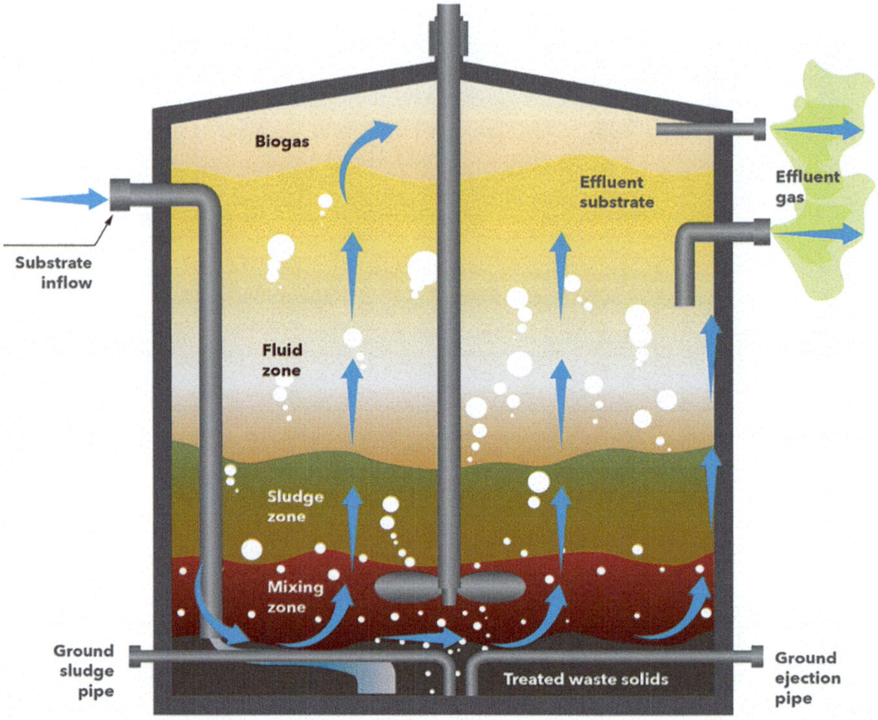

Fig. 1 Anaerobic digestion processes. (Source: Peter Knot, 2021)

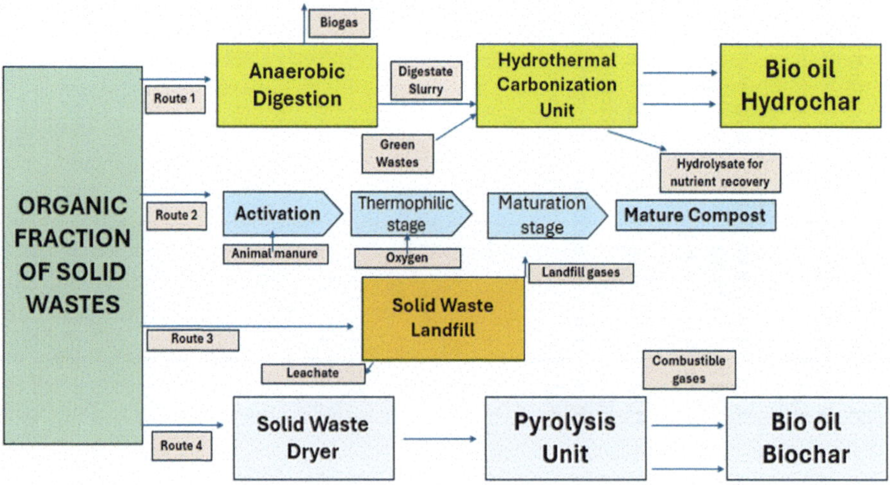

Fig. 2 Organic fraction of solid wastes to resources (route 1—anaerobic digestion, route 2—composting, route 3—land filling, route 4—pyrolysis)

need for chemical fertilizers and contributes to waste diversion, reducing landfill pressure and greenhouse gas emissions (Nair et al., 2019). Composting, a natural process of aerobic decomposition of organic matter, is one of the most effective ways to manage organic waste and improve soil health. This method has been employed extensively in the Global South, particularly in agricultural regions, to convert food waste and crop residues into nutrient-rich compost. Composting reduces the volume of waste destined for landfills and promotes sustainable agricultural practices by enhancing soil fertility and reducing the need for synthetic fertilizers (Mekonnen & Hoekstra, 2020). In Bangladesh, community-based composting initiatives have been implemented to manage urban organic waste, contributing to improved food production in peri-urban areas. Composting not only enhances crop yields but also fosters environmental sustainability by reducing methane emissions from landfills. It is a biochemical process in which the soil organic carbon is broken and stabilized, with this it can be used as a fertilizer or soil conditioner. It consists of three stages, the initial activation, the thermophilic and mesophilic stages (Route 1, Fig. 2). The initial decomposition of compounds is done by mesophilic microorganisms in which the temperature is increased within the pile that is composting. Thermophilic microbes replace the mesophilic ones when the temperature is increased above 40 °C as they can survive this high temperature. At this stage of action of thermophilic microbes, they utilize oxygen and releases ammonia and carbon dioxide. Pathogens are destroyed at temperature range of 55–70 °C, seed viability in the organic wastes is reduced. Temperatures are best kept below 65 °C using aeration and mixing processes as high temperatures can result to death of microbes thereby limiting the rate of decomposition. The maturation phase occurs when the temperature in the compost decreases and exhaustion of organic fraction. In this phase, compost is produced has humic characteristics and produced from humidified and stabilized organic matter.

Compost is a major product from this process and a good source of organic matter useful as a fertilizer and can improve the soil's water retention capacity, structure, porosity, and buffering capacity. Parameters such as the particle size, moisture content, pH, temperature, time of aeration, heat insulation and type if inoculum can determine the characteristics of compost (Wang et al., 2018). The process of composting takes place naturally and composting is efficient when the odor and dust is controlled and a good agricultural fertilizer with appropriate quantity of nitrogen, phosphorus, potassium, and carbon are produced. In monitoring compost, the following indicators must be considered:

C/N Ratio: It refers to the rate of decomposition and when its high, the rate of decomposition is slow because of the reduction in activities of microbes. When C/N is low, failure of air diffusion through the pile may occur and result in conditions that are anaerobic. They typical values are usually between 25 and 30:1 (Wang et al., 2018).

Moisture Content: It is required by microbes to decompose organic matter, and levels between 50 and 60% are recommended (Wang et al., 2018).

Bulk Density: It describes the rate of air diffusion and flow of water through a pile of compost. The flow characteristics are better when the density is low, and density between 600 and 700 kg/m^3 is recommended.

The advanced methods of technology in composting are the duration of composting, the loss of nitrogen by ammonia and volatilization from emissions of greenhouse gases. The duration of compost varies because of the type of material and time can be hastened using microbial communities. An example is vermicomposting which is the addition of biochar and use of earthworms which can reduce greenhouse gas emissions. Zeolite and biochar can be added to reduce the loss of nitrogen during composting. Lastly, the improvements in composting are around improvement in the control and removal of pathogens when sludge are used as additional feedstock.

Bioenergy Production

Bioenergy production refers to the conversion of biomass into energy through various technologies such as combustion, gasification, and pyrolysis (Demirbas, 2009). Biomass-to-energy conversion technologies produce biofuels like bioethanol and biodiesel, which are derived from organic waste such as crop residues, municipal solid waste, and animal manure. These biofuels offer a renewable energy source and a means of reducing reliance on fossil fuels in the Global South. Bioenergy production, especially through biomass conversion, is a significant resource recovery strategy in the Global South. Biomass-to-energy technologies convert organic waste materials like crop residues, forestry by-products, and animal waste into energy in the form of biofuels (bioethanol, biodiesel) and biogas. Countries like Brazil and South Africa have established large-scale biofuel programs that convert sugarcane and maize into bioethanol, contributing to their renewable energy mix and reducing dependence on fossil fuels (Goldemberg et al., 2008). In Brazil, the ethanol industry, primarily based on sugarcane, has significantly contributed to energy security while reducing greenhouse gas emissions (Bailis et al., 2016). In South Africa, biodiesel production from sunflower oil and other crops is part of the national strategy to promote renewable energy (Davies & Pienaar, 2019).

Microbial Fuel Cells

Microbial fuel cells (MFCs) are bioelectrochemical systems that harness microbial processes to convert organic waste into electricity (Logan, 2008). Microbes within the MFC system break down organic matter, releasing electrons that generate electrical currents. This technology is an innovative approach to both waste treatment and energy generation, offering a sustainable solution for energy recovery in regions lacking access to conventional energy infrastructure (Pant et al., 2010). MFCs represent an emerging biobased technology that generates electricity from organic waste by exploiting the metabolic processes of microorganisms. This innovative technology holds great promise in the Global South, where energy access is limited, and waste management infrastructure is underdeveloped (Pant et al., 2020). MFCs not only generate clean energy but also help in wastewater treatment, making them suitable for use in off-grid and rural areas. Pilot projects in countries like India and Ghana have demonstrated the feasibility of MFCs in treating wastewater while providing electricity for small communities (Saravanan et al., 2021).

Biochar Production (Pyrolysis)

Biochar is a carbon-rich material produced through the pyrolysis of organic waste, such as agricultural residues, under limited oxygen conditions (Lehmann & Joseph, 2015). Biochar serves as a soil amendment, enhancing soil fertility and moisture retention, while also acting as a carbon sequestration tool by storing carbon in the soil for extended periods. This technique offers substantial environmental benefits, particularly in regions prone to soil degradation (Lehmann, 2007). Biochar has multiple environmental benefits, including enhancing soil fertility, sequestering carbon, and improving water retention in soils. In the Global South, biochar production offers a sustainable solution for managing agricultural waste and combating soil degradation, particularly in regions facing the impacts of climate change (Lehmann et al., 2011). In Ethiopia and Zambia, biochar initiatives have been implemented to improve crop yields, enhance soil health, and sequester carbon in agricultural lands (Woolf et al., 2021). Biochar not only provides environmental benefits but also contributes to food security by boosting agricultural productivity. It is the process of organic matter decomposition using thermal processes in the absence of oxygen, resulting to the production of biochar. Pyrolysis is of two types: the slow pyrolysis and the fast pyrolysis. The slow pyrolysis is done at temperature range of 400–600 °C and heat of 0.2 °C/s and the result is the maximization of biochar yield to 35%. The fast pyrolysis results into a 70% yield of bio-oil which is a major byproduct. It is performed at a temperature of 1000 °C and heat rate of about 500 °C. Before pyrolysis, drying is required because the heat is consumed, and the lower yield is induced by the moisture that is embedded in the input. A recommended moisture level of less than 7% is required to achieve a performance that is reasonable during pyrolysis. Other parameters that are required to produce biochar are operating pressure, the presence of minerals, temperature, residence time, characteristics of feedstock and size of particles. It is essential to complement these parameters with the pyrolysis mechanisms and interaction of the biomass main components.

There are three products that are obtained when solid wastes are converted using the process of pyrolysis, they are biochar (it is usually solid form and rich in carbon), bio-oil (it is gotten as liquid after volatile organic matter has condensed) and non-condensable gases (examples are carbon monoxide, carbon dioxide, methane, and hydrogen). The production of biochar has received global attention because of its stability, its wide use in treatment of wastewater, soil amendment and remediation, capture of carbon dioxide and storage of energy. However, further investigation is required in determining the effects of parameters in the biochar stability during the process of pyrolysis.

Some advances in this process are the catalyzation of pyrolysis processes and methods of heating. Catalyzation is achieved using the process of co-pyrolysis and hydro-pyrolysis. The process of adding other feedstocks (e.g., Coal, plastic, sludge) is the basis of co-pyrolysis, and was adopted to utilize the chemical interactions that may occur between feedstocks. Hydro-pyrolysis occurs when there is an occurrence of fast biomass pyrolysis and catalytic hydrodeoxygenation in combination, with the presence of hydrogen to produce aromatic hydrocarbons. The yield of bio-oil has been reported to be enhanced by catalysts such as zeolites, metal oxides or inorganic salts.

Membrane Separation

Membrane separation is a biobased recovery technique that uses semipermeable membranes to separate valuable products from biomass. This approach is used for biomass fractionation (e.g., separating proteins, sugars, and phenolics) and bioactive compound extraction (e.g., antioxidants, flavors, and fragrances). This approach has high product yields and quality and reduced energy and chemical consumption which is effective for minimized waste generation. Membrane separation is a versatile biobased recovery technique, offering a sustainable and efficient way to extract valuable products from biomass, contributing to a circular economy and renewable resource management.

Bioleaching

Bioleaching is a biobased recovery technique that uses microorganisms to extract valuable metals and minerals from biomass. This approach offers several advantages like the reduced chemical consumption and selective extraction of target metals. Moreover, bioleaching is cost effective, and the process of bioleaching involves biomass preparation (e.g., crushing, grinding), inoculation with microorganisms (e.g., bacteria, archaea), leaching (extraction) of metals and recovery and purification of extracted metals. The approach has a reduced environmental impact and increased metal recovery efficiency. Bioleaching is a promising biobased recovery technique, offering a sustainable and efficient way to extract valuable metals and minerals from biomass, contributing to a circular economy and renewable resource management (Jaiswal & Srivastava, 2024).

Hydrolysis

Hydrolysis is a biobased recovery technique that involves the use of water to break down organic materials into valuable products. Hydrolysis offers several benefits as a biobased recovery technique as it has gentle operating conditions preserve sensitive molecules, having high product yields and selectivity with low energy requirements. Hydrolysis has been applied in biofuels and biochemicals production, food and feed processing, pharmaceutical and cosmetic industries, biorefineries and biotechnology and waste management and recycling. Hydrolysis is a versatile biobased recovery technique, offering a sustainable and efficient way to extract valuable products from biomass, contributing to a circular economy and renewable resource management.

Solvent Extraction

Solvent extraction is a biobased recovery technique that involves using a solvent to selectively extract valuable products from biomass. This approach offers several advantages such as high extraction efficiency and selective extraction of target compounds. Solvent extraction is used for bioactive compounds (e.g., phenolics, flavonoids, terpenes), amino acids, sugars and carbohydrates, proteins, fatty acids, and lipids. This approach is applied in food and feed processing (e.g., protein extraction, sugar production), biorefineries and biotechnology pharmaceutical and cosmetic industries and waste management and recycling. Solvent extraction is a versatile biobased recovery technique, offering a sustainable and efficient way to extract valuable products from biomass, contributing to a circular economy and renewable resource management.

Algae-Based Resource Recovery

Algae-based resource recovery leverages algae's ability to absorb nutrients from wastewater, thereby treating water while simultaneously producing biomass (Christenson & Sims, 2011). The harvested algae can be converted into biofuels, such as biodiesel or biogas, or used as a feedstock for biobased products like bioplastics and biofertilizers. This technique has gained attention for its potential to integrate waste treatment and bioenergy production, particularly in coastal and wastewater-intensive regions (Singh & Olsen, 2011). Algae-based resource recovery is a promising technology that uses algae to treat wastewater and produce valuable by-products such as biofuels, biofertilizers, and animal feed. Algae cultivation can absorb nutrients from wastewater, effectively reducing water pollution while producing biomass for bioenergy production (Muñoz & Guieysse, 2006). In Thailand and Brazil, algae-based systems have been used to treat wastewater from agriculture and aquaculture, producing biofuels that can be used for energy generation (Razzak et al., 2013). Algae-based technologies not only address environmental challenges but also create economic opportunities through the production of high-value bioproducts.

Bioplastics Production from Organic Waste

Bioplastics production from organic waste offers a sustainable alternative to conventional plastics derived from fossil fuels. The Global South, particularly in regions like Southeast Asia and Latin America, faces a growing challenge of plastic waste pollution due to high plastic consumption and inadequate waste management infrastructure (Rujnic-Sokele & Pilipovic, 2017). Bioplastics, derived from organic materials such as agricultural residues, food waste, and other biomass, present a promising solution to this issue. Innovations in bio-polymer synthesis have allowed for the development of bioplastics that decompose more quickly than conventional plastics, thereby reducing environmental pollution (Narancic et al., 2020). For example, in countries like Brazil, sugarcane bagasse, a byproduct of sugar production, has been used to produce polylactic acid (PLA), a biodegradable plastic. This not only reduces plastic pollution but also promotes the circular economy by repurposing agricultural waste. Bioplastics production also supports rural development and job creation by creating value-added products from organic waste streams (Wang et al., 2018).

Waste-to-Bioenergy Conversion

Waste-to-bioenergy conversion is a critical biobased recovery technique that transforms organic waste into renewable energy sources, such as biogas, bioethanol, and biodiesel. In the Global South, where access to reliable energy remains a significant challenge, waste-to-energy technologies can address both waste management and energy access issues (Karekezi et al., 2012). For instance, countries like Kenya and India have implemented biogas production facilities that convert agricultural residues, animal waste, and food waste into biogas, which can be used for cooking, heating, and electricity generation (Njenga et al., 2019a, b). Advancements in anaerobic digestion and gasification technologies have improved the efficiency of biomass-to-energy conversion processes, making them more viable for large-scale deployment in the Global South (Saidur et al., 2011). The production of biofuels such as biodiesel from used cooking oil and other organic waste materials is also gaining traction in countries like India and Indonesia, where local governments and private enterprises are investing in biofuel production to reduce reliance on imported fossil fuels (Sorda et al., 2010). This approach not only mitigates environmental impacts but also supports energy security and rural livelihoods.

Bioremediation

Bioremediation is a biobased recovery technique that uses microorganisms, plants, or fungi to degrade, detoxify, or remove pollutants from contaminated environments. This method is particularly relevant in the Global South, where industrial pollution, poor waste management practices, and agricultural runoff have led to significant environmental contamination. Bioremediation offers a cost-effective and environmentally friendly approach to addressing pollution in soil, water, and air. For example, in India, the use of microbial consortia to degrade petroleum hydrocarbons in oil-contaminated soils has shown promising results (Das & Chandran, 2011). Similarly, in sub-Saharan Africa, phytoremediation, which involves using plants to absorb heavy metals from contaminated soils, has been applied to restore degraded lands affected by mining activities (Garbisu & Alkorta, 2001). These bioremediation techniques not only restore environmental health but also contribute to sustainable agriculture by improving soil fertility and reducing the need for chemical inputs. The adoption of bioremediation technologies in the Global South is growing, driven by both environmental regulations and the need for sustainable land management practices (Atagana, 2014). However, the success of bioremediation projects depends on local environmental conditions, the nature of the pollutants, and the availability of trained personnel to manage the processes effectively (Kumar et al., 2011). Collaborative efforts between governments, research institutions, and international organizations are essential to scaling these technologies across the Global South.

Solid Waste Management Techniques

It is a process and activities that are involved in managing wastes from its generation to disposal. It involves the aspects of waste generation, collection, transportation, treatment, disposal, monitoring, and regulations (Fig. 2). Effective solid waste management techniques aim at reducing the wastes that are generated, increasing recycling, and reusing, minimizing the environmental impacts of wastes, support the natural resources and sustainable development. Some of the techniques are:

- Anaerobic digestion.
- Composting (Aerobic digestion).
- Hydrothermal Carbonization.
- Pyrolysis.
- Land filling.

- *Anaerobic Digestion*: The AD of solid waste is a waste that is biological in nature, and it involves the breakdown of organic matter into compounds that are smaller in the presence of microorganisms that are anaerobic in nature. The AD process involves some stages namely, hydrolysis, acidogenesis, acetogenesis, and methanogenesis.
- *Hydrolysis stage*: It is the process in which compounds that are simpler and soluble and are formed from organic compounds (amino acids and sugars), that are complex in nature such as proteins, carbohydrates, and fatty acids. The enzymes that are required for this stage are produced by microbes found in the anaerobic digester. The hydrolysis of some lignocellulosic materials is very slow, and this step is therefore referred to as a rate-limiting step in the process of AD.
- *Acidogenesis stage*: It is the process by which products from hydrolysis are converted into alcohols, volatile fatty acids (VFA) and hydrogen sulfide (H_2S), carbon dioxide (CO_2), ammonia (NH_3) and hydrogen (H_2). This stage is activated by the presence of anaerobic bacteria that are facultative and obligate in nature.
- *Acetogenesis Stage*: It is the process in which products from acidogenesis is converted to produce carbon dioxide, acetate, and hydrogen. AD process stability is monitored and evaluated by the quantity of VFAs produced at the stages of acidogenesis and acetogenesis. Although, accumulation of VFA is an issue that is major in the process of AD, and it results primarily from the ratio of acid from acidogenesis stage producers to the consumed acid in the methanogenesis stage.
- *Methanogenesis Stage*: It is the final step in solid waste management which is a process in which organic materials are degraded, leading to the production of biogas. The presence of methanogens that are acetotrophic and hydrogenotrophic in nature accomplishes this stage. About 70% methane are produced from acetotrophic methanogens, and limited quantities of hydrogen are supplied by AD systems. Methane is produced by the acetotrophic methanogens and hydrogenotrophic bacteria by degrading acetate and utilization of hydrogen and carbon respectively.

Hydrothermal Carbonization

It is a conversion that is thermochemical in which organic substrates that is wet with moisture of greater than 80% by weight reacts under hot pressurized water. The processing conditions are mild as temperature is between 180 and 250 °C and pressure between 10 and 40 bars, to maintain water in a liquid state. Hydrochar is the main product formed from HTC process, and it is a carbonaceous solid with high content of carbon which is greater than 80% by weight and its composition is like lignite or peat. In comparison with other thermochemical processes, the investment and operating cost for HTC is low and easier to provide solutions to problems of safety. The resistance time is low, and volume of equipment used is reduced.

The hydrochar produced from HTC can be used as improver of the soil, solid fuel and adsorbent. The characteristics and yield of hydrochar is related to the properties of feedstocks and the condition in which it is operated. The pathway of degradation is guided by the contents of carbon and minerals as well as the feedstock structure. Solid wastes are composed of high volatile matter percentage and low fixed carbon content. The application of HTC ensures conversion of solid wastes and carbonization at low temperatures.

The production of biofuel from HTC is useful in mitigating the effects of algae proliferation resulting from eutrophication. In the process of liquid extraction, the microalgal biomass is exposed to hydrothermal conditions to produce crude oil that can be used for generation of biodiesel (245 °C and 10 min). The fuel extracted from this process has a high heating value of 41.10 MJ/kg. Heilmann et al. reported the application of HTC at conditions of 200 °C and 0.5 h, and char produced from the biomass of algae had energy and consumption that was like coal with combustion heat value of 31 MJ/kg. Brown et al. proposed the use of hydrochar as a solid fuel in the production of biomethane through AD.

A major disadvantage in the process of AD is the handling of the digestate that is produced from the digestion. A major issue is the environmental problem that is caused by the release of high ammonia concentration, despite its use as a fertilizer. Also, the high content of water makes the use of digestate unsuitable in thermal treatments. The cost for disposal of the sludge is high as the slurry is bulky, making transportation of digestate expensive. Besides, HTC produces a by-product which is hot and compressed liquid and can be removed by dewatering. The issues of handling and watering in digestate have led to studies focused on the integration of AD and HTC to produce hydrochar from the digestate and improved generation of biogas.

5 Challenges in Implementing Biobased Resource Recovery in the Global South

The Global South faces unique challenges in implementing biobased resource recovery techniques, largely due to economic, infrastructural, policy, social, and technological factors. These challenges hinder the widespread adoption of sustainable practices and limit the potential of biobased recovery systems to promote environmental and economic sustainability.

Economic and Infrastructure Barriers: The implementation of biobased resource recovery in the Global South is often constrained by limited financial resources. Many countries in Africa, Latin America, and Southeast Asia struggle with inadequate funding for waste management infrastructure and technology. The high capital costs associated with setting up resource recovery systems, such as biogas plants, composting facilities, and bioenergy systems, make it difficult for governments and private entities to invest in these technologies (Nzila et al., 2012). Moreover, insufficient infrastructure for waste collection, sorting, and treatment adds another layer of complexity, as many rural areas lack the basic facilities required for effective resource recovery (Godfrey et al., 2019).

Policy and Regulatory Hurdles: The absence of cohesive policies and weak enforcement of existing regulations serve as significant barriers to biobased resource recovery in the Global South (Oberlin, 2011). In many countries, waste management is not a priority, and regulatory frameworks are often fragmented and poorly implemented. A lack of coordinated policies on resource recovery and the circular economy results in an inconsistent approach to sustainability efforts (Kosoe et al., 2023; Akudugu & Ogwu, 2024). Furthermore, ineffective collaboration between governmental bodies, local authorities, and private stakeholders exacerbates these challenges, leading to regulatory uncertainty and a slow adoption of recovery technologies (Simelane & Mohee, 2012).

Social and Cultural Factors Influencing Adoption: Cultural perceptions and social norms also influence the adoption of biobased resource recovery techniques. In some regions, waste is viewed as a public health issue rather than a potential resource, leading to reluctance to engage with waste recovery systems (Rouse, 2006; Agoh et al., 2024). In addition, public awareness of the environmental and economic benefits of resource recovery is often low, and there is limited community participation in waste management initiatives (Simelane & Mohee, 2012). Education and awareness campaigns are necessary to shift societal attitudes and increase acceptance of these systems.

Technological Limitations and Innovation Gaps: Technological limitations in the Global South further impede the implementation of biobased resource recovery. Many countries lack access to advanced recovery technologies, such as anaerobic digestion and bioenergy systems, which are necessary for efficient waste-to-resource conversion. The cost of importing or developing these technologies is often prohibitive, while local technological capacity is limited by a shortage of skilled workers and technical expertise (Kaza et al., 2018). Moreover, research and development in biobased recovery technologies is underfunded, leading to innovation gaps that prevent the adaptation of these technologies to local contexts (Nzila et al., 2012).

6 Technological Innovations and Pathways for Resource Recovery

The advancement of technological innovations in biobased resource recovery is crucial for enhancing efficiency, scaling solutions, and addressing the growing waste management challenges in the Global South. This section highlights emerging technologies, the role of biotechnology and bioengineering, innovations in waste-to-energy processes, and the integration of digital tools and monitoring systems to optimize resource recovery efforts.

Recent advancements in biobased resource recovery technologies are significantly improving the efficiency and scalability of waste management systems. One such innovation is anaerobic digestion (AD), a biological process that converts organic waste, such as agricultural residues and food waste, into biogas and digestate (Mata-Alvarez et al., 2014; Nizzy & Ogwu, 2024; Ogwu et al., 2024). This technology has the potential to address the energy needs of rural communities while reducing greenhouse gas emissions. Similarly, advancements in composting techniques, such as in-vessel composting, have enabled the rapid processing of organic waste into nutrient-rich compost for agricultural use, promoting soil health and reducing chemical fertilizer dependency (Said-Pullicino et al., 2016).

Biotechnology and bioengineering play a pivotal role in advancing biobased resource recovery, particularly in optimizing microbial processes and enhancing the efficiency of bioconversion technologies. Genetic engineering of microbial strains used in anaerobic digestion and composting has shown promise in increasing biogas yields and reducing process time (Lazarovits et al., 2020). Furthermore, bioengineering advancements have improved enzymatic breakdown of lignocellulosic biomass, making biofuels like bioethanol more accessible and cost-effective (Bhatia et al., 2017). The development of biobased catalysts and microbial fuel cells further contributes to the transformation of organic waste into valuable energy products, such as electricity and biohydrogen (Pandey et al., 2019).

Waste-to-energy (WtE) technologies have evolved significantly, providing an effective means of addressing organic waste streams and contributing to energy security. Innovations such as gasification and pyrolysis can convert agricultural and industrial waste into bioenergy, biochar, and syngas, which can be used as renewable energy sources or soil enhancers (Lehmann & Joseph, 2015). Additionally, advancements in microbial fuel cells (MFCs) offer the potential to generate electricity directly from organic waste, utilizing microbial activity to drive the process. The value-added products derived from these processes, such as biochar, fertilizers, and bioplastics, further enhance the economic viability of resource recovery initiatives.

Digital tools and smart monitoring systems are revolutionizing the field of biobased resource recovery by enabling real-time tracking of waste streams, optimizing resource recovery processes, and reducing operational costs. Technologies like Internet of Things (IoT) sensors and blockchain have been employed to monitor waste generation, collection, and treatment in urban and rural settings (Manoj & Hoang, 2020). These systems allow for data-driven decision-making, enabling operators to adjust recovery processes to maximize efficiency and output. Moreover, digital platforms facilitate knowledge sharing and collaboration among stakeholders, fostering innovation in biobased resource recovery technologies.

7 Sustainability and Resource Efficiency

Biobased resource recovery plays a crucial role in promoting sustainability and resource efficiency by transforming waste into valuable products, thus contributing to environmental, social, and economic benefits. One of the key environmental benefits of biobased resource recovery is the reduction of greenhouse gas emissions. Techniques like anaerobic digestion and composting help mitigate methane emissions from organic waste, while converting agricultural and food waste into biofuels reduces the reliance on fossil fuels (Mata-Alvarez et al., 2014; Ogwu, 2019). Additionally, waste minimization is achieved through the reuse and recycling of biomass, food waste, and other organic materials, which reduces the burden on landfills and conserves natural resources (Lehmann & Joseph, 2015). Biobased resource recovery techniques offer significant socioeconomic benefits, particularly in rural areas where agricultural waste is abundant. The implementation of resource recovery projects can create job opportunities in waste collection, processing, and bioenergy production (Albrecht et al., 2020). Furthermore, the development of biobased industries in rural areas can stimulate local economies and contribute to rural development by improving energy security and generating additional income streams for farmers through bio-product sales.

Biobased resource recovery contributes to the circular economy by creating closed-loop systems where waste is repurposed as a resource. By converting organic waste into valuable products such as compost, bioenergy, and bioplastics, biobased recovery enhances resource efficiency and reduces the need for virgin materials (Zhang et al., 2021). This circular approach not only minimizes environmental impacts but also ensures the sustainable management of resources, aligning with global sustainability goals.

7.1 Case Studies of Success

Several successful biobased resource recovery projects across the Global South have demonstrated the potential for transforming waste into valuable resources, with significant impacts on local communities and economies. One such example is the Kenya Biogas Program, which has implemented anaerobic digestion to convert organic waste into biogas for cooking and heating. This initiative has provided a cleaner energy source for rural households, reducing reliance on firewood and improving indoor air quality (Njenga et al., 2019a, b). In India, agricultural waste is repurposed through composting and bioenergy production. The Punjab Biomass Power Project, for example, utilizes crop residues such as rice husks and wheat straw to generate electricity. This project not only addresses waste management issues but also supplies renewable energy to rural areas, enhancing energy security

(Sinha & Yadav, 2020). In Brazil, sugarcane bagasse, a byproduct of sugar production, is used for bioenergy production. The Raízen Energy Project is a large-scale initiative that converts sugarcane residues into bioethanol, providing a sustainable fuel alternative. The project has reduced greenhouse gas emissions and created jobs in rural communities (Amorim et al., 2021).

These successful projects have highlighted key lessons, including the importance of stakeholder engagement, community involvement, and government support. For instance, the Kenya Biogas Program demonstrated the need for local capacity-building and technical training to ensure long-term sustainability. The Punjab Biomass Power Project underscored the value of integrating agricultural waste management with energy production, offering a replicable model for regions with abundant crop residues. Another critical lesson is the importance of policy frameworks that incentivize biobased resource recovery initiatives. The success of Brazil's Raízen Energy Project has shown that government support through subsidies and favorable policies can drive the adoption of bioenergy technologies and attract private investment (Amorim et al., 2021).

The impact of these projects on local communities has been transformative. In Kenya, the biogas program has improved energy access for rural households, reducing deforestation and improving health by replacing traditional cooking methods with cleaner alternatives (Njenga et al., 2019a, b). In India, the biomass power project has created employment opportunities, particularly in rural areas, and contributed to local economic development through the sale of electricity and byproducts (Sinha & Yadav, 2020). Similarly, Brazil's bioethanol industry has provided jobs, stimulated the rural economy, and positioned the country as a leader in bioenergy production (Amorim et al., 2021).

8 Policy and Governance Frameworks for Sustainable Biobased Resource Recovery in the Global South

Supportive policies are critical to scaling biobased resource recovery in the Global South. Without robust governance and regulatory frameworks, the adoption of sustainable practices is limited by a lack of incentives and enforcement mechanisms. Policies that promote biobased resource recovery can address gaps in waste management, reduce environmental impacts, and foster the circular economy (Lehmann et al., 2020). These policies often include tax incentives, subsidies, and research and development support for eco-friendly technologies, encouraging investment and innovation in the sector.

International organizations play a pivotal role in promoting biobased resource recovery through funding, technical assistance, and knowledge sharing. Agencies like the United Nations Environment Programme (UNEP) and the Food and Agriculture Organization (FAO) have initiated programs aimed at fostering resource-efficient and environmentally sustainable practices in developing countries. Regional cooperation, such as within the African Union or ASEAN, is also vital for cross-border collaboration on policy frameworks, especially in areas where shared ecosystems and waste streams require coordinated management efforts (FAO, 2021). Brazil stands as a leading example in the adoption of effective biobased resource recovery policies. The Brazilian National Solid Waste Policy promotes waste minimization and recycling, particularly in the agro-industrial sector. Through comprehensive legislation, Brazil has established a robust infrastructure for bioethanol production from sugarcane bagasse, creating both environmental and economic benefits (Amorim et al., 2021). Another notable example is South Africa, which has implemented progressive waste management policies, including the National Waste Management Strategy. This strategy encourages waste-to-energy projects and the recovery of resources from agricultural and municipal waste, particularly through public–private partnerships (Godfrey et al., 2017).

To enhance policy frameworks for biobased resource recovery, governments in the Global South must focus on:

Incentivizing Innovation: Governments can provide financial incentives, such as tax credits or subsidies, for companies that develop or adopt biobased technologies. This would stimulate private-sector investment and encourage small and medium-sized enterprises to engage in biobased resource recovery.

Strengthening Regulatory Enforcement: Effective policy implementation requires strong enforcement mechanisms. Governments should ensure compliance through stricter monitoring and penalties for non-compliance with waste management regulations.

Promoting Public Awareness and Stakeholder Engagement: Public education campaigns are essential to shift perceptions and practices regarding waste management. Governments should work closely with communities, businesses, and civil society to create an inclusive and cooperative approach to biobased resource recovery.

Regional Collaboration: Nations within the Global South should collaborate on shared environmental challenges by harmonizing waste management policies, exchanging technical expertise, and developing joint resource recovery initiatives (UNEP, 2021).

9 Future Direction: Collaboration and Knowledge Sharing

Collaborative efforts between the Global North and South, as well as within the Global South, are critical to advancing biobased resource recovery. North–South collaborations can facilitate technology transfer, access to funding, and capacity building for biobased innovations. South–South collaborations, on the other hand, can promote the exchange of region-specific knowledge, innovations, and best practices. For example, countries like Brazil and India, which have pioneered resource recovery initiatives, can share expertise with other nations in Africa and Southeast

Asia to address common challenges such as waste management and bioenergy production (Siqueira et al., 2021; Smith et al., 2020).

Academic institutions and research organizations play a pivotal role in generating the knowledge needed to advance biobased resource recovery. Universities and research centers can partner with industry stakeholders and governments to explore new technologies and sustainable business models for resource recovery. Public–private partnerships (PPPs) are essential in implementing pilot projects and scaling up proven technologies, as they pool resources, expertise, and funding. PPPs can also foster innovation by creating platforms for collaboration between scientists, entrepreneurs, and policymakers, thus accelerating the commercialization of biobased technologies (Pereira et al., 2019).

Capacity building initiatives, including training programs and workshops, are essential for scaling biobased resource recovery in the Global South. These programs should target key stakeholders, such as local communities, entrepreneurs, policymakers, and technical personnel, to equip them with the skills and knowledge necessary to implement resource recovery techniques effectively. Knowledge exchange platforms, facilitated by international organizations and NGOs, can also play a significant role in disseminating best practices, case studies, and technological advancements (Chavez-Tafur & Zylstra, 2020).

The growing demand for sustainable solutions in agriculture, energy, and waste management presents a unique opportunity to scale biobased resource recovery across the Global South. Scaling biobased technologies requires leveraging local resources, creating supportive policies, and investing in infrastructure. Furthermore, innovations in digital tools, such as remote sensing, data analytics, and blockchain, can be used to monitor resource recovery activities, optimize supply chains, and ensure compliance with environmental standards (Patel et al., 2021; Obahiagbon & Ogwu 2023a, b, 2024a, b).

Despite the progress made in biobased resource recovery, several research gaps remain. These include the need for more efficient technologies to recover resources from complex waste streams, the development of cost-effective solutions for smallholder farmers and low-income communities, and an improved understanding of the socioeconomic impacts of biobased recovery. Additionally, more research is needed on the lifecycle impacts of biobased products, including their potential to contribute to circularity in specific sectors (Murray et al., 2020). Policymakers should focus on creating enabling environments by providing incentives for biobased industries, developing regulatory frameworks, and fostering international collaboration. Industry stakeholders should invest in research and development, collaborate with academic institutions, and engage in public–private partnerships to scale sustainable technologies. Researchers should prioritize interdisciplinary studies that address both technical and socioeconomic challenges in biobased recovery and develop scalable models tailored to the Global South. International organizations should facilitate knowledge exchange, capacity building, and funding for collaborative projects aimed at advancing biobased resource recovery.

10 Conclusion

Biobased resource recovery is essential for achieving sustainability goals, as it promotes circular economy principles by converting waste into useful products, reducing greenhouse gas emissions, and enhancing resource efficiency. Furthermore, these techniques contribute to rural development and job creation, particularly in regions where waste management infrastructure is lacking. The environmental benefits, such as carbon sequestration and waste minimization, further underscore the importance of scaling bio-based recovery processes in the Global South. Biobased resource recovery offers significant potential to address environmental, economic, and social challenges in the Global South. By leveraging waste materials from agriculture, food, and other organic sources, biobased recovery processes such as anaerobic digestion, composting, bioenergy production, and microbial fuel cells can reduce waste, generate clean energy, and produce valuable by-products. However, successful implementation of these technologies depends on overcoming numerous challenges, including economic constraints, policy barriers, and technological gaps.

Achieving widespread adoption of biobased resource recovery requires a concerted global effort. Governments, private sectors, international organizations, and research institutions must collaborate to foster technological innovations, create enabling policy environments, and invest in capacity building. North–South and South–South collaborations can accelerate the transfer of knowledge and resources, while public–private partnerships can drive innovation and scale successful models. To fully unlock the potential of biobased resource recovery, it is imperative to invest in research, infrastructure, and policy frameworks that support the sustainable transition to a circular bioeconomy. These efforts will not only address waste and resource management challenges but also contribute to global sustainability and economic development goals.

References

Agoh, E. C., Ogwu, M. C., Chukwuemeka, O. S., & Ekeledo, P. I. (2024). Environmental and human health effects of cassava processing and processing waste. In M. C. Ogwu, S. C. Izah, A. A. Cunha Alves, & S. C. Babu (Eds.), *Plant biology, sustainability and climate change, sustainable cassava* (pp. 203–219). Academic. https://doi.org/10.1016/B978-0-443-21747-0.00001-1

Akudugu, M. A., & Ogwu, M. C. (2024). Sustainable development policies and interventions: A bibliometric analysis of the contributions of the academic community. *Journal of Cleaner Production*, 139919. https://doi.org/10.1016/j.jclepro.2023.139919

Albrecht, J., Fortes, P., Pina, A., & Santos, H. (2020). Resource recovery in the circular economy: A systematic approach to waste valorization. *Journal of Cleaner Production, 256*, 120442.

Amigun, B., Sigamoney, R., & von Blottnitz, H. (2017). Commercialisation of biofuels: A strategic perspective for Sub-Saharan Africa. *Journal of Cleaner Production, 86*, 247–257.

Amorim, C. E., Carvalho, R. J., & Rodrigues, M. A. (2021). Sugarcane bagasse for bioenergy in Brazil: Challenges and opportunities. *Renewable Energy, 164*, 678–689.

Atagana, H. I. (2014). Bioremediation of contaminated soils in the Global South. *Sustainable Industrial Development, 12*, 95–113.

Bailis, R., Solomon, B. D., & Marnay, C. (2016). Energy, biofuels, and the Global South. In *Bioenergy and sustainability: Bridging the gaps* (pp. 51–75). Springer.

Bernal, M. P., Alburquerque, J. A., & Moral, R. (2017). Composting of animal manures and chemical criteria for compost maturity assessment: A review. *Bioresource Technology, 100*(22), 5444–5453.

Bhatia, S. K., Joo, H. S., Yang, Y. H., & Gurav, R. (2017). Microbial biosynthesis of ethanol: Role of biotechnology and bioengineering for the production of biofuels. *Biotechnology Advances, 35*(7), 1075–1088.

Börjesson, P., & Gustavsson, L. (2020). Bioenergy from agriculture: Greenhouse gas balances and land use implications. *Biomass and Bioenergy, 63*, 153–157.

Bourguignon, D. (2016). *Circular economy in the European Union*. European Parliamentary Research Service.

Chavez-Tafur, J., & Zylstra, M. (2020). Knowledge sharing for sustainable development: Best practices and opportunities. *Sustainability, 12*(8), 3432.

Chime, A. O., Aiwansoba, R. O., & Ogwu, M. C. (2018). Pathological status of plant germplasm and sustainable crop production and conservation. *Journal of Energy and Natural Resources Management, 1*, 17–21.

Christenson, L., & Sims, R. (2011). Production and harvesting of microalgae for wastewater treatment, biofuels, and bioproducts. *Biotechnology Advances, 29*(6), 686–702.

Das, N., & Chandran, P. (2011). Microbial degradation of petroleum hydrocarbons. *Indian Journal of Experimental Biology, 49*(4), 225–235.

Davies, G., & Pienaar, J. J. (2019). The role of bioenergy in South Africa's energy future. *Renewable and Sustainable Energy Reviews, 105*, 95–102.

Demirbas, A. (2009). Biofuels securing the planet's future energy needs. *Energy Conversion and Management, 50*(9), 2239–2249.

Egboduku, W. O., Egboduku, T., Golohor, O. M., Imarhiagbe, O., & Ogwu, M. C. (2024). Cassava as raw material for sustainable bioeconomy development. In M. C. Ogwu, S. C. Izah, A. A. Cunha Alves, & S. C. Babu (Eds.), *Plant biology, sustainability and climate change, sustainable cassava* (pp. 57–73). Academic. https://doi.org/10.1016/B978-0-443-21747-0.00022-9

FAO. (2021). *Promoting circular economy and sustainable resource management*. Food and Agriculture Organization.

Garbisu, C., & Alkorta, I. (2001). Phytoremediation: A cost-effective plant-based technology for the removal of metals from the environment. *Bioresource Technology, 77*(3), 229–236.

Girotto, F., Alibardi, L., & Cossu, R. (2015). Food waste generation and industrial uses: A review. *Waste Management, 45*, 32–41. https://doi.org/10.1016/j.wasman.2015.06.008

Godfrey, L., Oelofse, S., & Phukubye, R. (2017). Opportunities for enhancing innovation in waste management in South Africa. *Waste Management & Research, 35*(3), 236–246.

Godfrey, L., Ahmed, M. T., Gebremedhin, K. G., Katima, J. H., Oelofse, S., Osibanjo, O., et al. (2019). Waste management practices in Africa: Current status and future prospects. *Waste Management, 105*, 153–161.

Goldemberg, J., Coelho, S. T., Nastari, P. M., & Lucon, O. (2008). Ethanol learning curve – The Brazilian experience. *Biomass and Bioenergy, 26*(3), 301–304.

Gómez, S., Rittmann, B. E., & Lassaux, C. (2021). Emerging technologies for bio-based resource recovery in developing countries: Challenges and opportunities. *Renewable and Sustainable Energy Reviews, 141*, 110801.

González, L. O., & Chauhan, C. (2020). Waste management and resource recovery in Latin America. *Journal of Cleaner Production, 265*, 121505.

Hochman, G., & Zilberman, D. (2016). The economics of biofuels and food vs. fuel debate. *Annual Review of Resource Economics, 6*, 273–293.

Holm-Nielsen, J. B., Al Seadi, T., & Oleskowicz-Popiel, P. (2009). The future of anaerobic digestion and biogas utilization. *Bioresource Technology, 100*(22), 5478–5484.

Hoornweg, D., & Bhada-Tata, P. (2012). *What a waste: A global review of solid waste management*. World Bank.

Islam, M. S., & Hossain, M. I. (2021). Community-based composting as a waste management strategy in Bangladesh. *Waste Management & Research, 39*(12), 1230–1240.

Jain, S., Basak, B. B., & Patra, D. D. (2020). *Recycling of organic wastes for sustainable agriculture* (pp. 42–49). Springer.

Kalmykova, Y., Sadagopan, M., & Rosado, L. (2018). Circular economy – From review of theories and practices to development of implementation tools. *Resources, Conservation and Recycling, 135*, 190–201.

Karekezi, S., Lata, K., & Coelho, S. T. (2012). Traditional biomass energy: Improving its use and moving to modern energy use. *Renewable Energy, 44*, 7–14.

Kaza, S., Yao, L. C., Bhada-Tata, P., & Van Woerden, F. (2018). *What a waste 2.0: A global snapshot of solid waste management to 2050.* The World Bank.

Kosoe, E. A., & Ogwu, M. C. (2024). Sustainable cassava processing: Processes, techniques, wastes, and waste streams. In M. C. Ogwu, S. C. Izah, A. A. Cunha Alves, & S. C. Babu (Eds.), *Plant biology, sustainability and climate change, sustainable cassava* (pp. 257–272). Academic. https://doi.org/10.1016/B978-0-443-21747-0.00011-4

Kosoe, E. A., Achana, G. T. W., & Ogwu, M. C. (2023). Regulations and policies for herbal medicine and practitioners. In S. C. Izah, M. C. Ogwu, & M. Akram (Eds.), *Herbal medicine phytochemistry. Reference series in phytochemistry.* Springer. https://doi.org/10.1007/978-3-031-21973-3_33-1

Kumar, M., Leon, V., Materano, A. S., & Ilzins, O. A. (2011). Bio-augmentation of petroleum hydrocarbon degradation by gamma-proteobacteria isolated from hydrocarbon-contaminated soil. *Journal of Environmental Science and Health, 46*(9), 1051–1059.

Kumar, S., Gaikwad, S. A., & Patel, S. (2019a). Renewable energy-based biogas plants for rural energy needs. *Renewable Energy, 142*, 1827–1835.

Kumar, A., Sahota, S., & Sharma, A. (2019b). Agricultural waste management in India: A review. *Waste Management & Research, 37*(6), 539–550.

Lazarovits, G., Khachatourians, G. G., Tewari, J. P., & Lawrence, T. L. (2020). Microbial bioengineering: Innovative approaches to maximize bioenergy production. *Biotechnology Advances, 43*, 107567.

Lehmann, J. (2007). A handful of carbon. *Nature, 447*(7141), 143–144.

Lehmann, J., & Joseph, S. (2015). *Biochar for environmental management: Science, technology and implementation.* Routledge.

Lehmann, J., Rillig, M. C., & Thies, J. (2011). Biochar effects on soil biota: A review. *Soil Biology and Biochemistry, 43*(9), 1812–1836.

Lehmann, M., Lee, M., & Rapoport, E. (2020). Policy frameworks for a circular economy: Challenges and opportunities. *Journal of Environmental Management, 271*, 110800.

Lehtomäki, A., Viinikainen, T. A., & Kokko, M. (2021). The role of biogas in sustainable energy transitions in Sub-Saharan Africa. *Renewable and Sustainable Energy Reviews, 150*, 111–119.

Logan, B. E. (2008). *Microbial fuel cells.* Wiley Online Library.

Lomillos, J. M., & De la Fuente, J. M. (2020). Circular bioeconomy: Policies and global development trends. *Resources Policy, 67*, 101658.

Manoj, B. S., & Hoang, D. H. (2020). Blockchain-based smart contract platforms for sustainable supply chain management in agriculture. *Journal of Cleaner Production, 284*, 124707.

Mata-Alvarez, J., Dosta, J., Romero-Güiza, M. S., Fonoll, X., Peces, M., & Astals, S. (2014). A critical review on anaerobic co-digestion achievements between 2010 and 2013. *Renewable and Sustainable Energy Reviews, 36*, 412–427.

McCormick, K., & Kautto, N. (2013). The bioeconomy in Europe: An overview. *Sustainability, 5*(6), 2589–2608.

Mekonnen, M., & Hoekstra, A. Y. (2020). The environmental footprint of composting as a solution for waste management in Africa. *Journal of Environmental Management, 268*, 110–119.

Muñoz, R., & Guieysse, B. (2006). Algal–bacterial processes for the treatment of hazardous contaminants: A review. *Water Research, 40*(15), 2799–2815.

Murray, A., Skene, K., & Haynes, K. (2020). The circular economy: An interdisciplinary exploration of the concept and application in a global context. *Journal of Business Ethics, 140*(3), 369–380.

Mutungwe, P., Waweru, E., & Maina, J. (2021). Biogas production and utilization in Kenya: Status, opportunities, and challenges. *Renewable and Sustainable Energy Reviews, 137*, 110611.

Nair, J., Sekiozoic, V., & Anda, M. (2019). Waste composting for sustainable agriculture. *Waste Management, 29*(2), 1045–1053.

Narancic, T., Verstichel, S., Reddy, A., et al. (2020). Biodegradable plastics: Where is the gap between the global policy and the sustainability? *Current Opinion in Green and Sustainable Chemistry, 21*, 45–50.

Nizzy, A. M., & Ogwu, M. C. (2024). Valorization of cassava processing by-products into biofuel for a sustainable environment. In M. C. Ogwu, S. C. Izah, A. A. Cunha Alves, & S. C. Babu (Eds.), *Plant biology, sustainability and climate change, sustainable cassava* (pp. 291–309). Academic. https://doi.org/10.1016/B978-0-443-21747-0.00010-2

Njenga, M., Mendum, R., & Iiyama, M. (2019a). Biogas technology: Opportunities and challenges in East Africa. *Renewable Energy, 143*, 743–752.

Njenga, M., Mendum, R., & Iiyama, M. (2019b). Small-scale biogas digesters in Africa: A synthesis of technologies, practices, barriers, and incentives. *Renewable and Sustainable Energy Reviews, 102*, 212–225.

Nzila, C., Dewulf, J., Spanjers, H., Tuigong, D., Kiriamiti, H., & Van Langenhove, H. (2012). Multi criteria sustainability assessment of biogas production in Kenya. *Applied Energy, 93*, 496–506.

Obahiagbon, E. G., & Ogwu, M. C. (2023a). Sustainable supply chain management in the herbal medicine industry. In S. C. Izah, M. C. Ogwu, & M. Akram (Eds.), *Herbal medicine phytochemistry. Reference series in phytochemistry*. Springer. https://doi.org/10.1007/978-3-031-21973-3_64-1

Obahiagbon, E. G., & Ogwu, M. C. (2023b). Consumer perception and demand for sustainable herbal medicine products and market. In S. C. Izah, M. C. Ogwu, & M. Akram (Eds.), *Herbal medicine phytochemistry. Reference series in phytochemistry*. Springer. https://doi.org/10.1007/978-3-031-21973-3_65-1

Obahiagbon, E. G., & Ogwu, M. C. (2024a). Organic food preservatives: The shift towards natural alternatives and sustainability in the Global South's markets. In M. C. Ogwu, S. C. Izah, & N. R. Ntuli (Eds.), *Food safety and quality in the Global South* (pp. 299–329). Springer. https://doi.org/10.1007/978-981-97-2428-4_10

Obahiagbon, E. G., & Ogwu, M. C. (2024b). The nexus of business, sustainability, and herbal medicine. In S. C. Izah, M. C. Ogwu, & M. Akram (Eds.), *Herbal medicine Phytochemistry. Reference series in phytochemistry*. Springer. https://doi.org/10.1007/978-3-031-21973-3_67-1

Oberlin, A. S. (2011). The role of households in solid waste management in East Africa capital cities. *Waste Management & Research, 29*(7), 727–738.

Ogwu, M. C. (2019). Understanding the composition of food waste: An "-omics" approach to food waste management. In A. P. Gunjal, M. S. Waghmode, N. N. Patil, & P. Bhatt (Eds.), *Global initiatives for waste reduction and cutting food loss* (pp. 212–236). IGI Global. https://doi.org/10.4018/978-1-5225-7706-5.ch011

Ogwu, M. C., Ahana, C. M., & Osawaru, M. E. (2018). Sustainable food production in Nigeria: A case study for Bambara groundnut (*Vigna subterranean* (L.) Verdc. Fabaceae). *Journal of Energy and Natural Resources Management, 1*, 68–77.

Ogwu, M. C., Iyiola, A. O., & Izah, S. C. (2023). Overview of African biological resources and environment. In S. C. Izah & M. C. Ogwu (Eds.), *Sustainable utilization and conservation of Africa's biological resources and environment. Sustainable development and biodiversity* (Vol. 888, pp. 1–34). Springer. https://doi.org/10.1007/978-981-19-6974-4_1

Ogwu, M. C., Odozi, I. P., Ahonsi, O. C., Uleanya, K. O., & Odozi, E. B. (2024). Organic acid production from cassava. In M. C. Ogwu, S. C. Izah, A. A. Cunha Alves, & S. C. Babu (Eds.), *Plant biology, sustainability and climate change, sustainable cassava* (pp. 395–418). Academic. https://doi.org/10.1016/B978-0-443-21747-0.00009-6

Oteng-Ababio, M., Arguello, J. E. M., & Gabbay, O. (2020). Recycling in the informal sector: Examining waste pickers' resilience in Accra, Ghana. *Journal of Environmental Management, 272*, 110881.

Pandey, A., Larroche, C., Ricke, S., Dussap, C. G., & Gnansounou, E. (Eds.). (2019). *Biorefineries: Integrated biochemical processes for liquid biofuels*. Elsevier.

Pant, D., Van Bogaert, G., Diels, L., & Vanbroekhoven, K. (2010). A review of the substrates used in microbial fuel cells (MFCs) for sustainable energy production. *Bioresource Technology, 101*(6), 1533–1543.

Pant, D., Singh, A., Van Bogaert, G., & Verstraete, W. (2020). Bioelectricity production from microbial fuel cells: Current challenges and future prospects. *Biotechnology Advances, 39*, 107–125.

Patel, R., Gupta, P., & Sharma, A. (2021). Digital technologies for enhancing resource recovery: Opportunities and challenges in developing countries. *Journal of Environmental Management, 285*, 112074.

Pereira, L., Campos, P., & Vicente, M. (2019). Public-private partnerships for waste-to-energy: A viable solution for sustainable development in developing countries. *Renewable and Sustainable Energy Reviews, 102*, 376–389.

Power Knot. (2021, March 1). *6 reasons anaerobic digesters aren't as environmentally friendly as you think*. Power Knot. https://powerknot.com/2021/03/01/6-reasons-anaerobic-digesters-arent-as-environmentally-friendly-as-you-think/

Razzak, S. A., Hossain, M. M., & de Lasa, H. I. (2013). Biofuels from algae. *Progress in Energy and Combustion Science, 39*(4), 353–380.

Rouse, J. (2006). Planning for sustainable waste management in Africa: The roles of civil society and the private sector. *Habitat International, 30*(4), 789–796.

Rujnic-Sokele, M., & Pilipovic, A. (2017). Challenges and opportunities of biodegradable plastics: A mini-review. *Waste Management & Research, 35*(2), 132–140.

Said-Pullicino, D., Erriquens, F. G., & Gigliotti, G. (2016). Changes in soil structure and organic carbon quality induced by compost amendments: Evidence from physical fractionation and chemical analysis. *Soil Biology and Biochemistry, 48*, 210–219.

Saidur, R., Abdelaziz, E. A., Demirbas, A., Hossain, M. S., & Mekhilef, S. (2011). A review on biomass as a fuel for boilers. *Renewable and Sustainable Energy Reviews, 15*(5), 2262–2289.

Saravanan, V., Roy, A., & Sundaramoorthy, M. (2021). Microbial fuel cells for wastewater treatment and energy recovery in India: A review. *Renewable and Sustainable Energy Reviews, 141*, 110–155.

Scarlat, N., Dallemand, J. F., Monforti-Ferrario, F., & Nita, V. (2015). The role of biomass and bioenergy in a future bioeconomy: Policies and facts. *Environmental Development, 15*, 3–34.

Simelane, T., & Mohee, R. (2012). *Future directions of municipal solid waste Management in Africa*. Africa Institute of South Africa.

Singh, J., & Olsen, S. I. (2011). A critical review of algal biofuels production, life cycle analysis, and sustainability. *Journal of Renewable and Sustainable Energy Reviews, 14*(9), 2588–2596.

Sinha, S., & Yadav, S. K. (2020). Biomass power projects and their impact on rural communities in India: A case study from Punjab. *Journal of Cleaner Production, 258*, 120871.

Siqueira, D., Silveira, C., & Nogueira, A. (2021). Bio-based economy in Brazil: Policy frameworks and market dynamics. *Energy Policy, 152*, 112249.

Smith, P., Davis, S. J., & Creutzig, F. (2017). A framework for assessing the role of bioenergy with carbon capture and storage (BECCS) in climate change mitigation. *Energy and Environmental Science, 10*(12), 1501–1522.

Smith, J., Wilson, S., & Johnson, K. (2020). South-South cooperation for sustainable development: Lessons from the bioeconomy sector. *Development Policy Review, 38*(2), 219–234.

Sorda, G., Banse, M., & Kemfert, C. (2010). An overview of biofuel policies across the world. *Energy Policy, 38*(11), 6977–6988.

Sovacool, B. K., Lant, P., Hook, A., & Baker, L. (2021). The energy and food nexus: Costs, benefits, and security of bioenergy. *Energy Research & Social Science, 79*, 102125.

Tauseef, S. M., Abbasi, T., & Abbasi, S. A. (2013). Anaerobic digestion for global warming control and energy generation – An overview. *Renewable and Sustainable Energy Reviews, 20*, 338–345.

UNEP. (2021). *Waste management in developing countries: Challenges and strategies*. United Nations Environment Programme.

Verma, A., Vinoda, K. S., Papireddy, M., & Gowda, A. N. (2019). Toxic pollutants from plastic waste – A review. *Procedia Environmental Sciences, 35*, 701–708.

Wang, X., Cai, Y., Peng, Z., & Wang, X. (2018). The green cycle of biomass waste: Biorefinery and waste valorization. *Energy Procedia, 153*, 99–104.

Wilson, D. C., Rodic, L., Modak, P., Soos, R., Carpintero, A., Velis, C. A., et al. (2015). *Global waste management outlook*. United Nations Environment Programme.

Woolf, D., Lehmann, J., & Lee, D. R. (2021). Biochar as a tool to enhance soil fertility and sequester carbon in Africa. *Nature Communications, 12*(1), 1039.

Zhang, H., Wang, H., & Li, X. (2021). Enhancing the circular economy through bio-based resources: A review of current practices and future perspectives. *Journal of Environmental Management, 294*, 113012.

Zilberman, D., Gordon, B., Hochman, G., & Wesseler, J. (2018). Economics of sustainable development and the bioeconomy. *Applied Economic Perspectives and Policy, 40*(1), 22–37.

Index

M. C. Ogwu et al. (eds.), *Sustainable Bioeconomy Development in the Global South*, https://doi.org/10.1007/978-981-96-0305-3

C

D

E

The manufacturer's authorised representative in the EU is Springer Nature Customer Service Centre GmbH, Europaplatz 3, 69115 Heidelberg, Germany. If you have any concerns regarding our products, please contact ProductSafety@springernature.com

Printed and bound by CPI Group (UK) Ltd, Croydon, CR0 4YY
01/10/2026
02225937-0016